MULTI-WAVELENGTH CONTINUUM EMISSION OF AGN

INTERNATIONAL ASTRONOMICAL UNION

UNION ASTRONOMIQUE INTERNATIONALE

MULTI-WAVELENGTH CONTINUUM EMISSION OF AGN

PROCEEDINGS OF THE 159TH SYMPOSIUM OF THE
INTERNATIONAL ASTRONOMICAL UNION,
HELD IN GENEVA, SWITZERLAND, AUGUST 30–SEPTEMBER 3, 1993

EDITED BY

T.J.-L. COURVOISIER

and

A. BLECHA

Observatoire de Genève, Sauverny, Switzerland

KLUWER ACADEMIC PUBLISHERS
DORDRECHT / BOSTON / LONDON

A C.I.P. Catalogue record for this book is available from the Library of Congress.

ISBN 0-7923-2744-6

*Published on behalf of
the International Astronomical Union
by
Kluwer Academic Publishers, P.O. Box 17, 3300 AA Dordrecht, The Netherlands.*

*Kluwer Academic Publishers incorporates
the publishing programmes of
D. Reidel, Martinus Nijhoff, Dr W. Junk and MTP Press.*

*Sold and distributed in the U.S.A. and Canada
by Kluwer Academic Publishers,
101 Philip Drive, Norwell, MA 02061, U.S.A.*

*In all other countries, sold and distributed
by Kluwer Academic Publishers Group,
P.O. Box 322, 3300 AH Dordrecht, The Netherlands.*

Printed on acid-free paper

All Rights Reserved
© *1994 International Astronomical Union*

No part of the material protected by this copyright notice may be reproduced or utilized in any form or by any means, electronic or mechanical including photocopying, recording or by any information storage and retrieval system, without written permission from the publisher.

Printed in the Netherlands

Table of Contents

Foreword 1

Multi–wavelength Continuum Emission of AGN — 3

The Origin of Continuum Emission in Active Galactic Nuclei 5
Joel N. Bregman

The UV–Optical–near–IR emission of BL Lacertae Objects 17
R. Falomo, E. Pian, R. Scarpa, A. Treves

Misdirected Quasars in distant Radio Galaxies 21
Sperello di Serego Alighieri, Andrea Cimatti

Extreme Quasars: Observations and Constraints 25
Martin Elvis

Particle Acceleration in Extragalactic Jets and Implications for the High–Energy Emission from Blazars 29
R. Schlickeiser, C. D. Dermer

Thermal or Non–Thermal X–Rays from AGNs ? 33
Gabriele Ghisellini, Francesco Haardt

New Observations of AGN with Specific Instruments — 37

Compton Observatory Observations of AGN 39
J. D. Kurfess

EGRET High-Energy Gamma-ray Observations of AGN: Energy Spectra and Time Variability 49
D. J. Thompson, D. L. Bertsch, B. L. Dingus, J. A. Esposito, C. E. Fichtel, R. C. Hartman, S. D. Hunter, P. Sreekumar, J. Chiang, J. M. Fierro, Y. C. Lin, P. F. Michelson, P. L. Nolan, G. Kanbach, H. A. Mayer-Hasselwander, C. von Montigny, H.-D. Radecke, D. A. Kniffen, J. R. Mattox, E. J. Schneid

ROSAT Observations of Active Galactic Nuclei 53
W. Brinkmann

Review of GRANAT Observations of AGNs 63
E. Churazov, M. Gilfanov, A. Finoguenov, R. Sunyaev, M. Chernyakova, Yu. Apalkov, S. Grebenev, I. Lagunov, E. Jourdain, J. P. Roques, P. Mandrou, L. Bouchet, J. Ballet, A. Claret, P. Laurent, F. Lebrun

GINGA Observation of Active Galactic Nuclei 73
H. Inoue

Hubble Space Telescope Observations of AGN 83
F. Macchetto

Observations of AGN with the Extreme Ultraviolet Explorer 105
Herman L. Marshall

Variability   111

High–Energy Continuum Variability in Active Galactic Nuclei 113
Rick Edelson

The X-ray Emission of Quasars as Observed by *Ginga* 123
A. J. Lawson, M. J. L. Turner

Solving the Mystery of the Periodicity in the Seyfert Galaxy NGC 6814 127
Greg M. Madejski, Chris Done, T. Jane Turner, Richard F. Mushotzky, Peter Serlemitsos, Fabrizio Fiore, Marek Sikora, Mitchell C. Begelman

Multifrequency Variability of AGN's 131
J. Clavel

Variability in the Millimeter and Radio Domains 145
Esko Valtaoja

Contemporaneous Multiwaveband Observations of Blazars 155
Alan P. Marscher, Steven D. Bloom, Yun Fei Zhang, Walter K. Gear

Multiwavelength Experiments on AGN: Reverberation Experiment 159
W. Wamsteker, R. Vio

The Structure of the Broad Emission Line Region as Shown by Variability
    Monitoring 163
Julian H. Krolik

The Amplitude–Time Lag Relation for Emission–Line Flares of AGN 173
Ivan I. Shevchenko

Intensive Spectroscopic Monitoring of NGC 5548 with HST and IUE 177
B. M. Peterson, K. T. Korista

AGN Variability and VLBI 181
L. B. Bååth

MM–VLBI: Jets in the Vicinity of Galaxy–Cores 187
T. P. Krichbaum, K. J. Standke, D. A. Graham, A. Witzel, C. J. Schalinski, J. A. Zensus

High Resolution Radio Observations of Nearby Galaxies 189
D. J. Saikia, A. Pedlar

Correlated X-ray and Millimetre Variability of the Blazar 3C273 193
I. M. McHardy, I. Papadakis, C. M. Leach, E. I. Robson, W. Junor, R. Staubert, H. Steppe, W. K. Gear

High S/N ROSAT–PSPC Observations of the Quasar PG 1116+215: Power
    Law Shape of the Soft X-ray Spectrum 197
M.-H. Ulrich, S. Molendi

Correlations between Emission Components 201

Emission Components in Seyfert Galaxies and Quasars 203
Thierry J.-L. Courvoisier

X-ray Emission and Reflection in AGNs 213
Hagai Netzer

The Soft X-ray Excess in Quasars and Deep X-ray Surveys 217
Ajit Kembhavi, A. C. Fabian

On the Broad Band Energy Distribution of Blazars 221
Laura Maraschi, Gabriele Ghisellini, Annalisa Celotti

Modeling Bent Radio Jets: Another Probe of the Central Engine 233
P. A. Hughes, M. F. Aller, H. D. Aller

AGN Physics and Models 237

Models for Variability in AGNs 239
Martin J. Rees

Magnetized Accretion Disks Driving Jets 249
G. Pelletier, J. Ferreira, F. Rosso

Variability and Slim Accretion Disks 253
F. H. Wallinder

Structure and Emission of Parsec–Scale Jets in Quasars 257
Max Camenzind

Emission of Accretion Disks 261
B. Czerny

Spectra of Quasars with Extreme Continuum Properties 271
Smita Mathur

The Palomar Observatory Dwarf Seyfert Survey 275
Luis C. Ho, Alexei V. Filippenko, Wallace L. W. Sargent

Quasar Formation in Hierarchical Structure Formation Models 279
M. G. Haehnelt

Unified Schemes and Relations with other Types of Objects 283

On the Difference Between Radio Loud and Radio Quiet AGN 285
Karl Mannheim

Hubble Space Telescope Observations of NGC 4151: Implications for the Unified
Model of AGN 289
Z. Tsvetanov, I. N. Evans, G. A. Kriss, H. C. Ford

Quasi–Steady State Cosmology 293
G. Burbidge, F. Hoyle, J. V. Narlikar

Unified Models: Religion and Science 301
Robert Antonucci

X–ray Luminous Non–Seyfert Galaxies 311
M. J. Ward, D. H. Hughes, J. S. Dunlop, P. N. Appleton

Poster contributions: Continuum studies   315

Modelling the Soft X-ray Excess in E1615+061 317
M. Bałucińska-Church, L. Piro, H. Fink, F. Fiore, M. Matsuoka, G. C. Perola, P. Soffitta

HST, IUE and ROSAT Observations of High Z QSOs 318
D. L. Band, R. D. Cohen, P. R. Blanco, V. T. Junkkarinen, E. M. Burbidge, R. E. Rothschild, G. A. Reichert

Ground-based Observations of PKS2155-304 in November 1991 319
A. Blecha, T. J.-L. Courvoisier, H. D. Aller, M. F. Aller, P. Bouchet, P. Bratschi, M. T. Carini, M. Donahue, E. D. Feigelson, A. V. Filippenko, I. S. Glass, J. Heidt, P. A. Hughes, R. I. Kollgaard, T. Matheson, H. R. Miller, J. C. Noble, P. S. Smith, S. Wagner

Electron Energy Distributions of AGNs in the Thin Synchrotron Limit. II. Peaked Electron Energy Spectra of Blazars and OVV Quasars 320
Irene Cruz-González, Luis Salas, Luis Carrasco

Simultaneous Optical to Near-IR Observations of Blazars 321
R. Falomo, M. Bersanelli

Observations of PKS 2155–304 with the Extreme Ultraviolet Explorer 322
A. Fruscione, C. S. Bowyer, T. E. Carone, S. M. Kahn, A. Königl, H. L. Marshall

The Multiwavelength Spectra of CEN A, NGC 4151, 3C273, 3C279 323
Sten Odenwald, Neil Gehrels, Sethanne Howard

Submillimetre Spectral Indices of Radio–Quiet Quasars 324
D. H. Hughes, J. S. Dunlop, E. I. Robson, W. K. Gear

EUVE Observations of NGC 5548 325
J. S. Kaastra, R. Mewe, J. Heise, F. J. M. Alkemade, C. J. Schrijver, T. Carone

Instantaneous Flux Spectra Measurements of the about 100 AGNs at 6 Frequencies from 1 Ghz to 22 Ghz 326
N. A. Nizelski, Yu. A. Kovalev, A. B. Berlin

Simultaneous Observations of the Continuum Emission of the Quasar 3C 273 from Radio to γ–ray Energies 327
G. C. Lichti, T. Balonek, T. J.-L. Courvoisier, N. Johnson, M. McConnell, C. von Montigny, W. Paciesas, E. I. Robson, A. Sadun, C. Schalinski, A. G. Smith, R. Staubert, H. Steppe, B. N. Swanenburg, M. J. L. Turner, M.-H. Ulrich, O. R. Williams

Simultaneous Long Term Millimetre and Sub-millimetre Monitoring of Blazars . 328
S. J. Litchfield, J. A. Stevens, E. I. Robson

Near IR - Observations of the Jet in M87 329
M. Neumann, K. Meisenheimer, H.-J. Röser, M. Stickel

Simultaneous Observations of Seyfert 1 Galaxies with IUE, ROSAT and GINGA   330
A. Orr, R. Walter, T. J.-L. Courvoisier, H. H. Fink, F. Makino, C. Otani, W. Wamsteker

UV Variability of a Large Sample of AGN 331
S. Paltani, T. J.-L. Courvoisier

Multiwavelength Energy Distributions of Ultraluminous IRAS Galaxies 332
D. Rigopoulou, A. Lawrence

Multifrequency Observations of the Jet of 3C 273 333
H.-J. Röser, K. Meisenheimer, M. Neumann, R. G. Conway

The Nature of the NIR Emission from Seyfert 1 Galaxies: NGC 4051 335
M. Salvati, G. Calamai, L. K. Hunt, G. Del Zanna, E. Giannuzzo, F. Mannucci, R. M. Stanga, M. Kidger, W. Wamsteker

The Origin of AGN-IR Emission 336
L. G. Stenholm

On the Correlation of the FIR and Radio Radiation of Spiral Galaxies 337
H. M. Tovmassian

The Soft X-ray Excesses Observed with ROSAT 338
R. Walter, H. H. Fink

Poster contributions: Emission Processes   339

The Structure of Relativistic MHD Jets 341
Stefan Appl, Max Camenzind

Anisotropic Effects in Thomson Cascade Models of Active Galactic Nuclei 343
Matthew G. Baring

Free-free Emission and the Big Blue Bump 344
Richard Barvainis

Radiative Signatures of Neutron Beams in AGN 345
Ifeanyi E. Ekejiuba

Interaction of Jets with Clouds in Extragalactic Radio Sources 346
V. Fedorenko, A. Zentsova, T. J.-L. Courvoisier, S. Paltani

A Study of Gamma Spectral Break in AGN 347
A. Marcowith, G. Henri, G. Pelletier

The Hard X–ray Reflection on Cold Matter 348
E. Jourdain, J. P. Roques

Quasi-Spherical Accretion onto the Black Hole: The Virial Regime 349
V. S. Berezinsky, I. Lapidus

The Emission Spectra of Radioweak Quasars. I. The FIR Emission 350
Martina Niemeyer, Peter L. Biermann

The Nucleus of Our Home Galaxy: A Remnant of an Active or a Starburst Galaxy ? 351
Leonid M. Ozernoy

Dust Distribution in IRAS Seyfert Galaxies 355
M. G. Pastoriza, Charles Bonatto, Eduardo Bica, T. Storchi-Bergmann

Auger Process Following 1s–Photoionization: Ne III and Ne IV Line Production . 356
D. Petrini, F. X. De Araújo

Do Fluid Waves Propagate in Mildly Relativistic Thermal Pair Plasmas ? 357
P. Pietrini, J. H. Krolik

Electron Energy Distributions of AGNs in the Thin Synchrotron Limit. I. The Method 358
Luis Salas, Irene Cruz-González, Luis Carrasco

A Simple MHD Model for One-Sided Jets 360
G. Bodo, E. Trussoni, G. Chagelishvili

Spectra of Distant Quasars and Verification of Possible Variation of Fundamental Constants over Cosmological Time-scales 361
D. A. Varshalovich, A. Y. Potekhin

Poster contributions: X-Rays and Higher Energies 363

X-ray Loud AGN with Optical Starburst or Seyfert 2 Properties 365
N. Bade, S. Schaeidt

ROSAT X-Ray Observations of Pair MRK 474/NGC 5682 366
Hongguang Bi

ROSAT Observations of an Optical Quasar Survey Field 368
H. Brunner, T. Dörrer, P. Friedrich, G. Lamer, R. Staubert

ROSAT Spectra of Quasars 369
P. Bühler, T. J.-L. Courvoisier, R. Staubert, H. Brunner, G. Lamer

ROSAT Observations of Einstein EMSS AGNs 370
M. Cappi, G. G. C. Palumbo, R. Della Ceca, T. Maccacaro

Supernova Fragments and the Origin of the Rapid X-ray Variability 371
R. Cid Fernandes, R. Terlevich, G. Tenorio-Tagle, J. Franco, M. Rozyczka

ROSAT Selected Interacting Galaxies with Narrow Emission Lines 372
D. Engels, N. Bade, J. Studt, H. Fink

Soft X-ray Emission of Quasars 373
N. Schartel, R. Walter, H. H. Fink

ROSAT All Sky Survey AGN Spectra: Constraints on Accretion Disks 374
P. Friedrich, T. Dörrer, H. Brunner, R. Staubert

X-ray Detection of the Nuclear Source in the Cygnus A Galaxy 375
D. E. Harris, R. A. Perley, C. L. Carilli

ROSAT Observations of Bright BL Lacertae Objects 377
G. Lamer, H. Brunner, R. Staubert

Can AGN (Active Galactic Nuclei) Alone Make the Cosmic X-ray Background? 378
Darryl Leiter, Elihu Boldt

Acceleration Efficiency in Nonthermal Sources and the Soft Gamma-rays from
NGC 4151 Observed by OSSE and SIGMA 379
Andrzej Maciołek-Niedźwiecki, Andrzej A. Zdziarski, Alan P. Lightman

X-rays from Photoionized Accretion Discs 380
G. Matt, A. C. Fabian, R. R. Ross

Gamma Rays and Jets in Active Galactic Nuclei 381
Donald Meyer

Spectral Variability of 3C 273 at soft X-rays 382
R. Staubert, T. J.-L. Courvoisier, H. Fink, M.-H. Ulrich, H. Brunner, S. Friedrich, K. Otterbein

X-ray Observations of Blazars with *Ginga* and *ASCA* 383
M. Tashiro, K. Makishima, Y. Kohmura, T. Ohashi, C. Otani, T. Kii, R. Fujimoto, F. Makino, Ginga Team, ASCA Team

ROSAT Spectra and Lightcurves of Bright BL Lacertae Objects at Low Interstellar Absorption 384
H.-C. Thomas, H. H. Fink

New X-ray Results on Radio Galaxies 385
D. M. Worrall, M. Birkinshaw

Poster contributions: Variability   387

Centiarcsecond Structure and Variability of AGN 389
D. R. Altschuler, L. I. Gurvits, W. Alef, D. Graham, B. Dennison, J. E. Carson, A. S. Trotter

3C 345: The Periods in the Optical Variability and Further Confirmation of a Connection between Optical Outbursts and Superluminal Components of Radiojet 390
M. K. Babadzhanyants, E. T. Belokon'

The Recent Lightcurve of 3C 345 391
K. J. Schramm, U. Borgeest, J. v. Linde, S. J. Wagner, J. Heidt

Variable IR and Optical Sources in Blazars 392
V. A. Hagen-Thorn

Optical Microvariability in Complete Samples of BL Lac Objects 393
Jochen Heidt

Long-Term Optical Behavior of BL Lac Objects 394
S. Kikuchi

Line Profile Variations in AGN 395
W. Kollatschny, M. Dietrich

Structure Function Analysis Applied to the Metsähovi Sample 396
Markku Lainela, Esko Valtaoja

A Jet Model Interpretation of Multi Frequency Flux Observations of Radio
  Outbursts in the AGN 0235+16 397
Y. Y. Kovalev, G. M. Larionov

Time Scale of Radio Sources Activity from the Statistical Model of Variability . 398
P. Magdziarz, J. Machalski

How Large are Quasar Broad-Line-Regions ? Results from a Program to Monitor the PG Quasars 399
Dan Maoz, Buell T. Jannuzi, Paul S. Smith, Shai Kaspi, Hagai Netzer

Spectral Variability of Six Bright BL Lac Objects in the Near IR 400
E. Massaro, R. Nesci, G. C. Perola, D. Lorenzetti, L. Spinoglio

Spectral Variability of NGC 4151 in 1972-1991 401
V. L. Oknyanskij, V. M. Lyutyi, K. K. Chuvaev

Long Term X-ray Variability of NGC 4151 402
I. E. Papadakis, I. M. McHardy

Variability of the Ultraviolet Continuum and Emission Lines of NGC 3783 403
G. A. Reichert

Coordinated Multifrequency Observations of OJ 287 and MK 421 (March 1993) . 404
A. Sillanpää, L. O. Takalo, E. Valtaoja, H. Teräsranta, Yu. S. Efimov, N. Shakhovskoy, J. Heidt, H. Bock

NGC 4151: An Accurate Upper Limit to the Dimensions of the Central Continuum Source ? 405
Nicolaos H. Solomos

Variability of NGC 3783: First Results from an Intensive Ground–Based Monitoring Campaign 407
Giovanna M. Stirpe

Two Year UV–Optical Monitoring of the Seyfert 1 Galaxy Markarian 335 408
Wei-Hsin Sun, Chris R. Shrader, Tracey J. Turner, Matthew A. Malkan, Bradley M. Peterson, Paul M. N. Hintzen, Yoji Kondo, Sung-Nan Lin, Ting-Chang Lin, Remington P. S. Stone

OJ 287: A Blazar with Everything 409
L. O. Takalo

Long Term Monitoring of AGN with the Metsähovi and SEST Telescopes 410
H. Teräsranta, M. Tornikoski, E. Valtaoja

Connection Between Optical and High Frequency Radio Variability in AGN . . . 411
M. Tornikoski, E. Valtaoja, A. G. Smith, A. D. Nair

Optical Variability of Faint QSOs and AGNs 412
D. Trèvese, R. G. Kron

Multifrequency Variability of Blazars 3C 279 and 3C 345 413
James R. Webb

Optical Microvariability in Radio Quiet Quasars 414
Paul J. Wiita, Gopal-Krishna, Ram Sagar

An Eight-Month Monitoring Campaign on a Sample of AGN 415
Cláudia Winge, Bradley M. Peterson, M. G. Pastoriza, T. Storchi-Bergmann, J. Baldwin

Poster contributions: Radio Emission (Maps)   417

Optically Quiet Quasars – Radio and Optical Investigations 419
Chidi E. Akujor, R. W. Porcas, A. R. Patnaik, A. Ardeberg

Correlated Flux Density Outbursts and Structural Variations in a Sample of
   Low Frequency Variable Radio Sources 420
M. Bondi, L. Padrielli, R. Fanti, L. Gregorini, F. Mantovani, J. D. Romney, N. Bartel, K. W. Weiler, G. D. Nicolson

Relativistic Jets with Stellar Wind Entrainment 421
Mark Bowman

High Resolution HI Observations of NGC 1068 422
Elias Brinks, Evan D. Skillman, Roberto J. Terlevich, Elena Terlevich

Astrophysical Results From Geodetic VLBI Campaigns 423
S. Britzen, A. Witzel, A.-M. Gontier, C. J. Schalinski, J. Campbell

On the Age of GPS Radio Sources 424
Joel C. Carvalho

VLBI Morphology of GHz-Peaked Spectrum Radio Galaxies 425
D. Dallacasa, C. Fanti, R. Fanti

The Milliarcsecond Structure of Four Seyfert Galaxies at λ 18 426
T. Ghosh, R. T. Schilizzi, G. K. Miley, A. G. deBruyn, M. J. Kukula, A. Pedlar, D. Graham, D. J. Saikia

Radio Source Structure and Unified Models 427
J. B. Hutchings, S. G. Neff

Directivity Pattern Simulation of the Jet Radio Emission in an AGN Model ... 428
Y. Y. Kovalev

Neutral Hydrogen In The Seyfert Galaxy NGC 3227 429
Carole G. Mundell, Alan Pedlar, Dave J. Axon

MERLIN Observations of Seyfert Nuclei 430
A. Pedlar, M. J. Kukula

The Central Region of NGC 1365. SEST and VLA Observations of CO and the
   Radio Continuum 431
Aa. Sandqvist, S. Jörsäter, P. O. Lindblad

Helical Motion and Non-Adiabatic Expansion in the Jet of 3C345 432
W. Steffen, T. P. Krichbaum, A. Witzel, J. A. Zensus, S. J. Qian

Interpretation of Some Properties of Extragalactic Jets in the Context 433
   of a Two Component Model ....
Lourdes Vicente

Poster contributions: Line Studies   435

Bloated Stars as BLR Clouds: Numeric Results 437
Tal Alexander, Hagai Netzer

Type Transitions in Starbursts Powered AGN 438
Itziar Aretxaga, Roberto Terlevich

Variation of Broad Optical Emission Lines in AGN's................ 439
N. G. Bochkarev, A. I. Shapovalova, L. S. Nazarova

HST Images of the Seyfert Galaxy NGC 5929 and Its Companion NGC 5930 .. 440
G. A. Bower, A. S. Wilson, J. S. Mulchaey, G. K. Miley, T. M. Heckman, J. H. Krolik

Velocities Within 1 Arcsec of the Nucleus of NGC 4151 as Revealed 441
   by the HST Faint Object Camera ....
R. M. Catchpole, A. Boksenberg, The Faint Object Camera Team

Spectroscopic Studies of Emission Line Galaxies 442
M. S. Chun, E. C. Sung, H. K. Moon, Y. I. Byun

Emission Line Variations of BLRG 444
M. Dietrich, W. Kollatschny

High Spatial Resolution 2D Spectrography of ENLR in Seyfert Galaxies 445
E. Pecontal, P. Ferruit

Kinematics of the BLR Clouds in AGNs and Quasars 446
P. M. Gondhalekar

Infrared Coronal Emission Lines and the Possibility of their Laser Emission in
   Seyfert Nuclei 447
Matthew A. Greenhouse, Uri Feldman, Howard A. Smith, Marcel Klapisch, Anand K. Bhatia, Avi Bar-Shalom

Nuclear Activity in the Seyfert Galaxy NGC 1365 448
M. Hjelm, P. O. Lindblad, S. Jörsäter

Composite Models for Narrow Line Regions 449
Stefanie Komossa

Optical, HST, and ROSAT Observations of BAL QSOs 450
Michael Kopko, Jr., David A. Turnshek, Brian R. Espey

Wolf-Rayet Stars as Tracers of the Recent History of the Star Formation Rate . . 451
G. Meynet

Modelling of Double-Peaked Emission Lines in AGN's 452
L. S. Nazarova

Time Dependent BLR Photoionization Models 453
P. T. O'Brien, M. R. Goad

The Nuclear Region of the Sbc Spiral Galaxy NGC 5055: A Mildly Active Nucleus 454
P. Pişmiş, M. Manteiga, A. Mampaso, E. Recillas-Cruz, G. Cruz-González

Subarcsecond-scale Optical and Radio Structure Correlations in Seyfert Galaxies 455
R. W. Pogge, M. M. DeRobertis

Influence of the Gravitational Field on the Shape of Spectral Lines in Spectra
of Seyfert Galaxies and Quasars 456
L. Č. Popović, I. Vince, A. Kubičela

Size of the Broad Line Region in NGC 3227: Results from the LAG Campaign . 457
I. Salamanca, T. Baribaud, D. Alloin

A Unified View of NGC 4151 458
Hartmut Schulz

Implications of Nonlinear Line Response in Variable Seyfert Nuclei 459
Joseph C. Shields, Gary J. Ferland, Bradley M. Peterson

Star-Formation and Nuclear Activity in Three Galaxies with Nuclear Rings . . 460
Thaisa Storchi-Bergmann, Andrew S. Wilson, Jack A. Baldwin

High Resolution NIR Imaging of the Starburst Galaxy NGC 1808 461
L. E. Tacconi-Garman, A. Krabbe, A. Sternberg, R. Genzel

Infrared Line Shapes in Active Galactic Nuclei 462
Rodger I. Thompson

FeII, FeI Emission Lines from Accretion Disks: An Explanation for "FeII Problem" in AGNs ? 463
Chunyan Wei, Fuzhen Cheng, Junhan You

The Anisotropic Radiation Field in NGC 3516 464
I. Yankulova, V. Golev, T. Bonev, K. Jockers

Poster contributions: Polarisation 465

Internal Faraday Rotation in Compact Radio Sources (CRS) 467
Claes-Ingvar Björnsson

Scattered Light in a Distant Radio Galaxy 468
Andrea Cimatti, Sperello di Serego Alighieri

Optical and VLBI Polarization Measurements of AGN 469
D. C. Gabuzda, M. L. Sitko

Polarized Optical and Infrared Emission from High Redshift Radio Galaxies . . . 470
Buell T. Jannuzi

Multifrequency Polarimetry of CSS Sources 471
Everton Lüdke, Chidi E. Akujor, Simon T. Garrington

Frequency-Dependent Polarization in Comptonization Models for AGN 472
Juri Poutanen

A Tentative Sketch for BL Lacertae Objects 473
Hélène Sol, Lourdes Vicente

The Intranight Variability of Optical Polarization in PKS 0109+224 474
L. Valtaoja

Poster contributions: Disk Structure and Emission 475

On the Accretion Disk Models with Stationary and Non-Stationary Shock Waves 477
Sandip K. Chakrabarti

The Vertical Structures of Accretion Disks in AGN 478
H. Riffert, T. Dörrer, R. Staubert, H. Ruder

The Nuclear Disk of NGC 4261: HST Images and WHT Spectra 479
Laura Ferrarese, Holland C. Ford, Walter Jaffe, Frank van den Bosch, Robert W. O'Connell

Physics of Magnetized Accretion Disks in AGN 480
George Field, Robert Rogers

AGN Accretion Disks with Finite Turbulent Prandtl Number 482
O. M. Heinrich

Structure of Outer Regions of Accretion Disks in AGN 483
Jean-Marc Huré, Suzy Collin, Guillaume Pineau des Forêts

Structure and Emission Line Spectrum of an X-ray Heated Accretion Disk in AGN 484
Yuan-Kuen Ko, Timothy R. Kallman

Hot Accretion Disks with Magnetic Field and Thermal Cyclo–Synchrotron Radiation 485
Masaaki Kusunose, Andrzej A. Zdziarski

Galactic Dynamo Models 486
N. P. Moore, J. S. Panesar, A. H. Nelson

Variability of Black Hole Accretion Disks 487
Michael A. Nowak, Robert V. Wagoner

The Structure and Stability of Two-Temperature Accretion Disk 488
Myeong-Gu Park

Reprocessing of UV and Line Emission in AGN Accretion Discs 489
Evlabia Rokaki, Catherine Boisson

Simultaneous Implosive Accretion and Jet Formation in Quasars: Correlation of Optical Outbursts with VLBI Jets 490
M. M. Romanova, R. V. E. Lovelace

Hybrid Accretion–Disks in AGN 491
Amri Wandel

Spectrum of a Magnetized Accretion Disk AGN Model 492
Yi Wang

On Illumination Effect in AGN Disks 493
Insu Yi

Poster contributions: Statistical Studies and Evolution   495

Cm–Wavelength Flux Variability: Constraints on AGN Modeling 497
M. F. Aller, H. D. Aller, P. A. Hughes

K–Correction Biases and the Quasar Luminosity Function 498
A. C. Baker, P. C. Hewett

Nuclear Activity in Interacting Galaxies 499
Tapan K. Chatterjee

Spatial Distribution of Quasars in the LBQS Sample 500
Yaoquan Chu, Jianhui Tao

On the Unification of Radio Galaxies and Quasars 501
Krzysztof T. Chyzy

Principal Component Analysis of Multiwavelength Properties of Seyfert Galaxies   502
Deborah Dultzin-Hacyan, Carlos Ruano

Optical Differences Between Radio-Loud and Radio-Quiet QSOs 503
Paul J. Francis

On the Way AGN's Turn Off 504
D. Friedli

Multifrequency Spectra of EXOSAT Blazars 505
K. K. Ghosh, S. Soundararajaperumal

Radio to X-ray Energy Distribution of BL Lacertae Objects 506
P. Giommi, S. G. Ansari, A. Micol

Optical Properties of Soft X-ray Selected Bright New ROSAT AGN 508
D. Grupe, K. Beuermann, K. Reinsch, H.-C. Thomas, H. H. Fink

The Redshift Distribution of Ly$_\alpha$ Forest Lines in Spectra of QSOs 509
Huang Keliang, Zhou Hongnan

Do Luminous QSOs Have Softer UV Spectra ? 511
Thorsten Köhler

A Possibility of Reveal of Parent Populations for the Objects with Active Nuclei
   on the Basis of Comparing their Spatial Correlation Functions 512
B. V. Komberg, A. V. Kravtsov

Activity as the Result of Merging 513
V. M. Kontorovich, A. V. Kats, D. S. Krivitsky

The CfA Seyfert Sample at 8.4 GHz 514
Marek J. Kukula, Alan Pedlar, S. Baum, C. O'Dea, S. Unger

The Influence of Selection Effects on the Properties of BL Lacs 515
M. J. M. Marchã, I. W. A. Browne

The QED (Quasar Energy Distributions) Atlas 516
J. McDowell

A ROSAT Survey of a Spatially and Magnitude Complete Quasar Sample . . . 517
K. Molthagen, H. J. Wendker, U. G. Briel

Some Peculiarities of Variable Optical Spectra of 11 Low-Redshift Quasars 518
I. I. Pronik

A Lower Limit to the Excess of Companions Among Seyfert Galaxies 519
P. Rafanelli, M. Violato

Are Redshifts of Seyfert Galaxies Cosmological ? 520
A. K. Sapre, P. S. Parihar

AGN as a Result of Evolution of Binary Gravimagnetic Rotators 521
Olga K. Sil'chenko, Vladimir M. Lipunov

Buried Quasars In Radiogalaxies ? 522
C. J. Simpson, M. J. Ward, D. L. Clements, S. Rawlings, A. S. Wilson

A Comparison of the Centimetre-to-Submillimetre Continuum Spectra of BL Lacertae
   Objects and Flat Spectrum Radio Quasars 523
J. A. Stevens, S. J. Litchfield, E. I. Robson, W. K. Gear, D. H. Hughes

Spectrophotometrical Models for AGN 524
Doru Marian Suran, Nedelia Antonia Popescu

Subject Index 525

Author Index 529

List of Participants

Akujor C. E. Nuffield Radio Astronomy Observatory GB–Macclesfield, Cheshire SK11 9DL, U.K.
Albrecht P. Universitäts-Sternwarte D–37083 Göttingen, Germany
Alexander T. School of Physics and Astronomy Tel–Aviv University Tel–Aviv 69978, Israel
Aller H. D. Astronomy Department University of Michigan Ann Arbor, MI 48109–1090, U.S.A.
Aller M. F. Astronomy Department University of Michigan Ann Arbor, MI 48109–1090, U.S.A.
Alloin D. Observatoire de Paris–Meudon DAEC, 5 Place Jules Janssen, F–92195 Meudon Cedex, France
Almaini O. Dept. of Physics University of Durham GB–Durham, DH1 3LE, U.K.
Altschuler D. R. Arecibo Observatory, P.O. Box 995, Arecibo 00613, Puerto Rico
Antonucci R. Space Telescope Science Institute Baltimore, MD 21218, U.S.A.
Appl S. Landessternwarte Königstuhl D–69117 Heidelberg, Germany
Aretxaga I. Dpto. de Física Teórica, C–XI. Universidad Autónoma de Madrid. E–28049 Madrid, Spain
Arp H. C. Max–Planck–Institut für Astrophysik D–80802 Garching, Germany
Auh B. Yuseonggu Daejeon, 303–348, Korea
Bååth L. Onsala Space Observatory Göteborg University S–43992 Onsala, Sweden
Babadzhanyants M. Astronomical Institute St. Petersburg University SU–198904 St. Petersburg, Russia
Bade N. Hamburger Sternwarte Universität Hamburg, Gojenbergsweg 112, D–21029 Hamburg, Germany
Baker A. C. Institute of Astronomy, Madingley Road GB–Cambridge, CB3 0HA, U.K.
Baker J. C. Dept. of Astrophysics The University of Sydney Sydney, NSW 2006, Australia
Balbus S. A. University of Virginia Charlotteville, VA 22903–0818, U.S.A.
Balucinska–Church M. School of Physics & Space Research, Birmingham Univ., Birmingham B15 2TT, U.K.
Band D. L. Center for Astrophysics & Space Sciences La Jolla, CA 92093–0111, U.S.A.
Baring M. G. NASA–Goddard Space Flight Center, LHEA, Greenbelt, MD 20771, U.S.A.
Barvainis R. Haystack Observatory Westford, MA 01886, U.S.A.
Baryshev Y. Astronomical Institute St. Petersburg University SU–198904 St. Petersburg, Russia
Bednarek W. Max–Planck–Institut für Kernphysik D–6900 Heidelberg, Germany
Bel N. Observatoire de Paris–Meudon F–92195 Meudon Cedex, France
Bergeron J. A. Institut d'Astrophysique F–75014 Paris, France
Bi H. Beijing Astronomical Observatory Chinese Academy of Sciences Beijing, 100080, China
Bjoernsson C. Stockholm Observatory S–13336 Saltsjöbaden, Sweden
Blecha A. Geneva Observatory 1290 Sauverny, Switzerland
Blom J. J. SRON–Leiden 2333 CA Leiden, The Netherlands
Bochkarev N. Sternberg Astronomical Institute 119899 Moscow, Russia
Boksenberg A. Royal Greenwich Observatory GB–Cambridge, CB3 0EZ, U.K.
Bond I. A. Institute of Phys. and Chem. Res. Cosmic Radiation Laboratory Wako, Saitama 351-01, Japan
Bondi M. Nuffield Radio Astronomy Laboratories Jodrell Bank GB–Macclesfield, SK11 9DL, U.K.
Borgeest U. Hamburger Sternwarte Universität Hamburg D–2050 Hamburg 80, Germany
Bower G. A. Space Telescope Science Institute Baltimore, MD 21218, U.S.A.
Bowman M. Univ. of Manchester Nuffield Radio Astron. Lab. GB–Macclesfield, Cheshire SK11 9DL, U.K.
Bregman J. N. Astronomy Department University of Michigan Ann Arbor, MI 48109–1090, U.S.A.
Bremnes T. R. University of Tromso, Astrophysics Group, P.O. Box 953, 9001 Tromso, Norway
Brinkmann W. Max–Planck–Institut für Extraterrestrische Physik D–85740 Garching, Germany
Brinks E. National Radio Astronomy Observatory Socorro, NM 87801, U.S.A.
Britzen S. Max–Planck–Institut für Radioastronomie D–5300 Bonn 1, Germany
Brunner H. Astronomisches Institut Universität Tübingen D–7400 Tübingen, Germany

Buehler P. Observatoire de Genève 1290 Sauverny,
Burbidge G. Center for Astrophysics and Space Sciences La Jolla, CA 92093–0111, U.S.A.
Burbidge M. Center for Astrophysics and Space Sciences La Jolla, CA 92093–0111, U.S.A.
Camenzind M. Landessternwarte Königstuhl D–699117 Heidelberg, Germany
Cappi M. Università di Bologna Dipartimento di Astronomia I–40126 Bologna, Italy
Carone T. E. Space Sciences Laboratory University of California Berkeley, CA 94720, U.S.A.
Carvalho J. C. E–08013 Barcelona, Spain
Catchpole R. M. Royal Greenwich Observatory GB–Cambridge, CB3 OEZ, U.K.
Celotti A. Institute of Astronomy, Madingley Road GB–Cambridge, CB3 0HA, U.K.
Chakrabarti S. K. Tata Institute of Fundamental Research Bombay 400 005, India
Chatterjee T. K. Facultad de Ciencas Fisico-Matemáticas Universidad Autónoma de Puebla Puebla, Mexico
Chu Y. International Center for Theoretical Physics (ICTP) I–34100 Trieste, Italy
Chun M. Dept. of Astronomy Yonsei University Seoul 120–749, Korea
Churazov E. Russian Academy of Sciences Space Research Institute 117810 Moscow, Russia
Church M. J. School of Physics & Space Research, Univ. of Birmingham GB–Birmingham B15 2TT, U.K.
Chyzy K. T. Astronomical Observatory Jagiellonian University 30–244 Krakow, Poland
Cid Fernandes R. Institute of Astronomy University of Cambridge GB–Cambridge, CB3 OHA, U.K.
Cimatti A. European Southern Observatory D–8046 Garching, Germany
Clavel J. European Space Agency ESTEC Code SAJ NL–2200 AG Noordwijk, The Netherlands
Clements D. Astrophysics Oxford University GB–Oxford, OX1 2RH, U.K.
Collin–Souffrin S. Institut d'Astrophysique de Paris F–75014 Paris, France
Contini M. School of Physics and Astronomy Tel–Aviv University Tel–Aviv 69978, Israel
Courvoisier T. Observatoire de Genève 1290 Sauverny, Suisse
Cruz–Gonzalez I. Instituto de Astronomia, UNAM Ciudad Universitaria Mexico, D.F. 04510, Mexico
Czerny B. Nicolaus Copernicus Astronomical Center PL–00716 Warszawa, Poland
Dallacasa D. Radiosterrenwacht Dwingeloo NL–7990 AA Dwingeloo, The Netherlands
Di Serego Alighieri S. Osservatorio Astrofisico di Arcetri I–50125 Firenze, Italy
Dietrich M. Universitäts–Sternwarte D–3400 Göttingen, Germany
Doerrer T. Astronomisches Institut Universität Tübingen D–7400 Tübingen, Germany
Dubois P. Observatoire de Strasbourg F–67000 Strasbourg, France
Dultzin–Hacyan D. Instituto de Astronomia, Univ. Nac. Auton. de Mexico Mexico, D.F. 04510, Mexico
Edelson R. NASA Goddard Space Flight Center Code 666 Greenbelt, MD 21230, U.S.A.
Ekejiuba I. E. Dept. of Physics & Astronomy Georgia State University Atlanta, GA 30303–308, U.S.A.
Elvis M. Harvard–Smithsonian Center for Astrophysics Cambridge, MA 02138, U.S.A.
Engels D. Hamburger Sternwarte Universität Hamburg D–2050 Hamburg 80, Germany
Falomo R. Osservatorio Astronomico di Padova I–35122 Padova, Italy
Fedorenko V. A.F. Ioffe Physical Technical Institute SU–194021 St. Petersburg, Russia
Ferrarese L. Space Telescope Science Institute Baltimore, MD 21218, U.S.A.
Ferreira J. Laboratoire d'Astrophysique Observatoire de Grenoble F–38041 Grenoble Cedex, France
Ferruit P. Observatoire de Lyon F–69561 St Genis Laval Cedex, France
Field G. Osservatorio Astrofisico di Arcetri I–50125 Firenze, Italy
Fraix–Burnet D. Laboratoire d'Astrophysique F–31400 Toulouse, France
Francis P. J. Steward Observatory University of Arizona Tucson, AZ 85721, U.S.A.
Friedli D. Observatoire de Genève 1290 Sauverny,
Friedrich P. Astronomisches Institut Universität Tübingen D–7400 Tübingen, Germany
Friedrich S. Astronomisches Institut Universität Tübingen D–72076 Tübingen, Germany
Fruscione A. Center for EUV Astrophysics University of California Berkeley, CA 94720, U.S.A.

Gabuzda D. Dept. of Physics and Astronomy University of Calgary Calgary, AB T2N 1N4, Canada
Ghisellini G. Observatory of Torino I-10025 Pino Torinese, Italy
Ghosh K. K. Indian Inst. of Astrophysics Vainu Bappu Observatory Alangayam, N.A.A., TN 635 701, India
Ghosh T. Arecibo Observatory Arecibo, 00613, Puerto Rico
Giommi P. ESIS / ESRIN I-00044 Frascati, Italy
Gondhalekar P. Astrophysics Group Rutherford Appleton Laboratory Chilton, OXON, OX11 0QX, U.K.
Green A. Department of Physics The University GB-Southampton, SO9 5NH, U.K.
Greenhouse M. A. Lab. for Astrophys. MRC321 Nation. Air & Space Museum Washington, DC 20560, U.S.A.
Grupe D. Universitaets-Sternwarte Göttingen D-3400 Göttingen, Germany
Haehnelt M. Ecole d'Eté de Physique Théorique F-74310 Les Houches, France
Hagen-Thorn V. A. Astron. Observatory St. Petersburg State Univ. SU-198904 St. Petersburg, Russia
Harris D. MS-3, Center for Astrophysics Cambridge, MA 02138, U.S.A.
Heidt J. Landessternwarte Heidelberg D-6900 Heidelberg, Germany
Heinrich O. M. Institut für Theoretische Astrophysik D-6900 Heidelberg, Germany
Henri G. Laboratoire d'Astrophysique Observatoire de Grenoble F-38041 Grenoble Cedex, France
Hjelm M. Stockholm Observatory S-13336 Saltsjöbaden, Sweden
Ho L. C. Dept. of Astronomy University of California Berkeley, CA 94720, U.S.A.
Howard S. BDM – 409 Washington, DC 20024, U.S.A.
Huang K. Dept. of Physics Nanjing Normal University Nanjing 210024, China
Hughes D. H. Astrophysics Oxford University GB-Oxford, OX1 3RH, U.K.
Hughes P. A. Astronomy Department University of Michigan Ann Arbor, MI 48109-1090, U.S.A.
Hure J. M. Observatoire de Meudon DAEC F-92195 Meudon Principal Cedex, France
Hutchings J. B. Dominion Astrophysical Observatory Victoria, B.C. V8X 4M6, Canada
Iijima T. Osservatorio Astrofisico di Asiago I-36012 Asiago (Vicenza), Italy
Inoue H. Institute of Space and Astronautical Science Kanagawa 229, Japan
Jackson N. Sterrewacht Leiden 2300 RA Leiden, Netherlands
Jannuzi B. Institute for Advanced Study Princeton, NJ 08540, U.S.A.
Jourdain E. Centre d'Etude Spatiale des Rayonnements F-31089 Toulouse Cedex, France
Köhler T. Hamburger Sternwarte Universität Hamburg D-2050 Hamburg 80, Germany
Kaastra J. S. SRON Leiden 2300 RA Leiden, The Netherlands
Kallman T. NASA-Goddard Space Flight Center, LHEA, Greenbelt, MD 20771, U.S.A.
Kembhavi A. Inter-University Centre for Astronomy and Astrophysics Pune 41107, India
Kikuchi S. National Astronomical Observatory Tokyo, 181, Japan
Ko Y. Laboratory for High Energy Astrophysics Greenbelt, MD 20771, U.S.A.
Kollatschny W. Universitäts-Sternwarte D-3400 Göttingen, Germany
Komberg B. V. Astro Space Center of Lebedev Physical Institute 117810 Moscow, Russia
Komossa S. Astronomisches Institut Ruhr-Universität D-4630 Bochum, Germany
Kontorovich V. Institute of Radioastronomy Kharkiv 310002, Ukraine
Kopko M. Department of Physic & Astronomy University of Pittsburg Pittsburgh, PA 15260, U.S.A.
Kovalev Y. A. Astro Space Center of Lebedev Physical Institute 117810 Moscow, Russia
Kovalev Y. Y. Astro Space Center & Physical Dept. Moscow State University 117810 Moscow, Russia
Krichbaum T. Max-Planck-Institut für Radioastronomie D-5300 Bonn, Germany
Krolik J. H. John Hopkins University Dept. of Physics + Astronomy Baltimore, MD 21218, U.S.A.
Kukula M. J. Nuffield Radio Astronomy Laboratory GB-Macclesfield, Cheshire SK11 9DL, U.K.
Kurfess J. D. Naval Research Laboratory Code 4150 Washington, DC 20375, U.S.A.
Kusunose M. Copernicus Astronomical Center PL-00716 Warsaw, Poland
Lüdke E. University of Manchester Nuffield Radio Astr. Lab. GB-Macclesfield, Cheshire SK11 9DL, U.K.

Lainela M. J. Tuorla Observatory SF–21500 Piikkiö, Finland
Lamer G. Astronomisches Institut Universität Tübingen D–7400 Tübingen, Germany
Lapidus J. I. Institute of Astronomy University of Cambridge GB–Cambridge, CB3 OHA, U.K.
Larionov G. M. Moscos State University Physical Department 119899 Moscow, Russia
Lawson A. University of Leicester GB–Leicester LEI JRH, U.K.
Leach C. M. Dept. of Physics Southampton University GB–Southampton, Hampsh. S09 5NH, U.K.
Lehnert M. D. IGPP Lawrence Livermore National Labs. Livermore, CA 94588, U.S.A.
Leighly K. Laboratory f. High Energy Astrophysics Greenbelt, MD 20771, U.S.A.
Leiter D. NASA–Goddard Space Flight Center Greenbelt, MD 20771, U.S.A.
Lichti G. Max–Planck–Institut für extraterrestrische Physik D–8046 Garching, Germany
Lindblad P. O. Stockholm Observatory S–13336 Saltsjöbaden, Sweden
Lipunov V. Sternberg Astronomical Institute 119899 Moscow, Russia
Litchfield S. J. Univ. of Central Lancashire Dept. of Physics & Astronomy GB–Preston, PR1 2HE, U.K.
Macchetto F. Space Telescope Science Institute Baltimore, MD 21218, U.S.A.
Macchetto F. D. Space Telescope Science Institute Baltimore, MD 21218, U.S.A.
Machalski J. Jagiellonian University Astronomical Observatory PL–30244 Krakow, Poland
Maciołek–Niedźwiecki A. Nicolaus Copernicus Astronomical Center PL–00716 Warsaw, Poland
Madejski G. M. NASA–Goddard Space Flight Center Greenbelt, MD 20771, U.S.A.
Malkan M. Dept. of Astronomy UCLA Los Angeles, CA 90024–1562, U.S.A.
Mannheim K. Universitäts–Sternwarte D–3400 Göttingen, Germany
Maoz D. Institute for Advanced Study Princeton, NJ 08540, U.S.A.
Maraschi L. Istituto di Fisica I–20133 Milano, Italy
Marcha M. M. J. Univ. of Manchester NRAL Jodrell Bank GB–Macclesfield, Cheshire SK11 9DL, U.K.
Marcowith A. Laboratoire d'Astrophysique Observatoire de Grenoble F–38041 Grenoble Cedex, France
Marscher A. P. Dept. of Astronomy Boston University Boston, MA 02215–1401, U.S.A.
Marshall H. L. Center for Space Research Massachusetts Inst. of Technology Cambridge, MA 02139, U.S.A.
Massaro E. Istituto Astronomico Università di Roma "La Sapienza" I–00161 Roma, Italy
Mastichiadis A. Max–Planck–Institut für Kernphysik D–6900 Heidelberg, Germany
Mathur S. Harvard–Smithsonian Center for Astrophysics Cambridge, MA 02138, U.S.A.
Matt G. Institute of Astronomy University of Cambridge GB–Cambridge, CB3 OHA, U.K.
Mc Dowell J. C. Center for Astrophysics Cambridge, MA 02138, U.S.A.
Mc Hardy I. Astronomy & Space Physics Group Southampton Univ. GB–Southampton, SO9 5NH, U.K.
Menon T. Max–Planck–Institut für Radioastronomie D–5300 Bonn, Germany
Mewe R. SRON Space Research Organization Netherlands NL–3584 CA Utrecht, The Netherlands
Meyer D. Dept. of Physics University of Michigan Ann Arbor, MI 48109–1090, U.S.A.
Meynet G. Observatoire de Genève 1290 Sauverny, Switzerland
Molthagen K. Hamburger Sternwarte Universität Hamburg D–21029 Hamburg, Germany
Moore E. Dept. of Astronomy Boston University Boston, MA 02215, U.S.A.
Moore N. P. Dept. of Physics & Astronomy Univ. of Whales Coll. of Cardiff GB–Cardiff, CF2 3YB, U.K.
Mulchaey J. S. Space Telescope Science Institute Baltimore, MD 21218, U.S.A.
Mundell C. G. NRAL GB–Macclesfield, Cheshire SK11 9DL, U.K.
Naundorf C. E. Max–Planck–Institut für Radioastronomie D–5300 Bonn 1, Germany
Nazarova L. Royal Greenwich Observatory GB–Cambridge, CB3 OEZ, U.K.
Netzer H. School of Physics and Astronomy Tel–Aviv University Tel–Aviv 69978, Israel
Neumann M. Max–Planck–Institut für Astronomie D–6900 Heidelberg 1, Germany
Niemeyer M. Max–Planck–Institut für Radioastronomie D–5300 Bonn, Germany
Nolze W. Astronomisches Institut Westfälische Wilhelms–Univ. D–4400 Münster, Germany

Nowak M. A. Canadian Institute of Theoretical Astrophysics Toronto, Ontario M5S 1A7, Canada
O'Brien P. T. Department of Physics & Astronomy Univ. College London GB–London, WC1E 6BT, U.K.
Okayasu R. Nobeyama Radio Observatory Nagano, 384–13, Japan
Oknyanskij V. L. Sternberg Astronomical Institute 119899 Moscow, Russia
Orr A. Geneva Observatory 1290 Sauverny, Switzerland
Ozernoy L. M. Lab. de Radioastron. Millim. Ecole Normale Supérieure F–75231 Paris Cedex 05, France
Pacini F. Osservatorio Astrofisico di Arcetri I–50125 Firenze, Italy
Paltani S. Institut d'Astronomie de l'Université de Lausanne 1290 Chavannes–des–Bois,
Palumbo G. G. Università di Bologna Dipartimendo di Astronomia I–40126 Bologna, Italy
Papadakis I. Astronomy & Space Physics Group Southampton Univ. GB–Southampton, S09 5NH, U.K.
Park M. Kyungpook Natl. University Dept. of Astronomy & Meteorology Taegu 702–701, South Korea
Pastoriza M. G. Instituto de Fisica UFRGS Avenida Bento Goncalves 9500 91500 Porto Alegre, RS, Brasil
Pecontal E. Observatoire de Lyon F–69561 St Genis Laval Cedex, France
Pedlar A. University of Manchester NRAL, Jodrell Bank GB–Macclesfield, Cheshire SK11 9DL, U.K.
Pelletier G. Laboratoire d'Astrophysique Observatoire de Grenoble F–38041 Grenoble Cedex, France
Perez–Olea D. E. Dpto. de Física Teórica, C–XI. Universidad Autónoma de Madrid. E–28049 Madrid, Spain
Peterson B. M. Dept. of Astronomy Ohio State University Columbus, OH 43210, U.S.A.
Petrini D. Observatoire de la Côte d'Azur F–06304 Nice Cedex, France
Pian E. Scuola Internazionale Sup. di Studi Avanzati I–34014 Trieste, Italy
Pietrini P. Dipartimento di Astronomia Universita di Firenze I–50125 Firenze, Italy
Piro L. Istituto Astrofisica Spaziale, CMNR Via E. Fermi 21 I–00044 Frascati, Italy
Pişmiş P. Instituto de Astronomia Universidad Nac. Auton. de Mexico Mexico, D.F. 04510, Mexico
Pogge R. W. Dept. of Astronomy Ohio State University Columbus, OH 43210–1106, U.S.A.
Poll A. Institut für Astrophysik und extraterrestrische Forschung D–53121 Bonn, Germany
Popescu N. A. Astronomical Institute of the Romanian Academy R–75212 Bucharest 28, Romania
Popovic L. Astronomical Observatory 11050 Belgrade, Yougoslavia
Poutanen J. Observatory and Astrophysics Laboratory, Univ. Helsinki SF–00014 Helsinki, Finland
Pronik I. 334413 Crimea, Ukraine
Röser H. Max–Planck–Institut für Astronomie D–6900 Heidelberg 1, Germany
Rafanelli P. Department of Astronomy University of Padova I–35122 Padova, Italy
Raikov A. St. Petersburg Indep. University SU–196135 St. Petersburg, Russia
Rantakyrö F. T. Onsala Space Observatory Göteborg University S–43992 Onsala, Sweden
Rees M. J. Institute of Astronomy GB–Cambridge, CB3 OHA, U.K.
Reichert G. A. Univ. Space Research Assoc. & NASA GSFC Greenbelt, MD 20771, U.S.A.
Reinsch K. Universitäts–Sternwarte D–37083 Göttingen, Germany
Rigopoulou D. Queen Mary and Westfield College Dept. of Physics London E1 4NS, U.K.
Robson I. Joint Astronomy Centre Hilo, HI 96720, U.S.A.
Rodriguez P. M. IUE Observatory E–28080 Madrid, Spain
Rokaki E. Queen Mary and Westfield College Dept. of Physics GB–London, E1 4NS, U.K.
Romanova M. M. Space Research Inst. of the Russian Academy of Sciences 117810 Moscow, Russia
Saikia D. J. TATA Institute of Fundamental Research Pune 411007, India
Salamanca I. Observatoire de Meudon DAEC F–92195 Meudon Principal Cedex, France
Salas–Casales L. Instituto de Astronomia UNAM San Diego, CA 92143–9027, U.S.A.
Salvati M. Arcetri Astrophysical Observatory I–50125 Firenze, Italy
Sanders D. B. Institute of Astronomy Honolulu, HI 96822, U.S.A.
Sandqvist A. Stockholm Observatory S–13336 Saltsjöbaden, Sweden
Santos–Lleo M. Observatoire de Paris–Meudon DAEC F–92195 Meudon Principal Cedex, France

Sapre A. K. Reader, School of Studies in Physics Pt. R.S. Shukla University Raipur–492010, India
Schalinski C. IRAM F–38406 Saint-Martin-d'Hères Cedex, France
Schlickeiser R. Max–Planck–Institut für Radioastronomie D–5300 Bonn, Germany
Schramm K. Institut d'Astrophysique Université de Liège B–4000 Liège, Belgique
Schulz H. Ruhr–Universität Astronomisches Institut D–4630 Bochum 1, Germany
Selvelli P. CNR Osservatorio Astronomico di Trieste I–34131 Trieste, Italy
Setti G. Istituto di Radioastronomia I–40126 Bologna, Italy
Shevchenko I. I. Inst. of Theoret. Astronomy Russian Academy of Sciences SU–191187 St. Petersburg, Russ
Shields J. C. Dept. of Astronomy Ohio State University Columbus, OH 43210, U.S.A.
Sillanpaa A. Tuorla Observatory SF–21500 Piikkiö, Finland
Simpson C. Dept. of Astrophysics Oxford University GB–Oxford, OX1 3RH, U.K.
Sol H. Observatoire de Meudon F–92195 Meudon Principal Cedex, France
Solomos N. Hellenic Naval Academy Dept. of Physics GR–18503 Piraeus, Greece
Staubert R. Astronomisches Institut Universität Tübingen D–7400 Tübingen, Germany
Steffen W. Max–Planck–Institut für Radioastronomie D–5300 Bonn 1, Germany
Stenholm L. Uppsala Observatory S–75120 Uppsala, Sweden
Stevens J. A. University of Central Lancashire Dept. of Physics and Astronomy Preston PR1 2HE, U.K.
Stift M. J. Institut für Astronomie voir adresses vacances aout dans dossier A–1180 Wien, Austria
Stirpe G. M. Osservatorio Astronomico di Bologna I–40126 Bologna, Italy
Storchi–Bergmann T. Instituto de Fisica UFRGS Av. Bento Goncalves 9500 91500 Porto Alegre, RS, Bras
Sun W. Institute of Astronomy Natl. Central University Chung–Li, 320 ROC, Taiwan
Suran D. M. Astronomical Institute of the Romanian Academy R–75212 Bucharest 28, Romania
Svensson R. Stockholm Observatory S–13336 Saltsjöbaden, Sweden
Swings J. Institut d'Astrophysique Université de Liège B–4000 Liège, Belgique
Tacconi–Garman L. E. Max–Planck–Institut für extraterrestrische Physik D–8046 Garching, Germany
Takalo L. O. Tuorla Observatory SF–21500 Piikkiö, Finland
Tanaka Y. Institute of Space and Astronautical Science Kanagawa-Ken 229, Japan
Tanzi E. Istituto di Fisica Cosmica CNR I–20133 Milano, Italy
Tashiro M. Dept. of Physics University of Tokyo Tokyo, 113, Japan
Teräsranta H. Metsähovi Radio Research Station SF–02540 Kylmälä, Finland
Thomas H. Max–Planck–Institut für Astrophysik D–85740 Garching, Germany
Thompson D. J. NASA Goddard Space Flight Center Greenbelt, MD 20771, U.S.A.
Thompson R. Steward Observatory University of Arizona Tucson, AZ 85721, U.S.A.
Tornikoski M. Metsähovi Radio Research Station SF–02540 Kylmälä, Finland
Torricelli G. Observatoire de Genève 1290 Sauverny,
Tovmassian H. Instituto Nacional de Astrofisica Optica y Electronica 72000 Puebla, Mexico
Trevese D. Istituto Astronomico Università di Roma "La Sapienza" I–00161 Roma, Italy
Trussoni E. Osservatorio Astronomico di Torino I–10025 Pino Torinese, Italy
Tsvetanov Z. Center for Astrophysical Sciences John Hopkins University Baltimore, MD 21218, U.S.A.
Tytler D. University of California, San Diego La Jolla, CA 92093–0111, U.S.A.
Ulrich M. European Southern Observatory D–8046 Garching, Germany
Valtaoja E. Metsähovi Radio Research Station SF–02540 Kylmala, Finland
Valtaoja L. NORDITA Nordisk Institut for Teoretisk Fysik DK–2100 Copenhagen 0, Denmark
Van Groningen E. Astronomiska Observatoriet S–75120 Uppsala, Sweden
Varshalovich D. A. A.F. Ioffe Inst. Phys. and Technology, SU–194021 St. Petersburg, Russia
Vicente L. Observatoire de Meudon DARC F–92195 Meudon Cedex, France
Vio R. Villa Franca del Castillo Satellite Tracking Station E–28080 Madrid, Spain

Vogel S. Hamburger Sternwarte Universität Hamburg D-2050 Hamburg 80, Germany
Von Linde J. Hamburger Sternwarte Universität Hamburg D-21029 Hamburg, Germany
Wagner S. J. Landessternwarte Königstuhl D-6900 Heidelberg, Germany
Wallinder F. NORDITA Nordisk Institut for Teoretisk Fysik DK-2100 Copenhagen, Denmark
Walter R. Max-Planck-Institut für Extraterrestrische Physik D-8046 Garching, Germany
Wamsteker W. VILSPA ESA IUE Observatory E-28080 Madrid, Spain
Wandel A. Observatoire de Paris F-92195 Meudon, France
Wanders I. Astronomiska Observatoriet S-75120 Uppsala, Sweden
Wang Y. Acton, MA 01720, U.S.A.
Ward M. J. Oxford University Astrophysics GB-Oxford, OX1 3RH, U.K.
Warwick R. S. Dept. Physics and Astronomy University of Leicester GB-Leicester, LEI 7RH, U.K.
Webb J. R. Dept. of Physics Florida Intl. University Miami, FL 33199, U.S.A.
Wei C. Center for Astrophysics University of Science and Technology 230026 Hefei, Anhui, China
Wiita P. J. Dept. of Physics and Astronomy Georgia State University Atlanta, GA 30303-3083, U.S.A.
Winge C. Dept. of Astronomy The Ohio State University Columbus, OH 43210, U.S.A.
Wisotzki L. Hamburger Sternwarte Universität Hamburg D-2050 Hamburg 80, Germany
Woltjer L. Observatoire de Haute-Provence F-04870 St Michel-l'Observatoire, France
Worrall D. M. Harvard-Smithsonian Center for Astrophysics Cambridge, MA 02138, U.S.A.
Yankulova I. M. Dept. of Astronomy Sofia University BG-1126 Sofia, Bulgaria
Yi I. Harvard-Smithsonian Center for Astrophysics, MS 51 Cambridge, MA 02138, U.S.A.
Zdziarski A. Copernicus Astronomical Center PL-00716 Warsaw, Poland
Zensus A. N.R.A.O. Socorro, NM 87801, U.S.A.
Zycki P. Nicolaus Copernicus Astronomical Center PL-00716 Warszawa, Poland

Foreword

Active Galactic Nuclei radiate over the electro-magnetic spectrum from radio waves to gamma rays. Understanding the physics of these objects therefore requires the synthesis of results from many different domains of Astronomy. It was the aim of the conference "Active Galactic Nuclei across the Electromagnetic Spectrum" to provide a forum where this exchange could take place.

Some 300 astronomers participated to the conference, 250 of them presented results either as oral papers or in the form of posters. Observations in all domains of the electro-magnetic spectrum in which astronomical observations can be made from the ground or from space were presented. Many theoretical contributions were also given.

There has been a tremendous growth in the number and quality of Astronomical observations in many spectral domains over the past several years. Students of Active Galactic Nuclei have been particularly keen to make use of the available facilities (both space born and on the ground), often in a very organised way, in order to obtain repeated simultaneous data covering large bands of the spectrum. This approach has produced a qualitatively new set of data for understanding the physics of Active Galactic Nuclei. The task of the meeting was to review this data in a coherent way.

Models of Active Galactic Nuclei were often devised in the past from observations in a single spectral domains. These models can now be confronted with much more comprehensive sets of data and often showed the need for changes or improvements. As a result there was a very open debate during the conference between different viewpoints for example on the physical origin of the optical and ultraviolet emission.

The meeting was hosted by the Observatory of Geneva in one of the buildings of the University of Geneva in the center of the city. It took place August 30 to September 3 1993. There were no parallel sessions, in line with the aim of the meeting to provide a synthesis of present knowledge rather than a set of particular views. Half of the posters were displayed from Monday to Wednesday, while the second half were presented from Wednesday to Friday. The quality of the posters was quite impressive.

The State of Geneva offered the participants a reception in the Palais Eynard, a magnificent building few hundred yards from the University. M. D. Föllmi, member of the Government of the State of Geneva gave a very warm welcome to all participants, insisting on the value of Astronomy for Peace in the world.

The conference dinner took place during a cruise on the lake of Geneva. Prof. L. Woltjer gave the after dinner speech, showing how (little) our fundamental knowledge on Active Nuclei has evolved in the 30 years since their discovery.

These proceedings contain almost all the presentations given at the conference both as posters and as oral invited and contributed papers. We offered all participants the possibility to publish their contribution so that the subjects reviewed during the conference are all presented and no arbitrary choices had to be made by the organisers. As a consequence, and to keep this volume to a reasonable size, the poster contributions are limited for most of them to one page. This implies that the posters must be viewed as a hint to a new development and/or as the pointer towards more detailed descriptions of the work presented here.

The meeting was supported by the Observatory of Geneva, the University of Geneva, the University of Lausanne, the Swiss National Science Foundation, the Swiss Academy of

Sciences, the Swiss Bank Corporation, the International Science Foundation the European Astronomical Society and the International Astronomical Union. We are grateful for this support which allowed us to welcome a large number of participants from all over the world in excellent conditions. Many of the invited speakers could arrange for independent funding of at least part of their expenses. This permitted us to support a larger number of astronomers who could not have participated otherwise.

The meeting could not have taken place without the generous help of a large number of people. Our secretary, Ms E. Teichmann, worked very hard and successfully to handle the large amount of administrative tasks related to the organisation. G. Simond and D. Mégevand devised numerous informatic tools to help us handle efficiently the data we needed from registrations to the publication of these proceedings. Several Ph.D. Students and Post Docs of the Geneva Observatory spent time helping at the registration desk and handling slides and microphones. All these people and several more are to be thanked for the smooth running of the meeting.

T. J.-L. Courvoisier   *A. Blecha*

Scientific Organising committee

R.D. Blandford
J.N. Bregman
S. Collin-Souffrin
T. J.-L. Courvoisier (Chairman)
H. Inoue
A. Kembhavi
J. Krolik
E.I. Robson
R. Staubert
R. Sunyaev
E. Tanzi
E. Valtaoja

Local Organising Committee

A. Blecha (Chairman)
G. Burki
J.-F. Bopp
M. Grenon
B. Hauck
A. Maeder
A. Orr
P. Bartholdi
L. Weber

Multi–wavelength Continuum Emission of AGN

THE ORIGIN OF CONTINUUM EMISSION IN ACTIVE GALACTIC NUCLEI

JOEL N. BREGMAN
Department of Astronomy, University of Michigan

Abstract. The general understanding of the continuum emission from AGN has changed from the picture where nonthermal processes were responsible for all of the emission. The current body of observation indicates that there are two types of objects, one being the blazar class (or blazar component), where nearly all of the emission is nonthermal, due primarily to synchrotron and inverse Compton emission. Variability studies indicate that the emitting region decreases with size from the radio through the X-ray region, where the size of the X-ray region is of order a light hour. More than two dozen of these radio-loud AGNs have been detected at GeV energies (one source at TeV energies), for which the radiation mechanism may be inverse Compton mechanism.

In the other class, the radio-quiet AGN (component), the emission is almost entirely thermal, with radiation from dust dominating the near infrared to submillimeter region. The optical to soft X-ray emission is often ascribed to black body emission from an opaque accretion disk, but variability studies may not be consistent with expectations. Another attractive model has free-free emission being responsible for the optical to soft X-ray emission. The highest frequencies at which these AGN are detected is the MeV range, and these data should help to determine if this emission is produced in a scattering atmosphere, such as that around an accretion disk, or by another model involving an opaque pair plasma.

1. Introduction

There has been steady progress in understanding the continuum emission from active galactic nuclei (AGNs), although the present picture is, in some respects, considerably different than it was a decade ago. This review examines how the models have changed, presents my view of the current picture, and outlines a few of the remaining issues facing the field.

Prior to about 1980, it was widely believed that, aside from a variety of emission lines and some line recombination radiation, all of the emission from AGN was nonthermal in origin. This point of view developed from the strong evidence that there was abundant nonthermal emission from radio-loud objects, combined with the philosophical belief that all AGNs were fundamentally similar. This philosophical belief of a single type of object, which might be attributed to Occam's Razor, was shown to be too simple by a wealth of data that accumulated in the 1980's.

2. The Radio Quiet AGNs

2.1. EMISSION FROM DUST

The two most important changes that occurred in the 1980's were the detailed measurements of the submillimeter through infrared region, and the study of the continuum emission over a very large wavelength region. In the application of these techniques, perhaps the greatest surprise occurred in the study of the radio-quiet

AGNs, such as the Seyfert galaxies and the radio-quiet quasars. These objects are not entirely radio-silent, as they are detected by the sensitive radio telescopes of today. However, the ratio of the radio to optical fluxes is orders of magnitude less than for the radio-loud AGNs, such as the BL Lac objects and the radio-loud quasars. Despite the relative weakness of their radio emission, the radio-quiet AGNs were often powerful sources of infrared emission, even at the longest wavelengths accessible to the *Infrared Astronomical Satellite* (IRAS), 100 μm. The transition from a far infrared flux density (3×10^{12} Hz), measured in Jy, to the radio emission (10^{10} Hz), measured in mJy, necessitates a rapid decrease in the flux density with decreasing wavelength.

A similar sharp drop from powerful far infrared emission to weak radio emission was seen in an entirely different class of extragalactic object, starburst galaxies (e.g., Soifer, Houck, and Neugebauer 1987; Telesco 1988). However, the emission from these galaxies is successfully modeled by thermal dust emission of absorbed optical and ultraviolet light rather than by nonthermal mechanisms. In the thermal dust model, the sharp decrease in the flux density occurs because dust becomes transparent at wavelengths longward of 50-100 μm and no longer emits as a blackbody (Draine 1990). The falloff can be extremely rapid, although the precise shape depends on grain composition, size, and temperature.

The synchrotron process can also produce a sharp falloff in the continuum emission, so it is necessary to try to distinguish between the dust and synchrotron models. For synchrotron emission, the low frequency falloff occurs as the plasma becomes opaque to its own radiation. In a homogeneous plasma, the opaque region has a slope of $F_\nu \propto \nu^{5/2}$ below some critical frequency that depends upon magnetic field, density and size of the region. Since these properties might be expected to vary amongst individual objects, the critical frequency should be seen by observers over a range of frequencies in an ensemble of sources.

These predictions were tested by observations that spanned the radio, millimeter, submillimeter, and far infrared regions (e.g., Chini *et al.* 1988, 1989; Barvainis and Antonucci 1989; Hughes *et al.* 1993; also, these proceedings). These researchers found that the spectral slope in the falloff region was often steeper than 5/2 (average slope of 3.75; Hughes *et al.* 1993), which violates the most extreme model with synchrotron emission (inhomogeneous synchrotron emitting regions without a sharp outer boundary always would produce a slope shallower than 5/2). Furthermore, the falloff occurred at approximately the same rest wavelength, for sources with a wide range of luminosities, and this falloff region was at approximately the same wavelength as in starburst galaxies. A characteristic turnover frequency is predicted from the dust model (near 100 μm), where dust becomes transparent. Such studies firmly established that thermal dust emission was the primary continuum mechanism in the submillimeter through infrared range for the radio-quiet AGNs.

This understanding led to renewed interest in developing models for dust emission. A general property of these models is that there must be a range of temperatures in the emitting dust region in order to fit the observed continuum. The dust temperature is determined by the balance between the absorbed optical-UV light from the AGN nucleus and the radiative rate of the dust grains (provided that the grains are not very small; Rees et al. 1969; Penston et al. 1974). Consequently, the warmest dust grains are closest to the central source (typically 0.1-1 pc distance), while the coolest dust grains, those responsible for the 60-100 μm emission, must lie 1-10 kpc from the nucleus (Barvainis 1987). This far infrared emitting region is so large, that it should have appeared to be non-variable during the year when the IRAS satellite was making measurements. An analysis of the 60-100 μm IRAS fluxes from radio-quiet AGNs confirmed this prediction of no variability (Clement et al. 1987; Edelson and Malkan 1987). At the shorter wavelengths, variability would be detectable, but the amplitude would be expected to decrease with increasing wavelength, which was seen by Cutri et al. (1985). Also, there would be a time delay between an outburst in the core region (the optical-UV region) and the reradiated emission (the near IR region). This time delay is of particular value because it provides valuable information on the size of the emitting region, for which there is a prediction. Such time delays are observed: in Fairall 9 the dust is thought to lie in the 0.3-1.3 light year region (Clavel, Wamsteker, and Glass 1989; Barvainis 1992); in NGC 3783, and 80-90 day lag is seen between the near-IR and UV regions (Glass 1992); in GQ Comae, a 250 day time delay is seen between the optical and near infrared regions (Sitko et al. 1993). These time delays are close to expectations, further supporting the dust model and the predicted spatial distribution.

Multifrequency observations of radio-quiet AGNs revealed a second important feature, a relative minimum in the power per logarithmic bandpass occurring at a rest frame wavelength of about 1 μm (e.g., Sanders et al. 1989; Hughes et al. 1993). This feature fit into the dust picture naturally, as it is linked to the sublimation temperature for dust. At temperatures exceeding 1000 K, dust begins to sublimate, a process that proceeds rapidly at temperatures closer to 2000 K. Dust close enough to the central source is heated to these temperature and quickly destroyed, so the emission from them is expected to be minimal. At these temperatures, dust behaves like a blackbody, so 2000 K corresponds to 1.5 μm, the approximate wavelength at which the minimum in the power occurs.

There is an alternative view of the origin of the near infrared emission in radio-quiet AGNs, a nonthermal power-law origin, although any support for this view is shrinking. Such a power-law is desirable to those fitting accretion disk models to the optical-UV emission (e.g., Sun and Malkan 1989), and it gathers its only real support by the near infrared to X-ray correlation that is seen (Edelson and Malkan 1986; Carleton et al. 1987). However, the near infrared emission is contaminated by galaxy light, so the correlation should improve if the galaxian contribution were removed. Such a test was performed by Kotilainen et al. (1992), who found that, in Seyfert 1 galaxies, the correlation did not improve, in conflict with expectation for the power-law picture.

2.2. THE BIG BLUE BUMP

The emission from the optical to the beginning of the soft X-ray emitting region is often referred to as the Big Blue Bump because, when plotting power per logarithmic bandwidth, there is a prominent local maximum in the ultraviolet region, with a noticeable falloff at shorter and longer wavelengths (e.g., Kolman et al. 1993). Usually, this is not just a local maximum, but an absolute maximum in the power output of these systems. If one defines this Big Blue Bump as extending from 1 μm to 0.5 keV, it contains typically one-third or one-half of the entire observed power from the object; the second most important energetic region is usually the infrared to submillimeter region. Some of the emission in the Big Blue Bump region is due to broad and narrow emission lines, a forest of Fe II emission lines, plus hydrogen recombination radiation. However, these contribute only modestly to the total power, most of which is attributed to some other feature or structure, such as an accretion disk.

Thermal emission from an accretion disk seemed to be the natural explanation for the Big Blue Bump, although recent observations call into question this explanation. One of the most powerful arguments for an accretion disk is theoretical in that it is should occur naturally and is an efficient mechanism through which energy can be extracted as material falls into the gravitational potential of a putative black hole (Rees 1984). Efforts to test this model observationally are concerned with comparing data to predictions of emission from the opaque surface, or to a possible atmosphere above the surface of the accretion disk. The observed Big Blue Bumps can be fit with thermal emission from an accretion disk, provided that one assumes an underlying nonthermal power-law for the near infrared emission (Sun and Malkan 1989; Malkan 1991). Since the near infrared emission is due to dust, the assumed nonthermal power-law continuum should be discarded. It may be possible to fits accretion disk models along with dust emission models, but current efforts seem to require a dust temperature that may be in excess of the maximum expected value (Malkan 1989).

Other tests of emission from accretion disks fail to provide confirmation of the model, but they do not rule out the model either. Accretion disks are not solid bodies, but have a vertical extent with a decreasing gas density distribution, creating a type of photosphere for which the actual structure is highly model-dependent. Hydrogen continuum absorption shortward of 912 A would be expected for a wide range of models, but this phenomenon is not detected (Antonucci, Kinney, and Ford 1989). If this disk and atmosphere are observed in any orientation other than face-on, the continuum emission should be polarized. Polarization at the 1-2% range is observed, and while that was originally believed to be below the expected values, more accurate predictions, including the effects of general relativity, indicate that it is consistent with the data (Laor, Netzer, and Piran 1990; Netzer 1991).

An important inconsistency between models of accretion disks and the data deals with the time delay between variations seen in various wavelength bands. In all opaque accretion disk models, the emitting region at a particular wavelength is associated with a particular radius, with the hottest regions (shortest wavelength) closest to the center. Then, a disturbance near the center of the accretion disk would cause an outburst at short wavelengths, and this disturbance might propagate radially outward at the local sound speed, causing an outburst at longer wavelengths at some later time. For the typical black hole masses and accretion disks discussed for AGNs, the time delay between UV and optical variations should be weeks. However, in the few systems for which measurements of the delays have been attempted, such as NGC 5548, no time delay has been measured, with upper limits of a few days (Couroisier and Clavel 1991; Molendi, Maraschi, and Stella 1992).

Efforts to find periodic behavior in the light curve, associated with rotation of or through the accretion disk, has been dealt a severe blow with recent ROSAT observations of NGC 6814 (see Madjeski, these proceedings). They show that the 12,000 second periodic variability is not intrinsic to the central engine of NGC 6814 but is due to an X-ray binary star system in the Galaxy that was close on the sky to NGC 6814 so that the two systems were confused when observed with previous X-ray instruments.

The only other evidence in support of orbital motion in the cores of AGNS is the double-peaked line profiles, a signature of rotating material, seen in generally radio-loud objects, such as 3C390.3 or 3C 332 (Chen, Halpern, and Fillipenko 1989; Halpern 1990). The double- peaked lines are seen to vary in 3C390.3, and continued observations are warranted to determine if there is any validity to the 10.4 year "sinusoidal" variations reported by Veilleux and Zheng (1991).

Given the difficulties with sustaining the standard accretion disk model, perhaps the model should be strongly modified or entirely discarded (e.g., Celotti, Fabian, and Rees 1992). One of the most attractive alternative suggestions is that optically thin free-free radiation dominates the observed optical-ultraviolet continuum (detailed discussion in Barvainis 1993; also Ferland, Korista, and Peterson 1990; Malkan and Sargent 1982). In this model, the shape of the 1 μm to 0.1 μm continuum is adequately fit without requiring an underlying power-law or very hot dust. Also, this model solves the time-delay problem, discussed above. There would be no time delay between an ultraviolet and optical outburst, because free-free emission occurs over a broad wavelength band, unlike blackbody radiation.

For this model to be successful, the continuum emitting region must be small enough to be consistent with the observed variations. However, it appears possible to make the emitting region small enough and still retain the transparent nature of the emission, with an optical depth that may be as great as 0.1.

A complete "cartoon" of the emitting region must be able to relate the behavior at various wavebands, and critical to this picture are the X- rays. As discussed by Clavel (these proceedings) for the case of NGC 5548, the optical-UV region does not show strong variability on timescales of less than 1 day, while the X-ray emission can vary strongly to a timescale of 0.1 days. This suggests that the X-rays come from a region smaller (interior to) the optical-UV region. Furthermore, the two regions are closely related as seen in the correlated variability between the X-ray and optical-UV bands.

Then, in our "cartoon", the X-ray emission lies closest to the black hole and may be largely opaque. Exterior to this emitting region is the free-free emitting region, which may be a dense extended atmosphere around an accretion disk. The free-free emitting region is heated by the photons and possibly fast particles from the X-ray region. Depending upon the geometry, the free-free region (as an accretion disk atmosphere) may be transparent when seen face-on, but nearly opaque as seen edge-on, which would be its orientation to the X-ray emission region, making absorption of the X-rays an efficient process.

2.3. The X-Ray Emission

The X-ray emission, which spans three orders of magnitude in energy from 0.1-100 keV is a complex region with several phenomena present (see review by Mushotzky, Done, and Pounds 1993). At the low energy end of this band, the 0.1-1 keV region, the spectrum is falling rapidly and is probably the high frequency extension of the Big Blue Bump (e.g., Masnou et al. 1992; Turner et al. 1993). This feature, referred to as the "soft excess", appears to be present in most, if not all radio-quiet AGNs (Walter and Fink 1993; Shastri et al. 1993).

The emission above 1 keV is adequately fit by a power-law component, Fe K emission, plus Fe reflected continuum emission (Pounds, Nandra, and Stewart 1992). Upon removing the contribution from Fe, the underlying power-law has a slope (energy index) of 0.9-1.0. This power-law continuum extends to about 100 keV, at which point, a spectral break is seen (see summary of OSSE observations by Kurfess, these proceedings).

Two distinct models for this power-law continuum have become popular, a pair cascade model and a Compton scattering model (reviewed by Zdziarski, these proceedings). In the first model, the emission emerges from a region that is at least partly opaque to photon-photon collisions, which mediate electron-positron pair production (e.g., Svensson, R. 1987; Done and Fabian 1989; Zdziarski et al. 1990). In the second model, the X-ray emission is produced by Compton scattering of soft photons by a thermal population of hot electrons, such as might occur in an accretion disk corona (Maraschi and Molendi 1991; Walter and Courvoisier 1992; Haardt and Maraschi 1993). The latter model was first developed to explain certain X-ray properties of Galactic black hole binaries, for which there are some spectral similarities to AGNs (Tanaka 1992). The critical test of the model will come from studies at energies near the electron-positron annihilation line (0.511 MeV), where the two models make significantly different predictions for continuum shape and annihilation line strength.

3. Blazars

3.1. Is It All Nonthermal Emission?

In contrast to the radio-quiet AGNs, where our understanding changed radically during the past decade, recent observations of radio-loud objects have reaffirmed the picture that developed in the mid-1970s (Jones, O'Dell, and Stein 1974). That is, nonthermal synchrotron emission dominates the radio through ultraviolet region, and inverse Compton scattering probably dominates the hard X-ray and gamma-ray regions.

To assure ourselves that nonthermal processes are required by the observations, we examine the "purest" of the radio-loud AGNs, the BL Lac objects. The reason that the radio data are believed to be nonthermal synchrotron emission is that the brightness temperature is commonly $10^{11} - 10^{12}$ K in the core, requiring relativistic electron (e.g., Gabudza, Wardle, and Roberts 1989). Furthermore, the emission is highly polarized, power-law in shape, and there is evidence for ordered magnetic fields, all consistent with synchrotron theory; finally, relativistic bulk motion is required. The radio spectrum extends smoothly into the submillimeter, infrared, and optical region, suggesting that they have an origin in synchrotron emission (e.g., Browne et al. 1989a,b; Bregman et al. 1990). In the far infrared, an estimate of the brightness temperature can be derived from the limited variability observations made with IRAS, which has a characteristic timescale shorter than typical radio variations but longer than optical variations (Edelson and Malkan 1987). These observations suggest a brightness temperature greater than 10^6 K, well above the temperature that dust can radiate at (recall that dust is the source of the far infrared emission in the radio-quiet AGNs), but consistent with a synchrotron origin. At higher frequencies, optical data exhibit rapid variability and high polarization (Carini and Miller 1992; Carini et al. 1992; Moore and Stockman 1981, 1984), and the soft X-ray data has shown variability so rapid that it occurs on a time scale comparable to or shorter than the light crossing time of a massive black hole, suggesting the presence of relativistic motion (Doxsey et al. 1983; Feigelson et al. 1986; Giommi et al. 1986). These observations of BL Lac objects (and Blazars in general) provide either direct evidence for, or consistency with the synchrotron process over approximately ten orders of magnitude in frequency space (more detailed review in Bregman 1990).

The identification of continuum features was a key to understanding radio-quiet AGNs, so it is worth looking for such features in the blazar class of objects. The general nature of the continuum is that the power per logarithmic bandwidth rises from the radio, peaking somewhere in the millimeter to ultraviolet band. The peak is generally quite broad and its location peak has an enormous range in the rest wavelength (Landau et al. 1986; Browne et al. 1989a,b; Bregman 1990). This range in continuum characteristics suggest a sizable range in the underlying properties of the emitting relativistic plasma.

A distinctive feature has been seen in the infrared-UV region of some BL Lacs, a sharp high frequency turnover in the continuum that has been attributed to an upper energy cutoff to the electron distribution (Rieke et al. 1979, 1982; Beichman et al. 1981a,b; Bregman et al. 1981). The identification of this feature is best achieved in the infrared-UV range, where instrumental sensitivity is good, but for some objects, this feature may occur in the extreme UV or X-ray region as well. Giommi (these proceedings) suggests that the difference between radio and X-ray selected BL Lacs is that for the X-ray selected BL Lacs, the cutoff does not occur until the X-ray or higher frequency regions; the other, and possibly more common view, is that the two groups of objects possess different orientations of their relativistically boosted beams with respect to our line of sight (Maraschi et al. 1986; Padovani and Urry 1992; Padovani 1992).

The BL Lac objects show virtually no evidence for thermal emission from gas (except a few exceedingly weak emission lines), but some of the objects in the blazar class possess the same spectral features as were seen in the radio-quiet AGNs (e.g., Sanders et al. 1989). These objects seem to be a combination of the pure nonthermal objects (BL Lacs) and the radio-quiet thermally emitting sources. The relationship between these two type of sources is one of the major problems that remains to be clarified. Such crossover objects are the most complex to study, and it is somewhat unfortunate that the brightest quasar in the sky, 3C 273, is this type of object. The study of AGN continuum emission might have progressed in a more direct manner if the brightest and first objects studied had been a classical BL Lac object (i.e., OJ 287, BL Lac) and a typical radio-quiet AGN (i.e., the Seyfert 1 galaxy NGC 5548).

3.2. Relativistic Jets

Early VLBI observations of the central core of the radio emission established the asymmetric nature of the central emitting region, and subsequent observations have shown that the emission is confined to a narrow channel (jet-like) on lengthscales spanning orders of magnitude ($1-10^5$ pc; Bridle and Perley 1984). These observations led to the development of models for the radio region, which also have been applied to the higher frequency emission, where a relativistic shock propagates along an inhomogeneous plasma jet. One of the most impressive successes of this model has been the ability to reproduce the radio outbursts, seen in several frequencies, in objects such as BL Lac (Hughes, Aller, and Aller 1989a,b; Hughes, these proceedings).

The optical and X-ray emitting regions are believed to originate in a jet, but on a size scale smaller than the radio emitting region (Ghisellini et al. 1986, 1993; Marscher, Gear, and Travis 1992). The relative sizes are suggested by the shape of the continuum, but more convincingly by the variability studies. The typical timescale of substantial variability is months-years in the radio region, days in the optical region, and hours in the X-ray region (review in Bregman 1990).

The relative sizes of one emitting region with respect to another can be investigated by examining correlated variability and time delays. In the transparent region, such as the infrared through X-ray wavebands, the synchrotron emission from an ensemble of electrons is broad-band, which may make it difficult to determine time delays. In a variability study of PKS 2155-304 (Edelson et al. 1994; these proceedings), the X-ray variations are correlated with and leads those in the ultraviolet by 2 hours, clearly establishing the connection between the two regions and the estimate for the size. Formally, no lag is seen between the UV, optical, and infrared regions, with an upper limit to the lag of 1 day (a similar limit is found for BL Lac; Bregman et al. 1990), although there is evidence that the ultraviolet variations precede those in the optical region (Urry et al. 1993). There is a modest correlation between optical and radio variations, with a time delay that is typically a year, but determining such correlations is hampered because of the different statistical nature of the variability in the two bands (Hufnagel and Bregman 1992).

3.3. HIGH ENERGY EMISSION

When the X-ray emission appears to be an extension of the infrared- optical-ultraviolet continuum, it is likely to be synchrotron emission from the inner part of the jet (i.e., PKS 2155-304; Edelson et al. 1994). However, for those objects where the X-ray emission is not an extrapolation of the lower frequency synchrotron continuum, another emission mechanism must be active, such as the inverse Compton process.

There is, as yet, no definitive proof that the inverse Compton process produces such high frequency emission, but there are some observations that offer support. A good correlation might be expected between the scattered (X-ray) and the seed photons (infrared-millimeter), and a good correlation is seen (Owen, Helfand, and Spangler 1981). Additionally, the fluxes of the seed and scattered photons would be expected to vary together, and such correlated variations are seen, in a limited data set, for BL Lac (Bregman et al. 1990; Kawaii et al. 1991).

Perhaps the most exciting new development at high energies comes from the EGRET instrument on the *Compton Gamma-Ray Observatory*, which can measure fluxes in the GeV photon range. As of the time of this meeting, EGRET has detected 26 AGNs, and every one is a radio-loud object (one of these sources, Mrk 421 is detected in the TeV range by the Whipple telescope; Punch et al. 1992); not a single radio- quiet object is detected in the GeV range (see the contributions by Kurfess and Thompson, these proceedings). The best correlation between the GeV emission and the continuum at lower frequencies is with the radio emission: all detected sources are at least 0.5 Jy at 5 GHz. This astonishing correlation is suggestive of an upscattering process, such as the inverse Compton mechanism, which is a promising explanation (Marscher and Bloom 1992; Bloom and Marscher 1993).

Variability has been seen at these gamma-ray energies in 3C 279 (Kniffen et al. 1993), and the timescale of variation (a few days) is characteristic of the optical-ultraviolet region, although it would be impossible to detect very rapid variations at gamma-ray energies because of the extremely low count rates.

In the distribution of the emitted power, it is difficult to determine whether the gamma-rays may is the dominant power source or only a modest contributor. If the Doppler boosting factor is the same at all frequencies so that the observed flux density distribution is a true representation of the power distribution, then some of the sources detected by EGRET are dominated by gamma-ray emission (e.g., 3C454.3; Hartman et al. 1993). However, the sources detected in gamma- rays may represent only transient outburst events. It is equally important to examine which of the optically and radio bright sources were not detected and the implications for the power distribution. For example, the bright source BL Lacertae has not been detected by EGRET, which implies that the gamma-ray contribution to the total power is less than that from the submillimeter through optical region. Finally, a definitive discussion of the power distribution is hampered by the uncertainty of the magnitude of the Doppler boosting parameter in the gamma-ray region, relative to the lower frequency emitting regions. With the continued operation of GRO, we look forward to what should be rapid progress in this field and an elucidations of many of these important issues.

JNB wishes to acknowlege support from NASA under grant NAGW-2135.

References

Antonucci, R.R.J., Kinney, A.L., and Ford, H.C. 1989, *Ap. J.*, **342**, 64.
Barvainis, R. 1987, *Ap. J.*, **320**, 537.
Barvainis, R. 1992, *Ap. J.*, **400**, 502.
Barvainis, R., and Antonucci, R. 1989, *Ap. J. Sup. Ser.*, **70**, 173.
Barvainis, R. 1993, *Ap. J.*, **412**, 513.
Beichman, C.A., Neugebauer, G., Soifer, B.T., Wootten, H.A., Roellig, T., and Harvey, P.M. 1981a, *Nature*, **293**, 711.
Beichman, C.A., Provdo, S.H., Neugebauer, G., Soifer, B.T., Matthews, K., and Wootten, H.A. 1981b, *Ap. J.*, **247**, 780.
Bloom, S.D., and Marscher, A.P. 1993, in *Proceedings of the Compton Observatory Symposium*, ed. N. Gehrels (New York: AIP), in press.
Bregman, J.N. 1990, *Astr. Ap. Rev.*, **2**, 125.
Bregman, J.N., et al. 1981, *Nature*, **293**, 714.
Bregman, J.N., et al. 1990, *Ap. J.*, **352**, 574.
Bridle, A.H., and Perley, R.A. 1984, *Ann. Rev. Astr. Ap.*, **22**, 319.
Browne, L.M.J., et al. 1989a, *Ap. J.*, **340**, 129.
Browne, L.M.J., et al. 1989b, *Ap. J.*, **340**, 162.
Carini, M.T., and Miller, H.R. 1992, *Ap. J.*, **385**, 146.
Carini, M.T., and Miller, H.R., Noble, J.C., and Goodrich, B.D. 1992, *Astr. J.*, **104**, 15.
Carleton, N.P., et al. 1987, *Ap. J.*, **318**, 595.
Celotti, A., Fabian, A.C., and Rees, M.J. 1992, *M.N.R.A.S.*, **255**, 419.
Celotti, A., Maraschi, L, and Treves, A. 1991, *Ap. J.*, **377**, 403.
Chen, K., Halpern, J.P., and Filippenko, A.V. 1989, *Ap. J.*, **339**, 742.
Chini, R., Steppe, H., Kreysa, E., Krichbaum, T., Quirrenbach, A., Schalinski, C., and Witzel, A. 1988, *Astr. Ap.*, **192**, L1.
Chini, R., Kreysa, E., and Biermann, P.L. 1989, *Astr. Ap.*, **219**, 87.
Clavel, J., Wamsteker, W., and Glass, I.S. 1989, *Ap. J.*, **337**, 236.
Clement, R., Sembay, S., Hanson, C.G., and Coe, M.J. 1988, *M.N.R.A.S.*, **230**, 117.

Couroisier, T.J.-L, and Clavel, J. 1991, *Astr. Ap.*, **248**, 389.
Cutri, R.M., Wisniewski, W.Z., Rieke, G.H., and Lebofsky, M.J. 1985, *Ap. J.*, **296**, 423.
Done, C., and Fabian, A.C. 1989, *M.N.R.A.S.*, **240**, 81.
Doxsey, R. *et al.* 1983, *Ap. J.*, **264**, L43.
Draine, B.T. 1990, in *The Interstellar Medium in Galaxies*, ed. H.A. Thronson, and J.M. Shull (Kluwer: Dordrecht),p. 483.
Edelson, R.A., and Malkan, M.A. 1986, *Ap. J.*, **308**, 59.
Edelson, R.A., and Malkan, M.A. 1987, *Ap. J.*, **323**, 516.
Edelson, R.A., *et al.* 1994, *Ap. J.*, in press.
Feigelson, E.D., *et al.* 1986, *Ap. J.*, **302**, 337.
Ferland, G.J., Korista, K.T., and Peterson, B.M. 1990, *Ap. J.*, **363**, L21.
Gabuzda, D.C., Wardle, J.F.C., and Roberts, D.H. 1989, *Ap. J.*, **336**, L59.
Ghisellini, G., Maraschi, L., Tanzi, E., Treves, A. 1986, *Ap. J.*, **310**, 317.
Ghisellini, G., Padovani, P., Celotti, A., and Maraschi, L. 1993, *Ap. J.*, **407**, 65.
Giommi, P., *et al.* 1986, *Ap. J.*, **303**, 596.
Glass, I.S., 1992, *M.N.R.A.S.*, **256**, 23P.
Haardt, F., and Maraschi, L. 1991, *Ap. J.*, **380**, L51.
Halpern, J.P. 1990, *Ap. J.*, **365**, L51.
Hartman, R.C. *et al.* 1993, *Ap. J.*, **407**, L41.
Hufnagel, B.R., and Bregman, J.N. 1992, *Ap. J.*, **386**, 473.
Hughes, D.H., Robson, E.I., Dunlop, J.S., and Gear, W.K. 1993, *M.N.R.A.S.*, **263**, 607.
Hughes, P.A., Aller, H.D., Aller, M.F. 1989a, *Ap. J.*, **341**, 54.
Hughes, P.A., Aller, H.D., Aller, M.F. 1989b, *Ap. J.*, **341**, 68.
Jones, T.W., O'Dell, S.L., and Stein, W.A. 1974, *Ap. J.*, **188**, 353.
Kawai, N. *et al.* 1991, *Ap. J.*, **382**, 508.
Kniffen *et al.* 1993, *Ap. J.*, **411**, 133.
Kolman *et al.* 1993, *Ap. J.*, **402**, 514.
Kotilainen, J.K., Ward, M.J., Boisson, C., DePoy, D.L., and Smith, M.G. 1992, *M.N.R.A.S.*, **256**, 149.
Landau, R., *et al.* 1986, *Ap. J.*, **308**, 78.
Laor, A., Netzer, H., and Piran, T. 1990, *M.N.R.A.S.*, **242**, 560.
Malkan, M. 1989, in *Theory of Accretion Disks*, ed. F. Meyer, W. Duschl, J. Frank, and E. Meyer-Hofmeister (Kluwer: Dordrecht), p. 19.
Malkan, M. 1991, in *Structure and Emission Peoperties of Accretion Disks*, ed. C. Bertout, S. Collin, J.-P. Lasota, J. Tran Thanh Van, (Gif sur Yvette: Ed. Frontiers), p. 165.
Malkan, M.A., and Sargent, W.L.W. 1982, *Ap. J.*, **254**, 22.
Maraschi, L., Ghisellini, G., Tanzi, E., Treves, A. 1986, *Ap. J.*, **310**, 325.
Maraschi, L., and Molendi, S. 1991, *Ap. J.*, **368**, 138.
Marscher, A.P., and Bloom, S.D. 1992, in *Compton Observatory Science Workshop*, p. 346.
Marscher, A.P., Gear, W.K., and Travis, J.P. 1992, in *Variability of Blazars*, ed. E. Valtaoja and M.J. Valtonen (Cambridge: CUP), p. 85.
Masnou, J.L., Wilkes, B.J., Elvis, M., McDowell, J.C., and Arnaud, K.A. 1992, *Astr. Ap.*, **253**, 35.
Molendi, S., Maraschi, L., and Stella, L. 1992, *M.N.R.A.S.*, **255**, 27.
Moore, R.L., and Stockman, H.S. 1981, *Ap. J.*, **243**, 60.
Moore, R.L., and Stockman, H.S. 1984, *Ap. J.*, **279**, 465.
Mushotzky, R.F., Done, C., and Pounds, K.A. 1993, *Ann. Rev. Astr. Ap.*, **31**, 717.
Netzer, H. 1991, in *Structure and Emission Properties of Accretion Disks*, IAU Colloquium No. 129, p. 177.
Owen, F.N., Helfand, D.J., and Spangler, S.R. 1981, *Ap. J.*, **250**, L55.
Padovani, P. 1992, *Astr. Ap.*, **256**, 399.
Padovani, P., and Urry, C.M. 1992, *Ap. J.*, **387**, 449.
Penston, M.V., Penston, M.J., Selmes, R.A., Becklin, E.E., and Neugebauer, G. 1974, *M.N.R.A.S.*, **169**, 357.
Pounds, K.A., Nandra, K., and Stewart, G.C. 1992 in *Ginga Memorial Symposium*, ed. F. Makino

and F. Nagase (ISAS: Tokyo), p. 45.
Punch, M. et al. 1992, Nature, **358**, 477.
Rees, M.J., Silk, J.I., Werner, M.W., and Wickramsinghe, M.C. 1969, Nature, **223**, 788.
Rees, M.J. 1984, Ann. Rev. Astr. Ap., **22**, 471.
Rieke, G.H., Lebofsky, M.J., and Kinman, T.D. 1979, Ap. J., **232**, L151.
Rieke, G.H., Lebofsky, M.J., Wisniewski, W.Z. 1982, Ap. J., **263**, 73.
Sanders, D.B., Phinney, E.S., Neugebauer, G., Soifer, B.T., and Matthews, K. 1989, Ap. J., **347**, 29.
Shastri, et al. 1993, Ap. J., **410**, 29.
Sitko, M., Sitko, A.K., Siemiginowska, A., and Szczerba, R. 1993, Ap. J., **409**, 139.
Soifer, B.T., Houck, J.R., and Neugebauer, G. 1987, Ann. Rev. Astr. Ap., **25**, 187.
Sun, W.-H., and Malkan, M.A. 1989, Ap. J., **346**, 68.
Svensson, R. 1987, M.N.R.A.S., **227**, 403.
Tanaka, Y. 1992 in *Ginga Memorial Symposium*, ed. F. Makino and F. Nagase (ISAS: Tokyo), p. 19.
Telesco, C.M. 1988, Ann. Rev. Astr. Ap., **26**, 26.
Turner, J. et al. 1993, Ap. J., **407**, 556.
Urry, C.M., et al. 1993, Ap. J., **411**, 614.
Veilleux, S., and Zheng, W. 1991, Ap. J., **377**, 89.
Walter, R., and Courvoisier, T.J.-L. 1992, Astr. Ap., **266**, 65.
Walter, R., and Fink, H.H. 1993, Astr. Ap., **274**, 105.
Zdziarski, A.A., Ghisellini, G., George, I.M., Svensson, R., Fabian, A.C., Done, C. 1990, Ap. J., **363**, L1.
Zdziarski, A.A., and Krolik, J.H. 1993, Ap. J., **409**, L33.

THE UV–OPTICAL–NEAR–IR EMISSION OF BL LACERTAE OBJECTS.

R. FALOMO
Osservatorio Astronomico di Padova, v. Osservatorio 5, 35122, Padova, Italy

E. PIAN
Scuola Internazionale Superiore di Studi Avanzati, via Beirut 2-4, 34014 Trieste, Italy

R. SCARPA
Dipartimento di Astronomia dell'Università di Padova, v. Osservatorio 5, 35122 Padova, Italy

and

A. TREVES
Scuola Internazionale Superiore di Studi Avanzati, via Beirut 2-4, 34014 Trieste, Italy

Abstract. Results from UV, optical and near–IR simultaneous observations for 11 BL Lac objects are reported. We find that for all but one source the spectral flux distribution can be described by a single power law ($f_\nu \propto \nu^{-\alpha}$) plus, where relevant, the contribution of the host galaxy. The comparison of the optical-near-IR and UV spectral indices for two samples of BL Lacs suggests the same picture for a larger sample of objects.

Introduction

The overall spectral flux distribution (SFD) of BL Lac objects, is usually interpreted as due to the synchrotron or synchrotron self-Compton processes. Depending on the considered energy range, complex forms, like broken power laws or a curve that steepens with increasing frequency are used to describe the SFD rather than a single power law (e.g. Landau *et al.* 1986, Impey and Neugebauer 1988). Spectral "breaks" are reported to occur between near-IR and optical or between optical and UV frequencies (*e.g.* Ghisellini *et al.* 1986). These can possibly be due to the lack of simultaneity among observations in different bands, or to the reddening which introduces a steepening of the continuum at optical-UV frequencies, or to the contribution of starlight from the host galaxy, which produces a steepening of the energy distribution in the optical and a flattening in the near-IR. If instead the observed spectral breaks are intrinsic to the emission they may be significant to constrain the radiation process.

We report here on quasi simultaneous ($\Delta t \lesssim 1$ day) UV, optical and near-IR observations of 11 BL Lac objects obtained in the course of our 10 years systematic multifrequency study of BL Lacs (see *e.g.* Falomo *et al.* 1993 and references therein). In two cases (see Table 1) simultaneous IR-optical and optical-UV data taken at different epochs have been combined. UV observations were obtained using both cameras (SWP and LWP) onboard of the *International Ultraviolet Explorer* (IUE) while optical spectrophotometry and J,H,K, L photometry of the sources were obtained at the European Southern Observatory (ESO). Table 1 gives program objects, along with observation date, redshift, galactic extinction, radio or

X-ray selection, UV-to-IR spectral slope, percentage contribution and absolute magnitude of the host galaxy.

Table 1: Blazars Observed in UV-Opt-IR

Objects	Date	z	A_V	R/X	α	% Gal	$M_V^\dagger$
0048 − 097	87 Jan 7,8	...	0.22	R	0.93 ± 0.02	...	...
0118 − 272	89 Aug 10	0.559	0.09	R	1.20 ± 0.01	...	...
0301 − 243	89 Aug 9	(0.2)	0.11	R	0.79 ± 0.02	11	-22.2
0323 + 022	89 Aug 10	0.147	0.50	X	0.23 ± 0.04	30	-22.2
0414 + 009	89 Feb 15	0.287	0.51	X	0.54 ± 0.05	3	-21.3
0422 + 004	88 Jan 9,10	(0.1)	0.42	R	1.20 ± 0.05	10	-21.5
0521 − 365	87 Jan 8	0.055	0.21	R	1.43 ± 0.03	51	-21.7
1538 + 149	88 Aug 2	0.605	0.20	R	1.34 ± 0.02	...	...
1553 + 113	88 Aug 2,5	...	0.22	X	††	...	...
2005 − 489	86 Sep/89 Aug	0.071	0.33	X	0.57 ± 0.02	12	-22.2
2155 − 304	86 Sep/89 Aug	0.116	0.10	X	0.47 ± 0.02	3	-22.2

† Values are not K-corrected and computed assuming $H_0=50$; $q_0=0$.

†† IR-Optical region: $\alpha = 0.78 \pm 0.02$; UV region: $\alpha = 1.56 \pm 0.04$.

Results

A composite (SFD) was constructed for each object from quasi simultaneous observations. Data were corrected for interstellar reddening using A_V of Table 1. In 7 cases we attempted a decomposition of the SFD in terms of a power law plus a standard giant elliptical galaxy. For 2 objects of unknown redshift a rough estimate of z has been assumed based on fit optimization of the galaxy contribution.

In Fig. 1 we report as example the cases of 0048-097, 0323+022, 0422+004, 1553+113. We found that in all but one case the observed SFD is well accounted for either by a single power law or by a power law plus an elliptical galaxy. The exception is PG 1553+11 whose emission shows a spectral break at $\nu \sim 10^{15}$ Hz.

The average spectral index of the non thermal component is $<\alpha> = 0.87 \pm 0.41$ with a marked tendency for X-ray selected objects to be flatter than radio selected ones (respectively $<\alpha> = 0.45 \pm 0.15$ and $<\alpha> = 1.15 \pm 0.24$).

The constancy of the UV-to-near-IR spectral shape has been tested (Table 2) comparing the spectral indices in the optical-near-IR region for 33 blazars (Falomo et al. 1993) with the UV spectral indices of 24 sources (Pian and Treves 1993). This constancy is observed also if only common objects (14) are considered and is still maintained subdividing them in X-ray and radio selected subsamples.

The main result that no significant change is present in the shape of BL Lac spectra between their IR-optical part and the UV one is at variance with the conclusions reached by other authors who report spectral breaks in the near-IR optical or UV region (*e.g.* Cruz-Gonzalez and Huchra 1984; Ghisellini *et al.* 1986; Smith *et al.* 1987; Ballard *et al.* 1990). We suggest that this is related to our use of simultaneous data, appropriate correction for reddening and for the host galaxy contribution. The absence of any relevant spectral break indicates that a non thermal synchrotron process from a unique radiating volume is likely responsible for the observed emission.

Table 2: Average Spectral Indices

	$<\alpha_{IROP}>$	$1-\sigma$	N_{obj}	$<\alpha_{UV}>$	$1-\sigma$	N_{obj}
	Whole samples					
All	1.08	0.34	33	0.97	0.41	24
X-ray selected	0.68	0.28	8	0.66	0.30	10
Radio selected	1.20	0.30	25	1.20	0.33	14
	Common objects					
All	0.92	0.41	14	0.98	0.46	14
X-ray selected	0.52	0.15	6	0.66	0.39	6
Radio selected	1.22	0.22	8	1.22	0.36	8

References

Ballard, K.R., Mead, A.R.G., Brand, P.W.J.L. and Hough, J.H.: 1990, MNRAS, 243, 640
Cruz-Gonzalez, I. and Huchra, J.P.: 1984, AJ, 89, 441
Falomo, R., Bersanelli, M., Bouchet, P. and Tanzi, E. G.: 1993, AJ, 106, 11
Ghisellini, G., Maraschi, L., Tanzi, E.G. and Treves, A.: 1986, ApJ, 310, 317
Impey, C.D. and Neugebauer, G.: 1988, AJ, 95, 307
Landau, R. *et al.*: 1986, ApJ, 308, 78
Pian, E. and Treves, A.: 1993, to appear in ApJ
Smith, P. S., Balonek, T. J., Elston, R. and Heckert, P. A.: 1987, ApJS, 64, 459

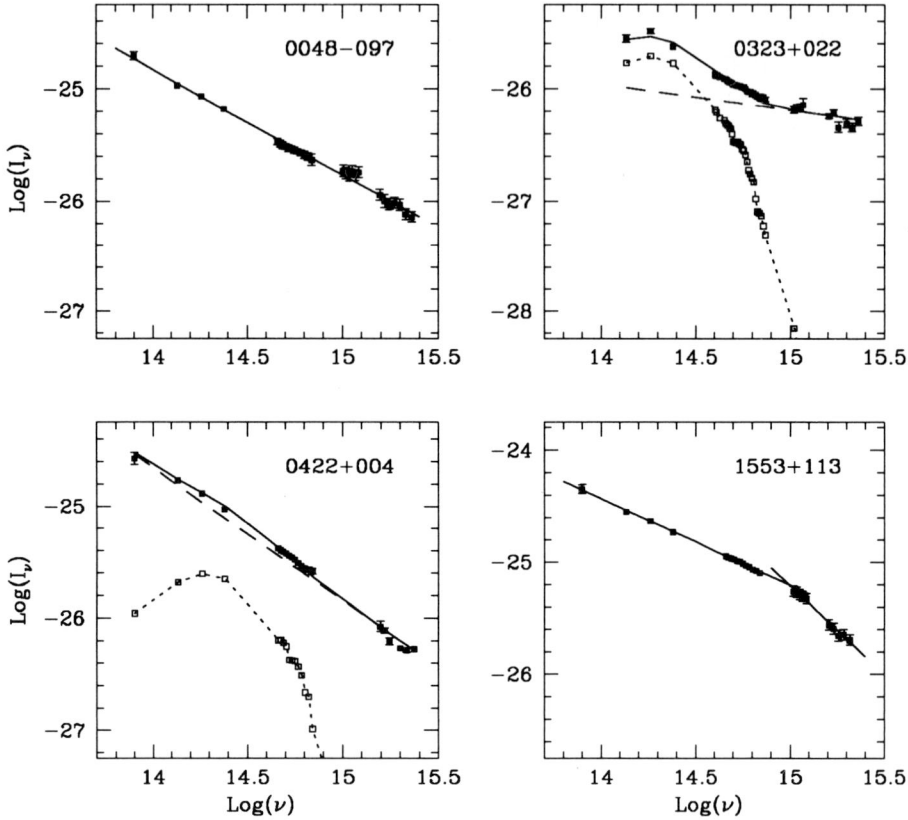

Fig. 1. SFDs of BL Lac objects observed simultaneously at UV, optical and near-IR frequencies. Data (*filled squares*) are corrected for interstellar extinction. The solid line is the best fit model which is either a single power law or the combination of a host galaxy (*dotted line*) and a power law component (*dashed line*). Open squares represent the spectrum of the standard elliptical.

Misdirected Quasars in distant Radio Galaxies

Sperello di Serego Alighieri
Osservatorio Astrofisico di Arcetri
Largo E. Fermi 5, I-50125, Firenze, Italy

and

Andrea Cimatti
European Southern Observatory
Karl-Schwarzschild Str.2, D-8046, Garching bei München, Germany

Abstract. We present the results of recent spectro–polarimetry and imaging–polarimetry of distant radio galaxies which show: **(1)** broad polarized permitted emission lines, **(2)** narrow unpolarized forbidden emission lines, **(3)** a flat (in f_λ) polarized UV continuum and **(4)** an absorption feature, probably interstellar. The direction of the E vector of polarization is always perpendicular to the optical/radio axis. These observations are strong evidence that these objects harbour a quasar, which is visible only through scattering by the interstellar medium of the galaxy. The continuum polarization drops to the red of the 4000Å break, suggesting dilution by an evolved stellar population. A two–component model made of a dust scattered quasar and an evolved stellar population reproduces well the polarization and the spectral energy distribution, including the IR data.

1 Introduction

The unified models of Active Galactic Nuclei (AGN, see [1] for a recent review) foresee that a type 1 spectrum, consisting of a featureless continuum and broad emission lines, is emitted anisotropically from the nuclear regions in two opposite cones and is visible directly only if our line of sight falls in the cones. For misdirected objects an indirect view of the central regions is provided by scattering, as has been beautifully demonstrated by the presence of a polarized type 1 spectrum for a few nearby AGN (e.g. NGC 1068, [11]). For the more distant radio loud AGN the unified model is based on the evidence for relativistic beaming in quasars — which may well not be the only source of anisotropy — and on a statistical analysis of the angular size of a complete sample of extragalactic powerful radio sources [2], leaving it as an open question whether the unbeamed parent population is made of radio galaxies.

The discovery of perpendicular polarization in distant radio galaxies ([7] and [4] for a recent review) is a strong indication of the presence of scattered nuclear radiation, but, since it results from broad–band imaging polarimetry, it gives little information on the scattered light, not allowing, for example, to decide whether it has a quasar– or a blazar–like spectrum. We have therefore started a programme of spectro–polarimetry of distant radio galaxies, with the aim of studying the nature of the scattered radiation, understanding the scattering mechanism and disentangling the spectral energy distribution of the stellar radiation. We present here the first results for two radio galaxies with redshift around one.

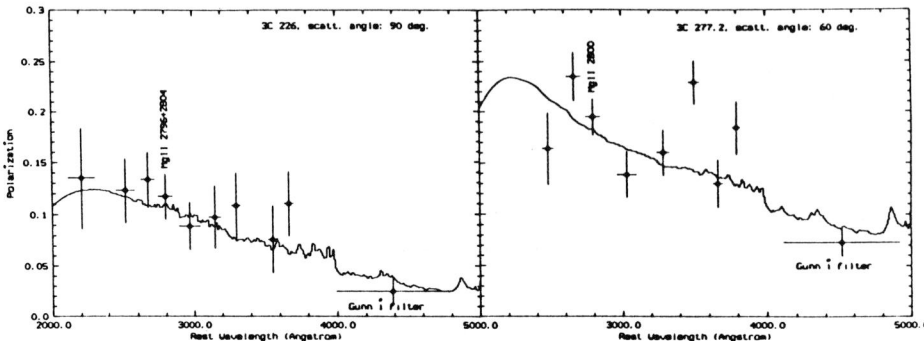

Fig. 1. The degree of polarization of the continuum radiation. The data for a band across the broad MgIIλ2800 line are also shown. The horizontal "error" bars represent the band over which the measurements have been performed. The continuous line shows the polarization foreseen by our two–component model.

2 Observations and Results

The results of recent imaging– and spectro–polarimetry of 3C 226 (z=0.818) and 3C 277.2 (z=0.766), obtained with the ESO Faint Object Spectrograph and Camera, are the following:

The MgIIλ2800 line is clearly broader than the forbidden lines, like [NeV]λ3426, with a relative broadening of 3700 km/s (FWHM). The width of the line corresponds well to that measured for the average spectrum of radio loud quasars [5]. The continuum is almost flat in f_λ. A clear absorption feature is present at 2600Å in all individual spectra.

The polarization of the continuum was measured on the spectra for wavelength bins carefully selected to avoid emission lines and strong sky features, and is shown in Fig. 1 together with the polarization obtained in the Gunn i band with imaging polarimetry. The polarization is high (10–20%) and almost flat between 2000 and 4000Å, and suddenly drops beyond 4000Å to a much lower but still significant value. The polarization over a band across the broad MgIIλ2800 line is also shown in Fig. 1: it is high, equal to that of the surrounding continuum. Since the equivalent width of the MgIIλ2800 line is large ($\sim$ 50Å), the line itself must be polarized at a level close to that of the continuum. The direction of the E vector is constant with wavelength and perpendicular to the optical/radio axis. The narrow forbidden lines are not polarized for both galaxies.

3 Discussion

The results presented here are strong evidence that the two distant radio galaxies observed harbour a quasar, which is not visible directly, but shines its Type 1 spectrum into two opposite cones along the radio axis, where this radiation is scattered in our direction by the interstellar medium of the host galaxy, thereby confirming the beaming/scattering idea [12] to explain the alignment effect in distant radio galaxies, and already supported by the results of imaging polarimetry ([4] and references therein). The main evidence comes from the presence of a polarized type 1 spectrum, which has a continuum approximately flat in f_λ and a broad MgIIλ2800 line. In addition the direction of the E vector is perpendicular to the radio/optical axis at all wavelengths observed. Our results provide strong support for the radio loud quasars/radio galaxies unification.

The scattering medium could be dust [6] or electrons [8] or a mixture of both. Fabian [8] has foreseen that scattering by electrons in a hot halo would broaden the incident spectrum by about 30000 km s^{-1}, effectively smearing out any spectral feature. The fact that broad polarized lines are observed at about the expected width for a quasar spectrum therefore rules out the possibility that scattering is due to hot electrons. Also the observed polarization is too low for electron scattering, while it is consistent with dust scattering.

The sudden drop in the degree of polarization observed both in 3C 226 and 3C 277.2 beyond 4000Å points to the presence of diluting radiation with a strong 4000Å break. We interpret this drop as due to dilution by the light of the host galaxy which contains an evolved stellar population. The possible presence of young stars is severely limited by the fact that the dilution at 2800Å must be small since the MgII line has a polarization close to that of the continuum.

The polarization of the MgII line indicates that, at least in this case, anisotropy must be procuced by obscuration, the broad line region being inside the obscuring material. On the other hand, the less dens gas producing the unpolarized narrow lines must emit isotropically and therefore be outside the obscuring material.

We interpret the absorption feature at 2600Å as due to interstellar FeIIλ2600.18, which is the strongest UV interstellar absorption line with MgIIλ2796+2804 [9].

We have modelled the spectral energy distribution of the two galaxies, including the infrared photometric data [10], with a two component model made of a dust scattered quasar and an evolved stellar population. We have used the typical spectrum of a radio loud quasar [5] scattered by dust with a galactic composition and grain size distribution, adopting Mie single scattering [4]. For the stellar component we use the synthetic spectra of simple stellar populations (instantaneous burst) [3] and Salpeter IMF over a range of 0.1–125 M$_\odot$. We have also introduced a certain amount of dust extinction, to decrease the UV light. Nevertheless at least part of this extinction is probably due to the fact that the scattering cross section that we have used has a strong 2200Å feature, while recent data show that this feature is only present in absorption, but not in scattering.

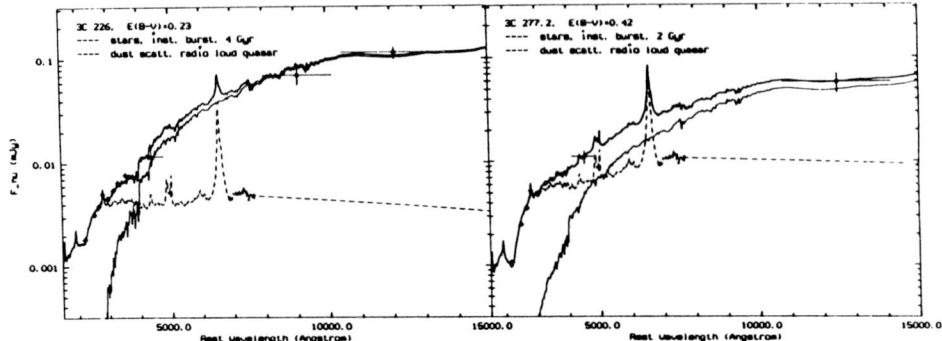

Fig. 2. The SED of 3C 226 and 3C 277.2, compared with our two-component model.

The results of our spectral models are shown in Fig. 2, which give also the amount of extinction and the age of the stellar population. Clearly the model fits the data of both galaxies very well (χ^2_{red} = 1.17 and 1.31 for 3C 226 and 3C 277.2 respectively). The age of the stellar population that give an acceptable fit is limited to be larger than 3 Gyr for 3C 226 and to be between 0.5 and 3 Gyr for 3C 277.2. The presence of an evolved stellar population is also consistent with the drop of the polarization beyond 4000Å. The polarization predicted by our model for the total light is shown in Fig. 1 and is consistent with the observed values. With the advent of 8–10 m class telescopes similar work can be extended to a larger sample and will provide a more precise definition of the incident quasar spectrum, of the scattering parameters, and of the age of the stellar population.

References

1. Antonucci R., 1993, *ARAA*, **31**, 473
2. Barthel P.D., 1989, *ApJ*, **336**, 606
3. Bruzual G., Charlot S. 1993, *ApJ*, **405**, 538
4. Cimatti A., di Serego A. S., Fosbury R.A.E., Salvati M., Taylor D. 1993, *MNRAS*, in press
5. Cristiani S., Vio R., 1990, *A&A*, **227**, 385
6. di Serego Alighieri S., Binette L., Courvoisier T., Fosbury R.A.E., Tadhunter C.N., 1988, *Nature*, **334**, 591
7. di Serego Alighieri S., Fosbury R.A.E., Quinn P.J., Tadhunter C.N., 1989, *Nature*, **341**, 307
8. Fabian A.C., 1989, *MNRAS*, **238**, 41p
9. Kinney A.L., Bohlin R.C., Calzetti D., Panagia N., Wise R.F.G., 1993, *ApJ Suppl.*, in press
10. Lilly S.J., Longair M.S., 1984, *MNRAS*, **211**, 833
11. Miller J.S., Goodrich R.W., Mathews W., 1991, *ApJ*, **378**, 47
12. Tadhunter C.N., Fosbury R.A.E., di Serego Alighieri S., 1989, in Maraschi L., Maccacaro T. & Ulrich M.H., eds., *BL Lac Objects*, Springer-Verlag, Berlin, p. 79

EXTREME QUASARS: OBSERVATIONS AND CONSTRAINTS

MARTIN ELVIS

Harvard-Smithsonian Center for Astrophysics, Cambridge MA 02138 USA

Abstract. The range of continuum shape the quasar phenomenon covers is larger than is usually thought, and is limited by observational sensitivity. between 1/3 and 1/2 of all quasars are "extreme" in at least one band. We have embarked on a program of observations to explore extreme spectral shapes, and to push to high redshift/luminosity. This paper presents X-ray and multi-wavelength SEDs for 12 $z\sim 3$ quasars.

1. Extreme Quasars

Over the past decade Spectral Energy Distributions (SEDs) have been compiled from the radio to X-ray bands for several dozen AGN and quasars (Malkan and Sargent 1982, Elvis et al. 1986, Neugebauer et al. 1987, Sanders et al. 1989, Elvis et al. 1994). All of the quasars with SEDs however are 'moderate', i.e. they have middling values of luminosity, low redshifts and are quite flat in their SEDs. This moderation is the result of the sensitivity of the IRAS, IUE and *Einstein* missions. Because each of these satellites had a limiting sensitivity only about a factor 10 below the flux of the brightest sources only sources with comparable $\nu f\nu$ across all three bands could have SEDs constructed. Any SED that strongly deviated from a flat distribution would not be detected. Thus the instruments define a 'Spectral Window Function' through which all observed SEDs must pass (Brissenden 1989, Elvis and Brissenden 1994). Even so, the observed SEDs show as much variation as is allowed by the constraints of the Spectral Window Function, about a factor 10-20 (see McDowell, these proceedings). Other considerations (Elvis and Brissenden 1994) also suggest that the Spectral Window Function is limiting our understanding of the true range of quasar SEDs, so that perhaps 1/3-1/2 of all quasars are extreme in at least one band.

With the advent of ROSAT, ASCA, HST and improved infrared detectors this window is being opened wider. We have taken the opportunity to search for 'Extreme Quasars' and build SEDs for them. Our hope is that extreme conditions will highlight the basic dependencies of the many observed quasar properties, and so lead us to the physics. Mathur (these proceedings) presents SEDs for an 'X-ray quiet quasar', and a 'Red Quasar'. Here we present SEDs of high redshift quasars.

2. Physical Evolution Of Quasars

Quasars evolve strongly with redshift such that the break point, L*, of their luminosity function shifts to higher luminosities by a factor of 40-50 between z=0.1 and z=2 (Boyle et al 1987) The physical interpretation of this *population* evolution for *individual* quasars is not clear. There are two basic options: long-lived and short-lived. In the long-lived case quasars are intrinsically rare and their present day remnants are massive. Individual quasars are born at high luminosities at z~ 2 − 3 and decline in luminosity gradually so that at the present epoch they emit at no more than 1-2% of L_{Edd}. As the quasar black holes grow larger, and become more starved of accreting matter, the continuum they produce is likely to change in form. In the opposite, short-lived, case quasars are common (so that most large galaxies had one at one time), but each is active only for a short time ($\sim 10^8$ yr). In order not to require too much mass these quasars must be accreting close to their Eddington limits while they are active. In this case all quasar spectra are likely to be similar. While these expectations are qualitative, because there is no detailed physical evolution model of quasars at present, they do indicate how high z quasar SEDs could constrain and guide such modeling efforts.

In the past few years a dozen or so bright (m<18) high redshift (z>2.8) quasars have been discovered that are prime targets for SED construction.

3. Quasar X-Ray Spectra at z=3

We observed a sample of 14 z=3 quasars with ROSAT (detecting 12), radio, thermal infrared (where 10μm ground-based bolometers can go almost 10 times fainter than typical IRAS observations), near infrared (where new arrays can give spectra in the rest frame optical band), and optical (rest frame ultraviolet) (Elvis et al. 1994, Bechtold et al. 1994). ROSAT observes from 0.2-2.4 keV, and so covers the ~1-10keV emitted band covered at low redshifts by *Ginga* (Williams et al. 1992). These quasars also match the *Ginga* quasars in *evolved* luminosity, at 10^{46}-10^{47} erg s^{-1} in the *emitted* 2-10 keV band (assuming isotropic emission), c.f. 10^{45}erg s^{-1} for *Ginga*.

Figures 1a,b show X-ray spectral slope against luminosity and redshift. The radio-quiet slopes must be treated with caution because of the small number of counts (~25). The mean spectral index of the radio-quiet quasars ($if N_H \equiv N_H(\mathrm{Gal})$) is 1.1±0.04, and that of radio-loud quasars is 0.7±0.08. We note: (1) The low redshift slope difference between radio-loud and radio-quiet quasars (Wilkes and Elvis 1987, Williams et al. 1992) persists at z~3, radio-quiet quasars having steeper slopes; (2) neither radio-quiet ($\Delta\overline{\alpha} < 0.7$, 99% confidence) nor radio-loud ($\Delta\overline{\alpha} < 0.3$) quasars show any strong change in 2-10keV spectral slope with redshift, or with luminosity. The absence of any change in spectral index at early epochs when L* is ~ 40L*(z=0) argues in favor of short lifetime quasar models. However, if radio-quiet quasars are intrinsically absorbed (see below) then they may evolve by $\Delta\overline{\alpha} = 0.85 \pm 0.25$. Determining their absorption is thus crucial.

3.1. X-ray Absorption toward High z Quasars

The ROSAT spectra of high z quasars showed unexpected strong absorption. For three out of four radio-loud quasars showed X-ray absorption at a level of $\sim 10^{22}$ atoms cm^{-2} (z=z(quasar), Elvis et al. 1994) so excess absorption is common, but not universal, at z=3. The absorber redshifts are unknown- they may be internal

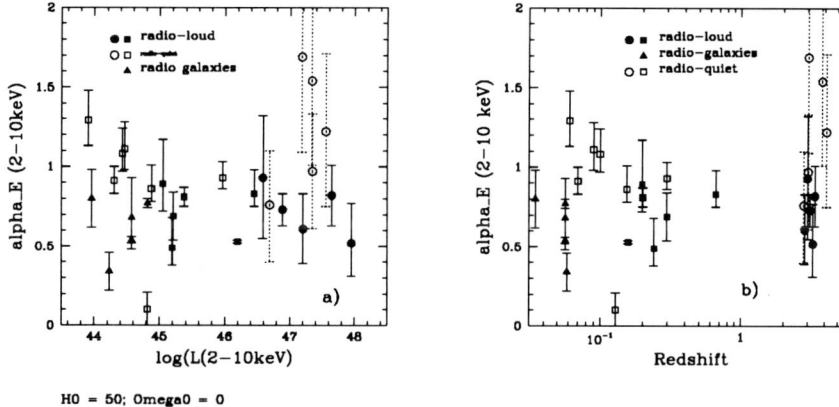

Fig. 1. X-ray spectral slope ($f_\nu \propto \nu^{-\alpha}$, 2-10 keV) vs. (a) luminosity, (b) redshift.

(perhaps related to the Gigahertz Peaked nature of some of these radio sources), or intervening (possibly 'Damped Lyman-α' systems at z<2). The hardness ratios of the radio-quiet quasars suggest no strong absorption and so argues, weakly, for an intrinsic origin for the absorption in radio-loud quasars, since intervening absorption would be as likely for radio-quiet and radio-loud quasars. We refer the interested reader to Elvis et al (1994) for details.

4. Quasar IR to UV SEDs at z=3

Combining optical, near and thermal infrared we can produce SEDs for z=3 quasars such as the one shown in figure 2 (Kuhn et al 1994). Clearly this (and other) z=3 quasars look similar to low z quasars, again favoring short-lived models (Bechtold et al 1994). The MMT 10μm bolometer point precludes any excess of 1000K dust. Large amounts of hot dust are expected in some models (e.g. Haehnalt and Rees, 1993, these proceedings). Radio-quiet high z, high luminosity quasars are more X-ray quiet (i.e. have larger α_{OX}) than those at low z and/or luminosity, but we cannot distinguish the two dependencies (Bechtold et al 1994). More subtle changes in the shape of the UV Bump may be present. The effects of reddening are now being investigated. Accretion disk models predict changes with z for long-lived quasars. First attempts to fit these models (Bechtold et al 1994) fail for large universes (H_0=50, Ω_0=0) in one case, but small universes (H_0=100, Ω_0=1) produce the observed UV slopes, and match the mass and accretion rate values derived solely from total luminosity requirements (Turner 1991).

5. Conclusions

It is now possible to gather SEDs for redshift 3 quasars across many bands in the electromagnetic spectrum. First indications are that no strong changes in the

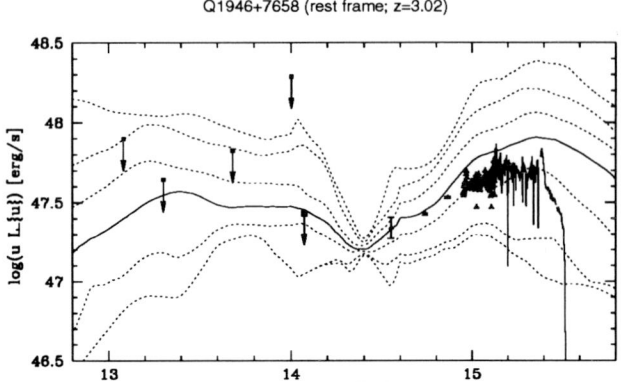

Fig. 2. IR to UV SED for the z=3.02 quasar HS1946+7658. The solid line is the mean SED for a low z sample; the dashed lines are the 68% 90% and 99% ranges normalized at the near-IR inflection (see McDowell these proceedings).

continuum shape even while L* changes by a factor of $\sim$50. This favors short-lived quasar models. More subtle effects may be present.

Acknowledgements

Jill Bechtold, Roc Cutri, Fabrizio Fiore, Olga Kuhn, Smita Mathur, Jonathan McDowell, Marcia Rieke, Aneta Siemiginowska, and Belinda Wilkes have all contributed to the present investigation. This work was supported in part by NASA grants NAGW-2201 (LTSARP), NAG5-1872 and NAG5-1536 (ROSAT), and NASA contracts NAS5-30934 (RSSDC), NAS5-30751 (HEAO-2) and NAS8-39073 (ASC).

References

Bechtold J., et al., 1994, Ap.J., submitted.
Boyle B., Fong R., and Shanks T., 1987, MNRAS, 227, 717.
Brissenden R.J.V., 1989, PhD Thesis, Australian National University.
Elvis M. et al., 1994, Ap.J.Suppl., submitted.
Elvis M. and Brissenden R.J.V. 1994, Ap.J., in preparation
Elvis M., et al., 1986, Ap.J., 310, 291.
Haehnalt M.G. and Rees M.J., 1993, MNRAS, 263, 168.
Kuhn O., Bechtold J., Cutri R., and Elvis M., 1994, Ap.J.Letters, submitted.
Neugebauer G., et al., 1987, Ap.J. Suppl., 63, 615.
Malkan M.A., and Sargent W.L.W., 1982, Ap.J., 254, 22.
Sanders D., et al 1989, Ap.J., 347, 29.
Turner E.L., 1991, A.J., 101, 5.
Wilkes B.J. and Elvis M., 1987, Ap.J., 323, 243.
Williams O.R., et al, 1992, Ap.J., 389, 157.

PARTICLE ACCELERATION IN EXTRAGALACTIC JETS AND IMPLICATIONS FOR THE HIGH-ENERGY EMISSION FROM BLAZARS

R. SCHLICKEISER

Max-Planck-Institut für Radioastronomie, Auf dem Hügel 69, 53121 Bonn, Germany

and

C. D. DERMER

E. O. Hulburt Center for Space Physics, Naval Research Laboratory, Code 7653, Washington, DC 20375-5352, USA

September 15, 1993

Abstract. We demonstrate that the prevalence of superluminal sources in the sample of γ-ray blazars and the peak of their luminosity spectra at γ-ray energies can be readily explained if the γ-rays result from the inverse Compton scattering of the accretion disk radiation by relativistic electrons in outflowing plasam jets. Compton scattering of external radiation by nonthermal particles in blazar jets is dominated by accretion disk photons rather than scattered radiation to distances of $\sim 0.01 - 0.1$ pc from the central engine for standard parameters. The size of the γ-ray photosphere and the spectral evolution of the relativistic electron spectra constrain the location of the acceleration and emission sites in these objects.

Key words: γ-rays – blazars – emission process – particle acceleration

1. Introduction

Since its launch in April 1991, the EGRET experiment on board the Compton Gamma Ray Observatory has increased the number of known ($> 5\sigma$) extragalactic sources of >100 MeV γ-radiation from formerly one (3C273) to 27 (status August 1993, Thompson 1993). With the exception of the LMC, all these sources are radio-loud active galactic nuclei (AGNs), of which 5 are classified as BL Lacs and 17 as quasars. Eight of them (0234+285, 0836+710, Mrk 421, 1156+295, 3C273, 3C279, 1633+382, 3C454.3) have exhibited apparent superluminal ($V_{app} > c$) motion components; in a few other cases (0235+164, 0528+134), superluminal motion is very likely. The γ-ray emission in general is variable from time scales as short as 2 days to several months or years. Mrk 421 is the only source that has also been detected at TeV energies (Punch et al. 1992). Truly remarkable are both the high fraction ($> 30\%$) of known superluminal sources in this sample and the dramatic peak of the luminosity spectra νF_ν at γ-ray frequencies.

2. A model for the high-energy blazar emission

Recently, we (Dermer et al. 1992, Dermer and Schlickeiser 1993) have developed a model for the blazar γ-ray emission based on the relativistic beaming hypothesis (Blandford and Rees 1979) and the unified scenario for AGNs. Simplifying many details, its basic features read as follows: (i) AGNs are powered by accretion of matter onto a black hole; (ii) in the case of radio-loud AGNs, collimated relativistic jets composed of shocks or plasma blobs move with speed Bc and bulk Lorentz

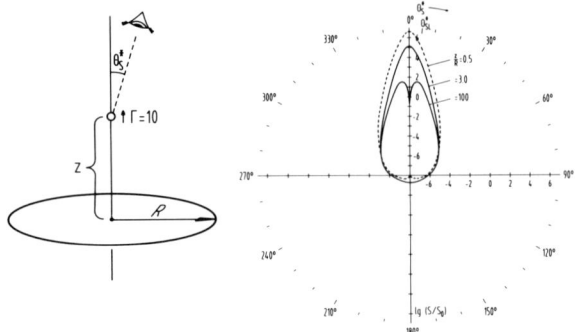

Fig. 1. Diagram illustrating jet model for γ-ray emission from AGNs (left figure) and radiation beaming pattern (right figure), showing the ratio of scattered luminosity to soft photon luminosity as a function of observing angle θ_s^*. Different ratios of blob height z to radius R of the soft-photon source are given for a relativistically outflowing plasma blob with bulk Lorentz factor $\Gamma = 10$.

factor $\Gamma = (1 - B^2)^{-1/2}$ outward along the axes of the accretion disk (see left diagram in Fig.1); and (iii) the line-of-sight of the observer determines the AGN classification.

We assume that
(a) a spherical blob of radius r_b contains a homogeneous magnetic field of strength H_o and relativistic electrons and positrons with an isotropic angular and power-law energy distribution, given by

$$N(\gamma) = N_e \gamma^{-s} \text{ for } 1 \leq \gamma_1 \leq \gamma \leq \gamma_2 \qquad (1),$$

in the rest frame of the blob;
(b) the relativistic particles are confined to the blob by scattering off intrinsic MHD turbulence;
(c) the blob is of low compactness, and the inverse Compton optical depth $\tau = \sigma_T N_e r_b < 1$ is smaller than unity.

3. γ-ray production

Among the many possible physical processes in such a system (synchrotron self-Compton, inverse Compton scattering of ambient 2.7K microwave background and optical photons, etc.), we have demonstrated (Dermer and Schlickeiser 1993) that *the 100 MeV-10 GeV γ-rays result from Compton scattering of the ultraviolet and soft X-ray accretion disk photons traversing the outflowing blob by the relativistic electrons and positrons in the blob.* An important feature of this process is the strong beaming of the scattered γ-ray emission. Due to Doppler boosting and the strong anisotropy of the accretion disk photons in the comoving fluid frame, the angular distribution of the inverse Compton scattered radiation is very anisotropic and peaks near the superluminal direction $\mu_{SL}^* = \cos\theta_{SL}^* = B$ once the blob is at distances from the central object $z(\text{pc}) > \Gamma R = 0.003(\Gamma/10)(R/10^{15}\text{ cm})$, where R denotes the effective size of the accretion disk. From the diagram on the right of Fig. 1, we see that the beaming pattern peaks in the forward direction at values of $z/R \ll \Gamma$.

On the basis of estimates for the total electron loss rates in the various interaction processes (see Dermer and Schlickeiser 1993 for details), we find that for parameters appropriate to 3C279, the inverse Compton scattering of accretion disk photons dominates the production of γ-rays by

- microwave scattering if $z(\text{pc}) < 360 \frac{(L_{UV}/10^{46}\text{ erg/s})^{1/2}}{(\Gamma/10)^2(1+z_{redshift})^2}$,
- Compton scattering of stellar optical photons if $z(\text{pc}) < 4.3 \frac{(L_{UV}/L_{opt})^{1/2}}{(\Gamma/10)}$,
- the synchrotron-self Compton process if $z(\text{pc}) < 0.1 \frac{(L_{UV}/10^{46}\text{erg/s})^{1/2}}{\Gamma(H_o/1\text{ G})}$,
- Compton scattering of reprocessed radiation in emission line clouds (Sikora et al. 1994) if $z(\text{pc}) < 0.06 \frac{(M_8)^{1/3} R_{ELC}^{2/3}(1\text{ pc})}{\tau_{-2}^{1/3}}$, where M_8 denotes the central black hole mass in units of 10^8 solar masses, R_{ELC} the distance in units of pc of the emission line clouds from the central black hole, and τ_{-2} the electron scattering optical depth in emission line clouds in units of 10^{-2}. Note that the lack of an observed soft X-ray turnover due to photoelectric absorption in contemporaneous GINGA observations of 3C279 (Makino et al. 1993) constrains the value $\tau_{-2} \leq 0.25$. Moreover, we established (see Dermer and Schlickeiser 1994) that for parameters characteristic of 3C279, $\gamma - \gamma$ pair annihilation of the produced 1 GeV γ-radiation is negligible

- with the accretion disk radiation if $z(\text{pc}) > 4 \cdot 10^{-4} \frac{(L_{UV}/10^{46}\text{ erg/s})^{1/2}(M_8)^{1/2}}{(k_B T_A/50\text{ eV})^{1/2}}$,
- with reprocessed radiation from emission line clouds at all z, since its pair annihilation optical depth is

$$\tau(1\text{ GeV}) \cong 0.06 E_{\text{GeV}}^{0.7}(L_{UV}/10^{46}\text{ erg/s})\tau_{-2}\ln(z/z_i) \ll 1 \qquad (2).$$

Here T_A denotes the temperature of the accretion disk and z_i is the distance of the base of the jet from the central source. Estimate (2) is a factor 200 smaller than the estimate of Blandford (1993) because, according to contemporaneous GINGA observations, 3C279 was in a low-state (1-10 keV) during the γ-ray flare.

The above constraints place the location of the γ-ray emission site and the particle acceleration site at distances 10^{15} cm $\leq z \leq 10^{17}$ cm from the central object, where the lower limit is comparable to the radial extent of the accretion disk and is determined by the obvious absence of $\gamma - \gamma$ pair annihilation in the observed γ-ray spectra, while the upper limit is set by the dominance of other interaction processes which could, however, also contribute to the more slowly-varying gamma ray emission. Our proposed model of γ-ray production by inverse Compton scattering of accretion disk photons by outflowing relativistic electrons and positrons in a relativistic jet is very economical: for favorable viewing conditions of the observer with $\theta_{obs}^* \simeq \theta_{SL}^*$, we estimate the total energy in relativistic electrons and positrons in the blob to be

$$E_{tot} \geq 5 \cdot 10^{46}(r_b/10^{15}\text{ cm})^2(\gamma_1/10)\text{ erg} \qquad (3).$$

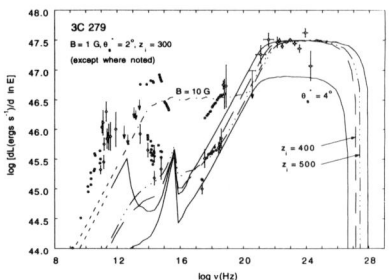

Fig. 2. Model results for the time-integrated luminosity spectrum compared to the multi-wavelength observations of 3C279. The radio-through-optical synchrotron emission, which is probably made farther out in the radio jet, is not modelled here.

Combined with the observed γ-ray luminosity $L(> 100 \text{ MeV}) = 10^{48} L_{48} f$ erg s^{-1}, where f is the beaming factor, Eq.(3) implies that the intrinsic relativistic electron and positron loss time

$$t_R = E_{tot}/L(> 100 \text{ MeV}) \geq 50 L_{48}^{-1}(f/10^{-3})^{-1} \text{sec} \qquad (4).$$

Coupled with the Doppler time-dilation effect, we see from Eq.(4) that very short γ-ray flare durations can easily be explained in this model if the acceleration time scale is sufficiently short.

The short radiation loss time (4) implies that the time evolution of the electron energy spectrum during the γ-ray flare is very important. We calculated the time-integrated γ-ray emission from a modified Kardashev-model by instantaneously injecting the power law electron distribution (1) at height z_i, calculating its modification with height in the relativistically outflowing blob due to the operation of the various inverse Compton energy loss processes, and finally integrating the hard X-ray and γ-ray emissivities over height (Dermer and Schlickeiser 1993). Comparison with multiwavelength observations of 3C279 is shown in Fig.2. One notices the good fit to the high-energy emission from this source, including the break at MeV energies which results from incomplete relativistic electron cooling. Moreover the peak of the luminosity spectrum at γ-ray energies is well reproduced by the model.

References

Blandford, R.D., Rees, M.J.: 1979, *Pittsburg Conference on BL Lac Objects, ed. A. M. Wolfe, Pittsburg, Pittsburg Univ. Press, 328* ,
Blandford, R.D.: 1993, *Proceedings of the Compton Symp., ed. N. Gehrels & M. Friedlander* , in press
Dermer, C.D., Schlickeiser, R., Mastichiadis, A.: 1992, *AA* **256**, L27
Dermer, C.D., Schlickeiser, R.: 1993, *ApJ* **416**, in press
Dermer, C.D., Schlickeiser, R.: 1994, *ApJS* **89**, in press
Makino, F. et al.: 1993, *International Symposium on Neutrino Astrophysics, Takayam/Kamioka, Japan, October, 19-22, 1992* , in press
Punch, M., et al.: 1992, *Nat* **358**, 477
Sikora, M., Begelman, M.C., Rees, M.: 1994, *ApJ* , in press
Thompson, D.J.: 1993, *this conference* ,

Thermal Or Non-Thermal X-Rays From AGNs?

Gabriele Ghisellini
Osservatorio di Torino, Strada Osservatorio 20, 10025 Torino, Italy

and

Francesco Haardt
ISAS/SISSA, Via Beirut 2-4, 34013 Trieste, Italy

Abstract. Recent data from OSSE on CGRO and SIGMA on GRANAT challenge the non-thermal interpretation of the origin of the high energy emission of AGNs, showing that the hard X-ray spectra of several Seyfert AGN are steep like those of Galactic black hole candidates. Thermal models are therefore favoured. Two-phase models, in which a hot corona is placed above a relatively cold accretion disk can account for the observed X-ray spectra and the correlated variability in the UV and X-ray bands. Cold matter, both in the vicinity of the nucleus, and located further away in the torus surrounding the nucleus, may modify substantially the spectrum with important consequences on the expected variability and spectral shape.

Key words: X-rays, Active Galactic Nuclei, Comptonization

1 Introduction

The hard X-ray and γ-ray satellites GINGA, SIGMA and CGRO have greatly changed our views on the high energy emission of Active Galactic Nuclei, deeply influencing our understanding of the main mechanisms responsible for this emission. With the improved spectral resolution of GINGA it was discovered the presence of a fluorescence K_α iron line at 6.4 keV, and that the spectra flatten above 10 keV. Both features are interpreted as the contribution of the so called Compton reflection component (Pounds et al. 1990): cold matter, possibly in an accretion disk, is illuminated by the power law X-ray spectra, and reflects part of it by scattering, degrading high energy photons by Compton recoil and absorbing low energy photons by the photoelectric effect (Guilbert & Rees 1988; Lightman & White 1988). The resulting reflected spectrum has a broad peak at an energy of 30–50 keV. Fitting the data with a power law plus the Compton reflection hump yields a steeper power law index, with $\Delta\alpha_x \sim 0.2$ with respect to a fit with a power law only.

These GINGA observations revived theoretical models in which the continuum is produced by non-thermal electron-positron pairs, because they predict $\alpha_x \sim 0.9 - 1$ in a large region of the parameter space (Zdziarski et al. 1990). However, these models are now challenged by the new observations of SIGMA on GRANAT and especially of OSSE on CGRO. The observed spectra are steep spectrum above 50 keV in several Seyfert 1 galaxies (Cameron et al. 1993). This indicates that there must be a break in the spectrum at $\sim$ 30–100 keV. In NGC 4151 the existing data (Jourdain et al. 1992; Maisack et al. 1993) seems to exclude a pure non-thermal pair plasma model (Zdziarski, Lightman & Maciolek, 1993).

2 Thermal models and the 'universal' X–ray spectrum

Haardt and Maraschi (1991, 1993) considered the interplay between a hot, optically thin corona and a cold accretion disk. The hot plasma in the corona is thermal, and the main radiation mechanism is Comptonization. If all the power is released in the corona, it will emit half of the luminosity in the upward direction, and half down toward the disk (or more, if the Compton rocket effect makes the emission anisotropic, see Ghisellini et al. 1991 and Haardt 1993). The cold disk absorbs the incoming power, reemitting it in the UV, and thus producing the seed soft photons to be comptonized in the hot corona. Therefore, the UV and the X–ray radiation have approximately the same power, corresponding to an average spectral index of unity, and to a Comptonization parameter $y \sim 1$. Addition of the Compton reflection hump yields $\alpha_x \sim 0.7$.

In this model, luminosity balance suffices to yield the right spectral index and a relation between the optical depth τ_T and kT. Furthermore, if the hot plasma is pair dominated (for large compactnesses), kT depends only on the compactness, which becomes the only free parameter. Note that:

• The compactness of the source does not need to be large, the model works for any value of the compactness (but for low compactnesses, also τ_T must be specified).

• Pair production limits the temperature of the corona and very few γ–rays are emitted. In this model, as in all thermal steady plasma (Svensson 1984), no annihilation line should be visible.

• Since all the power is released in the corona, the UV is reprocessed radiation and should approximately have the same luminosity as the X–rays, and should follow changes of the overall X–ray power.

A major problem of thermal models concerns the thermalization mechanism able to mantain a Maxellian distribution even during the very fast variation of the luminosity we observe. It must be more efficient than electron–electron collisions. One possibility is that the thermalization is achieved through the exchange of photons (Ghisellini & Svensson 1990): in the presence of an equipartition magnetic field, all the cyclo–synchrotron radiation produced by the hot plasma is self absorbed. Self-absorption acts as a very fast thermalizing mechanism, able to produce an equilibrium quasi–maxwellian distribution (a Maxwellian plus an high energy tail) in few synchrotron cooling times, even if the primary power in injected in the source in the form of monoenergetic particles at, say, $\gamma_{max} \sim 10$.

Another possibility is that the particle distribution, albeit not perfectly Maxwellian, produce a thermal–like spectrum, as shown by Ghisellini, Haardt & Fabian (1993). What is important, in models where the X–rays are made by multiple Compton, is in fact not the shape of individual scattering orders (which indeed resemble the underlying particle distribution), but their sum, which gives indistinguishable spectra in very different cases (i.e. thermal or power law particle distributions) as long as the mean quadratic energy is similar.

3 Patchy (in space and time) coronae

There are two main objections to the two-phase model discussed above: i) it predicts that $L_{UV}/L_X \sim 1$, while observations indicate a larger ratio in many objects; ii) UV radiation should always correlate with X–rays, even at short timescales.

To overcome these problems, we (Haardt, Maraschi & Ghisellini, work in progress) suggest that the corona may not be extended, but patchy. In other words, liberation of gravitational energies is localized in small scale blobs, where accumulated energy (e.g. in magnetic fields) is suddendly released, resulting in flare activity. In this case we allow the accretion disk to emit not only reprocessed radiation, but also an important fraction of the gravitational power.

Under an X–ray blob (if the blob is sufficiently small and/or the energy dissipation time scale is short), the reprocessed radiation far exceeds the power produced locally by the disk, and this fixes the X–ray spectral shape as in the case of the extended corona discussed above.

4 The role of absorbing tori in the X–ray spectra

The popular unification scheme of Seyfert 1 and 2 galaxies assumes that an obscuring torus blocks the UV and soft X–ray radiation, hiding the Seyfert nucleous and the broad line regions at large inclination angles (e.g. Antonucci & Miller 1985). We have calculated the effects the torus has on the high energy spectrum, varying the inclination angle and the optical depth of the torus.

We were stimulated in this study by the suggestion that, if the torus is marginally Compton thick, than one can explain the X–ray background as due to Seyfert (mainly Seyfert 2) galaxies (Madau, Ghisellini & Fabian 1993). The results are presented in Fig. 1.

Compton thick tori ($N_H \gtrsim 10^{24}$ cm^{-2}) affect the overall emission even at small inclination angles, contributing to the 'reflection Compton component' and producing a fluorescence line emission at 6.4 keV, with an equivalenth width of $\sim$ 80–100 eV (Ghisellini, Haardt & Matt, 1993).

5 References

Antonucci, R.R., & Miller, J.S. 1985, ApJ, 297, 621
Cameron, R.A. et al., 1993, to be published in the proceedings of the Compton symposium, St. Louis, MO, ed. N. Gehrels
Ghisellini, G., George, I.M., Fabian, A.C. & Done, C., 1991, MNRAS, 248, 14
Ghisellini, G., Haardt, F. & Fabian, A.C., 1993, MNRAS, 263, L9
Ghisellini, G., Haardt, F. & Matt, G., 1993, submitted to MNRAS
Ghisellini, G. & Svensson, R., 1990, in Physical Processes in Hot Cosmic Plasmas, Eds. W. Brinkmann, A.C. Fabian & F. Giovannelli, Kluwer Acad. Publ., p. 395

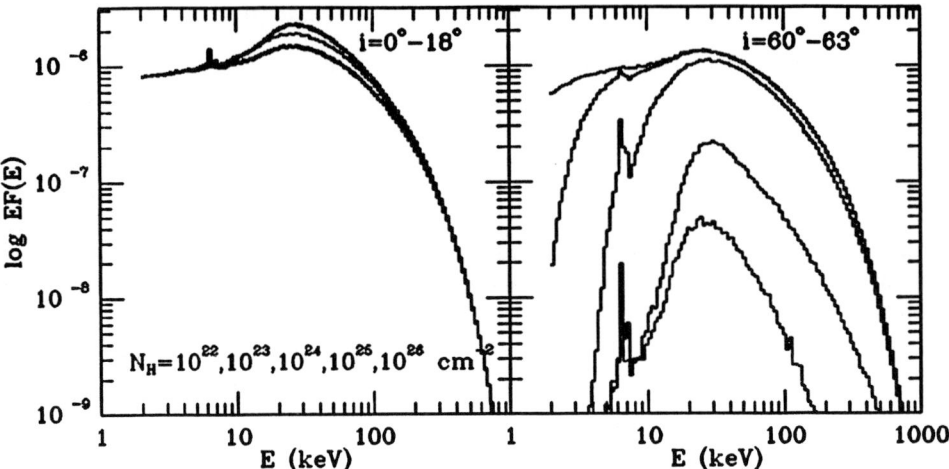

Fig. 1. The influence of the absorbing torus surrounding the nucleous of Seyfert galaxies on the X-ray spectrum. The left hand panel shows the spectra calculated for face on sources, for different values of the column density of the torus. Note that for $N_H \gtrsim 10^{24}$ cm^{-2} the funnel of the torus reflects in the line of sight some of the incident radiation, contributing to the 'reflection hump' at energies 10-50 keV. Furthermore, the funnel would produce fluorescence iron line at 6.4 keV. If the torus contributes to the observed emission, it dilutes the primary radiation, smearing fast variability at high energies. The right hand panel shows the spectrum for an inclination angle of about 60°. It shows the effects of the absorption of soft X-rays, and of the downscattering of hard X-rays as the column density of the torus increases. Not shown in the figure is the contribution of warm scattering material located above the torus, thought to contribute around 1 per cent of the primary flux. From Ghisellini, Haardt & Matt, 1993.

Guilbert, P.W. & Rees, M.J., 1988, MNRAS, 233, 475
Haardt, F. & Maraschi, L., 1991, ApJ, 380, L51
Haardt, F. & Maraschi, L., 1993, ApJ, 413, 507
Haardt, F., 1993, ApJ, 413, 680
Jourdain, E., et al., 1992, A.A 256, L38
Lightman, A.P. & White, T.R., 1988, ApJ, 335, 57
Madau, P., Ghisellini, G., & Fabian, A.C., 1993, ApJ, 410, L7
Maisack et al. 1993, ApJ, 407, L61
Pounds, K. A., Nandra, K., Stewart, G. C., George, I. M., & Fabian, A. C., 1990, Nature, 344, 132
Svensson, R., 1984, MNRAS, 209, 175
Zdziarski, A.A., Ghisellini, G., George, I.M., Svensson, R., Fabian, A.C. & Done, C., 1990, ApJ 363, L1
Zdziarski, A.A., Lightman, A.P. & Maciolek-Niedzwiecki, A., 1993, ApJ, 414, L93

New Observations of AGN with Specific Instruments

COMPTON OBSERVATORY OBSERVATIONS OF AGN

J. D. Kurfess
Naval Research Laboratory
Washington, DC 20375

ABSTRACT. The principal results on active galactic nuclei from the Phase 1 observations by the *COMPTON Gamma Ray Observatory* are presented. These include the detection of a new class of high-energy gamma ray sources by the EGRET instrument and extensive observations of Seyfert galaxies in low-energy gamma rays by OSSE. The identified EGRET sources are associated with core-dominated radio loud objects, OVV's and BL Lacs. EGRET has not detected any Seyfert galaxies. OSSE observes a thermal-like spectrum from NGC 4151, and the low-energy gamma ray spectra of other Seyferts are significantly softer than the spectra below 50 keV, suggesting that a thermal emission mechanism is characteristic of these objects. OSSE has not detected any positron annihilation radiation from any Seyfert, and neither OSSE nor COMPTEL have detected an MeV excess from these sources.

1. Introduction

Prior to the launch of the *COMPTON Observatory* there were only four extragalactic sources which had been detected above 100 keV: Cen A, NGC 4151, MCG 8-11-11 and 3C 273 (Levine et al. 1984). AGN observed in the 1-50 keV energy band by HEAO, EXOSAT and Ginga were generally well fit with a "canonical" power-law spectrum with an α spectral index of about -1.7, although Ginga analyses did provide evidence for iron line emission and an additional reflected component above 10 keV. It was clear that the spectra of Seyfert galaxies had to break above several hundred keV in order not to overproduce the diffuse gamma ray background. However, MeV emission consistent with the hard x-ray spectrum extending to above 1 MeV was reported for two Seyferts: NGC 4151 (Perotti et al. 1981a) and MCG 8-11-11 (Perotti et al. 1981b). Only one extragalactic source, the nearby quasar 3C 273, was detected above 50 MeV by COS-B (Swanenburg et al. 1978). Finally, observations dating from the late 1960's had detected a diffuse gamma ray background at energies from 20 keV to several hundred MeV. The origin of this gamma ray background has been the focus of intense theoretical work, but remains a mystery. Although AGN contribute to the diffuse gamma ray background, the fraction of this background which is attributable to AGN is unknown.

The *COMPTON Gamma Ray Observatory* was launched on 5 April 1991. It carries four instruments which undertake gamma ray observations from 20 keV to 20 GeV. The overall capabilities and characteristics of these instruments are described in a series of papers (Johnson et al. 1993, Thompson et al. 1993, Schoenfelder et al. 1993, Fishman et al. 1989). Briefly, the EGRET instrument observes gamma rays in the 30 MeV-20 GeV region by converting incident gamma rays into electron-positron pairs and uses the path of the pairs through a multiple spark chamber to determine the direction of the incident gamma ray. COMPTEL uses the compton scattering process to "image" 1-30 MeV gamma rays. Both EGRET and COMPTEL have broad fields-of-view and the first 18 months (Phase 1) of the mission were used to undertake a full sky survey with these two instruments. This was accomplished with a series of two and three week viewing periods (VP) during which the attitude of the observatory remained fixed. OSSE operates in the 50 keV -10 MeV region, has a 3.7° x 11.4° field-of-view (non-imaging), and is used to study one object at a time. The BATSE instrument consists of eight broad field-of-view detectors which provide all sky coverage for transient phenomena such a cosmic gamma-ray bursts and solar flares. Strong sources can be observed by BATSE in an Earth occultation mode, and the strongest active galactic nuclei are monitored on a nearly continual basis (Paciesas et al. 1993)

2. High Energy Sources

As noted earlier, 3C 273 was the only extragalactic high-energy gamma ray source detected prior to the launch of GRO. 3C 273 and the Virgo region were scheduled to be viewed in VP11, about 5 months into the mission. However, the occurrence of a Type 1a supernova in NGC 4527, SN 1991t, resulted in an early change to the viewing program which enabled the Virgo region to be viewed in VP3 (June 1991). Surprisingly, 3C 273 was not observed by EGRET as an intense source, but the nearby OVV quasar 3C 279 was. 3C 279 was the first quasar for which superluminal motion was observed, and it had been recognized as a good candidate for gamma ray emission. Remarkably, EGRET observed significant variability in 3C 279's high-energy gamma-ray emission during the two week observation (Hartman et al. 1991). The source was observed to increase by a factor of 3 over a period of one week, and then decreased by a factor of 3 over a period of only two days. At a redshift of 0.538, the apparent γ-ray luminosity of the source is 1 x 10^{48} erg/s if isotropic emission is assumed. This luminosity is an order of magnitude larger that the luminosity in any other spectral band. The short term variability and the luminosity strongly indicate the high-energy gamma ray emission originates in relativistic jets with the radiation beamed toward the Earth. The short term variability probably results from the emission originating relatively close (a few light days) to the central compact object, and the intrinsic gamma ray luminosity is reduced from the apparent luminosity by the beaming factor.

The energy spectrum of 3C 279 is observed to be a power law in the EGRET energy range. During June 1991 when the strong flaring activity occurred, the photon spectral index was -1.9. Several months later, when the source was in a somewhat lower intensity state, the spectrum was slightly steeper, spectral index = -2.1. Ginga observed this source several days prior to the EGRET observations in June, 1991, so contemporaneous x-ray observations are available. Broad-band *Compton* spectral results on this and other sources will be discussed later in this paper.

During Phase 1 of the *COMPTON* mission, EGRET has reported the detection of 22 extragalactic sources at greater that 5σ (Fichtel et al. 1993a). These are listed in Table 1. All of these sources belong to the class of radio loud active galactic nuclei. Fourteen are classified as flat spectrum quasars and four are BL Lacs. Six of the sources are associated with radio objects which exhibit superluminal characteristics. Identifications have been made with optical and/or radio objects in all but four cases. These identifications are based on the EGRET error box associated with each object and the most likely candidate from a pre-selected list of AGN (Fichtel et al. 1993b). Short term variability has been observed for several of these sources where the detection is sufficiently significant to perform temporal studies. In all cases, the spectra can be adequately

Table 1: AGN Detected by EGRET during Phase 1

Source	Type	Redshift (z)	Observ. Dates	Flux (E>100 MeV) (10^{-7} γ/cm^2-s)	Photon Index
PKS 0202+149 (4C+15.05)	Quasar		2/20 - 3/5 1992 4/23 - 4/28 1992 5/7 - 5/14 1992	0.25 ±0.08 <1.4 <0.4	2.5±0.1
PKS 0208-512	Quasar	1.003	9/5 - 9/12 1991 9/19 - 10/3 1991 11/7 -11/14 1991	0.17±0.07 1.07±0.08 0.53±0.12	1.7±0.1
0219+428			11/28-12/12 1991 8/11- 8/20 1992 9/1 - 9/17 1992	0.17±0.04 <0.2 <0.4	
0234+285			11/28-12/12 1991 8/11- 8/20 1992 9/1 - 9/17 1992	0.17±0.05 <0.6 <0.3	
PKS 0235+164 (OD+160)	BL Lac	0.94	11/28-12/12 1991 2/20 - 3/5 1992	<0.3 0.86±0.12	2.0±0.2
0420-014 (OA 129)	Quasar	0.92	5/16 - 5/30 1991 6/8 - 6/15 1991 2/20 - 3/5 1992 5/14 - 6/4 1992	<0.11 <0.11 0.44±0.13 <0.4	
PKS 0446+112	Quasar		5/16 - 5/30 1991 6/8 - 6/15 1991 8/11 - 8/20 1992 9/1 - 9/17 1992	<0.14 <0.14 0.96±0.18 <0.5	1.8±0.3
0454-234			5/14 - 6/4 1992	0.62±0.15	
PKS 0454-463		0.858	7/26 - 8/8 1991 9/19 - 10/3 1991 12/27/91-1/10/92 5/14 - 6/4 1992	0.27±0.07 <0.2 <0.3 <0.2	
PKS 0528+134	Quasar	2.06	5/16 - 5/30 1991 6/8 - 6/15 1991 8/11 - 8/20 1992	}0.84±0.10 <0.6	2.6±0.1
PKS 0537-441	BL Lac	0.894	7/26 - 8/8 1991 8/22 - 9/5 1991 12/27/91-1/10/92 5/14 - 6/4 1992	0.33+0.08 <0.2 <0.5 <0.3	2.0±0.2
0716+714	BL Lac		1/10 - 1/23 1992 3/5 - 3/19 1992 6/11 - 6/25 1992	0.18±0.04 0.52±0.13 <0.5	1.8±0.2 2.4±0.3
0836+710 (4C+71.07)	Quasar	2.17	1/10 - 1/23 1992 3/5 - 3/19 1992 6/11 - 6/25 1992	0.13±.03 0.43±0.11 <0.3	2.4±0.2 1.9±0.4
1101+384 (Mrk 421)	BL Lac	0.031	6/28 - 7/12 1991 9/17 - 10/8 1992	0.14±0.05 0.22±0.07	1.9±0.1
1226+023 (3C 273)	Quasar	0.158	6/15 - 6/28 1991 10/3 -10/17 1991	0.26±0.04 0.13±0.04	2.4±0.1

41

Source	Type	Redshift (z)	Observ. Dates	Flux (E>100 MeV) (10^{-7} γ/cm^2-s)	Photon Index
1253-055 (3C 279)	Quasar	0.538	6/15 - 6/28 1991 10/3 -10/17 1991 10/17 -10/31 1991 4/2 - 4/16 1992	2.6±0.1 0.85±0.07 1.02±0.27 0.56±0.20	1.9±0.1 2.1±0.1
1606+106 (4C +10.45)	Quasar	1.23	9/12 - 9/19 1991 12/12 -12/27 1991 4/2 - 4/23 1992	<0.3 0.53±0.12 0.29±0.07	2.2±0.3
1633+382 (4C +38.41)	Quasar	1.81	9/12 - 9/19 1991	0.95±0.08	1.9±0.1
2022-077 (NRAO 629)	Quasar		8/15 - 8/22 1991 10/31 - 11/7 1991 1/23 - 2/6 1992 2/6 - 2/20 1992 10/28 - 11/3 1992	0.67±0.12 <0.13 0.25±0.08 <0.16 <0.7	1.5±0.2
PKS 2052-474	Quasar	1.489	8/6 - 8/13 1992 8/27 - 9/1 1992 10/15 -10/29 1992	<1.1 <0.4 0.28±0.06	
2230+114 (CTA 102)	Quasar	1.037	1/23 - 2/6 1992 4/23 - 4/28 1992 5/7 - 5/14 1992 8/20 - 9/3 1992	0.25±0.05 <1.2 0.39±0.17 0.48±0.17	2.6±0.2
2251+158 (3C 454.3)	Quasar	0.859	1/23 - 2/6 1992 4/23 - 4/28 1992 5/7 - 5/14 1992 8/20 - 9/3 1992	0.78±0.08 1.1±0.3 0.43±0.13 1.23±0.18	2.2±0.1

described by a single power law, although there may be a suggestion for high energy steepening above 1 GeV in several of the sources. The range of redshifts extends to z > 2. This is also consistent with a subset from a population of objects in which the emission is beamed toward the observer.

This discovery of a class of high-energy gamma ray sources associated with active galactic nuclei is one of the early surprises of the *COMPTON* mission, and has enabled a new observational approach to the study of AGN. It has also made clear the importance of obtaining broad band coverage, from radio through high-energy gamma rays for these very intriguing sources. Results from such multi-wavelength campaigns are the subject of other talks at this conference.

3. Low Energy Sources

Unlike EGRET and COMPTEL, OSSE has a smaller non-imaging field-of-view and undertakes observations of isolated objects one at a time. OSSE observed 35 active galaxies during phase 1. These sources comprise 14 Seyfert Type 1's, 9 Seyfert Type 2's, 4 BL Lacs, 5 QSOs, 1 radio galaxy, and 2 starburst galaxies. Table 2 provides a list of the QSOs and BL Lacs which were OSSE targets. The rather small number of QSOs reflects the Phase 1 observation plan prior to the

EGRET discovery of the high energy AGN as well as the large number of Seyferts which had been observed with hard spectra in previous hard x-ray surveys (Rothschild et al. 1983).

Table 2: QSOs and BL Lacs Observed by OSSE during Phase 1

Source	Type	Observ. Dates	Detection Signif. (σ)	Flux @ 70 keV (10^{-3} γ/cm^2-s-MeV)	Photon Index (above 50 keV)
3C 273	QSO	6/15 - 6/28 1991	34.77	23.9±0.82	1.69±.05
		8/22 - 9/5 1991	6.86	76.2±1.29	1.59±0.23
		10/3 - 10/17 1991	6.66	5.94±0.9	1.63±0.23
		8/12 - 8/20 1992	6.41	23.6±3.94	2.33±.39
		9/1 - 9/17 1992	12.57	29.2±2.32	1.84±0.14
QSO 0736+016	QSO	6/15 - 6/28 1991	0.24		
3C 279	QSO	9/19 - 10/3 1991	9.34	10.2±1.26	2.10±0.25
		9/17 - 10/8 1992	0.66		
PKS 0528+134	QSO	10/8 - 10/15 1992	0.61		
		11/3 - 11/17 1992	1.03		
QSO 0834-201	QSO	10/8 - 10/15 1992	-0.22		
		11/3 - 11/17 1992	0.20		
Mrk 421	BL Lac	7/12 - 7/26 1991	-0.63		
		7/26 - 8/8 1991	-0.01		
		9/12 - 9/19 1991	-1.08		
PKS 0548-322	BL Lac	6/11 - 6/25 1992	2.08	6.82±3.72	2.92±1.64
4C 04.77	BL Lac	8/20 - 8/27 1992	0.84		
PKS 2155-304	BL Lac	10/15-11/17 1992	4.96	7.47±1.67	2.01±0.44

Table 3 lists the Seyfert galaxies which were OSSE Phase 1 targets. The strongest Seyfert galaxy observed by OSSE is NGC 4151. These results have been reported by Maisack et al. (1992). The spectrum for NGC 4151 observed in July 1991 is shown in Fig. 1. This spectrum is well fit by thermal spectra (thermal bremsstrahlung or Sunyaev-Titarchuk model) but is not consistent with a single power-law. A single power law fit to the data gives a spectral index of -2.72 ± 0.07 in the 65-800 keV region (Maisack et al. 1993), which is much harder than that observed at lower energies. This clear evidence for a spectral break and the best fit thermal temperature of about 40 keV supports thermal models of Seyfert sources, and is similar to the spectra observed for galactic black hole candidates such as Cyg X-1 and GRO J0422+34. No evidence for a positron annihilation feature or an excess MeV emission is observed raising doubts about non-thermal models that were developed for these sources following earlier reports of MeV emission for NGC 4151 and MCG 8-11-11. The OSSE and COMPTEL upper limits in the 1-10 MeV region are more than an order of magnitude below the level reported by Perotti et al. (1981a) from a balloon-borne observation.

The OSSE observation of NGC 4151 raises the question whether other Seyfert galaxies have similar spectra, and if so, what is the implication for the contribution of Seyferts to the diffuse gamma ray background. NGC 4151, detected at a sensitivity level of about 60σ, is the only Seyfert observed thus far for which a distinction between spectral models can clearly be made

(Maisack et al. 1993). Johnson et al. (1994) have studied the summed spectra of many of the lower intensity OSSE Seyferts and find that the summed spectrum is consistent with a thermal model, and not consistent with a power law model. This can be seen, in part, by inspecting Table 3 where the spectral indices for each of the Seyfert observations are given. Note that by excluding NGC 4151, the average spectral index of the six Seyferts (3C 111, 3C 390.3, IC 4329A, MCG - 6-30-15, MCG 8-11-11, and NGC 4388) detected with a significance above 6σ is 2.50. This is significantly harder that the 'canonical' spectral index observed for these sources below 50 keV, suggesting that a thermal-like spectrum may be typical of Seyfert galaxies.

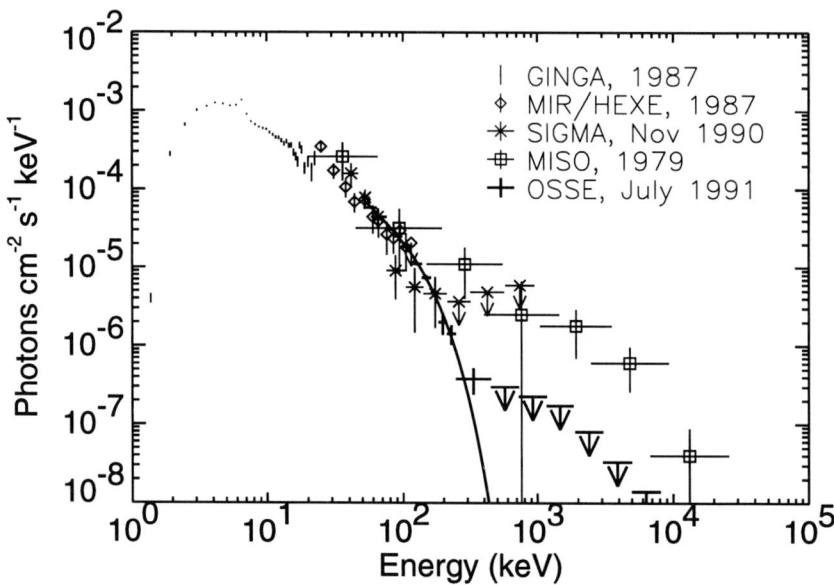

Figure 1. OSSE spectrum of NGC 4151 observed in July, 1991. The best fit Sunyaev-Titarchuk thermal comptonization model is shown. Previous x-ray and gamma ray results are also shown.

4. Broad-Band *COMPTON Observatory* Spectra

One of the objectives of the *Compton Observatory* is to obtain broad band γ-ray coverage of a variety of sources. Brown et al. (1994) have presented preliminary spectra for several AGN over the gamma ray band from 0.1 MeV to 10 GeV. These are shown in Figure 2 and include spectra for 3C 273, 3C 279, PKS 0528-134 and Mrk 421, which have been detected by EGRET at high energies and for which OSSE and/or COMPTEL observations have been made or for which significant low-energy upper limits exist. By broad-band spectra, we limit these to GRO data only. Other papers in these proceedings go into considerably more detail on the very broad band observations of AGN, from radio to gamma rays, which is the focus of this meeting .

Table 3: Seyfert Galaxies Observed by OSSE During Phase 1

Source	Type	Observ. Dates	Detection Signif. (σ)	Flux @ 70 keV (10^{-3} γ/cm^2-s-MeV)	Photon Index (above 50 keV)
3C 111	SY 1	6/28 - 7/12 1991	6.15	5.23±0.97	2.03±0.37
		5/14 - 6/4 1992	1.89		
3C 390.3	SY 1	10/17-10/31 1991	6.09	6.86±1.20	2.60±0.45
		5/14 - 6/4 1992	4.89	5.41±1.09	2.30±0.46
Mrk 279	SY 1	3/5 - 3/19 1992	4.01	4.12±1.13	2.38±0.62
IC 4329A	SY 1	10/8 - 10/15 1992	3.67	8.64±2.33	2.25±0.59
		11/3 - 11/17 1992	8.26	11.4±1.47	2.85±0.38
MCG -6-30-15	SY 1	10/8 - 10/15 1992	3.38	4.80±1.70	2.28±0.77
		11/3 - 11/17 1992	6.91	6.72±1.12	2.39±0.40
Mrk 509	SY 1	10/28 - 11/3 1992	3.21	6.14±2.08	3.83±1.28
3C 120	SY 1	5/14 - 6/4 1992	2.62	5.08±1.89	1.59±0.58
		6/4 - 6/11 1992	-0.31		
		7/2 - 7/16 1992	3.14	3.32±1.16	1.81±0.57
MCG +8-11-11	SY 1	6/11 - 6/25 1992	7.65	6.15±0.90	2.70±0.41
Mrk 841	SY 1	4/16 - 4/23 1992	-0.19		
NGC 3783	SY 1	6/25 - 7/2 1992	2.08	4.04±1.90	1.62±0.68
ESO 141-55	SY 1	8/27 - 9/1 1992	3.17	4.12±1.94	1.75±0.74
		10/15-10/29 1992	-0.19		
Mrk 335	SY 1	4/23 - 4/28 1992	-0.54		
		5/7 - 5/14 1992	0.92		
		8/20 - 9/3 1992	-0.17		
NGC 5548	SY 1.2	8/15 - 8/22 1991	3.47	6.49±2.00	3.36±1.05
		10/17-10/31 1991	2.38	8.42±2.99	3.68±1.37
		10/31 -11/7 1991	2.74	8.21±3.20	1.73±0.65
NGC 4151	SY 1.5	6/28 - 7/12 1991	64.41	41.5±0.83	2.42±.05
		4/2 - 4/9 1991	5.82	34.4±6.85	2.22±0.45
		4/9 - 4/16 1991	5.57	42.6±7.77	2.07±0.38
NGC 4593	SY 1.9	8/12 - 8/20 1992	-0.99		
		9/1 - 9/17 1992	-0.94		
NGC 7314	SY 1.9	4/28 - 5/7 1992	1.31		
NGC 7582	SY 2	12/12-12/27 1991	3.44	7.80±2.32	3.07±0.96
		4/2 - 4/9 1992	3.04	8.00±3.24	5.26±1.66
		4/9 - 4/16 1992	0.13		
		4/16- 4/23 1992	-0.31		
NGC 1275	SY 2	11/28-12/12 1991	2.81	2.75±0.92	2.49±0.83
NGC 2992	SY 2	6/4 - 6/11 1992	0.94		
		7/2 - 7/16 1992	0.19		
MCG -5-23-16	SY 2	8/6 - 8/12 1992	4.14	5.55±1.68	1.89±0.52
		8/12 - 8/20 1992	-0.62		
		8/27 - 9/1 1992	3.80	7.55±2.38	2.08±0.63
NGC 1068	SY 2	2/20 - 3/5 1992	0.07		
NGC 4388	SY 2	9/17 -10/8 1992	11.51	10.9±1.00	2.37±0.22
MCG +5-23-16	SY 2	8/12 - 8/20 1992	-0.51		
		9/1 - 9/17 1992	-0.22		

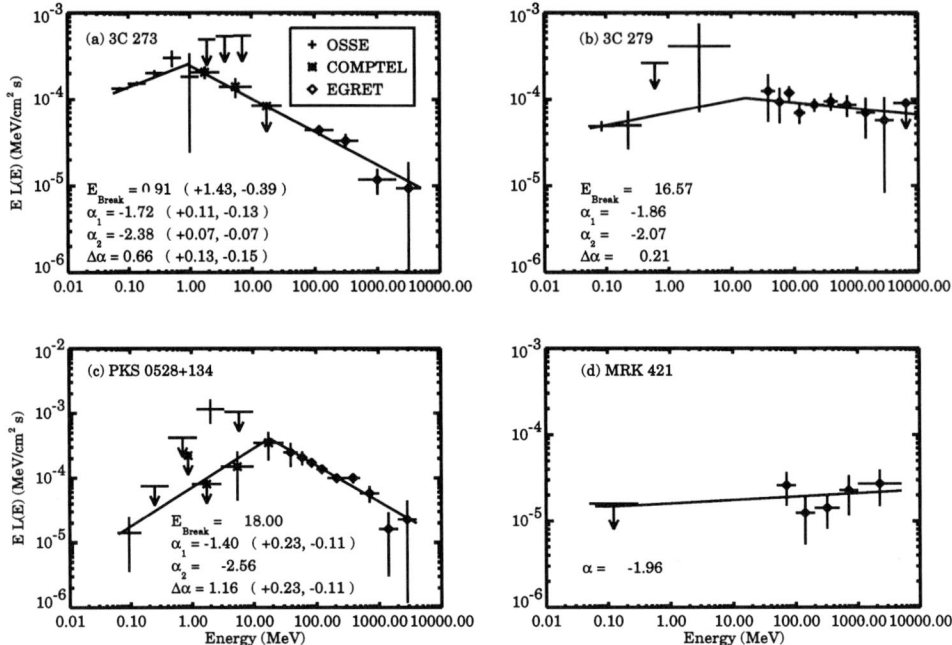

Figure 2: 0.1 MeV - 10 GeV spectra for four AGN: (a) 3C 273, (b) 3C 279, (c) PKS 0528+134, and (d) Mrk 421.

The GRO spectrum of 3C 273 in June, 1991 is shown in Figure 2a. During this observation, 3C 273 was relatively weak in high-energy gamma rays compared to the COS-B observation, and was also weak in low-energy gamma rays compared to historical data. The spectrum is characterized by a broken power law with a break near 2 MeV, and a change in spectral index of 0.66.

The spectrum for 3C 279 is shown in Figure 2b. The reader is cautioned that the data from the GRO instruments are not simultaneous. The EGRET data were taken in VP11 when OSSE was not observing the source. These are compared with OSSE data from VP10. It should also be noted that the high-energy gamma ray emission during VP3 showed clear variability, raising further caution about over-interpreting these data. Nevertheless, the EGRET data during VP3 and VP11 showed little change in intensity at the lower energy of the EGRET band (100 MeV). A broken power law fit to the VP10/VP11 data suggests only a small change in index in the GRO energy range, and the combined data are consistent with a single power law of index about -2.0

The spectrum for PKS 0528+134 is shown in Fig. 2c. This QSO, at a redshift of 2.08, has been detected by EGRET and COMPTEL, but OSSE only has upper limits for a non-contemporaneous observation. This source exhibits one of the softer spectra observed by EGRET ($\alpha = -2.6$). COMPTEL observes a weak signal, which requires a spectral break near 20 MeV. A spectral break of $\delta\alpha \geq 1.0$ is also required by OSSE limits in the 0.1-1.0 MeV region.

The spectrum of the nearby BL Lac, Mrk 421, is shown in Fig 2d. Mrk 421 is the only extragalactic source observed at TeV energies by ground-based gamma ray detectors (Punch et al. 1992). EGRET has obtained a spectrum with a power law index of -2.0 which agrees with the TeV data when extrapolated to higher energies. The OSSE upper limits at 100 keV are not constraining on the spectrum, so it is possible that this $\alpha = -2.0$ power law extends unbroken from 0.1 MeV to 1 TeV.

Even though the spectra of AGN observed by EGRET can be fit with power laws, broad band GRO data indicate that spectral breaks occur at energies ranging from less that 0.1 MeV to above 1 GeV. In the case of PKS 0528+134 the magnitude of the break is $\delta\alpha \geq 1.0$ and occurs near 20 MeV, while in 3C273 the spectral break, $\delta\alpha = 0.66$, occurs at lower energy, ~2 MeV. The energy and magnitude of the spectral break should provide information about the emission mechanism and/or geometry of the relativistic beams in these sources.

5. Future Plans for COMPTON/GRO

During Phase 1 the EGRET and COMPTEL instruments completed a full sky survey for gamma radiation above 1 MeV. Guest Investigations were started in Phase 2 and in the current phase (Phase 3) over one half of the observation time is devoted to GI observations. Starting with the next phase (November 1994) all observing time will be competed. Several key projects have been allocated time in the current phase for in-depth study of AGN. Several regions of the sky will be viewed by COMPTEL and EGRET to monitor a large number of the AGN which were discovered during the sky survey. New sources will also certainly be discovered based on the variability which seems to be commonplace in these objects. There is also a deep survey of the Virgo region to be undertaken in Phase 3 which will provide 8 weeks of coordinated observations with the OSSE, EGRET and COMPTEL instruments. Coordinated observations at other wavelengths are also being organized in connection with the Virgo survey. Finally, there will be a multi-wavelength campaign to observe NGC 4151 in December, 1993.

6. Summary

The COMPTON Observatory has made several important contributions to the understanding of AGN during the first 1-1/2 years of the mission. These include:

1. Discovery of a new class of high-energy gamma ray emitting AGN associated with radio loud, core-dominated sources. The short term variability, power-law spectra, and large apparent radiated power suggest that the emission arises in relativistic jets that are beamed toward the observer.

2. Measurements of the low-energy gamma ray spectra of Seyfert galaxies. These spectra, typified by NGC 4151, are much steeper above 50 keV than below 50 keV. No evidence for positron annihilation features or MeV excesses reported previously have been observed. This seems to suggest a thermal origin for the radiation, similar to that observed in several galactic black hole candidates.

3. Broad-band spectra of several radio loud objects with spectral indices and spectral break energy covering a wide range of these parameters. Further observations and characterization of these should help in understanding the emission mechanism and perhaps the geometry associated with these beamed sources. It is clear that much progress in understanding AGN will come from such multi-wavelength campaigns, and we expect that the *COMPTON Observatory* will contribute to a large number of such campaigns in the future.

Acknowledgments

The author wishes to thank C. Fichtel, V. Schoenfelder, N. Johnson and K. Brown for data provided for this paper.

References

Brown, K., et al., 1994, to be published in Proc. Second *COMPTON* Symposium.
Fichtel, C.E., et al., 1993a, presented at the 182nd AAS meeting, private communication.
Fichtel, C.E., et al., 1993b, AIP Conference Proceeding **280**, 461.
Fishman,G.J. et al. 1989, Proceedings of the GRO Science Workshop, 2-39.
Hartman, R.C. et al., 1992 Ap. J., **385**, L1
Johnson, W.N. et al., 1994, to be published in Proc. Second *COMPTON* Symposium.
Johnson, W.N., et al., 1993, Ap. J. Supp. **86**, 693.
Levine, A.M., et al., 1984, Ap. J. Supp., **54**, 581.
Maisack, M., et al., 1993 Ap. J. **407**, L61.
Paciesas, W.S., et al. 1993, AIP Conference Proceeding **280**, 473.
Perotti, F, et al., 1981a, Ap. J. **247**, L63.
Perotti, F., et al., 1981b, Nature, **292**, 133.
Punch, M., et al., 1992, Nature **358**,477.
Rothschild, R.E., et al., 1983. Ap. J. **269**, 423.
Swanenburg, B.N., et al., 1978, Nature **275**, 298.
Schoenfelder, V., et al., 1993, Ap. J. Supp. **86**, 657.
Thompson, D. J., et al., 1993, Ap. J. Supp. **86**, 629.

EGRET HIGH-ENERGY GAMMA-RAY OBSERVATIONS OF AGN: ENERGY SPECTRA AND TIME VARIABILITY

D.J. THOMPSON, D.L. BERTSCH, B.L. DINGUS*,
J. A. ESPOSITO*, C.E. FICHTEL, R.C. HARTMAN,
S.D. HUNTER, P. SREEKUMAR*
NASA/Goddard Space Flight Center
Greenbelt, Maryland 20771 USA

J. CHIANG, J. M. FIERRO, Y.C. LIN,
P.F. MICHELSON, P.L. NOLAN
Stanford University
Stanford, California, 94305 USA

G. KANBACH, H.A. MAYER-HASSELWANDER, C.V. MONTIGNY
H.-D. RADECKE
Max Planck Institut für Extraterrestrische Physik
Giessenbachstr, 85748 Garching Germany

D.A. KNIFFEN
Hampden-Sydney College
Hampden-Sydney, Virginia, 23943 USA

J.R. MATTOX
Compton Observatory Science Support Center,
operated by Astronomy Programs,
Computer Sciences Corporation
Greenbelt, MD 20770 USA

E.J. SCHNEID
Grumman Aerospace Corporation
Bethpage, New York 11714 USA

ABSTRACT. The Energetic Gamma Ray Experiment Telescope (EGRET) on the Compton Gamma Ray Observatory has detected more than 20 Active Galactic Nuclei (AGN) at photon energies above 30 MeV.

1. Introduction

Since the launch of the Compton Gamma Ray Observatory in April, 1991, EGRET has been mapping the high energy gamma ray sky. During the all-sky survey and in the

---

*USRA Research Associate

observations since then, EGRET has identified 26 gamma-ray-bright AGN. All were detected with statistical significance of 5 σ or greater.

2. Identifications

Although EGRET source error boxes are relatively large (ranging from a few arcminutes to more than a degree in diameter), the number of high-energy gamma ray sources at high latitudes is modest; therefore, source confusion is not a serious problem. Gamma rays in the 100 MeV energy range are also inherently nonthermal, requiring high energy particles as their progenitors. The search for counterparts can focus on likely sites of particle acceleration. In the case of the EGRET sources, a very high fraction (more than 70%) of the high galactic latitude error boxes contain one particular type of AGN. These are characterized by their radio properties, particularly in the 1-5 GHz band: radio bright (more than 0.5 Jy), flat or nearly flat spectrum (α > -0.6), and core dominated. An extrapolation of the Kühr et al. (1981) catalog indicates that there are fewer than 900 such sources in the sky. Such objects are generally thought to have jets, which provide the likely sites of particle acceleration to produce the gamma rays. Table 1 lists the EGRET detections, along with properties which indicate that many of these have characteristics of blazars.

3. Energy Spectra

Over at least the central part of the EGRET energy range (50 MeV to several GeV), all the AGN energy spectra can be well represented by power laws. As shown in Table 1, the photon number spectra range from -1.5 to -2.6. Combining the EGRET results with those from other frequencies (usually not contemporaneous) produces multifrequency spectra such as the example of Fig. 1. The gamma rays represent a major contributor to the total observed power of the source. In this example, the simultaneous EGRET (Hunter et al., 1993) and COMPTEL (Collmar, et al., 1993) observations show strong evidence for a spectral break near 10^{22} Hz.

4. Time Variability

Most of the EGRET-detected AGN show time variability on scales of days to months. An example is shown in Fig. 2. In addition to supporting the concept of a jet origin for the gamma radiation, the variability provides an important tool for modeling the physical processes in these jets. Correlations and time lags between gamma-ray flares and those seen at other wavelengths are crucial tests of models.

5. References

Bertsch, D.L. et al.: 1993, `Detection of Gamma-Ray Emission from the Quasar PKS 0208-512', *ApJ* **405**, pp. L21-L24.
Collmar, W. et al.: 1993, `Preliminary COMPTEL Results on the Quasar PKS 0528+134', *Proc. 23rd Inter. Cosmic Ray Conf. (Calgary)*, **1**, pp. 168-171.
Hunter, S.D. et al.: 1993, `Detection of High-Energy Gamma Rays from Quasar PKS 0528+134 by EGRET on the Compton Gamma Ray Observatory', *ApJ* **409**, pp. 134-138.
Kühr, H., Witzel, A., Pauliny-Toth, I.I.K., and Nauber, U.: 1991, `A Catalogue of Extragalactic Radio Sources Having Flux Densities Greater than 1 Jy at 5 GHz', *A&ASuppl* **45**, pp. 367-430.

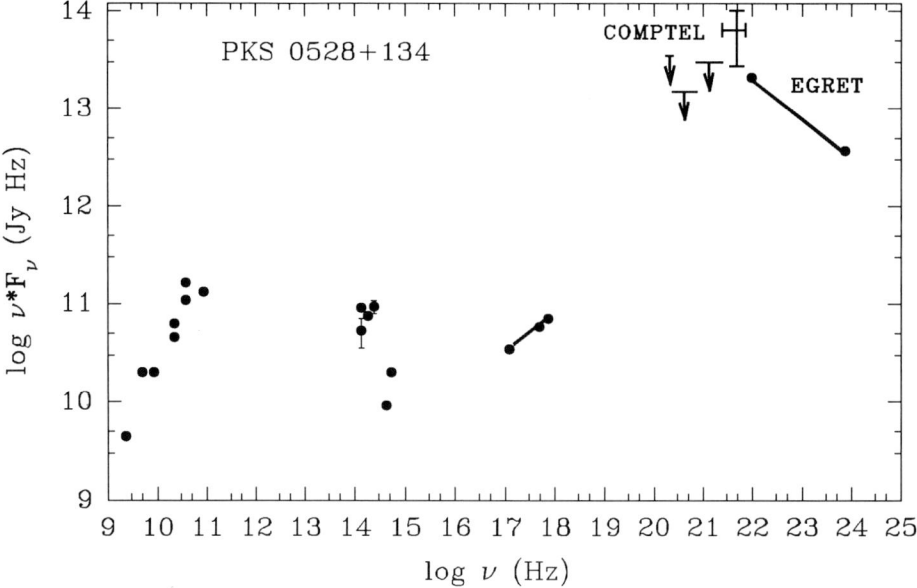

Fig. 1 -- Multifrequency spectrum of PKS 0528+134. The EGRET (Hunter et al., 1993) and COMPTEL (Collmar et al., 1993) observations were made in May, 1991. References to other observations are given by Hunter et al.

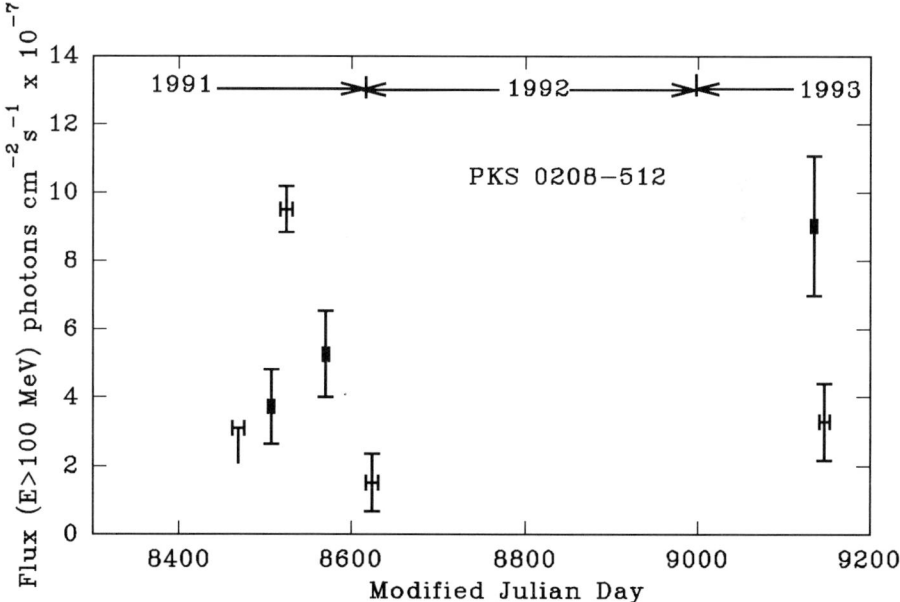

Fig. 2 -- Time variability of E > 100 MeV gamma rays from PKS 0208-512 as seen by EGRET. Data points from 1991-1992 are from Bertsch et al. (1993).

Table 1 -- EGRET High Energy Gamma Ray Observations of AGN

Source ID and characteristics		OVV	BL Lac	super lum.	radio bright	flat radio[1]	opt. pol.	flux (10^{-6} cm^{-2} s^{-1}) (E>100 MeV)	photon spectral index	z	relative luminosity[2]
0202+149	(4C+15.05)				√	√		0.3±0.1	-2.5±0.1	...	...
0208-512	PKS				√	√	√	0.4 to 0.9	-1.7±0.1	1.00	2
0219+428	(3C 66A)		√		√	√	√	0.17±0.05		0.444	0.03
0234+285				√	√	√	√	0.17±0.05		1.213	0.3
0235+164	(OD+160)		√	?	√	√	√	0.8±0.1	-2.0±0.2	0.94	2.0
0420-014	(OA 129)	√			√	√	√	0.4±0.1	-1.9±0.3	0.92	0.4
0446+11					√	√		1.0±0.2	-1.8±0.3		
0454-463	PKS	√			√	√		0.25±0.1	-2.6±0.1	0.86	0.3
0528+134	PKS				√	√	√	0.4 to ~3	-2.0±0.2	2.06	4 to 13
0537-441	PKS		√		√	√	√	0.3±0.1	-1.8±0.2	0.894	0.2
0716+714			√		√	√		0.20±0.06		...	...
0827+243		√		?	√	√		0.20±0.04		2.05	
0836+710	(4C+71.07)	√		√	√	√		0.15±0.04	-2.4±0.2	2.17	1.1
1101+384	(Mrk 421)		√		√	√	√	0.14±0.03	-1.9±0.1	0.031	0.0002
1156+295	(4C+29.45)	√		√	√	√		0.63±0.15		0.729	
1226+023	(3C 273)	√		√	√	√	√	0.30±0.05	-2.4±0.1	0.158	0.008
1253-055	(3C 279)	√		√	√	√	√	0.6 to 4.9	-2.0±0.1	0.54	0.3 to 2
1406-076	PKS				√	√		0.25±0.1		0.86	0.3
1606+106	(4C+10.45)				√	√		1.0±0.1		1.49	
1611+343		√			√	√		0.33±0.06		1.40	
1633+382	(4C+38.41)	√		√	√	√		0.4 to 1.4	-1.9±0.1	1.81	3 to 11
1739+522	(4C+51.37)				√	√		0.3 to 0.5		1.38	
2022-077	(NRAO 628)				√	√		0.7±0.1	-1.5±0.2		
2052-474					√	?		0.3±0.1		...	0.6
2230+114	(CTA 102)			?	√	√	√	0.24±0.07	-2.6±0.2	1.489	0.5
2251+158	(3C 454.3)	√		√	√	√	√	0.8±0.1	-2.2±0.1	1.037	0.5
										0.859	

1. Flat spectrum radio sources: $\alpha_r > -0.5$ (2-5 GHz band).
2. The source luminosity for the observed energy range (0.1 GeV<E<5 GeV), computed using the known redshift assuming $H_0 = 75$ km s^{-1} Mpc^{-1} and $q_0 = 1/2$, is equal to the relative luminosity times (10^{48} erg s^{-1})f, where f is an unknown beaming factor. Typically f is thought to be in the range from 10^{-2} to 10^{-3}.

ROSAT OBSERVATIONS OF ACTIVE GALACTIC NUCLEI

W. BRINKMANN
Max-Planck-Institut für Extraterrestrische Physik,
D-85740 Garching, F. R. G.

Abstract. The large number of Active Galactive Nuclei detected for the first time through their X-ray emission in the ROSAT All Sky Survey as well as the first measurements of the X-ray emission of many previously known AGN provide a new unprecedented large basis for the statistical and morphological exploration of these objects.
The soft energy range of the X-Ray Telescope, the good energy resolution of the PSPC detector, and the high sensitivity of the instrument further allows an investigation of the spectral properties of sources in this energetically important energy band.
A short overview is given of the actual ongoing research concentrating on the study of the soft X-ray class properties of the various types of AGN.

Key words: Active Galactic Nuclei, X-rays, Unification Schemes

1. Introduction

ROSAT is in orbit for more than three years and it is still working exceptionally well. After performing the first All Sky Survey in the soft X-ray band between August 1990 and February 1991 /1/ it is used for pointed observations in an international PI programme. The fact that about 30% of the proposed targets are Active Galactic Nuclei (AGN) demonstrates the large scientific interest and the observational potential of ROSAT in this field. Two properties make ROSAT especially well suited for the study of AGN:
The soft energy band: The very soft energy band (0.1 – 2.4 keV) together with its high sensitivity and the moderate spectral resolution of the PSPC /2/ allows the spectral study of AGN in a previously unaccessible energy range. Most pointed observations utilize these instrument characteristics. Many results of these investigations on previously known AGN are presented in this volume. In this review I will therefore concentrate on the second main aspect of the ROSAT mission:
The All Sky Survey: The ROSAT All Sky Survey (RASS) provides a large number of new X-ray identifications of AGN of all types and, in particular, it will yield an enormous number of previously unknown AGN.
So far most studies of the class properties of the AGN population are biased by strong selection effects and in many cases data from pointed observations of known objects had to be used. The RASS yields a large number of AGN homogeneously selected all over the sky. However, differences in the Survey exposure as well as spatial variations of the galactic N_H-values introduce some selection biases which have to be taken into account in any statistical study.
The vast majority of these objects remain to be identified. Only about 2000 objects are previously catalogued AGN but mostly without any X-ray detection so far. They will be used for studying the broad band spectral behaviour of these sources

and form the basis for the study of the evolution of different classes of AGN and their statistical properties.

2. The AGN Content of the Survey

Early estimates predicted that about 40% to 60% of the sources found in the RASS would be AGN, i.e., from the $\sim$ 50000 sources found so far there are $\sim$ 25000 new extragalactic objects, detected for the first time through their X-ray emission. The correctness of these estimates could be confirmed recently in several observational programs in which the complete optical identification of all RASS sources in certain limited areas of the sky was attempted.

The currently most advanced one is an ESO Key Project. 488 of the 688 ROSAT sources in the 'study areas' have been identified up to date. 259 objects turned out to be AGN, 31 were galaxies, only 40 were clusters of galaxies, 2 were CVs, 4 white dwarves, and the rest stars. As the sources are being identified according to their optical brightness it is felt that all stars have been found and the remaining, optically faint unidentified sources are predominantly of extragalactic origin.

From existing catalogues of optically known AGN, for example from the more than 8000 objects listed by Veron-Cetty and Veron /3/ only around 1500 sources were detected in the RASS. Basically, this leaves us with the same number of $\sim$ 25000 objects to be identified optically as AGN.

The above mentioned areas for which the complete identification of RASS sources are attempted cover only a few hundred square degrees of the sky. Other projects try to identify as many sources as possible from existing observational data without pretention to be complete.

2.1. THE HAMBURG QUASAR SURVEY

For this project digitized objective prism and direct plates are used obtained with the Calar Alto Schmidt telescope of the northern sky $\delta > 0°$ and $|b_{II}| > 20°$ /4/. So far $\sim$ 6200 ROSAT sources have been looked at, of which 63% could be identified. 29% were AGN, 7% galaxies and clusters, and 27% stars. About 30% of the objects were in fields with either too many candidates, with 'implausible' candidates, or on plates with errors. 7% of the fields were 'empty', i.e., no optical counterpart brighter than the plate limit ($B > 18.5^{mag}$) could be found.

This fact highlights one of the observational challenges for optical identification programs: a substantial fraction of the optical counterparts are rather faint. In Fig. 1 I give a histogram for the optically unidentified objects from the ROSAT - Molonglo correlation (see below). Plotted are the magnitudes m_J of those objects for which a 'plausible' optical candidate could be selected from the NRL / ROE digitized finding charts. As seen from figure 1, about half of the objects seem to be fainter than 19^{mag}.

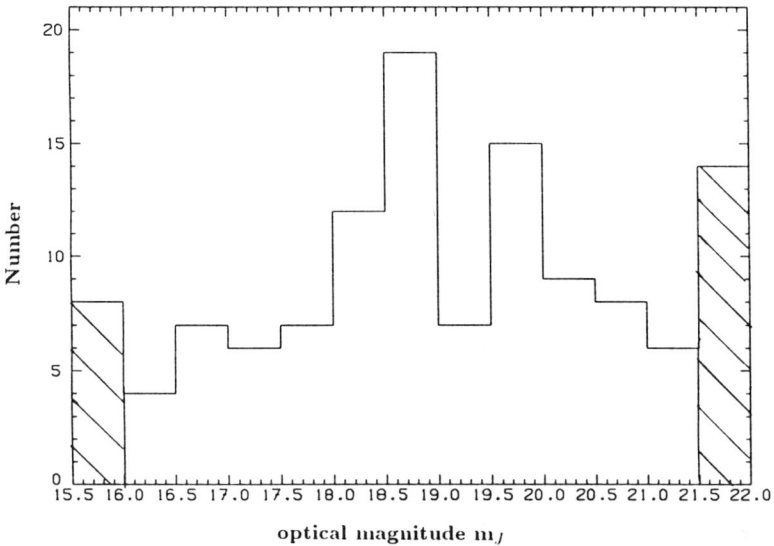

Fig. 1. Histogram of the optically unidentified objects from the ROSAT - MRC correlation. m_J magnitudes from the NRL/ROE digitized finding charts. Hatched areas denote all objects outside the respective magnitude range.

2.2. RADIO SURVEYS

The cross - correlation of the RASS source list with the Molonglo reference catalogue of radio sources (MRC /5/) and a point source catalogue generated from the Condon-Broderick-Seielstad 6cm survey of the northern hemisphere /6/ yielded a total of more than 2500 matches. A large fraction of them (more than 2/3 of the sources) remain optically unidentified. Extensive radio- and optical follow-up observation programs have been initiated to obtain more information on this large sample. First results from the statistical analysis of the optically identified subset of these correlations are presented in /7, 8, 9/ and some of the questions related to the class properties of AGN are discussed below.

To give an impression about the data to be expected, Fig. 2 shows the soft X-ray luminosity as function of redshift z for about 700 ROSAT detected AGN from the Veron-Cetty & Veron catalogue for which all necessary information was readily available. The luminosities are K-corrected according to their group-averaged spectral power law index. Marked with different symbols are different types of AGN. Diamonds represent objects with classifications not strictly belonging to the four mentioned categories or with questionable identification. For an average value of the galactic absorption and a power law index $\Gamma = -2.0$ the full line indicates a flux of 5×10^{-13} erg cm$^{-2}s^{-1}$, representative for the limiting sensitivity

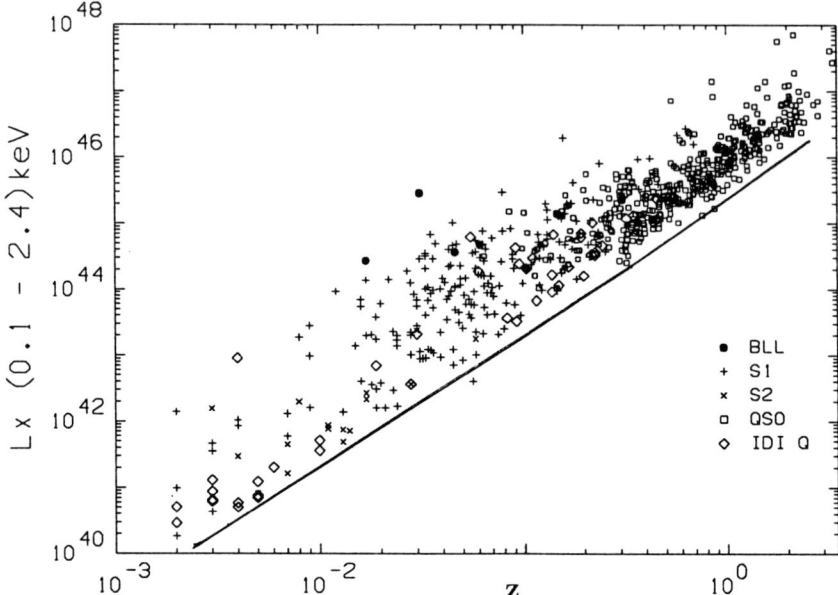

Fig. 2. Soft X-ray luminosities of optically known AGN as function of their redshift. Different types of objects are indicated by different symbols. The full line represents an incident X-ray flux of $5 \times 10^{-13} erg\ cm^{-2} s^{-1}$, characteristic for ROSAT's survey sensitivity.

of the Survey. Clearly visible at low redshifts are 'active galaxies' (like starbursts or radio galaxies), then Seyferts and BL Lac objects. At larger redshifts, $z \gtrsim 0.4$, the quasar component dominates.

The currently most distant object is a ROSAT detected quasar with $z = 4.32$ /10/ found in the NEP region with its very deep exposure. Correspondingly, it's inferred soft X-ray luminosity $L_{0.1-2.4keV} \sim 2 \times 10^{45}$ erg/s is rather low for a quasar. The second most distant object is, with $z = 3.87$, the object Q 1745+624, first seen by Einstein /11/, close to the NEP as well. The brightest object is PKS 1937-101 with $L_{0.1-2.4keV} \sim 3 \times 10^{47}$ erg/s at a redshift of z=3.78 /8/.

3. The X-ray AGN - a schematic view

Recently, it has become clear that the wide range of different appearances of AGN is only partially caused by truly different properties of the central machines. A great deal of the observed diversity can be attributed to the actual geometrical viewing conditions of the observer with respect to a preferred direction of emission /12/.

The by far largest number of AGN is radio-quiet. Using the source brightness as additional classification criterium, we find as radio-quiet, low luminosity objects the Seyfert galaxies, as high luminosity objects quasars.

The nearby and therefore bright Seyfert I galaxies show a very high detection probability in the Survey. Detailed comparative spectral studies of this class of objects can be found elsewhere in these proceedings. Only very few Seyfert II objects were seen with rather low countrates. This is in accordance with unification schemes which predict the obscuration of the central nuclear region by dense surrounding material.

It should be noted that in this region of the parameter space additional objects like starbursts and IRAS galaxies are found which form, at least with respect to their soft X-ray luminosity, a 'transition' between normal galaxies and AGN. They are detected only due to their relative proximity. Some of them, like NGC 4258, show characteristics which are indicative for a previous Seyfert activity in their nuclei /13/.

3.1. QUASARS: RADIO QUIET VERSUS RADIO LOUD

About 90% of the quasars are radio quiet /14/ and the available optical information suggests that they form a class quite different from the radio loud objects. Even in X-rays both types of quasars seem to have qualitatively different properties as well.

1. Detection probability: Quasars are the largest group of X-ray detected objects in the survey but with respect to their total number only about 16% of them are seen by ROSAT. However, while $\gtrsim 30\%$ of the catalogued radio-loud objects were discovered only $\lesssim 7\%$ of the large number of radio-quiet quasars were found.

2. Redshift distribution: Figure 3 shows the relative detection rate as function of redshift, i.e., the number of detected ojects in a redshift bin divided by the total number of objects in that bin. In Fig. 3a the data for the radio-loud quasars are plotted. The detection rate drops roughly like $1/z^2$ (indicated by the dots) due to flux limits of the telescope. Fig 3b shows the equivalent data for the radio-quiet quasars.

These results are in accordance with the well known fact that radio-loud quasars are on the average stronger X-ray emitters than the radio-quiet objects. The detection rate for radio-quiet quasars clearly drops much faster than that for the radio-loud objects. Apart from the different space density of the objects this is indicative for additional differences in the luminosity function of these two classes of objects.

3. Spectral parameters: The ROSAT soft X-ray (0.1 - 2.4 keV) spectra are in general much softer than those found at higher energies and show large intrinsic scatter. While radio-quiet quasars have an average power law photon index of $<\Gamma>\sim -2.4$, radio-loud quasars have a flatter spectrum with $<\Gamma>\sim -2.2$ (see /8/, Schartel et al. and Brunner et al., these proceedings).

4. Luminosity correlations: In figure 4 the soft 0.1-2.4 keV X-ray luminosity is plotted as function of the monochromatic optical luminosity for radio-loud quasars.

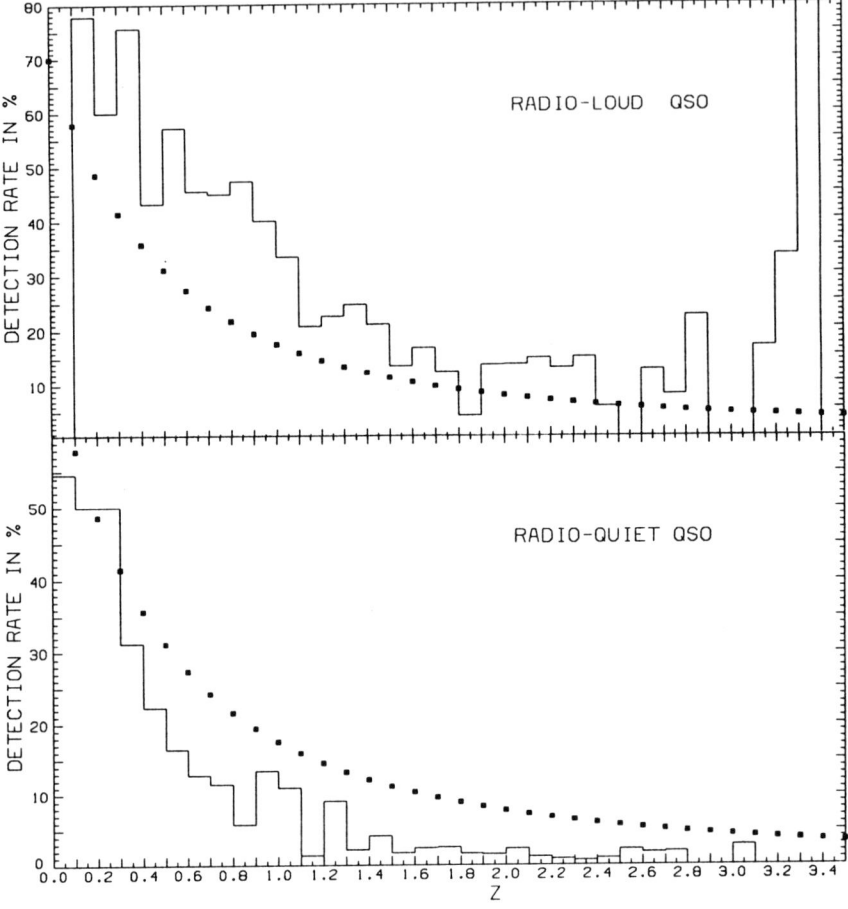

Fig. 3. Relative detection rate (in percent) for quasars in the ROSAT All Sky Survey as function of redshift. Upper panel, radio-loud QSOs, lower panel, radio-quiet objects. Please note the different scaling of the vertical axes. The dotted curves represent a $1/z^2$ sensitivity variation.

A regression analysis gives a correlation of the form $log(L_x) \sim 0.69 log(l_{opt})$ with nearly 100% confidence. A similar analysis for radio-quiet objects results in a different slope $log(L_x) \sim 0.89 log(l_{opt})$. The values for the slopes are securely out of their mutual errors.

5. Selection effects ? Finally, it must be noted that for the ROSAT X-ray selected quasars there seem to be strong selection effects. Even for the radio-brightest quasars the Survey detection rate is only around 50% without obvious exposure, redshift, or galactic absorption dependencies /8/. This indicates that all statistical

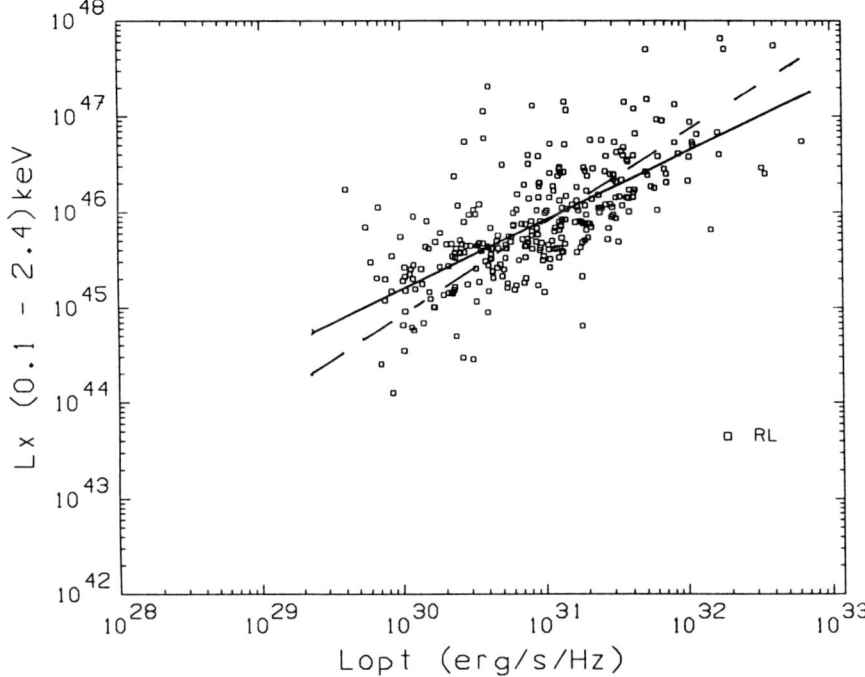

Fig. 4. Integrated soft X-ray luminosity as function of the monochromatic 5500 Å optical luminosity for radio-loud quasars. Full line is the linear regression curve, dashed line that for the radio-quiet quasars.

correlations are valid only for special, X-ray loud sub-samples of the quasar population.

On the other hand, there is a substantial number of objects without optical counterparts brighter than $\sim 20^{mag}$ but relatively luminous at radio- and X-ray wavelengths. It remains to be seen what the nature of these radio / X-ray bright but optically faint objects is. So far they have been regarded as being relatively rare as only very few of them have been found /15/.

4. Unification schemes

Only a small fraction of the order of 10% of known quasars are strong radio emitters /14/ and only very few radio galaxies have been detected in X–rays /16/. However, extragalactic radio sources have been studied in great detail with unsurpassed sensitivity and spatial resolution and they represent an important subgroup of

X-ray detected AGN. Further, radio quasars are the most luminous objects in the Universe and thus the distances and look back times are large enough that cosmological and evolutionary effects can be studied.

The largest sub-group of the cross - correlation of the ROSAT All Sky Survey source list with catalogues of radio sources (see sect. 2.2) with about 40% of the objects are quasars, followed by galaxies (17%), clusters of galaxies (3%), and BL Lacs (1%).

These samples are sufficiently large to allow to test current AGN unification models over a broad energy range. For more than a decade unification schemes have been proposed, which try to relate the various types of extragalactic objects /17, 18/, and references therein. In none of these schemes the X-ray properties of the various classes were taken into account. Therefore, a comparison of the properties of the above samples of radio selected extragalactic X-ray sources can give new information about the applicability of these schemes. First results indicate, that schemes based on purely "geometrically different" viewing conditions might be questionable:

i) Radio galaxies have an average photon index $\Gamma \sim 1.8$ in the soft X-ray range, quasars an index of $\Gamma \sim 2.25$

ii) About 25% of the radio galaxies of the MRC are in clusters or groups of galaxies. But from the more than 200 quasars only one object is found in a cluster.

iii) Quite a fraction of the objects in the various subclasses of AGN can be found in the IRAS catalog, i.e., they are infrared sources as well, but none of these sources is associated with a cluster environment.

At present it seems to be premature to draw strong conclusions from these observational facts. The influences of the various selection effects have not been evaluated yet and so far we have to rely on rather 'inhomogenous' catalogued data. Even more, we don't know what 'special' properties of the sources in an object class lead to their detection in the ROSAT All Sky Survey. These X-ray detected objects certainly form some not yet defined 'sub-groups' of their corresponding classes. Nevertheless, the above X-ray results strongly indicate an additional environmental or/and evolutionary difference between different types of objects. The real question to ask is perhaps not whether certain unification schemes are correct or not, but more, what fraction of the apparent source properties can be related to orientation dependent 'viewing conditions'.

Another question raised in the study of associations of galaxies with clusters is whether the X-ray emission attributed to the radio galaxies is predominantly cluster emission or whether the cluster environment is the evolutionary cause for active galaxies with strong X-ray and radio emission. In this case X-ray studies of distant clusters might be severely biased by a galaxy contribution. The spatial resolution of the PSPC in the Survey is usually insufficient to separate these two components and there seem to be examples for both possibilities /19/.

5. Conclusions

The ROSAT mission changed our approach of studying the AGN phenomenon considerably: The soft energy band allows an examination of the energetically vital region between the 'canonical power law' and the 'big blue bump' where spectral diversity promises new insights in the source conditions. The large number of newly detected AGN provide statistically sufficiently large samples for all classes of AGN. The results obtained from the ROSAT All Sky Survey presented in this paper must still be regarded as somehow preliminary as they are based on data obtained from the first processings of the Survey when the data were not complete.
Qualitatively, the results are not expected to change drastically, although the quantitative properties of individual sources might have to be revised slightly after a reprocessing of the data.
The following main findings and open questions will, however, remain unaffected:
* ROSAT multiplied the number of known X-ray emitting AGN
* Most AGN have a considerably **steeper** spectral slope in the ROSAT band than the canonical $\Gamma \sim 1.7$ found in the medium energy band
* The number of known X-ray emitting radio sources has been greatly increased
* What is the nature of the "unknown" radio sources ?
* What are the X-ray / radio - loud but optically "dull" objects ?
* X-ray observations provide new tests for Quasar/radio galaxy unification schemes

And, finally, the $\gtrsim 25000$ unidentified X-ray sources from the ROSAT All Sky Survey form an invaluable source of information both for observers and theoretical astronomers.

Acknowledgements

The ROSAT project is supported by the Bundesministerium für Forschung und Technologie (BMFT). I like to thank my colleagues at the Max-Planck-Institut für Extraterrestrische Physik for fruitful discussions and providing parts of the presented material.

6. References

1. Voges, W. 1992. in *Proc. of the ISY Conference "Space Science"*, ESA ISY-3, ESA Publications, p.9
2. Pfeffermann, E., Briel, U.G., Hippmann, H., Kettenring, G., Metzner, G., Predehl, P., Reger, G., Stephan, K.H., Zombeck, M.V., Chappell, J., Murray, S.S., 1986, Proc SPIE **733**, 519
3. Veron-Cetty, M.-P. and Veron, P., 1991, ESO Scientific Report No. 10.
4. Bade, N., Engels, D., Fink, H., Hagen, H.-J., Reimers, D., Voges, W., and Wisotzki, L., 1992, *Astr. Astrophys.*, **254**, L21

5. Large, M.I., Cram, L.E., & Burgess, A.M., 1991, *The Observatory* **111**, 72.
6. Condon, J.J., Broderick,J.J., and G.A. Seielstad, 1989, *Astron. J.* **97**, 1064.
7. Brinkmann, W., 1992. in: *Proc. of a Conference on "X-Ray Emission from AGN and the Cosmic X-Ray Background"*, eds. W. Brinkmann and J. Trümper, MPE Report 143.
8. Brinkmann, W., Siebert, J., and Boller, Th., 1993, *Astr. Astrophys.*, in press.
9. Brinkmann, W., and Siebert, J., 1993, *Astr. Astrophys.*, submitted.
10. Henry, J.P. et al , 1993, *Astron. J.*, submitted
11. Becker, R.H., Helfand, D.J., and White, R.L., 1992, *Astron. J.*, **104**, 531.
12. Antonucci, R., 1993, Ann. Rev. Astron. Astr.**31**, 473.
13. Pietsch, W., et al., 1993, *Astr. Astrophys.*, submitted
14. Kellermann, K.I., et al., 1989, *Astron. J.*, **98**, 1195.
15. Elvis, M., Schreier, E.J., Tonry, J., Davis, M., and Huchra, J.P., 1981, *Astrophys. J.*, **246**, 20.
16. Fabbiano, G., et al., 1984. *Astrophys. J.* **277**, 115.
17. Barthel, P.D., 1989. *Astrophys. J.* **336**, 319.
18. Scheuer, P.A.G., 1987. in *Superluminal Radio Sources*, eds. J.A. Zensus and T.J. Pearson, Cambridge Univ. Press, Cambridge, p. 104.
19. Pierre, M., Hunstead, R., and Unnewisse, A., 1993, in: Proc. of the NATO ASI 'Clusters of Galaxies', Münster, eds. Seitter et al.

REVIEW OF GRANAT OBSERVATIONS OF AGNS

E.CHURAZOV, M.GILFANOV, A.FINOGUENOV, R.SUNYAEV,
M.CHERNYAKOVA, YU.APALKOV, S.GREBENEV and I.LAGUNOV
Space Research Institute, Profsoyuznaya 84/32, 117810 Moscow, Russia

E.JOURDAIN, J.P.ROQUES, P.MANDROU and L.BOUCHET
Centre d'Etude Spatiale des Rayonnements, 9, avenu du Colonel Roche, BP 4346, 31029 Toulouse Cedex, France

and

J.BALLET, A.CLARET, P.LAURENT and F.LEBRUN
Service d'Astrophysique, Centre d'Etudes Nucleaires de Saclay, Orme des Merisiers, 91191 Gif-sur-Yvette Cedex, France

Abstract. Brief review of AGNs observations in the X-ray / soft gamma-ray bands with the orbital observatory GRANAT is presented.
   For three well known bright objects (3C273, NGC4151 and Cen A) broad band (3 keV - few hundreds keV) spectra have been obtained. Imaging capabilities allowed accurate (several arcminutes) identification of these objects with sources of hard X-rays.
   The spectrum of NGC4151 above $\approx$ 50 keV was found to be much steeper than that in most of the previous observations, while in standard X-ray band the spectrum agrees with observed previously. The comparison of the observed spectra with that of the X-Ray Background (XRB) indicates that sources similar to NGC4151 could reproduce the shape of XRB spectrum in 3-60 keV band.
   Cen A was observed in the very low state during most of observations in 1990-1993, except for two observations in 1991. The variability of the hard X-ray flux has been detected on the time scales of several days.

Key words: X-Rays: AGNs

1. Introduction

Two coded aperture imaging X-ray telescopes ART-P and SIGMA on board the GRANAT satellite have been operated since 1989 December 1. ART-P telescope consists of four identical modules, having 5' angular resolution over $3°.6 \times 3°.6$ fully coded field of view, and operates in standard (3-30 keV) energy range (Sunyaev et al., 1990). The SIGMA telescope, operating in hard X-ray / Soft gamma-ray band, provides $11°.4 \times 10°.6$ images (at half-sensitivity) with nominal angular resolution of $\sim 15'$ (Paul et al., 1991). More than 40 $\approx$ one-day observations of 3C273, NGC4151 and Cen A have been performed. These objects were unambiguously identified with sources of hard X-rays with several arcminutes accuracy (except for possible contamination of 3C273 flux due to GRS1227+02/1E1227+0224). Shown in Fig. 1 are the $4° \times 4°$ SIGMA images in 35-150 keV energy range averaged over most of observations of these three objects in 1990-1993.

The results of AGNs observations with GRANAT during first years of the mission are presented in a number of papers by Apalkov et al., 1992 (ART-P, observations in 1990, 3C273, NGC 4151, Cen A); Ballet et al., 1991, 1992a,b (SIGMA

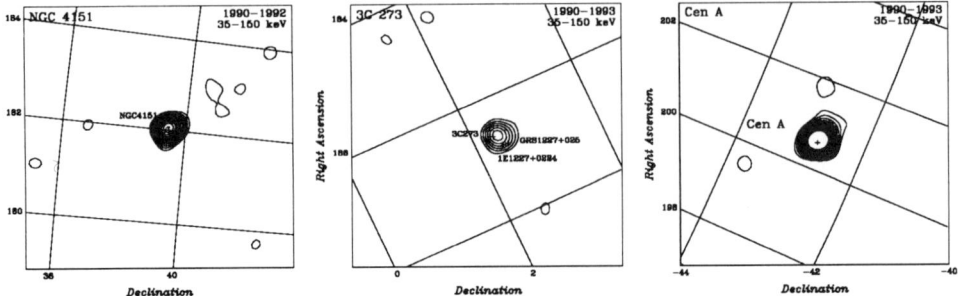

Fig. 1. Slices of SIGMA images in 35-150 keV band, averaged over most of observations in 1990-1993 of NGC4151 (a), 3C273 (b) and Cen A (c). Contours are 3,4,..,15σ levels. The images have been convolved with the detector Point Spread Function in order to increase the sensitivity to the point sources. As the result apparent angular resolution (width of the peaks on the images) is considerably worse than the best possible resolution.

and ART-P, 1990-1992, 3C273, NGC 4151, Cen A); Jourdain et al., 1992 (SIGMA, 1990, NGC4151); Bassani et al., 1992 (SIGMA, 1990, 3C273); Jourdain et al., 1993 (SIGMA, 1990-1991, Cen A); Lebrun et al., 1992 (SIGMA, 1990-1991, M87 and NGC4388); Bassani et al., 1993 (SIGMA, 1992, 3C273, NGC4151, Cen A), Finoguenov et al., 1993 (ART-P & SIGMA, 1990-1992, NGC4151). We will concentrate below on the data of observations in the hard X-ray / soft Gamma-ray band obtained by GRANAT/SIGMA.

2. NGC 4151

NGC 4151 is one of the nearest ($d \sim 20$ Mpc for $H_0 = 50$ $km\ s^{-1}\ Mpc^{-1}$) low-luminosity Seyfert galaxies. It has 2-10 keV photon index of $\alpha \sim 1.5 \pm 0.2$ (Yaqoob et al., 1993) somewhat harder than "canonical" value of $\alpha \sim 1.7$ for Sy I galaxies. In standard X-ray band the flux varies by at least factor of 5 with the fastest variability detected so far of the order of 1/2 day. Observations at higher energies detected power law spectrum with indices of ≈ 1.7 up to hundreds of keV or even MeV (see Perotti et al., 1991 for review) placing NGC 4151 among the hardest objects on the sky in this energy range.

The source has been observed by SIGMA 19 times in 1990–1992 (see Table 1 for the list of sessions with highest quality of the data). In contrast to previous high-energy observations the spectrum observed during first 6 sessions in 1990 was found to be very steep with photon index $\alpha \sim 3.1 \pm 1$ in the 30-200 keV energy band (Jourdain et al., 1992). On the other hand the observations of ART-P on 1990 November 11,12,16,19 and 20 gave the value of the 3-20 keV index of

TABLE I
The list of SIGMA/GRANAT observations of NGC 4151

Ses. #	Date (UT)	Exposure, (hours)
0122	12.579-13.473 07/90	17.78
0123	13.636-13.880 07/90	5.00
0190	15.576-16.301 11/90	14.40
0191	16.526-17.218 11/90	13.83
0193	19.664-20.593 11/90	18.39
0194	20.831-22.193 11/90	27.01
0304	29.457-30.523 06/91	21.01
0306	11.425-12.471 07/91	20.75
0367	19.470-20.297 11/91	16.38
0368	20.455-21.869 11/91	28.01
0470	24.648-25.547 06/92	17.99
0531	18.492-19.255 11/92	15.08
0532	19.377-20.230 11/92	17.00
0534	23.377-24.229 11/92	17.00

$\alpha = 1.48 \pm 0.03$ (Apalkov et al., 1992), i.e. typical for this source at these energies. All further SIGMA observations confirmed the steepness of NGC 4151 spectrum at high energies (Finoguenov et al., 1993). Observations of OSSE in 1991 (Maisack et al., 1993) also show similar very steep spectrum at energies above 65 keV, suggesting that steep state is now the dominant state for NGC 4151.

The spectrum of NGC 4151 averaged over all useful SIGMA observations in 1990-1992 is shown in Fig. 2. The assumption of power law spectrum in 35-300 keV range with photon index less than 2 can be rejected at the confidence of 99%. (Finoguenov et al., 1993). Comparison of the SIGMA spectrum with the ART-P data (Apalkov et al., 1992) suggests the break somewhere between 30 and 100 keV (Fig. 2). The position of the break in the broken power law approximation was estimated to be ~ 50 keV (best fit to the 3-300 keV GRANAT data in November 1990; Finoguenov et al., 1993). The similar value can be inferred from Fig. 2, but since shown ART-P and SIGMA spectra are not contemporaneous one should treat this result with caution.

The SIGMA data (in 35–200 keV energy range) can be satisfactory approximated by the power law with index $\alpha = 2.3 \pm 0.2$ indicating the steepening from standard to hard X-rays of at least $\Delta\alpha \sim 0.8$ (see Table 4) The SIGMA data are consistent with further steepening of the spectrum ($\alpha \approx 3.4$) above ~ 100 keV found by OSSE/CGRO (Maisack et al., 1993). So broad band spectrum over few keV – few hundred keV range (ART-P,SIGMA, OSSE) requires at least two breaks in broken power law approximation (Finoguenov et al., 1993). The broad band spectrum of NGC4151 can be satisfactory approximated by the models in-

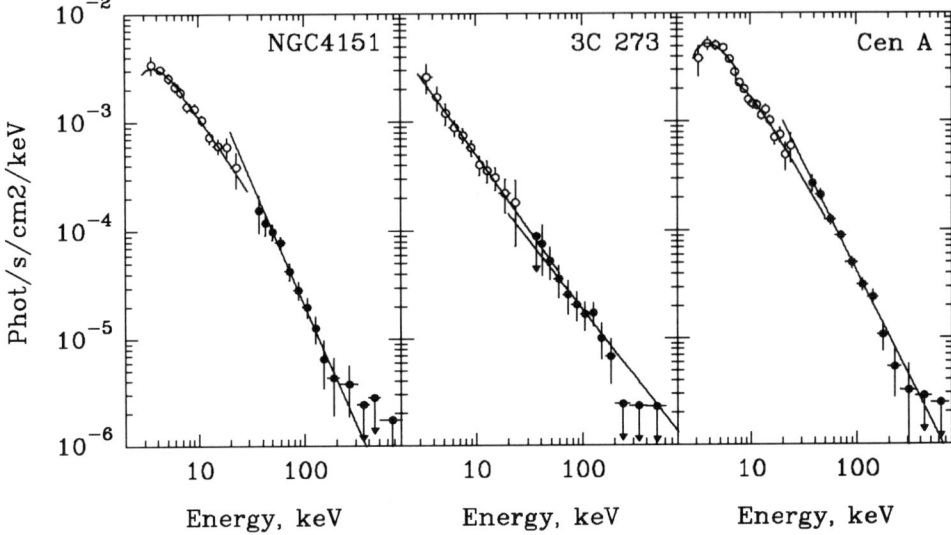

Fig. 2. The Spectra of NGC 4151, 3C273 and Cen A averaged over ART-P observations in 1990 and SIGMA observations in 1990-1993. The points below 1σ have been replaced with 1σ upper limits.

volving the exponential cutoff at high energies. Such a behavior, when the spectrum rapidly steepens with energy, argues in favor of the thermal models of origin of the X-ray spectra (see also Zdziarski et al., 1993).

The absorption corrected 2-10 keV flux $(20\pm1)\times10^{-11}$ $ergs\,cm^{-2}\,s^{-1}$ measured by ART-P in November 1990 is considerably higher than $\sim 11\times10^{-11}$ $ergs\,cm^{-2}\,s^{-1}$ expected for $\alpha = 1.48$ from index/luminosity relation obtained in the frame of nonthermal models considered by Yaqoob & Warwick (1991). On the other hand rather similar combinations of photon index and 2-10 keV flux have been observed by Ginga in 1991 January (c.f. Jan 91d from Table 3 of Yaqoob & Warwick ,1993).

Variability of the X-ray flux has been detected both in standard (Apalkov et al., 1992) and hard X-ray bands (Finoguenov et al., 1993). Shown in Fig. 3 is the 35-150 keV light curve of NGC4151 in 1990-1992 obtained by GRANAT. The probability (according to χ^2 test) that measurements correspond to constant source flux is about 3%.

The AGNs are known to contribute at least 30-40% to the X-Ray Background (hereafter XRB) at 2 keV. The shape of XRB spectrum in board 3-60 keV band can be well approximated by the emission of optically thin plasma with temperature $\approx 45 keV$ (Rothschild et al., 1983). Below ~ 30 keV it has power law slope of $\approx$

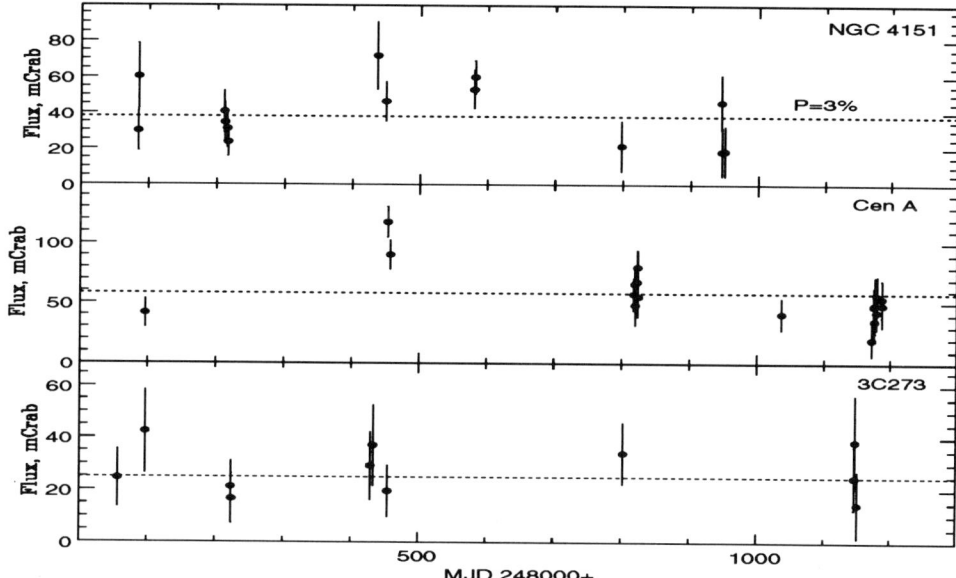

Fig. 3. The light curves of NGC4151, Cen A and 3C273 in 35-150 keV band according to SIGMA measurements in 1990-1993

1.4 (photon index) and steepens to index ≈ 2.4 in 30-200 keV range. However there is obvious lack of bright objects having the spectra of the similar shape. The assumption that dominant contribution in this energy band come from the objects with "canonical" power law spectra (photon index ≈ 1.7) extending to hundreds keV is in conflict (1987) with XRB shape – the latter is flatter at low energies and steeper at high.

Although the NGC4151 is probably rather unique source it is interesting that the spectrum of this object observed by GRANAT well satisfy the requirements mentioned above. Indeed NGC4151 has almost proper slope at low energies and steepens at high. Shown in Fig. 4 is the expected shape of background due to the objects with spectra similar to that of NGC4151 in comparison with measured spectrum of XRB. Two curves (with and without evolution of luminosities are shown for illustration). The fit to the spectrum of NGC4151 observed by GRANAT in 1990 November (Finoguenov et al., 1993) has been adopted ($I_\nu = E^{-\alpha} * exp(-E/kT)$, $\alpha = 0.32$, $kT = 79$ keV, $N_H = 5 \times 10^{22} cm^{-2}$).

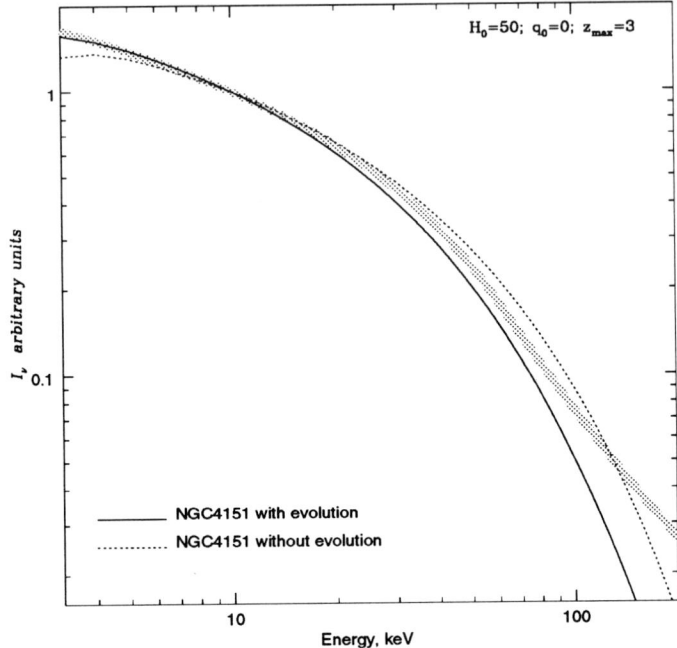

Fig. 4. The expected shape of the diffuse background due to the sources with spectra similar to that of NGC4151, observed by GRANAT. The following parameters were adopted: $z_{max} = 3$; $q_o = 0$; $H_0 = 50$. Two thin curves correspond to the case when all objects have the same luminosity and when luminosity evolves according to the model proposed by Maccacaro et al., 1991. Thick line is the empirical approximation of XRB spectrum (Gruber, 1992).

3. 3C 273

Numerous observations of this nearby ($z = 0.158$) QSO have detected the power law spectrum extending to hundreds keV with nearly the same photon index $\alpha = 1.4$.

The list of SIGMA observations of 3C273 with highest quality of data are given in Table 2. In the first observations the source was found in the state similar to that observed previously (Bassani et al., 1992). Shown in Fig. 2 is the spectrum of 3C273 in broad energy band observed with GRANAT. The ART-P data points have been averaged over the observations on 1990 (Apalkov et al., 1992); the SIGMA points have been averaged over all observations in 1990-1992. Unlike NGC 4151 there is no significant evidence for strong break of the spectrum from standard to hard X-ray bands. The photon index in 3-24 keV range ($\alpha = 1.4 \pm 0.05$) (Apalkov et al., 1992)

TABLE II
The list of SIGMA/GRANAT observations of 3C273

Ses.#	Date (UT)	Exposure,(hours)
0104	14.431-15.589 06/90	23.00
0130	25.467-26.118 07/90	13.01
0197	27.875-28.593 11/90	14.22
0198	28.805-29.708 11/90	18.00
0300	21.742-22.553 06/91	16.00
0302	25.752-26.148 06/91	8.01
0309	16.670-17.508 07/91	16.64
0474	29.612-30.363 06/92	15.00
0610	7.427- 8.463 06/93	20.62
0611	8.595- 9.041 06/93	9.00
0613	11.511-12.470 06/93	19.00
0614	12.598-13.346 06/93	11.50

is consistent with inferred from 35-200 keV SIGMA data ($\alpha = 1.3 \pm 0.32$). However the source is rather weak in SIGMA energy band and substantially steeper slopes in 35-200 keV band ($\alpha \approx 1.6 - 1.7$) can not be ruled out with SIGMA data.

Although ART-P and SIGMA observations are not simultaneous there is no strong mismatch of spectra at 30 keV. The whole set of SIGMA data points in 35-150 keV band is consistent with the assumption of constant flux. The marginal evidence for variability based on first few SIGMA measurements (see Fig. 3) has been mentioned by Bassani et al., 1992. The average flux at 100 keV of $(1.86 \pm 0.21) \times 10^{-5} ph/cm^2/s/keV$ found by SIGMA is in general compatible with previous measurements.

Another source GRS1227+025 has been reported 15′ apart from 3C273 (Jourdain et al., 1992). It was relatively bright only during short period of time (observations 197 and 198 in Table 2). Grindlay (1992) recently pointed out that Einstein HRI source 1E1227+0224 (Tananbaum et al., 1979) is located within the 5′ SIGMA error circle reported for GRS1227+025. Note that image of this field shown in Fig. 1 has been convolved with detector Point-Spread-Function which cause apparent broadening of the peaks. For detailed discussion see Jourdain et al., 1992; Grindlay, 1992.

4. Cen A

Cen A is radiogalaxy best studied in X-rays. This object has rather low luminosity ($L_{3-10\ keV} \approx 10^{42} ergs/s$), however due to small distance to the source ($\sim 5 Mpc$) Cen A is the brightest extragalactic source seen by GRANAT.

Cen A is known to be strongly variable both in standard and hard X-ray band.

TABLE III
The list of SIGMA/GRANAT observations of Cen A

Ses.#	Date (UT)	Exposure,(hours)
0129	24.388–25.261 07/90	17.26
0308	15.381–16.471 07/91	21.42
0311	19.507–20.596 07/91	21.57
0482	14.529–15.443 07/92	18.04
0483	15.574–16.422 07/92	16.91
0484	16.553–17.048 07/92	10.00
0485	18.475–19.464 07/92	19.59
0486	19.587–20.423 07/92	16.61
0487	20.545–21.142 07/92	12.00
0567	17.656–18.556 02/93	18.00
0625	2.515– 3.365 07/93	17.00
0626	5.463– 6.379 07/93	18.17
0627	6.511– 7.337 07/93	16.44
0628	7.457– 7.750 07/93	6.00
0629	9.615–10.525 07/93	18.03
0630	10.645–11.337 07/93	13.88
0635	17.696–18.461 07/93	15.18
0636	18.581–19.229 07/93	13.00

The variability of 35-150 keV flux is also apparent from SIGMA data (Fig. 3). In the most of observations the source was in very low state in hard X-rays comparable only with the HEXE/MIR observations in June 1987 (Maisack et al., 1991). The hard X-ray flux from Cen A seems to gradually decrease in 1992 to 1993. The Cen A was found in the relatively high state only during two observations in 1991 (observations 308 and 311 in Table 3). These two observations were separated by 4 days and the flux in the latter observation was significantly lower than in the first one (Jourdain et al.,1993). If the fast decrease of the flux between these two observations characterizes also the rise time, than it is possible that the source remained in the low state during whole period of SIGMA observations in 1990-1993 and two observations in 1991 July occasionally happened during relatively short flare of the source. Observations of ART-P in 1990 (Apalkov et al., 1992) in standard X-ray band also revealed the source in relatively low state in comparison with most of previous measurements (Morini et al., 1989).

According to preliminary analysis there is evidence for spectral variation in SIGMA data between the states with high and low flux at 100 keV (observations 308, 311 in 1991 and all observations in 1992 and 1993 respectively). The power law photon index in 35-200 keV is $\approx$ 1.76 for 1991 observations (flux at 100 keV $F_{100} \sim 6.7 \times 10^{-5} phot/s/cm^2/keV$) and $\approx$ 2.3 for the spectrum averaged over

TABLE IV
The power law approximations of NGC4151, 3C273 and Cen A spectra, averaged over observations of ART-P (1990) and SIGMA (1990-1993)

Source	N_H^a	Flux 1 keV ^{b}	3-24 keV index	Flux @ 100 keV ^{c}	3 5-200 keV index
NGC4151	6 ± 2	3.6 ± 0.2	1.48 ± 0.03	2.06 ± 0.21	2.3 ± 0.2
Cen A	9 ± 2	9.6 ± 0.5	1.69 ± 0.02	4.16 ± 0.23	1.98 ± 0.11
3C 273	0.018 ^{d}	1.2 ± 0.1	1.40 ± 0.05	1.86 ± 0.21	1.3 ± 0.3

Errors are 68% confidence limits for single parameter estimates
^{a} – in units of $10^{22} cm^{-2}$
^{b} – $10^{-2} ph/cm^2/s/keV$
^{c} – $10^{-5} ph/cm^2/s/keV$
^{d} – fixed

1992-1993 observations ($F_{100} \sim 3.0 \times 10^{-5} phot/s/cm^2/keV$). The spectral shape in 1992-1993 observations is probably more complicate than simple power law over 35-200 keV band – it is somewhat steeper below $\approx$50 keV and somewhat flatter in 50-150 keV bands. The analysis of the observations in 1992 and 1993 is in progress now and detailed results will be published elsewhere.

The spectrum of Cen A averaged over ART-P observations in 1990 (Apalkov et al., 1992) and SIGMA in 1990-1993 is shown in Fig. 2. In standard X-ray band the source had photon index $\approx$ 1.7 while in SIGMA 35-200 keV band it is $\approx$1.98 (see Table 4). There is evidence for steepening of the spectrum above $\approx$ 150 keV in agreement with results of OSSE/CGRO (Johnson et al., 1993).

References

Apalkov Yu. et al., 1992, in the Proc. 28th Yamada Conf. on Frontiers of X-Ray Astronomy, ed. Y.Tanaka, K.Koyama, p.251
Ballet J. et al., 1991, 26th Moriond conference on "The early observable universe from diffuse backgrounds", Rocca-Volmerange, Deharveng and Tran Thanh Van editors, Editions Frontieres, 43
Ballet J. et al., 1992a, Heidelberg conference on "Physics of Active Galactic Nuclei", ed. Duschl and Wagner, Springer Verlag, 67
Ballet J. et al., 1992b, Garching conference on "X-ray emission from Active Galactic Nuclei and the X-ray background", ed. Brinkmann and Truemper, MPE report 235, p 50
Bassani L. et al., 1992, ApJ, 396, 504
Boldt E., 1987, Phys. Rep., 146, 215
Finoguenov A.et al., 1993, in preparation
Grindlay J., 1992, A&AS, 97, 113
Gruber D.E., 1992, In The X-ray background, eds. X.Barcons & A.C.Fabian, Camb. Univ. Press.
Johnson W.N. et al., 1993, A&AS, 97, 21
Jourdain E. et al., 1992, A&A, 256, L38
Jourdain E. et al., 1993, A&A, 412, 586
Lebrun F. et al., 1992, A&A, 264, 22
Maccacaro T. et al., 1991, ApJ, 374, 117

Maisack M. et al., 1989, 23rd ESALB Symp., "X-Ray Astronomy", eds. J.Hunt & B.Battrick, 975.
Maisack et al., 1993, ApJ, 407, L67
Morini M., Anselmo F. & Molteni D., 1989, ApJ, 347, 750
Paul J. *et al.*, 1991, Adv.Space Res., 11, (8) 289
Perotti F. et al., 1991, ApJ, 373, 75
Rothschild R.E. et al., 1983, ApJ, 269, 423
Sunyaev R. *et al.*, 1990, Adv.Space Res., 10, (2) 233
Tananbaum H. *et al.*, 1979, ApJ, 234, L9
Yaqoob T. & Warwick R.S., 1991, MNRAS, 248, 773
Yaqoob T. et al., 1993, MNRAS, 262, 435.
Zdziarski A. et al., 1993, ApJ, 414, L93

GINGA OBSERVATIONS OF ACTIVE GALACTIC NUCLEI

H. Inoue
Institute of Space and Astronautical Science
3-1-1, Yoshinodai, Sagamihara
Kanagawa 229, Japan

Abstract. Ginga observed 116 AGNs during its operational life from 1987 to 1991: Among them, there were 55 Seyfert galaxies, 42 quasars, 9 BL Lac objects and 10 other AGNs, although the classification is not strict. From these AGN observations, a number of fruitful results were obtained. In this review, I briefly summarize the Ginga observations of AGNs and show several similarities between the Seyfert-type AGNs and the galactic black hole candidates. I also discuss two inverse correlations between the break energy of the power law spectrum and the flux observed from the black hole candidate GS2023+338 and between the equivalent width of the iron fluorescent line and the continuum flux observed from NGC4151.

1 Summary of the Ginga observations

The results of the Ginga observations are summarized as follows (for the general review of the X-ray observations of AGN, see Mushotzky, Done and Pines 1993):

1.1 SPECTRAL SLOPE

Statistically significant X-ray spectra have been obtained from many of the above AGNs by Ginga. The systematic study of the spectral properties of Seyfert galaxies (Awaki 1991; Nandra 1991) show that the 2-30 keV spectrum of the most of Seyfert galaxies can be approximated by a simple power law form and the specral energy indices distrubute in a fairly narrow range around 0.7 (see e.g. Mushotzky, Done and Pines 1993). The mean spectral index of QSOs observed with Ginga was separately investigated and found to be still higher than that of the cosmic X-ray background (Williams et al. 1992).

The determination of the mean spectral index of AGNs fainter than the Ginga detection limit can also be addressed by analyzing the source confusion noise in the LAC background (Hayashida 1990). From one pointing position to another, the integrated flux of these faint sources fluctuates in the LAC field of view. An analysis which obtained the mean spectral index of the source confusion noise has been performed and the result shows that the spectral energy-index is about 0.8. This spectral slope is still significantly steeper than that of the cosmic X-ray background. The mean flux of these objects is 5-10 $\times 10^{-13}$ ergs cm^{-2} s^{-1}.

1.2 REFLECTION COMPONENT IN SPECTRA

Ginga observations of AGN established that the presence of an iron emission line in the spectrum is common to almost all Seyfert galaxies. The line center energies obtained from the Ginga observations of Seyfert galaxies are almost all consistent with the K_α-line energy of neutral iron, after we take account of the cosmological

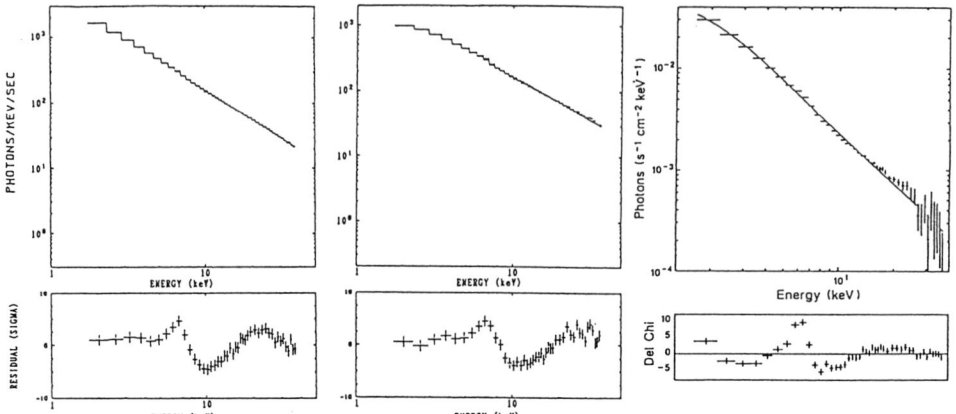

Fig. 1. Simple power law fits to the spectra in the hard state (upper panel) and residual plots of the data minus model results (lower panel) for Cyg X-1 (left), GS2023+338 (center), and an average of twelve active galactic nuclei (right, Pounds et al. 1990) observed with Ginga.

redshift of the line energy. If the iron-line-emitting matter subtends the central X-ray source with a large solid angle, X-rays reflected from the iron-line emitting, cold matter should be observed in the spectrum simultaneously with the fluorescent iron line. In fact, Ginga observations of AGNs revealed that the spectrum generally shows a shallow and broad absorption like feature at energies from 7 to 15 keV together with the emission line feature at 6-7 keV. Furthermore, the power law continuum tends to flatten at higher energies (Pounds et al. 1989). These features are very similar to those observed from black hole candidates as shown in Fig.1.

These spectral features can be reproduced very well by a superposition of a component reflected by cool Thomson-thick matter from a single power law. The emission line feature at 6-7 keV, the shallow and broad absorption feature in the 7-15 keV range and the excess above 15 keV are all consistent with the presence of a reflection component in these spectra.

1.3 SEYFERT I AND SEYFERT II GALAXIES

Seyfert II galaxies which Ginga observed generally showed evidence of a large obscuration of the central engine by cool matter. In particular, Ginga detected an intense iron line with an equivalent width as large as 1.3 keV from a Seyfert II galaxy, NGC1068 (Koyama et al. 1989). An equivalent width as large as 1-2 keV can be expected only when the direct X-rays from the central source are blocked by dense matter while the X-rays reprocessed by cold matter around the central engine are still observed.

1.4 IRON EMISSION LINES FROM QUASARS

Ginga has discovered the presence of iron emission lines in the spectra from two bright quasars, 3C273 (Turner et al. 1990) and 1E1821+643 (Kii et al. 1991).

The equivalent width of the line from a radio quiet quasar 1E1821+643 is 180 eV a value similar to that from Seyfert I galaxies. This distant ($z \simeq 0.3$) radio-quiet quasar probably has a geometry similar to those of nearby Seyfert galaxies as discussed above. On the other hand, significant evidence of an iron line from a radio loud quasar 3C273 was found only when 3C273 was very dim, and the equivalent width was only 50 eV. Since 3C273 is known to exhibit superluminal motion, the line of sight to the central engine in 3C273 can be considered to be very close to the direction of the mass ejection, according to the canonical interpretation of superluminal motion. The weak iron line from 3C273 would imply that the X-ray beaming in the direction of the mass ejection is relatively strong in radio-loud/OVV quasars exhibiting superluminal motion. In fact, none of the OVV quasars observed from Ginga shows a significant iron line.

1.5 BL LAC OBJECTS

A simultaneous ROSAT-Ginga observation has clearly shown that the X-ray spectrum of a BL Lac object, Mrk 421, has a concave shape: providing support for synchrotron interpretation (Makino et al. 1991). A systematic relation is found between the intensity and the spectral hardness during the flux variation in the sense that harder X-rays always precede softer X-rays: another support for direct synchrotron emission for steep spectrum (index > 1.5) BL Lac objects (Sembay et al. 1993; Tashiro 1993). The time lag has been measured for some sources. Lack of iron emission line has been confirmed for all of the measured objects with high statistics (Kii et al. 1992), which is consistent with synchrotron model with the continuum emission likely to be boosted by a relativistic motion.

2 Similarities between AGNs and Galactic Black Hole Candidates

There are several similarities between AGNs and galactic black hole candidates (BHC: for the review, see e.g. Tanaka 1992; Inoue 1991) as shown below.

2.1 POWER LAW SPECTRUM

The power law slope, 0.5 - 1, of the Seyfert galaxies is very similar to that of the BHC in the hard state (see e.g. Tanaka 1992). This suggests that there is a common physical mechanism to produce a power-law spectrum with an index of 0.5 - 1 in the central power houses, independently of the wide range of the mass of the central power house from 10 solar mass probably up to 10^8 solar mass.

The spectral slope also seems to be independent of the wide range of the accretion rate onto the central power house. In the case of the Galactic transient source, GS2023+338, the slope stayed almost constant despite the large flux change of over three orders of magnitude. Similarly, the spectral slope of the OVV quasar 3C279 did not change greatly even when the source flared up by a factor of > 10 (Makino et al. 1989).

2.2 REFLECTION COMPONENT IN SPECTRA

Ginga observations of BHC have revealed that the reflection component in the spectrum, as observed from AGN, is also generally observed from galactic black hole candidates (see Fig.1). Figure 1 shows residual plots of the observed spectrum minus a single power-law model for Cyg X-1 and GS2023+338; the presence of these features is evident.

The similarity of AGN to BHC in the presence of the reflected component in the spectrum strongly suggests that an accretion disk exists in the central engine of AGN. It is generally believed that accreted matter will form an accretion disk in binary X-ray sources and the outerpart of the disk will be responsible for the X-ray reflection.

2.3 SPECTRAL BREAK AROUND 100 KEV

Recent hard X-ray observations by Mir-Kvant, GRANAT-SIGMA and GRO-OSSE revealed that the power law spectrum of both Seyfert galaxies and galactic black hole candidates doesn't extend to the energies higher than 100 keV but breaks around there (see e.g. Maisack et al. 1993; Salotti et al. 1992).

2.4 TIME VARIABILITY

Similarities between AGN and BHC exist also in their time variabilities.

The power spectra of X-rays from BHC such as Cyg X-1 and GS2023+338 have the logarithmic slope of power density against frequency ~ 0 in the frequency range below about $10^{-1.5}$ Hz, ~ -1 in the range between $10^{-1.5}$ Hz and $10^{0.5}$ Hz, and ~ -2 in the range above about $10^{0.5}$ Hz (see e.g. Inoue 1988). On the other hand, those of AGNs such as NGC5506 and NGC4051 obtained by Pounds and McHardy (1988) have remarkably similar properties except that the "knee" frequencies are 10^4 - 10^5 times lower.

If the similarities in the power spectra among these sources come from a common energy generation mechanism in the central power houses, the differences of the "knee" frequencies may be interpreted as differences of the typical dimension of the central power house, which will be proportional to the central mass. The masses of AGNs estimated by scaling the "knee" frequencies from the mass of Cyg X–1 (Paczynski 1974) are roughly proportional to the X-ray luminosities (Pounds and McHardy 1988 and see also Barr and Mushotzky 1986; Wandel and Mushotzky 1986).

These similarities over the wide range of the mass of the central object suggest the following scalings:

The structure of the accretion flow should be similar to all AGNs and galactic black hole candidates if the distance-scale is normalized by the Schwartzschild

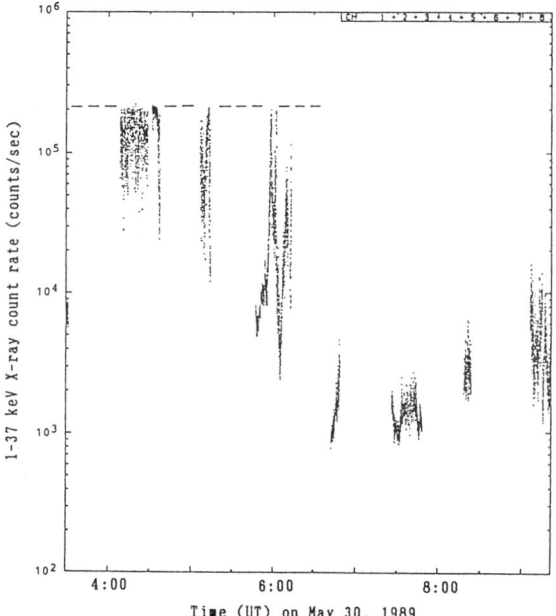

Fig. 2. X-ray light curve of GS2023+338 on May 30 in 1989 observed with Ginga. A dashed line indicates a possible luminosity saturation.

radius which is proportional to the central mass.

When the quantity, L/L_E, (the luminosity, L, normalized by the Eddington limit, L_E) is the same, the physical situation around the black hole would be the same over a wide range of the central mass.

If the particle energy is proportional to the depth of the potential energy, it will be independent of the central mass at the position close to the Schwartzschild radius. Then, the fact that the emergent photon spectrum has the same functional form of the energy independently of the central mass seems to suggest the photon emission mechanism to be thermal and optically thin.

The following discussions are consistent with these suggestions.

3 Inverse Correlation between Break Energy and Luminosity

Ginga discovered the bright transient source GS2023+338 (Kitamoto et al. 1989) and performed several observations during its decay. The follow-up optical observation revealed this source to be probably a black hole (Casares, Charles and Naylor 1992). As seen in the X-ray light curve of GS2023+338 on May 30, 1989, in Fig.2, the flux apparently saturated at the peak of the outburst. For the dis-

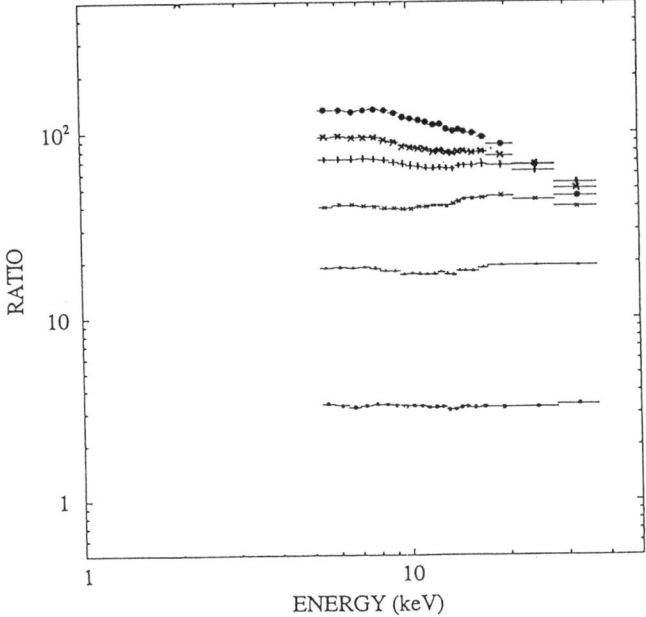

Fig. 3. Ratios of the spectra of GS2023+338 obtained at different flux levels to a typical power law spectrum at a certain flux level.

tance of 1–3 kpc evaluated from the optical observations (Casares et al. 1991), the saturation level is consistent with the Eddington limit of a several-solar-mass star. GS2023+338 showed a single power law spectrum with a reflection component throughout the decay by over three orders of magnitude, although the source sometimes suffered from heavy absorption and revealed a complicated spectrum during the first ten days after the discovery.

Figure 3 shows the ratios of the spectra obtained at different flux levels to a typical power law spectrum at a certain flux level. The spectrum at the highest flux level was obtained when the luminosity was very close to the saturation level probably corresponding to the Eddington limit. It is clearly seen in Fig.3 that the spectra when the flux is very bright reveals a break around 10 keV and the break energy seems to shift to the higher energy as the flux decreases. The shift of the break energy becomes clear by introducing the observations of this source by the Mir-Kvant Observatory (Sunyaev et al. 1991). The Kvant observations were done when the flux was about a 50th to a 200th of the highest flux of the Ginga observations. At those flux levels, the break was seen around 100 keV. The break energies at the various flux levels observed by Ginga and Mir-Kvant are roughly estimated and are plotted in Fig.4. The inverse correlation between the break

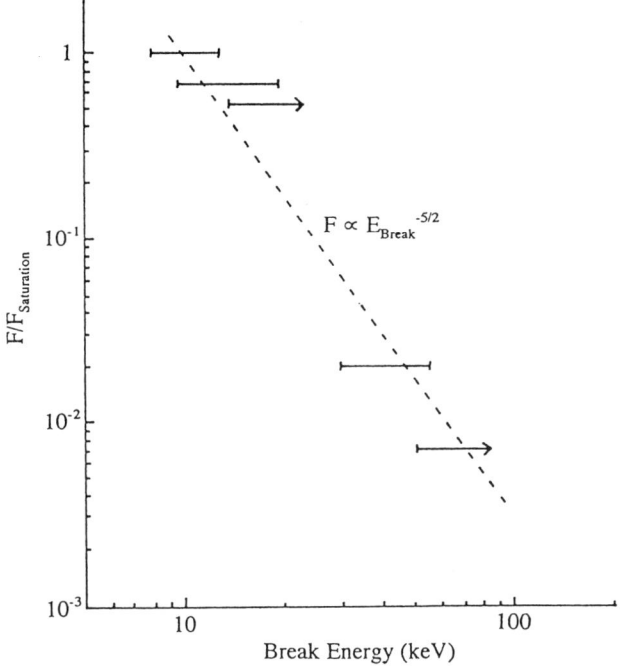

Fig. 4. Break energy of a power law spectrum observed at the various flux levels normalized by the saturation level of GS2023+338 by Ginga and Mir-Kvant (Sunyaev et al. 1991).

energy and the flux is obvious. This inverse correlation can be interpreted by a simple consideration as follows.

Let's consider a situation that an optically thin, geometrically thick accretion disk surrounds a central black hole. Here, the geometrically thick accretion disk is assumed to be an spherical hot plasma with the radius of R. The plasma will be heated up by the gravitational energy release of the accretion flow through some possible mechanism such as viscous heating or solar flare-like activity. Then the proton temperature, T_p, will approximately be given by $kT_p/m_p \simeq GM/R$, where k, m_p, G and M are the Boltzmann constant, the proton mass, the gravitational constant and the mass of the central black hole respectively. Since the energy loss rate of protons as discussed below is estimated to be sufficiently small compared to the energy input rate through the accretion, the proton temperature is assumed to be constant in the followings.

The protons will give their energy to electrons through two body collisions and the energy loss rate per unit volume is given as $(3/2)nkT_p/t_{pe}$ if $T_p \gg T_e$, where n and T_e are the number density and the electron temperature. t_{pe} is the relaxation time for protons and electrons to be in the thermal equilibrium and is proportional to $(kT_p/m_p + kT_e/m_e)^{3/2}/n$ (see e.g. Spitzer 1962), where m_e is the electron mass. This energy flow heats up electrons while the electrons loose their energy through Compton scatterings with ambient photons. The cooling rate per a unit volume is approximately given as $(4kT_e/(m_ec^2))fn\sigma$, where c is the velocity of light, f is the photon flux and σ is the electron scattering cross section. f is roughly related to the luminosity, L, as $f \simeq L\tau/(\pi R^2)$ (when $\tau > 1$) or $L/(\pi R^2)$ (when $\tau < 1$), where $\tau \simeq n\sigma R$.

This simple consideration for the energy flow from protons to photons can explain two important observational facts.

1) Since the luminosity should be equal to the total energy loss rate of the electrons in the volume of $4\pi R^3/3$, we obtain $y \simeq 1$ by equating $(4kT_e/(m_ec^2))fn\sigma(4\pi R^3/3)$ with L. Here, y is the so called Comptonization parameter defined as $y = (4kT_e/(m_ec^2))\tau$ (when $\tau < 1$) or $(4kT_e/(m_ec^2))\tau^2$ (when $\tau > 1$). If y is close to unity, the emergent photon spectrum is expected to be a power law with the energy index of about 1 (see e.g. Shapiro, Lightman and Eardley 1976; Sunyaev and Titarchuk 1980).

2) In the steady state, the energy flow rate from protons to electrons should balance with the flow rate from electrons to photons. By equating $(3/2)nkT_p/t_{pe}$ with $(4kT_e/(m_ec^2))fn\sigma$, we obtain a simple relation $T_e \simeq (L/L_E)^{-2/5}$ when $\tau > 1$ and $kT_e/m_e > kT_p/m_p$. This relation can roughly explain the inverse correlation between the break energy and the luminosity observed from GS2023+338. The weak dependency of the break energy on the normalized luminosity could also explain the recent observational evidence that the hard X-ray spectra of Seyfert galaxies as well as galactic black hole candidates seem to generally break from the simple power law spectrum around 100 keV, if the luminosities of these black hole candidates are well below the Eddington limit.

4 Inverse Correlation between Iron Line Equivalent Width and Flux

Iwasawa and Taniguchi (1993) have recently pointed out the inverse correlation between the equivalent width of the iron line and the luminosity for the Seyfert galaxies based on the Ginga observations. Figure 5 shows the relation of the equivalent width with the continuum flux for NGC4151 (Yaqoob and Warwick 1991) and the inverse correlation is obvious.

We shall assume that a geometrically thin, optically thick accretion disk surrounds the central hot region discussed above.

The equivalent width of the fluorescent iron line will be proportional to the solid angle of the line reproccessing region as viewed from the central X-ray emitter. If the central X-ray emitter has the height of R and the reprocessor has a negligibly small thickness, the solid angle of the disk further than a radius, r, is approximately given as $2\pi R/r$. The innermost radius, r, of the line reproccessing disk should be determined by the so called ionization parameter ξ. When ξ is sufficiently larger than 10^3 erg cm, the iron will be fully ionized and emits no line (Kallman and

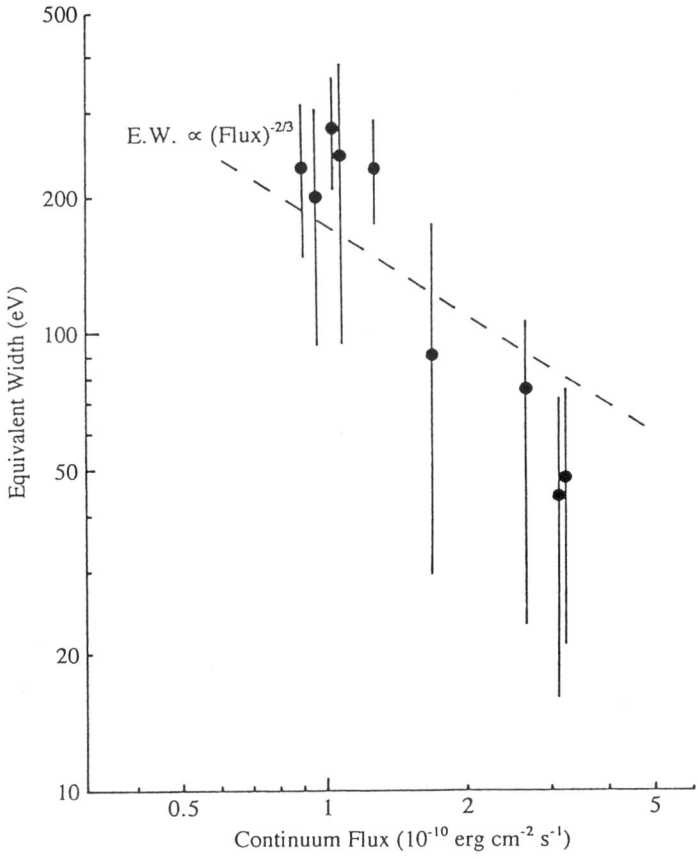

Fig. 5. Relation of the equivalent width of the fluorescent iron line with the continuum flux for NGC4151 (Yaqoob and Warwick 1991).

McCray 1982). Since X-rays graze the cool disk, the ξ should be redefined as $(L/nr^2)(R/r)$. Furthermore, the region responsible for the line reprocessing will be only a surface layer with the column density, nh, of about 10^{24} H atoms cm^{-2}, where h is the thickness of the surface layer. Hence, ξ can be further rewritten as $(L/nhr)(h/r)(R/r)$. Since h/r can be approximated by $(kT/m_p)^{1/2}r^{1/2}$, we finally obtain $\xi \propto T^{1/2}(R_S/R)^{1/2}(nh)^{-1}(L/L_E)(r/R)^{-3/2}$ (R_S is the Schwarzshild radius). If the innermost radius of the line reprocessing region is given at the place where ξ is about 10^{3-4} erg cm, T is about 10^{6-7} K and R is constant, (r/R) is in proportion to $(L/L_E)^{2/3}$. Since the equivalent width $\propto (R/r)$ as discussed above, the equivalent width of the reprocessed iron line is expected to be proportional to $(L/L_E)^{-2/3}$. This expectation is roughly consistent with the inverse relation between the equivalent width and the flux obtained from NGC4151 as seen in Fig.5. The non-dimensional form can also explain the common presence of the reflected component in the spectrum over the wide range of the mass of the central black hole. (For the recent more quantitative treatment for the surface layer of the line reprocessing geometrically thin disk, see Ross and Fabian 1993; Matt, Fabian and Ross 1993).

References

Awaki, H. 1991, *Ph.D.Thesis*, Nagoya Univ.
Barr, P. and Mushotzky, R.F.: 1986, *Nature*, **320**, 421.
Casares, J., Charles, P.A. and Naylor, T. 1992, *Nature*, **355**, 614.
Casares, J. et al. 1991, *M.N.R.A.S.*, **250**, 712.
Hayashida, K. 1990, *Ph.D.Thesis*, Univ. of Tokyo.
Inoue, H. 1988, in *"Two Topics in X-ray Astronomy"* , (ESA, SP-296), 783.
Inoue, H. 1991, in *"Frontiers of X-Ray Astronomy"* ed. Y.Tanaka and K.Koyama, (UAP, Tokyo), 291.
Iwasawa and Taniguchi 1993, *Ap. J. (Letters)*, **413**, L15.
Kallman and McCray 1982, *Ap. J. Suppl.*, **50**, 263.
Kii, T. et al. 1991, *Ap. J.*, **367**, 455.
Kii, T. et al. 1991, in *"Frontiers of X-Ray Astronomy"* ed. Y.Tanaka and K.Koyama, (UAP, Tokyo), 577.
Kitamoto, S. et al. 1989, *Nature*, **342**, 518.
Koyama, K. et al. 1989, *P.A.S.J.*, **41**, 731.
Maisack, M. et al. 1993, *Ap. J. (Letters)*, **407**, L61.
Makino, F. et al. 1989, *Ap. J. (Letters)*, **347**, L9.
Makino, F. et al. 1991, in *"Frontiers of X-Ray Astronomy"* ed. Y.Tanaka and K.Koyama, (UAP, Tokyo), 543.
Matt, G., Fabian, A.C. and Ross, R.R. 1993, *M.N.R.A.S.*, **262**, 179.
Mushotzky, R.F., Done,C and Pounds, K.A. 1993, *Ann. Rev. A. Ap.*, **31**, 717.
Nandra, K. 1991, *Ph.D.Thesis*, Univ. of Leicester.
Paczynski, B. 1974, *A. Ap.*, **34**, 161.
Pounds, K.A. and McHardy, I.M. 1988, in *"Physics of Neutron Stars and Black Holes"* ed. Y. Tanaka, (UAP, Tokyo), p.285.
Pounds, K.A. et al. 1990, *Nature*, **344**, 132.
Ross, R.R. and Fabian, A.C. 1993, *M.N.R.A.S.*, **261**, 74.
Salotti, L. et al. 1992, *A. Ap.*, **253**, 145.
Sembay, S. et al. 1993, *Ap. J.*, **404**, 112.
Shapiro, S.L., Lightman, A.P. and Eardley, D.M. 1976, *Ap. J.*, **204**, 187.
Spitzer, L.Jr. 1962, *Physics of Fully Ionized Gases* (Wiley, New York).
Sunyaev, R.A. and Titarchuk, L.G. 1980, *A. Ap.*, **86**, 121.
Sunyaev, R.A. et al. 1991, *Sov. A. Lett.*, **17(2)**, 123.
Tanaka, Y. 1992, in *"Proceedings of the Ginga Memorial Symposium"* ed. F.Makino and F.Nagase, (ISAS, Sagamihara), 19.
Tashiro, M. 1993, *Ph.D.Thesis*, Univ. of Tokyo.
Turner, M.J.L. et al. 1990, *M.N.R.A.S.*, **244**, 310.
Wandel, A. and Mushotzky, R.F. 1986, *Ap. J. (Letters)*, **306**, L61.
Williams, O.R. et al. 1992, *Ap. J.*, **389**, 157.
Yaqoob, T. and Warwick, R.S. 1991, *M.N.R.A.S.*, **248**, 773.

HUBBLE SPACE TELESCOPE OBSERVATIONS OF AGN[*]

F. MACCHETTO[**]
Space Telescope Science Institute

1. Introduction

Only little more than three years have passed since the launch of the Hubble Space Telescope (HST) and the wealth of results produced by astronomers using it, have already made fundamental contributions to our understanding of a variety of astrophysical processes. A considerable number of investigations have been, and are being, devoted to the study of the whole gamut of problems associated with activity in galaxies. These range from the very largest scales, namely those applicable to the study of the optical jets and galaxy mergers (10-100 kpc) to the smallest scales (1-10 pc) relevant to investigate the broad-line regions and the very center of the active galaxies. In all cases, the high-spatial resolutions, extended dynamic range and ultraviolet response, has made possible the study of a number of objects with a detail impossible without the HST.

In a broad sense, the results so far have generally confirmed, the "standard" unified picture of AGN, providing new observational constraints to the model; however, a number of notable exceptions or "complications" have been unveiled which will require considerable more theoretical work and even more precise observations with the refurbished HST to resolve.

In this review, I have followed the following scheme. The first two sections discuss the large-scale phenomena, jets and mergers, which are a manifestation of, or the cause of the AGN activity. In the subsequent sections I discuss some of the most recent and important results in the three subclasses of active galaxies proper and follow the "standard paradigm" as its backdrop, highlighting where appropriate the observations that do not conform to our current picture. In all cases, I have used only results of articles recently appeared in, or submitted to, the main journals or conference proceedings.

2. Optical Jets

The study of the optical counterparts to the radio jets has been the subject of a number of observing programs. We know that these jets play a fundamental role in transporting energy from the central source to the extended radio lobes. Observations at optical and ultraviolet wavelengths with the HST are essential to obtain spatial resolutions similar to, or better than, those achieved in the radio band

---

[*] Based on observations obtained with the NASA/ESA Hubble Space Telescope
[**] On assignment from the Space Science Department of ESA

and, thus, provide the possibility of directly comparing the sites and mechanisms responsible for the emission at these different wavelengths.

In all cases to date, the emission has been attributed to the synchrotron mechanism, and since the electron lifetime is a strong function of the observed frequency, observations at optical and ultraviolet wavelengths offer the possibility to determine the precise locations where particle acceleration occurs. Comparison of the radio and optical morphologies further allows the study of the confinement mechanisms and diffusion processes within the jet.

A number of important discoveries and observations that place the theoretical models on firmer observational grounds have been published or are about to be published: PKS 0521-36 (Macchetto et al 1991a), 3C66B (Macchetto et al 1991b), NGC 3862 (Crane et al 1993), M87 (Macchetto, 1991a,b; Boksenberg et al. 1992; Lauer et al. 1992) and have been reviewed in the past (Macchetto 1991a,b; 1992; 1993.) Particularly, interesting results have been published recently on 3C273 and M87.

3C273 is one of the nearest and brightest quasars known. Its jet has been extensively studied at radio and optical wavelength, the latter, however, limited by the low resolution attainable from the ground. Using the Faint Object Camera on HST, Thomson, Mackay and Wright (1993) have carried out high-resolution imaging polarimetric observations of the jet. Fig. 1 shows a ground-based image of 3C273 and its jet and the total intensity image of the jet as observed with the HST. The projected jet length is more than 70 kpc ($H_0 = 50 \,\mathrm{km\,s^{-1}\,Mpc^{-1}}$; $q_0 = 0.5$). The width is only a few tenths of an arcsecond ($\sim 0.5 - 1$ kpc). The optical emission is highly confined to the core of the radio jet. It runs along the ridge of the radio emission and is asymmetric compared to the radio.

The bright optical photo shows polarization contours in excess of 20% and the magnetic and polarization vectors are generally aligned along the jet near the quasar, but the orientation changes along the jet and they seem to twist by 40° and become parallel with the inclined inner structure of the outer knot. The HST data are consistent with pure synchrotron radiation as the source of both optical and radio emission, and is contrary to the suggestion that the optical emission is simply quasar light scattered in our direction by jet particles.

The authors interpret these results as a continuous fluid model in which a jet of ionized plasma burrows its way through the intergalactic medium, and where the outermost knot, and its rotated magnetic field angle, is the working surface of the jet. The asymmetrical optical and radio morphologies are interpreted as a transverse motion of the jet, due to either a processing jet or a transverse wind.

The authors also discuss the possibility that the jet is intrinsically one-sided, but confirmation of this model requires considerable more observational and theoretical work.

Optical and ultraviolet observations of **M87** have been carried out with HST and have been extensively reported (Macchetto 1991a,b, 1992, 1993; Boksenberg et al. 1992; Lauer et al. 1992; Macchetto, Biretta and Sparks 1993). The early

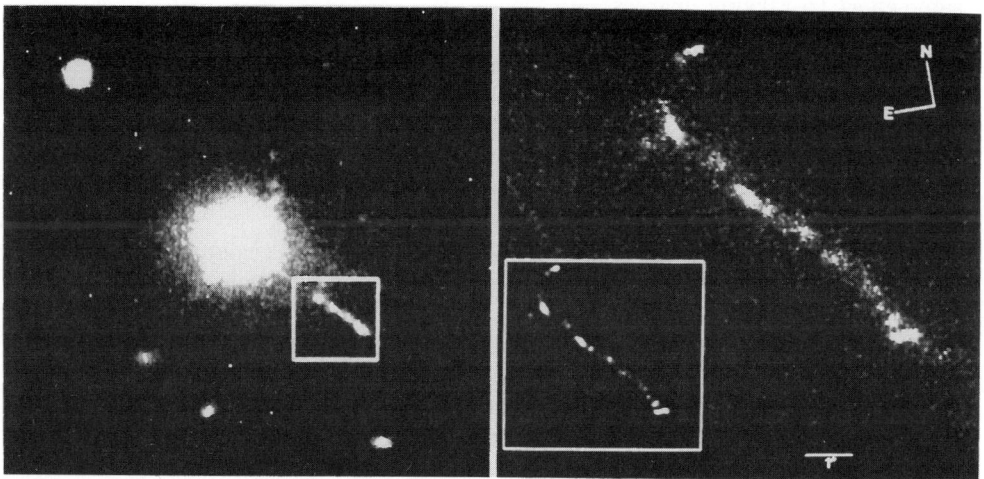

Fig. 1 Ground-based (left) and HST images of the 3C273 jet.

HST/FOC observations have demonstrated a remarkable similarity between the radio and optical morphologies. However, more detailed comparisons (Sparks, Biretta & Macchetto 1993) highlight the several systematic differences between these different wavelength bands.

The optical emission appears to be more concentrated in the knots than the radio emission. The ratio of knot to interknot brightness is higher for the optical data. This is evident in Fig. 2 where the radio image shows a diffuse component of emission which fills the inner jet (knots D, E, F, and I), while any such component is much weaker in the optical images. This is also apparent in the spectral index map, Fig. 3 where the centers of the knots have indices $\alpha \leq 0.6$ whereas the interknot regions are appreciably steeper with $\alpha \geq 0.7$. To confirm this effect, we have also compared the VLA and HST images with ground-based NTT, V and R band data. In both the HST and NTT data, the ratio of knot to interknot brightness is higher than in the radio.

The jet also appears narrower in the optical than in the radio band. This is most easily seen at knot A where the FWHM normal to the jet axis is about $1.13''$ in the VLA data, but only $0.85''$ in the HST FOC data. All of the knots show the same effect to varying degrees.

Besides these systematic trends, there are also localized regions which are relatively much brighter in the HST data than in the radio. This is most apparent in knots E and F of the inner jet.

The jet also appears narrower in the optical than in the radio band. This is most easily seen at knot A where the FWHM normal to the jet axis is about 1.13" in the VLA data, but only 0.85" in the HST FOC data. All of the knots show the same effect to varying degrees.

Besides these systematic trends, there are also localized regions which are relatively much brighter in the HST data than in the radio. This is most apparent in knots E and F of the inner jet.

There are two spectral effects which might account for the systematic variations in the radio to optical spectral index found here. A sharp spectral break where the spectrum steepens by $\triangle \alpha \sim 0.6$ is known to occur between the infrared and optical bands. Fluctuations in the frequency of this break would produce small changes in the radio-to-optical spectral index, even though all regions might have the same spectral shape. The observed spectral variations would indicate the magnetic fields are stronger at the jet center, and inside the knots, than elsewhere. While such a structure is possible, it is contrary to recent suggestions of magnetic confinement and evidence for a low-loss channel at the jet center.

An alternative possibility is that the interknot and jet-edge spectra are steeper throughout the entire electromagnetic spectrum. This picture gains some support from preliminary measurements of the interknot radio spectral index. Here the particle energy spectra would be steeper in these regions, perhaps due to variations in the acceleration process.

Future observations of the interknot radio spectral index, as well as deeper HST images giving the interknot optical spectra, will help decide between these pictures.

3. Mergers

As part of the standard paradigm, galaxy-mergers have become accepted as a key ingredient in either producing an AGN or in triggering activity in a "dormant" nucleus (e.g., Heckman 1992, and references therein). Numerical simulations have shown (Barnes & Haernquist 1991) that tidal interactions or mergers of galaxies can be responsible for supplying large amounts of gas and dust to the innermost regions of a galaxy and, thus, be responsible for resupplying the central engine (black hole?) with the necessary fuel to account for the observed activity. While direct evidence that AGN's are fueled via mergers or other accretion processes is not (yet!) overwhelming, the amount of indirect evidence continues to increase, and HST observations at high resolution are contributing to better define this picture.

Mrk 315 is a moderately luminous Seyfert 1.5 galaxy and has a redshift of $11,820 \text{ km s}^{-1}$; $M_V = -21.6$. The scale is $0.57 h^{-1}$ kpc arcsec^{-1}. It is a steep spectrum source with a diffuse morphology and a total extent of $2.9 h^{-1}$ kpc. This

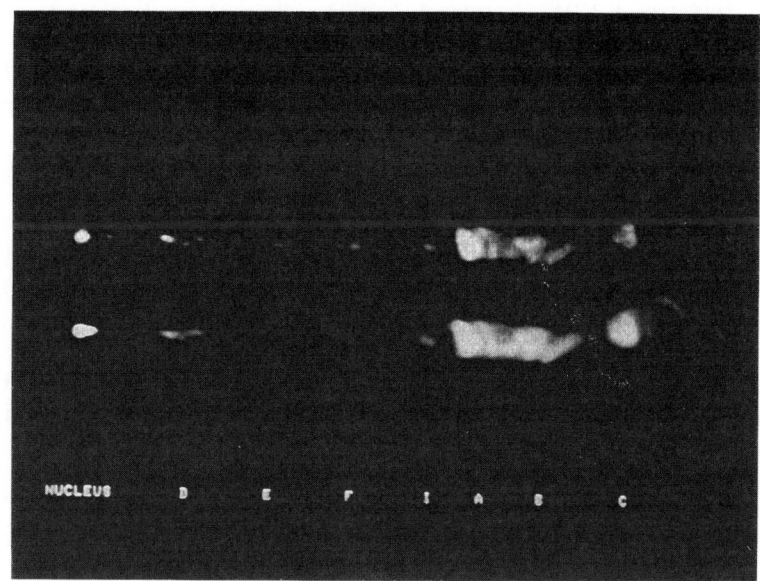

Fig. 2 The M87 jet as seen by the FOC (top) and the VLA (bottom). Note that the knot to interknot brightness ratio is higher in the optical data.

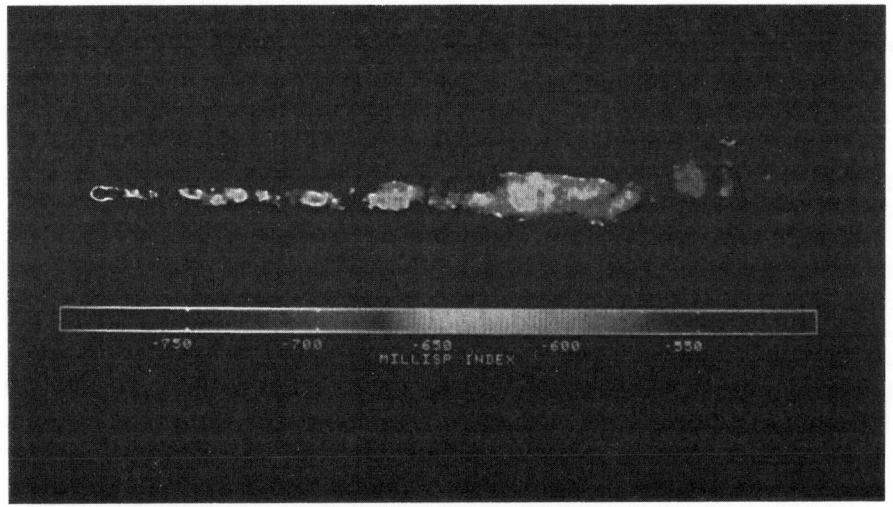

Fig. 3 Radio-to-optical spectral index map for the M87 jet.

structure is consistent with an extended starburst in the galaxy and the IRAS fluxes are also consistent with this interpretation. Mackenty (1986) discovered an 80 kpc streamer of ionized gas emerging from near the nucleus, extending in a straight line for 60 kpc then bending back in a hook. He suggested two possible origins for this feature: a tidal interaction or a dormant radio jet.

The HST observations, Fig. 4, (Mackenty et al. 1993) support the idea of a tidal origin and establish Mrk 315 as an extreme example of an active galaxy with a strong tidal disruption in the gas-rich material close to its nucleus.

The HST image shows a second, diffuse peak 2.27″ east of the diffraction limited, stellar Seyfert nucleus. This second peak is clearly resolved with a Gaussian FWHM of 0.66″ (corresponding to $380h^{-1}$ pc at the distance of Mrk 315). A fainter ring, or spiral-like structure is present 2.5″ – 3″ south of the nucleus ($\sim$ 1.4 kpc) opening towards the SW. The brighter of the two nuclei is the core of Mrk 315 and probably contains a massive black hole. The fainter nucleus is considered to be the surviving core of a galaxy that recently merged into Mrk 315.

The features seen in the data for Mrk 315 are consistent with an interpretation of the secondary continuum knot being a recently captured galaxy remnant. The knot is at the leading arc of what appears to be a stellar, spiral density enhancement. The inner ionized gas structures and the VLA 20 cm morphology are consistent with recent star formation, and subsequent supernova explosions in gas compressed by tidally induced cloud-cloud collisions. The asymmetrical velocity pattern in the gas kinematics suggests that the center of gravity for the system is displaced from the Seyfert nucleus. All of these different features are symptomatic of the type of gravitational forces which leads to mass inflow towards the nucleus.

NGC 7252 is the prototypical example of a merger between two disk galaxies. It shows two long tidal tails, one of which extends to $\sim$ 280″ (130 kpc) from the center of the galaxy. The outer regions show clear evidence of a recent interaction involving two disks, while the inner region of NGC 7252 appears to have relaxed into a relatively spheroidal distribution. Schweizer (1982); found that the luminosity profile follows a $r^{1/4}$ law, and discovered a counterrotating disk of ionized gas within 8″ of the center, with a 1″–2″ hole at its center.

Whitmore et al. (1993), report on observations made with the Planetary Camera of the HST, Fig. 5. They discovered a population of about 40 blue pointlike objects in this galaxy with a mean absolute magnitude of $M_V = -13$ mag; mean color $V - I = 0.7$ mag; and mean effective radius of 10 pc (for $H_0 = 50$ km s^{-1} Mpc^{-1}). The luminosities, colors, projected spatial distribution, and sizes are all compatible with the hypothesis that these object formed within the last 1 Gyr following the collision of two spiral galaxies, and that they are young globular clusters. They conclude that new globular clusters can be created during a merger between two gas-rich galaxies.

Whitmore et al. also discovered a bright spiral structure within 3.5″ (1.6 kpc) of the center, with dust lanes and weak spiral structure extending out to about 9″ (4.2 kpc). This structure closely corresponds to the counterrotating disk of

Fig. 4 HST view of the nucleus of Mrk 315. The diffuse peak is 2.3" east of the stellar Seyfert nucleus.

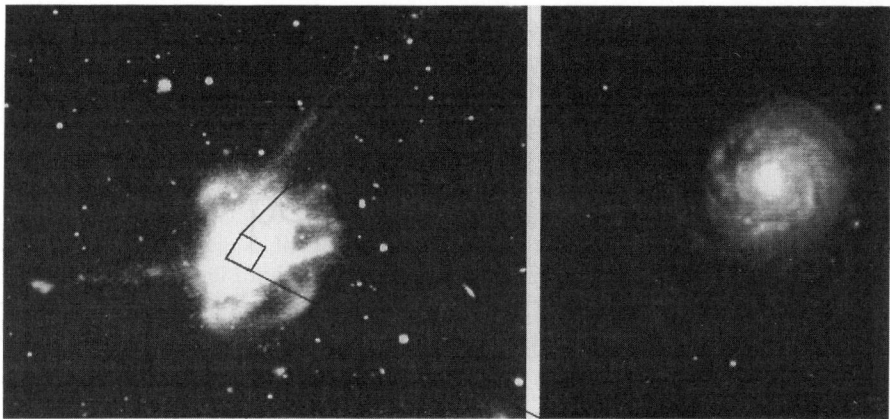

Fig. 5 Ground-based (left) and HST (right) images of NGC 7252. The inner spiral structure has a diameter of only 9" (4.2 kpc).

ionized gas observed by Schweizer (1982), and is presumably formed through the continued infall of gas into this central disk. This infall would be sufficient to trigger the activity of an AGN, if one existed in this galaxy.

Arp 220 is a low surface brightness galaxy with a well defined dust lane along its center. IRAS observations have shown that it emits $1.2 \times 10^{12} L_\odot$, at far infrared wavelengths, and thus, it is the prototype of ultraluminous IR galaxies, with a far-infrared luminosity that exceeds its optical luminosity by one order of magnitude. The best candidate mechanisms for this enormous amount of energy are either an AGN, or a large population of bright young stars. In both cases, the trigger is thought to be a galaxy merger in the recent past.

Planetary Camera images of Arp 220 were taken in the V, R, and I-band filters, Fig. 6 by Shaya et al. (1993). They show a very luminous object near the position of the western radio continuum source and 7 fainter objects within $2''$ of this position. The most luminous object appears to coincide with the radio source, however, an alternate alignment of maps shows the eastern radio source to coincide with one of the fainter objects and the OH radio sources to coincide with yet other objects. The authors favor the interpretation of those objects as massive young star associations with luminosities $10^9 - 10^{11} L_\odot$, but highly extinguished by intervening dust. These massive associations should fall into the nucleus on a timescale of 10^8 years. This implies that the associations are young and that dynamical friction is a very efficient mechanism to funnel large amounts of material into the obscured nucleus.

Assuming that all of these objects are young associations, between 10% - 20% of the far-IR flux could arise from these objects depending on how many remain totally obscured. In addition, if the diffuse starlight out to a radius of $8''$ is dominated by stars with typical ages of order 10^8 years, the time since the merger of two galaxies, then the reradiation of diffuse starlight contributes at least $3 \times 10^{11} L_\odot$ to the far-infrared flux, or $\geq 25\%$ of the total far-IR flux.

The additional bright objects ($M_V \approx -13$) located about $5''$ from the core are likely protoglobular clusters, but any of these could be recently exploded supernovae instead. The expected supernova rate, if the dominant energy source is young stars, is about one per month for the region where the anomalous far-infrared flux originates. Also, individual giant dust clouds are visible in these images. Their typical size is 300 pc ($1''$).

4. Active Galaxies

In the standard and, perhaps, simplistic unified model of AGN, the basic differences between the various classes of active galaxies can be explained as the result of varying orientations to the line of sight, (see Antonucci, 1993 for a review).

All active galaxies have a central energy source (e.g., black-hole) surrounded by the broad-line region and optical featureless continua. In turn, these are surrounded by an optically thick torus, which is oriented perpendicular to the radio jets (in

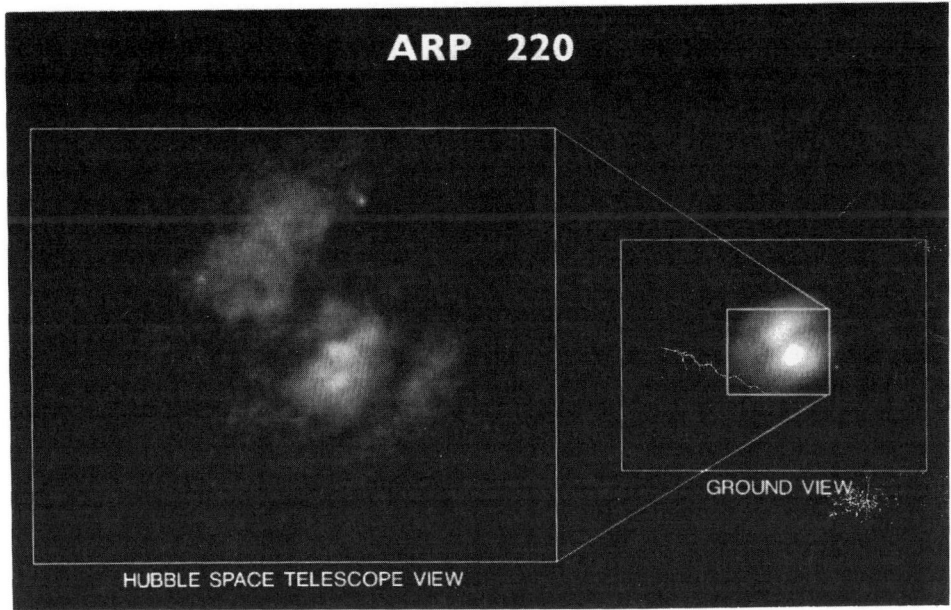

Fig. 6 HST (left) and ground-based (right) views of Arp 220. Note the resolved dust lane and the bright regions.

the case of radio-loud galaxies). All sources have, therefore, the same intrinsic geometry, although a wide range of luminosities is possible.

Those galaxies seen face-on are defined spectroscopically as Type 1 and have wide permitted lines in emission from the broad-line region, as well as narrow-emission lines, both permitted and forbidden from the narrow-line region. Type 1 galaxies also show strong featureless continua which are almost always observed to be variable.

Type 2 galaxies do not show emission from the broad-line region, nor a strong nuclear featureless continuum. HST observations of the different types of AGN seem to conform to this broad picture. However, in all cases, they considerably expand our understanding and in some cases (e.g., NGC 4151), they challenge the standard scenario.

4.1. TYPE 1 GALAXIES

A number of Seyfert 1 galaxies have been studied with HST, e.g., M51 (Ford et al 1993); NGC 4395 (Fillipenko et al 1993), etc. Of particular interest are recent observations of two radio-loud galaxies NGC 4261 and 3C449.

The radio source 3C270 has been mapped at various frequencies, and consists of a nuclear point source and two jets with outer lobes.

Observations of **NGC 4261** (3C270) with the HST have been reported by Jaffe et al (1993), Fig. 7.

Jaffe et al, discovered a large 2×10^{20} cm radius disk of cool dust and gas surrounding the bright unresolved nucleus of NGC 4261. They suggest that the bright point corresponds to thermal emission from the hot disk and that this is fueled by material flowing from the cool 'outer' accretion disk. The spin axis of the accretion disk is parallel to the direction of the radio jets.

The emission in the outer areas of the HST picture arises from normal early type stars; their density increases smoothly towards the nucleus. The centre of the image shows a well-defined disk that absorbs the light from the stars behind it. At the centre of the disk is the bright unresolved nucleus. The disk itself is elliptical, with axes of $1.71'' \times 0.74''$, and its major axis is at a position angle of $-16°$.

For an intrinsically circular disk, this implies that the plane of the disk is inclined $64°$ to the plane of the sky. Since the distance of NGC 4261 is 14.7 Mpc ($1'' = 71$ pc), the projected disk size is 121 pc $\times$ 51 pc.

The position angle of the disk's major axis coincides with that of the galaxy as a whole but, it is worth noting that the galaxy rotates about its major axis, rather than its minor axis.

The absorption in the disk is presumably caused by interstellar dust embedded in cool gas. The jet axis is parallel to the spin axis of the disk, and thus, is perpendicular to the rotation axis of the galaxy. This strongly suggests that the angular momentum of the disk is responsible for the orientation of the jets. The effect might be caused either by transfer of angular momentum from the disk to a central black hole, or by formation of a jet 'nozzle' in the inner disk by hydrodynamic forces. The disk observed in the HST picture is 10^5 - 10^7 times larger than the presumed inner accretion disk. Jaffe et al, estimate that the timescale for the viscous transport of 'fuel' into the central engine is about 10^8 yr., which is compatible with the timescale and energy requirements for the radio jet.

HST observations of the low luminosity radio-source **3C449** have been discussed by Capetti, Macchetto, Sparks & Miley (1994). This source has two extended ($> 15'$) radio jets, which display a remarkable mirror symmetry. Ground-based observations failed to detect any optical counterpart to the radio jets (Fraix-Burnet et al. 1991). Because of its relatively close distance ($z = 0.017$; $0.1'' = 50h_{50}^{-1}$ pc), this CD galaxy is ideal for high resolution studies with the HST. The observations were obtained with the Faint Object Camera, and show a 23 mag. nucleus surrounded by a ring whose projected radius is $0.4''$, Fig. 8. The ring center is coincident with the nucleus and has an ellipticity of 0.11 ± 0.09 and a position angle of its major axis of $\theta = 81 \pm 10$ degrees. Assuming cylindrical symmetry for the inner region, this corresponds to an angle between the axis of the ring and the line of sight of $\phi = 27 \pm 12$ degrees. The de-projected radius is $0.20'' < r < 0.23''$

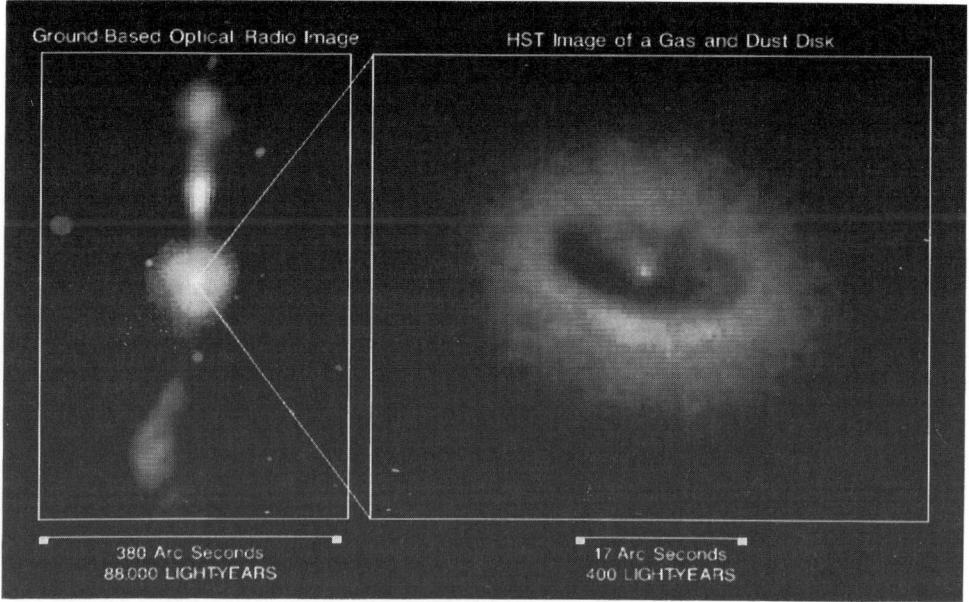

Fig. 7 Composite optical and radio image of NGC 4261 (left) and the HST image (right). The extended accretion disc is $1.7'' \times 0.7''$

corresponding to about 100 pc. The luminosity profile is well fitted by an $r^{1/4}$ law for $r > 0.3''$, about the radius of the ring, but in the inner region, the $r^{1/4}$ law overestimates the observed light profile.

The interpretation is that the nuclear region is strongly absorbed and the observed ring is actually produced by the high brightness gradient and the contrast with the inner absorbed region. The size of this region is consistent with it being due to cool material associated with an extended-accretion disk which partially absorbs the light of the underlying galaxy. This is also supported by the close alignment of the ring axis to the radio axis and by the presence of two bright and compact emitting regions found North and South of the nucleus.

The two bright knots are observed at a distance from the nucleus of about $0.55''$. The angle between the knots and the core is 167 degrees, about the same angle formed by the radio jets. There is an angular displacement of 15° between the jets and the line joining the optical knots and the nucleus, but this difference can be explained with lateral oscillations of the radio-jets in the inner 10", oscillations that are required by the observed radio morphology at larger distance. Therefore, it is very likely that the radio-jets are responsible for the formation of the optical knots. Most, if not all, of the flux from the knots appears to be continuum emission. This essentially rules out the possibility that the ionization is produced by the nuclear source, supporting the interpretation that the radio-jets are producing the knots.

Therefore, we conclude that the natural explanation for the peculiar morphology observed in 3C449 is that we are seeing material associated with an extended accretion disk obscuring the inner region of the galaxy. The jets originating from the nucleus at the center of the accretion disk produce two optical knots, probably associated with shock waves internal to the jets.

4.2. TRANSITION TYPE GALAXIES

When mother nature wants to remind us that our attempts at describing the physical phenomena with "standard paradigms" or "grand unified theories" are not to be taken too seriously, it provides us with objects such as NGC 4151 that "almost" fit our models. In this case, we call them "transition" objects (or Seyferts 1.5!), since they display characteristics of more than one class. We study these type of objects in great detail, because we can learn a great deal about the fundamental physics by how well they do or do not fit in detail our models.

NGC 4151 is the nearest ($\sim$ 30 Mpc) example of a (sometime) type 1 Seyfert galaxy. The broad emission component of Hβ, which is characteristic of Seyfert 1 galaxies, varies dramatically and in a low state can almost disappear. This led to NGC 4151 being reclassified as Seyfert 1.5. Both the permitted and the continuum emission show variations on time scales as short as days.

HST observations of NGC 4151 at several wavelengths in the optical and ultraviolet were obtained by the FOC (Boksenberg 1992, Boksenberg et al. 1993). These include observations obtained in the line and continuum around [O III] λ5007 Å. The UV observations including those around C IV and Lyα shows strong emission arising from the nucleus. However, there is no ultraviolet counterpart to the well-defined biconical structure abong PA 245 seen in [O III]. This consists of the bright nucleus and a number of distinct emission-line clouds with typical sizes of 10 pc, which lie within the NRL. The clouds are distributed within the biconical structure with a cone opening angle of $\sim$ 75° ± 10° and with apices coincident with the central point source Fig. 9. The cone is aligned with the extension of the nuclear VLB1 radio-source indicating that the same mechanism is responsible for the alignment of the optical ionizing radiation field and the parsec scale radio structure.

Long-slit spectroscopic observations carried-out with the FOC (Boksenberg et al. 1993a) show narrow-line emission components due to the bright knots within the cones and significant continuum and broad-line components, which arise from the intense nuclear emission. The radial velocity structure indicates an outflow of material along the jets, as well as material which shares in the general rotation of the galaxy; from these measurements we derive an upper limit of $4 \times 10^8 M_\odot$ for the total material contained within a radius of 30 pc of the nucleus. From a

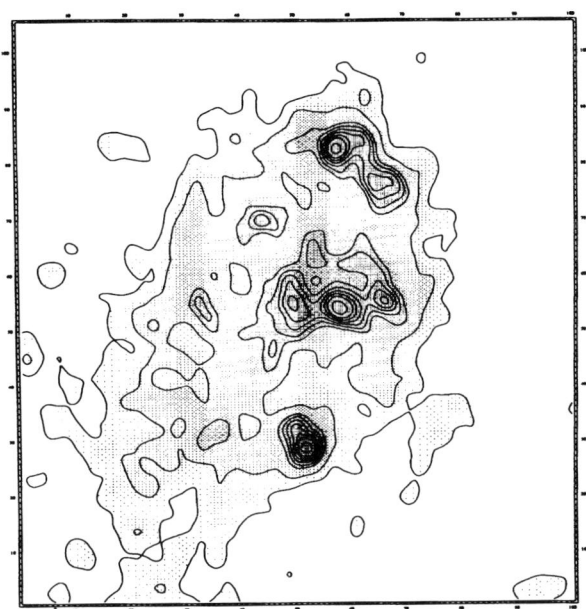

Fig. 8 In 3C449, the HST shows an extended accretion disc (0.2″ or 100 pc radius) and two bright knots aligned with the radio-axis.

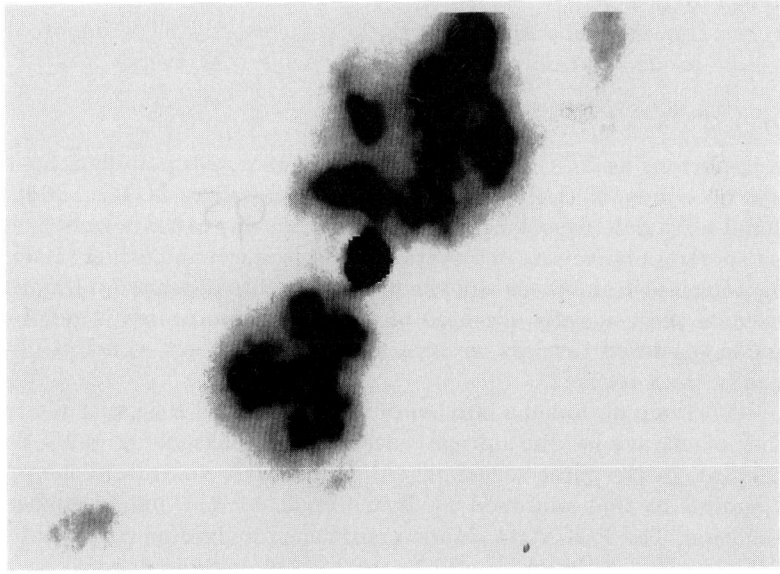

Fig. 9 This deconvolved HST image of NGC 4151 was taken in [O III] and shows a bright-nucleus and a well defined biconical structure. Typical cloud sizes are 10 pc.

detailed comparison of the optical and radio morphology and from the ultraviolet observations, Boksenberg et al, derive a rotation of the cone position axis as a function of the distance from the nucleus and conclude that the line of sight falls within, although barely, the ionization cone.

Different conclusions have been reached by Evans et al. (1993), which using similar data obtained with the PC observe the same emission-line morphology (although without any accompanying ultraviolet data). They propose that the geometry inferred from the data is incompatible with the simplest version of the unified model of AGN, namely, a very optically thick molecular torus surrounding the BLR to obscure the source of ionizing photons.

They conclude that for feasible model parameters, the line of sight must lie outside of the ionization cone, and since the very optically thick molecular torus model would predict that the central broad-line point source should not be visible directly, the model is not correct, or does not apply to NGC 4151.

On the other hand, obscuration models with effective neutral hygrogen column density along the line of sight of $10^{20} < N_H < 10^{21}\,\mathrm{cm}^{-2}$ are compatible with the observations.

Possible ways out include different mechanisms for collimating the radiation close to the central source: a very optically thick torus surrounded by a lower density atmosphere that collimates the ionizing radiation, but leaves the line of sight to the nucleus relatively unobscured; a disk geometry for the BLR clouds so that their opacity absorbs most of the ionizing radiation emitted in the disk plane; and finally, an optically thick, but geometrically thin accretion disk which surrounds a black-hole.

It is clear that more observations with the refurbished HST and more theoretical work will be needed to resolve this very interesting controversy.

4.3. TYPE 2 GALAXIES

Just as important as NGC 4151 to the development of a paradigm for the AGN, has been the study of the prototypical Seyfert 2 galaxy **NGC 1068**. While at optical and ultraviolet wavelengths, it appears like any other normal Seyfert 2, pioneering spectropolarimetric observations by Antonucci and Miller (1985) revealed a strong polarized component and the presence of broad-emission Balmer lines. In what became the generally accepted picture, they interpreted these observations as emission scattered towards us from a Seyfert 1 nucleus, which is obscured by an optically thick torus.

The FOC Team obtained a number of images of NGC 1068 with filters covering a number of ultraviolet and optical wavelengths. (Boksenberg 1992, Boksenberg et al. 1993b). In the filter containing [O III] 5007 Å the morphology Fig. 10 is clearly similar to that obtained by Evans et al. (1991), but at higher contrast and resolution. The FOC data shows a number of individual emission-line clouds, resolved with sizes of the order of $0.1'' - 0.2''$ ($\sim 6 - 12 h_{50}^{-1}$ pc) and which outline a clear biconical structure, with emission projecting also from South-West of the nucleus. The cone apex is coincident with the apex of the radio-emission suggesting that both collisional and excitation mechanisms are at play.

The ultraviolet images show high-contrast extended features clearly related to the [O III] structure, but considerably different in detail. They each show a dense diffuse concentration of emission near the nucleus which is of similar general appearance to the resolved bright feature in the optical continuum image, but which is not present in the [O III] 'pure' line image. This feature in the strong continuum component corresponding to the optical continuum emission and the result of scattering into our line of sight by the electron cloud. These observations considerably strengthen the case for the obscured-nucleus model.

Of considerable relevance are recent multiaperture spectropolarimetric observations caried-out with the FOS in the ultraviolet at wavelengths between 1575 Å and 3300 Å (Hurt, Antonucci & Miller, 1993), which allowed them to investigate the scattered, nuclear light without the complication of polarization dilution caused by strong unpolarized starlight.

The main results can be summarized as follows: the UV continuum polarization is constant (p $\sim$ 16% at $\theta \sim 97°$) from 1600 Å to 2800 Å. The value and wavelength independence of the polarization agree with other UV observations (Code et al 1993) and with expectations from optical spectropolarimetry (Miller & Antonucci 1983). The polarization decreases redward of 2800 Å reaching a value of $\sim$ 11.5% at 3300 Å. This can be explained by dilution from unpolarized starlight and implies a starlight fraction of $\sim$ 30% at 3300 Å. The narrow lines are much less polarized than the continuum, with typical polarizations of $<$ 2%. In polarized flux, the narrow lines are absent, and two lines, C III λ1909 and Mg II λ2798 have large equivalent widths of $\sim$ 25 Å and $\sim$ 14 Å respectively.

The relative fluxes in these three apertures show that the reflecting region is $\sim 1.0''$ which corresponds to $\sim$ 100 pc. Fig. 11 clearly distinguishes the polarized and unpolarized emission in NGC 1068. The starlight dilution at the red end is apparent. In addition, it is clear that both [C III] and Mg II have polarized components, while the forbidden line [Ne IV] λ 2423 is unpolarized.

These UV spectropolarimetric observations are also fully explained by the occultation/reflection model proposed by Antonucci & Miller (1985). Thus, the HST observations build up a model according to which NGC 1068 harbors a Seyfert 1 nucleus which is obscured along the line of sight by an opaque torus. The symmetry axis of the torus is aligned with the radio jet. Radiation from the continuum source and BLR can escape along the poles of the torus where it is scattered toward us, giving this radiation a partial linear polarization. The wavelength-independence of the polarization indicates electron scattering as the process responsible for the reflection.

The FOC ultraviolet observations confirm this geometry and locate the electron scattering medium close to and around the nucleus.

The quasar **OI 287** has a high and constant optical polarization but unlike most radio sources, is oriented parallel to the radio axis, a quiescent optical flux,

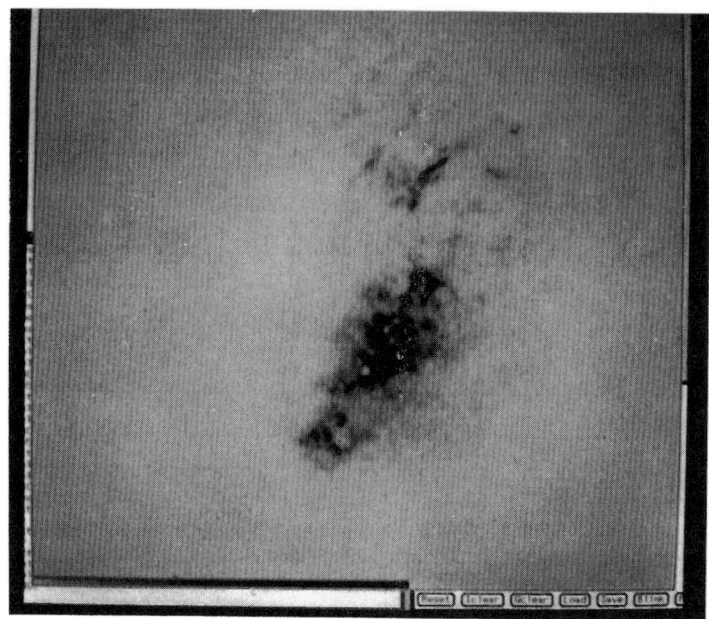

Fig. 10 This FOC image of NGC 1068 taken in [O III] shows the emission-line clouds arranged in a biconical structure. The cloud sizes are of the order of 10 pc.

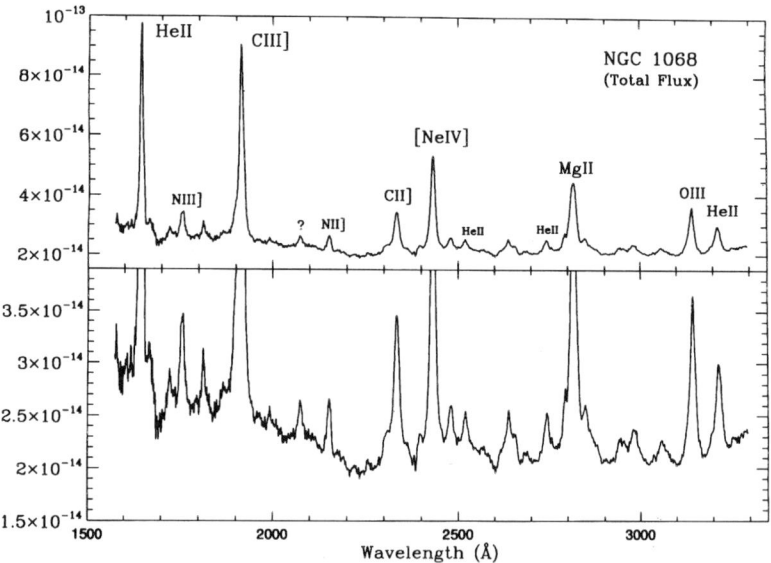

Fig. 11 Total and polarized spectra of NGC 1068. The starlight dilution at the red end of the total spectrum is apparent.

and a lobe-dominant radio source. It is a possible transition object.

Previous studies have led to the picture of an edge-on thin dusty torus occulting a featureless continuum (FC) source and a broad-line region. The FC source and the BLR are seen only in reflected polarized light. In the unified models, this makes OI 287 a "Quasar 2," analogous to the Type 2 Seyfert galaxies and the narrow line radio galaxies.

Observations of OI 287 were carried out with the FOS 4.3″ square aperture and covered the wavelength range 1200 Å – 2300 Å (Antonucci, Kinney & Hurt, 1993). The HST spectrum, Fig. 12, shows that OI 287 does have two spectrally resolved and statistically significant absorption features on the blue wing of the C IV line! The absorbers have substantial velocities relative to the quasar ($\sim$ 7100 km/s and $\sim$ 4000 km/s), and also substantial velocity ranges. These features are likely to represent material outflowing from the quasar. Thus, they are similar to BAL absorbers. The percent polarization is fairly constant with wavelength in OI 287. This wavelength-dependence is inconsistent with an origin due to transmission through dust grains like those in the Milky Way disk and in nearby galaxies, but can easily be accounted for by electron scattering in a region between the BLR and the NLR.

The data for OI 287 shows that it is probably an "edge-on" quasar, with the continuum and BLR sources occulted by an opaque, but geometrically thin dusty disk. By analogy, the UV spectrum of OI 287 also provides support for the notion that BAL objects may have obscuring disks relatively close to edge-on to our line of sight.

Observations of **Mrk 78** were carried out with the FOC at wavelengths including [O III], and at the nearby continuum (Capetti, Macchetto, Sparks & Boksenberg, 1994).

Two regions of high luminosity within a fainter extended halo are seen in the data, Fig. 13; they are located East and West of the galaxy's center and show smaller scale structures. The halo is more extended towards the West. In this region, it has several bright knots and two bright arcs with an overall biconical shape.

The cone has a full opening angle of 40 ± 5 degrees, and its axis is at a position angle of 67 ± 2 degrees from the North. The central region of the cone is completely obscured in an area of 0.8″ in size along the E-W direction and displaced from the center of the cone by 0.2″. Its angular size is 0.8″ (corresponding to 900 h_{50}^{-1} pc) in the direction perpendicular to the cone axis.

The biconical structure observed in the [O III] image, strongly indicates that the origin of the ionization is a central and compact source whose emission is restricted to a well defined solid angle by a parsec size thick accretion torus.

These results define the geometry of this region: obscuring material prevents the UV emission from the central ionizing source to escape along the line of sight; the same thing happens to any [O III] emission from the innermost gas. The continuum (starlight) emission on the other hand comes from an external region, and it is not

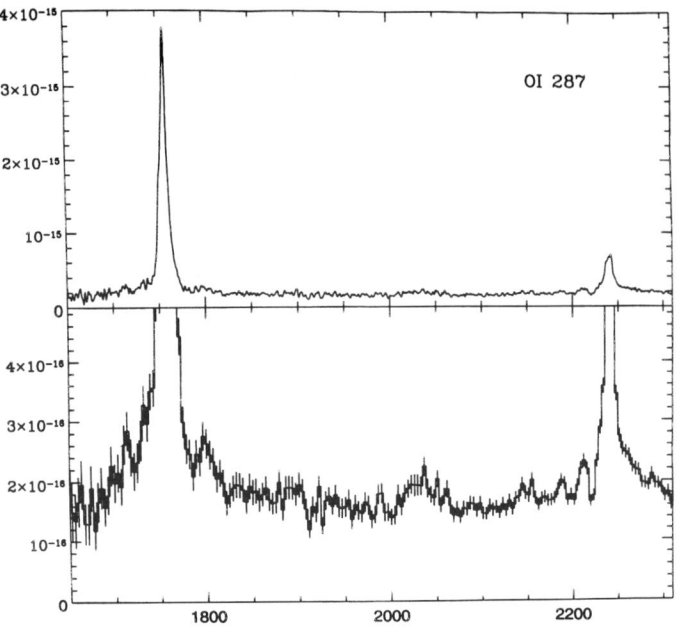

Fig. 12 The spectrum of OI 287 as observed with the FOS. Note the absorption features on the blue wing of C IV.

affected by the nuclear obscuring material.

The large size of the [O III] obscured region, $900h_{50}^{-1}$ pc, suggests that absorbing material (i.e., dust) is required on a much larger scale in addition to the torus (few parsecs size) obscuring the Seyfert 2 nucleus. The geometry of the obscuring material must be related to, or even determined by, the central torus, since otherwise we would require two different obscuration mechanisms, one for the nuclear ionizing source and the other for the central region.

This result is a strong confirmation of the occultation/reflection model for Seyfert 2 galaxies, with a compact central source as the origin for the ionization. The UV emission from this source is obscured, as confirmed by the total lack of far UV emission in the F130M data.

Dust surrounding the accretion torus is needed to explain the extended obscured central region. The origin for the overall geometric configuration is likely to be connected with the ionizing source and/or with outflowing gas.

The alignment of the extended [O III] emission with the cone of [O III] emission supports the idea of a common origin for the ionization. Nevertheless, the compact

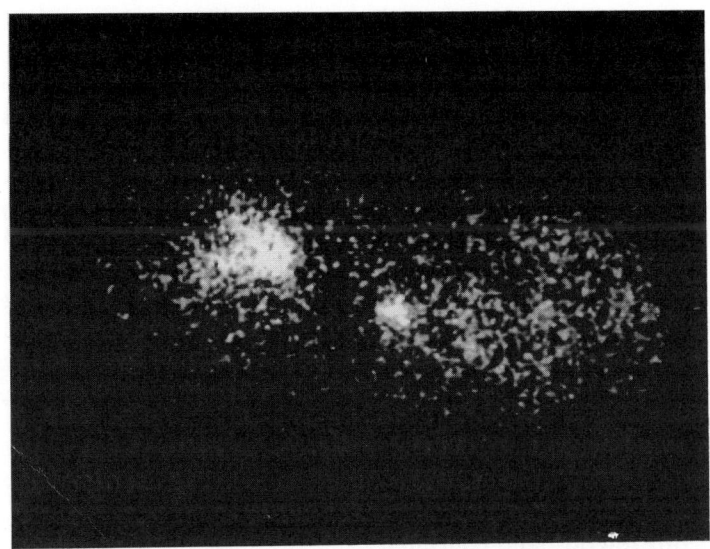

Fig. 13 FOC observations of Mrk 78 in [O III] showing two regions of high luminosity within a fainter extended halo, which has a biconical structure with 40° opening angle.

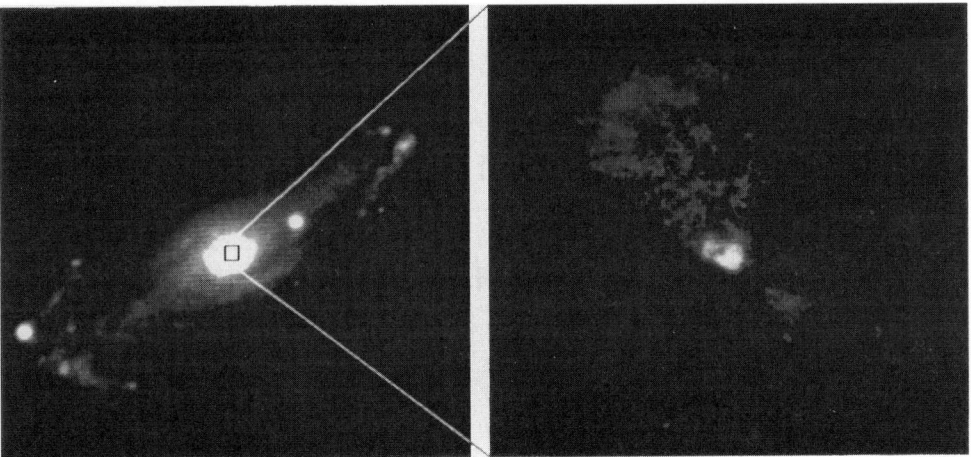

Fig. 14 Ground-based (left) and HST (right) images of NGC 5728. The HST data shows [O III] extended emission (8″) with a well defined biconical structure. The opening angle is 60°.

and extended emitting gas display different dynamic behaviours.

The radio emission is misaligned with respect to the [O III] emission by about 25 degrees; there is no direct morphological association between the radio and [O III] emission. This strengthens the interpretation of a pure photoionization from a nuclear source as the origin for the overall emission line region.

The observed UV flux is at least three order of magnitude smaller than the flux required to ionize the NRL, corresponding to a column density of $N_H > 0.5 \times 10^{22}$ cm^{-2}. For an isotropic source, the corresponding far infrared luminosity would greatly exceed the value obtained from the IRAS observations; thus, the dust must be heated by a highly anisotropic source, as expected in the case of an accretion torus, with reflecting funnel walls.

NGC 5728 is one of the galaxies which are known to exhibit ionization cone structure. Emission-line images (Pogge 1989) reveal a fan-shaped region of high excitation gas extending $\simeq 8''$ towards the SE from a bright core.

HST observations were made using the Planetary Camera by Wilson et al. (1993). Exposures with filters were taken in the Hα + [N II] $\lambda\lambda 6548, 6583$ emission, and in the [O III] $\lambda\lambda 4959, 5007$ emission. Exposures were also taken through nearby filters to record the off-band continua.

The dominant continuum feature is an off-centered, bar-shaped region which extends from a peak near the nucleus to another peak $2.2''$ (400 pc) to the east. A third peak lies $\simeq 1.5''$ to the NW with a prominent dark band separating the two. This dark band coincides with the apex of the cones.

The Hα+[N II] and [O III] continuum-subtracted images, Fig. 14, reveal that the high excitation gas near the nucleus of NGC 5728 is contained within two conical regions with a common apex. No [O III] emission is detected at the apex and the Hα+[N II] emission is weak there. The cone axes are colinear to within the uncertainties in defining them and the opening angles of the two cones are also very similar at $\simeq 55° - 65°$. Profiles of the [O III] and Hα brightness plotted perpendicular to the cone axis in the SE cone show flat-topped and double-peaked profiles $0.25'' - 1''$ SE of the line maximum. Flat-topped transverse spatial profiles can result when the illuminated gas lies in a disk, rather than being distributed with spherical symmetry. Double-peaked transverse profiles indicate a hollow cone and are also apparent in the ionization cone of Mrk 78 (Capetti et al. 1994); such profiles may result from "shadowing" of the ionizing radiation by optically thick gas closer to the nucleus along the axis of the cone.

The morphologies of the ionized gas, and the continuum emission can be well understood in the context of the torus model for Seyfert galaxies. The torus is located at the apex of the ionization cones, with the axis of the torus aligning with that of the cones.

5. Conclusions

The observations described in this review constitute far less than a quarter of the data taken with the HST in the field of AGN, and currently being analyzed. It is clearly an impressive array of exciting scientific results, which impose tight new observational constraints on our current physical models. The HST has the unique capability of probing the innermost regions of active galaxies, precisely where the most important physical phenomena are taking place. These observations will undoubtedly contribute to the solution of some of the more difficult problems, such as: what is the real nature of the central energy source; what are morphology and physical properties of the inner accretion tori; what is, and where is the region responsible for the featureless continuum, etc. To solve these and a number of other related problems will require, however, not only the precise observations of which HST is capable, but correlated observations at other wavelength ranges.

References

Antonucci, R.R.J., and Miller, J.S. 1985, *ApJ*, **297**, 621
Antonucci, R.R.J. 1993, *Annu. Rev. Astron & Astrophys*, **31**, 473
Antonucci, R.R.J., Kinney, A.L., & Hurt, T. 1993, *ApJ*, in press
Barnes, J. & Haernquist, L. 1991, *ApJ L*, **370**, L65
Boksenberg, A. 1992, in *Science with the Hubble Space Telescope*, ed. P. Benvenuti & E. Schreier, Munich: ESO, 61
Boksenberg, A., Macchetto, F., Albrecht, R., et al 1992, *Astron & Astrophys*, **261**, 393
Boksenberg, A., Macchetto, F., Catchpole, R., et al 1993a, *Astron & Astrophys*, submitted
Boksenberg, A., Macchetto, F., Catchpole, R., et al 1993b, *Astron & Astrophys*, submitted
Capetti, A., Macchetto, F., Sparks, W. B., & Miley, G.K. 1994, *Astron & Astrophys*, to be published
Capetti, A., Macchetto, F., Sparks, W.B., & Boksenberg, A. 1994, *ApJ*, in press
Code, A.D., et al 1993, *ApJ L*, **403**, L63
Crane, P., et al 1993, *ApJ L*, **402**, L37
Evans, I.N., Ford, H.C., et al 1991, *ApJ L*, **369**, L27
Evans, I.N., Tsvetanov, Z., Kriss, G.A., Ford, H.C., Caganoff, S., & Koratkar, A.P. 1993, *ApJ*, **417**, 82
Fillipenko, A. V., Ho., L.C. & Sargent, W.L.W. 1993, *ApJ L*, **410**, L75
Ford, H.C., et al 1993, *ApJ*, submitted
Fraix-Burnet, D., Golombek, D., Macchetto, F., Nieto, J.-L., Perryman, M.A.C. & Di Serego-Alighieri, S. 1991 *AJ*, **101**, 88
Heckman, T., 1992, AIP Conference Proceedings, ed Holt S., & Urry, M; **AIP**, 595
Hurt, T., Antonucci, R.R.J., & Miller, J.S. 1993, *Proc. 182nd AAS*, 789, in preparation.
Jaffe, W., Ford, H.C., Ferrarese, L., van den Bosch, F. & O'Connell, R. W. 1993, *Nature*, **364**, 213
Lauer, T.R., Faber, S.M., Lynds, C.R., et al 1992, *AJ*, **103**, 703
Macchetto, F. 1991a, *Heidelberg Conf. on Active Galactic Nuclei*, ed., S.J. Wagner, W.J. Duschl, Springer-Verlag: Berlin, **325**
Macchetto, F., 1991b, *7th IAP Conf. in Extragalactic Radio Sources from Beams to Jets*, et. Sol. A, **309**
Macchetto, F. 1992, in Science with the Hubble Space Telescope, ed. P. Benvenuti & E. Schreier, Munich: ESO, 73
Macchetto, F., et al 1991a, *ApJ L*, **369**, L55
Macchetto, F., et al 1991b, *ApJ L*, **373**, L55

Macchetto, F. 1993, *Mem. S.A. It.*, **64**, 11
Macchetto, F., Biretta, J.A., & Sparks, W.B., 1992, *Proc 182nd AAS in Comparison of HST and VLA Images of the M87 Jet, Arizona, BAAS*, **24**, 1183
Mackenty, J.W. 1986, *ApJ*, **308**, 571
Mackenty, J.W., Simkin, S.M., Griffiths, R. E. & Wilson, A.W. 1993, *ApJ*, submitted
Miller, J.S. & Antonucci, R.R.J. 1983, *ApJ L*, **271**, L7
Pogge, R.W. 1989, *ApJ*, **345**, 730
Sanders, D.B., Phinney, E.S., Neugbauer, G., Soifer, B.T., & Mathews, K. 1989, *ApJ*, **347**, 29
Schweizer, F. 1982, *ApJ*, **252**, 455
Shaya, E.J., Dowling, D.M., Currie, D.G., Faber, S.M., & Groth, E.J. 1993, *ApJ*, submitted
Sparks, W.B., Biretta, J.A., & Macchetto, F. 1993, *ApJ*, submitted
Thomson, R.C., Mackay, C.D. & Wright, A.E. 1993, *Nature*, **365**, 133
Whitmore, B.C., Schweizer, F., Leitherer, C., Borne, K., & Robert, C. 1993, *AJ*, **106**, 1354
Wilson, A.S., Breatz, J.A., Heckman, T.M., Krolik, J.H., & Miley, G.K. 1993, *ApJ L*, in press

OBSERVATIONS OF AGN WITH THE EXTREME ULTRAVIOLET EXPLORER

HERMAN L. MARSHALL
Eureka Scientific, Inc.
2452 Delmer St, Suite 100
Oakland, CA 94602

and

Massachusetts Institute of Technology
70 Vasser St., 37-667a
Cambridge, MA 02139

Abstract. The first results from surveys performed in the extreme ultraviolet (EUV) will be described in the context of studies of active galaxies and BL Lac objects. About a dozen extragalactic sources are known so far to emit sufficient EUV radiation that they are detectable even through the Galactic interstellar medium. These results are interpreted in the context of a model of EUV or soft X-ray excesses in the case of AGN. In the case of BL Lac objects, the detections indicate that the steep soft X-ray power law spectra continue into the EUV and that there is little intrinsic gas. Finally, there now exists EUV spectra from the Extreme Ultraviolet Explorer for one BL Lac, PKS 2155-304 and two AGN: Mk 478 and NGC 5548. The spectra show no significant spectral features; for AGN, it indicates that optically thin and emission line models may have a difficult time explaining the EUV and soft X-ray bumps.

Key words: AGN, BL Lac Objects

1. Introduction

When I first started on the *Extreme Ultraviolet Explorer* (EUVE) project, it was painfully aware to all extragalactic astronomers that it would be nearly impossible to observe galaxies or their active nuclei (AGN) in the extreme ultraviolet (EUV), sometimes defined as the range of wavelengths between the boron K edge at 68Å and the Lyman α edge at 1216Å. The basic reason was that the interstellar medium (ISM) was known to present a large optical depth longward of 100Å: for a typical *low* value of $N_H \sim 2 \times 10^{20}$ cm^{-2}, the optical depth out of the Galaxy would be about 7.

My personal view on this subject was turned around by two papers. First, Lockman, Jahoda, and McCammon (1986) showed that there are regions where the N_H out of the Galaxy is very small, less than 10^{20} cm^{-2} in some cases. Second, Wilkes and Elvis (1987) showed that AGN often appear to have soft X-ray excesses that compensate for ISM absorption. With these two observations and some quantitative representations about them, I made the first predictions at the EUV conference at Berkeley in January, 1989 (Marshall, 1991). Surprisingly, I found that of order 100 AGN would be found in the EUVE and ROSAT Wide Field Camera (WFC) all-sky surveys. Only the shortest wavelength bandpasses, with the Lexan/B filter

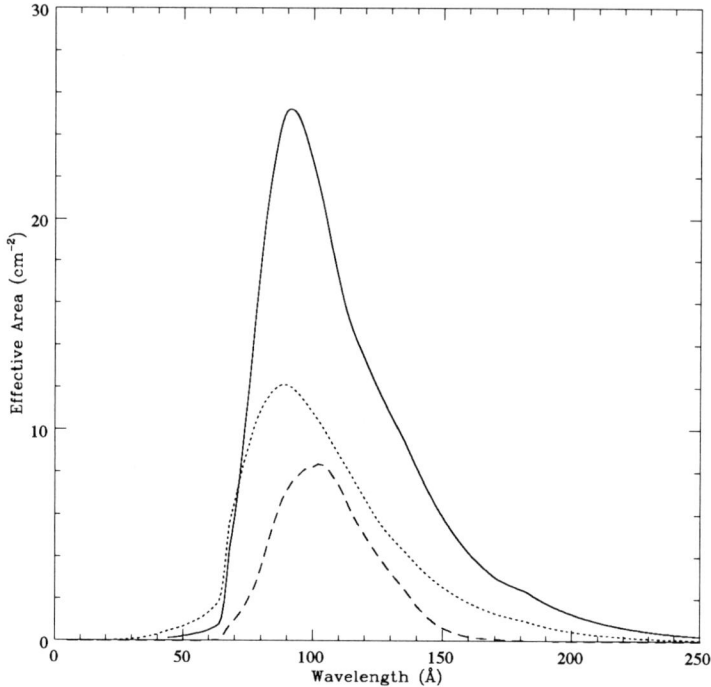

Fig. 1. Effective area functions for three short wavelength telescopes. *Solid line:* EUVE Deep Survey Lexan/B band. *Dotted line:* EUVE Scanner Lexan/B band. *Dashed line:* ROSAT WFC Lexan/B band. Note that the EUVE survey telescope has a much broader bandpass than the WFC.

(figure 1), are expected to allow detection of AGN because the ISM is too opaque longward of about 150Å.

2. EUV Survey Results

The ROSAT WFC all-sky survey was completed in January, 1991. There were about 380 sources, of which 4 were identified as AGN and 3 were BL Lac objects (Pounds, et al., 1993). In the EUVE all-sky survey, completed in January, 1993 (although gaps were filled until July 1993), 11 AGN and 5 BL Lacs were detected (Marshall, Carone, and Fruscione, 1994). See Table I for the list of known sources and their count rates in these surveys. Note that the WFC results from Pounds, et al. are for the complete survey, so sources have to be bright enough to satisfy the criterion for sources independent of the possible identification. There are data in the table from Balucinska-Church and Gondhalekar (1992) as well as Marshall, Carone, and Fruscione, who select targets on the basis of optical or X-ray properties and then determine the possible flux from the target; this procedure picks up sources at fainter flux levels than the surveys.

TABLE I
AGN and BL Lacs Detected in EUV Surveys

Name	Type[a] (Q, A, B)	R_{WFC} (cnt/s)	R_{EUVE} (cnt/s)
Q0239-591	Q		0.019
H0425-573[c]	A		0.050
BL1011+496	B		0.021
X12325[b]	A	0.021	0.017
Mk 421[b,c]	B	0.069	0.059
PG1116+215	Q		0.031
NGC 4051	A	< 0.010[e]	0.016
3C 273	Q	0.008[e]	0.036
Mk 279[c]	A		0.043
NGC 5548	A	0.010[e]	0.027[d]
IC 3599[b]	A	0.013	< 0.005
Mk 478[b]	A	0.052	0.048[d]
H1426+427[b]	B	0.013	0.029
Mk 501[c]	B		0.032
PKS2155-304[b,c]	B	0.166	0.249
NGC 7213	A	<0.008[e]	0.033
RE2248-511[b]	A	0.018	0.020

[a]A = AGN, Q = quasar, B = BL Lac object
[b]Source is in WFC bright source list (Pounds, et al., 1993).
[c]Source is in EUVE bright source list and satisfies the 6σ selection criterion (Malina et al., 1994).
[d]Source was observed during EUVE survey gap fill-in.
[e]Source data are taken from Balucinska-Church and Gondhalekar (1992).

There are a few features of the list to note. First, the BL Lacs that are bright in the hard X-ray band show up well in the EUVE bandpass. This is probably due to the steep spectra of BL Lacs and that they probably have very little neutral gas intrinsic to the host galaxy. Similarly, BL Lacs make up a relatively large fraction of the population, 30%, although they comprise less than 15% of the population at hard x-rays (Remillard, 1991).

Second, there are not as many AGN as expected before launch. There are two reasons for this, which are different for the two surveys. For the WFC, the Lexan/B effective area was about a factor of three lower than the pre-launch estimates. Furthermore, the average sensitivity was expected to be about 0.01 count/s and it turned out closer to about 0.02 count/s. Taking the sky coverage function into account (Pounds, et al., 1993), I repeated the prediction using the method from Marhshall (1991) to obtain 4.5 sources expected in the WFC survey. They observed 4, so this model apparently works. The model is relatively simple: a power law with $f_{nu} \propto \nu^{-\alpha}$ and $\alpha = 1$ is augmented with a thermal component with $T \sim 8 \times 10^5$K that would match UV spectra. The Galactic absorption was handled with a simple distribution function fitted to low N_H data (from F. J. Lockman, private communication; the data are also given in Marshall 1991).

In the case of the EUVE survey, however, the prediction is 9.7, given the actual sky coverage function in the bright source list (Malina et al., 1994). There is significant incompleteness, however, so the final value is expected to be much closer to the observed number: 2 AGN brighter than the 6σ threshold required for the EUVE bright source list. Again, the sensitivity of the survey was significantly worse than expectations, decreasing the predicted number of AGN.

3. Results from Observations of Individual Sources

3.1. PKS 2155-304

This bright BL Lac object was observed with the EUVE Deep Survey/Spectrometer during the EUVE in-orbit checkout phase in early July, 1992 (before the survey started). It was known to be the brightest extragalactic source in the WFC survey and was chosen because there would be very little long wavelength flux so that the second order response of the EUVE spectrometer could be verified. The source was not strongly variable during the EUVE observations and simultaneous optical data indicated that the overall spectral energy distribution decreased by about a factor of two between in the two years between the EUVE and WFC observations (Marshall, Carone, and Fruscione, 1993). Furthermore, with an accurate value of N_H, 1.36×10^{20} cm^{-2} (Lockman and Savage, 1993), they showed that the count rates of PKS 2155-304 were consistent with a spectral model with an ISM HeI/H ratio of 0.07 to 0.10 and $\alpha = 1.66$, as observed in the soft X-ray band (Fink et al., 1992). Using nearly the same spectral index and N_H values Fruscione et al. (1993) showed that additional neutral gas, $\sim 3 \times 10^{19}$ cm^{-2}, would be required to fit the EUVE spectrometer data if one requires HeI/H = 0.07 (which is favored by other EUVE observations).

3.2. NGC 5548

This AGN was detected by the WFC (Balucinska-Church and Gondhalekar, 1992) and was observed in March and May, 1993. The observations were coordinated with HST and ground-based optical observations so that correlations of the emission lines and the ionizing continuum could be obtained to look for lags. The source was extremely faint, however, so the planned exposure for the observation was 600,000 s, or a total observatory time of about 23 days.

A preliminary analysis has been presented at this conference (Kaastra et al., 1993). The source is so weak that the continuum is practically invisible. There is a hint of emission near 95Å but it is improbable that this flux would be detectable through the Galactic ISM, estimated at $N_H = 2 \times 10^{20}$ cm^{-2}, especially if there is no detection at 75Å or shorter.

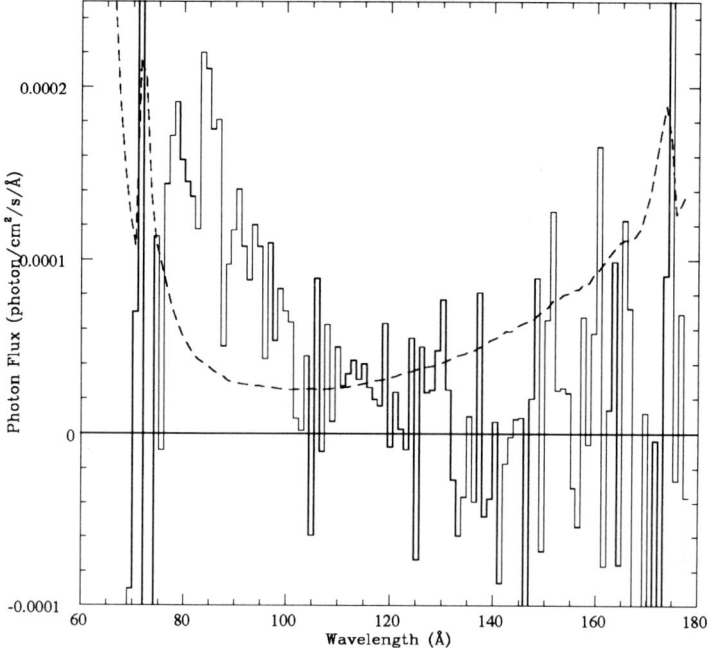

Fig. 2. EUVE spectrum of Mk 478. *Solid line:* Background subtracted photon flux in 1Å bins. *Dashed line:* Uncertainty in photon flux. There are no significant emission features at the resolution of the spectrometer, which is less than 1Å. As expected, the flux cuts off at long wavelengths due to the opacity of the Galactic ISM. Note that the uncertainties are very high shortward of 75Å, which is due to scattered geocoronal Lyman α, and that the continuum is marginally detectable in the 100-120Å range. The spike at 71.5Å is due to poor background subtraction at the bright detector rim.

3.3. Mk 478

The AGN Mk 478 ($z = 0.079$) was observed with the EUVE spectrometer in early April, 1993. It was proposed because it was the brightest AGN in the WFC survey and the spectrum shows that it was well detected (figure 2).

There are no significant features in the spectrum, so the first impression (the spectrum is only a week old) is that the EUV excesses are not due to a collection of emission lines, a possibility that may account for some soft X-ray excesses (Turner et al., 1991). Furthermore, there are no narrow emission features that might accompany a thin thermal plasma, if the UV bump and soft X-ray excesses are to be fitted with a bremsstrahllung model. Detailed analyses will place stringent limits on the amounts of gas that may be present with temperatures between 10^5 and $10^{6.5}$ K (Marshall et al., 1994).

4. Summary

In conclusion, current EUV instruments are capable of detecting AGN and BL Lac objects so that we have a new diagnostic. The spectroscopic data are especially exciting because we finally have sufficient resolution to test models of the soft X-ray excess.

Acknowledgements

I thank Antonella Fruscione for providing me the spectrum of PKS 2155-304 to present and Jelle Kaastra for two preliminary figures from the NGC 5548 analysis. This work has been supported by NASA: NAS5-32485 to Eureka Scientific, Inc., M.I.T. subcontract SV1-61010 under SAO contract NAS8-39073, and NAS5-30180 to U.C. Berkeley.

References

Balucinska-Church, M., and Gondhalekar, P. 1992, in *X-ray Emission from Active Galactic Nuclei and the Cosmic X-ray Background*, ed. W. Brinkmann, and J. Trümper, (Garching: MPE), 1992, 224.
Fink, H. H., Thomas, H.-C., Brinkmann, W., Okayashu, R., and Hartner, G. 1992, in *X-ray Emission from Active Galactic Nuclei and the Cosmic X-ray Background*, ed. W. Brinkmann and J. Trümper, (Garching: MPE), 1992, 202.
Fruscione, A., Bowyer, S., Konigl, A., Kahn, S. 1993, *Ap. J. (Letters)*, submitted.
Kaastra, J., et al. 1993, these proceedings.
Lockman, F.J., Jahoda, K., and McCammon, D., 1986, *Ap. J.*, **302**, 432.
Lockman, F.J. and Savage, B. 1993 *Ap. J. Suppl.*, submitted.
Malina, R.F., et al. 1994, *Astron. J.*, accepted.
Marshall, H. L. 1991, in *EUV Astronomy*, R. F. Malina and C. S. Bowyer, eds., (New York: Pergammon), p. 228.
Marshall, H.L., Carone, T.E., and Fruscione, A. 1993, *Ap. J. (Letters)*, **414**, L53.
Marshall, H.L., Carone, T.E., and Fruscione, A. 1994, *Ap. J.*, submitted.
Marshall, H.L., Carone, T.E., Shull, J.M., Malkan, M., Elvis, M., and Green, R.F. 1994, in preparation.
Pounds, K., et al. 1993, *M.N.R.A.S.*, 260, 77.
Remillard, R. 1991, MC-LASS catalog, unpublished.
Turner, T.J., Weaver, K.A., Mushotzky, R.F., Holt, S.S., and Madejski, G.M. 1991 *Ap. J.*, **381**, 85.
Wilkes, B.J. and Elvis, M. 1987, *Ap. J.*, **323**, 243.

Variability

HIGH-ENERGY CONTINUUM VARIABILITY IN ACTIVE GALACTIC NUCLEI

RICK EDELSON
Department of Physics and Astronomy; University of Iowa; Iowa City, IA 52242; U.S.A.

September 30, 1993

Abstract. *CGRO* and *IUE* observations suggest that the strong, aperiodic variability seen in the *Exosat* long-look observations of AGN extends over a much wider energy band. Some BL Lac objects (but no Seyfert 1 galaxies) have shown X-ray variations which were so rapid that they violate the assumptions of isotropy inherent in the Eddington limit. In the ultraviolet, Seyfert 1s as a class show an anti-correlation between the variability amplitude and luminosity, while BL Lacs show a positive correlation. Furthermore, Seyfert 1s show strong flux-correlated spectral variability, while BL Lacs show little or none. All of this suggests that the high-energy continua of BL Lacs are beamed towards us, while the ultraviolet continua of Seyfert 1s are emitted isotropically.

The November 1991 multi-waveband monitoring of the BL Lac PKS 2155-304 showed strong correlated variability, with the soft X-rays leading the ultraviolet by a few hours, and no measurable lag between the ultraviolet and optical down to a limit of $\lesssim 1.5$ hr. This indicates that the X-rays from this BL Lac are not produced by Compton upscattering, and that the ultraviolet does not come directly from a thermal source such as an accretion disk. This also strongly constrains the relativistic jet model, suggesting that all of the radiation is produced in a flattened region like a shock front.

Low temporal resolution ultraviolet/optical monitoring of the Seyfert 1 NGC 5548 in 1989 yielded a strong correlation with no measurable lag to a limit of $\lesssim 4$ days, casting some doubt on the standard model of thermal emission from an accretion disk in Seyfert 1s. Upcoming X-ray/ultraviolet/optical monitoring of the Seyfert 1 NGC 4151 in December 1993 will have much faster sampling, to permit a strong test of both this model and the competing reprocessing model.

Key words: BL Lacertae Objects, γ-Rays, Quasars, Seyfert Galaxies, Ultraviolet, Variability, X-Rays

1. Introduction

Variability is emerging as a powerful tools for constraining the physics of Active Galactic Nuclei (AGN). Early single-band variability studies yielded constraints on the source sizes and energy densities that led to the original black hole/accretion disk picture (Rees 1984). Also, the observation of rapid radio variability, combined with the relatively low X-ray fluxes provided the first strong evidence for beaming in flat spectrum radio sources, and led to predictions of superluminal motion that were later confirmed by VLBI (e.g., Pearson et al. 1981).

It has recently become possible to combine simultaneous observations taken at different wavebands (and with different telescopes), to construct single-epoch radio through X-ray (and, in some recent cases, through γ-ray) spectral energy distributions (SEDs) of AGN. These broadband "snapshots" have proven enormously valuable, suggesting what types of spectral components could contribute to different bands, thus providing a framework for formulating models of the physical conditions and emission mechanisms operating in the centers of AGN (see, e.g., the contributions by Bregman and Maraschi in this volume).

However, such single-epoch SEDs cannot by themselves provide strong tests of these models. Model fits are underconstrained because they require a large number of free parameters, and experience has shown that a wide variety of models can produce acceptable fits. For example, both synchrotron self-Compton (SSC) models (e.g., Ghisellini, Maraschi & Treves 1985) and accretion disk models (Wandel & Urry 1991) have successfully been fitted to the broadband SED of the BL Lac object PKS 2155–304. Likewise, a variety of accretion disk models (Czerny & Elvis 1987; Wandel & Petrosian 1988; Madau 1988; Laor & Netzer 1989; Sun & Malkan 1989, etc.) and free-free models (Barvainas 1993) provide adequate fits to Seyfert 1 SEDs.

Because the central regions of AGN are small and the spectral resolution and polarization capabilities of current satellites are limited, multi-waveband variability provides the best chance to resolve these ambiguities and constrain competing models. While this has long been recognized, only recently has there been the simultaneous availability of satellites operating at ultraviolet, X-ray and γ-ray wavelengths, the will to commit large amounts of telescope time, and the spirit of cooperation between large numbers of astronomers necessary to make it work.

Throughout this contribution, a distinction will be made between blazars (BL Lac objects and OVV quasars) and Seyfert 1 galaxies (and normal quasars). Also, while "high-energy" generally refers to the X-ray/γ-ray regime, the broadband nature of AGN emission means that the ultraviolet and occasionally optical wavebands must be considered as well.

2. Prevalence of Variability

Strong, rapid, aperiodic high-energy variability is a common feature of AGN, as seen in Figure 1. The most rapid variations have been seen at X-ray wavelengths, but this may be an artifact of the fact that *Exosat*, with its high Earth orbit (leading to long, uninterrupted light curves) and relatively high sensitivity, is particularly well-suited for studying AGN variability. By comparison, *IUE*, the workhorse of ultraviolet astronomy, can measure fluxes with high signal-to-noise ($\gtrsim 50/1$), but has a minimum cycle time of ~ 1 hr, and AGN have such low γ-ray count rates that *CGRO* is generally only sensitive to $\gtrsim 10\%$ variations on ~ 1 day time scales. Also, it is my impression that blazars show more rapid variations than Seyfert 1s, although no systematic comparison has ever been made.

The long-term ultraviolet light curves of blazars and Seyfert 1s look remarkably similar. This is probably due to the gross undersampling of the archival data, which are thus rendered insensitive to short time scale variability. Finally, with the discovery of a confusing source in the field of NGC 6814 (Madejski et al. 1993), there is no good evidence for periodicity in the light curve of **any** AGN.

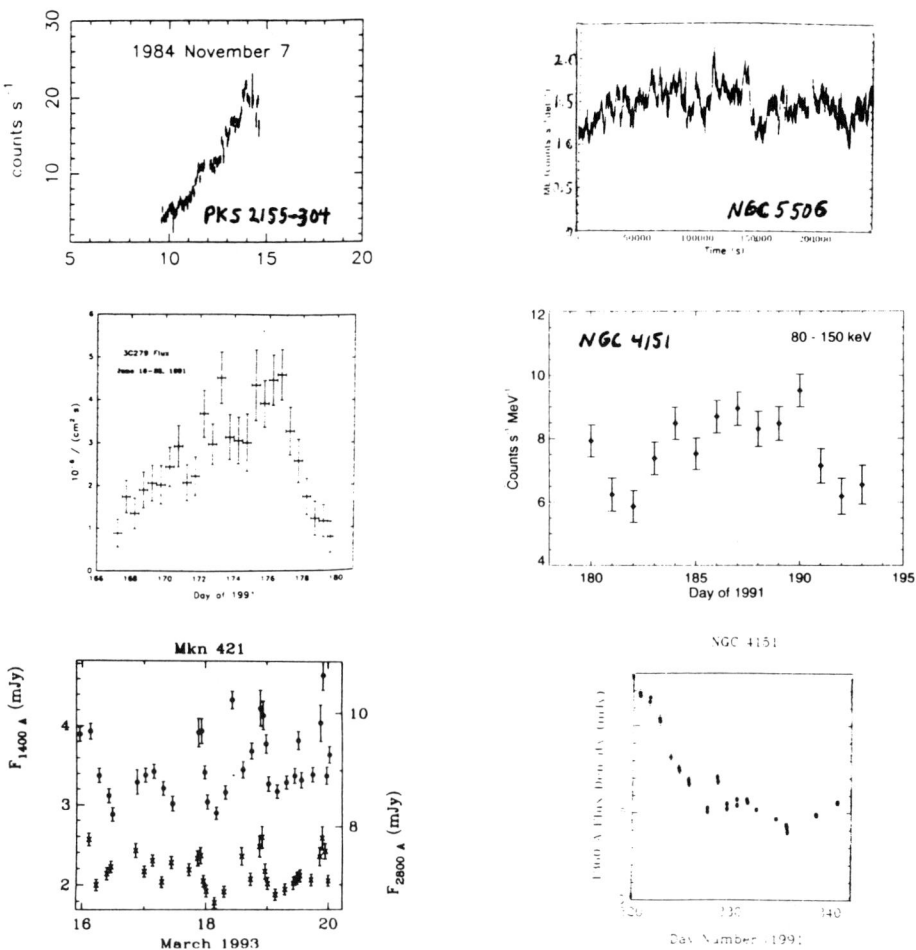

Fig. 1. Rapid variability of blazars (left) and Seyfert 1s (right) at X-ray (top), γ-ray (middle) and ultraviolet (bottom) energies. Data from Tagliaferri et al. (1991; *Exosat* ME observations of PKS 2155–304), Kniffen et al. (1993; *CGRO* EGRET, 3C 279), the ultraviolet archives (*IUE* SWP and LWP, Mkn 421), McHardy & Czerny (1985; *Exosat* ME, NGC 5506), Maisack et al. (1993; *CGRO* OSSE, NGC 4151), the ultraviolet archives (*IUE* SWP, NGC 4151).

3. Trends of Variability with Luminosity

The earliest variability tests involved comparing the observed "doubling time" ($t_{1/2} = dt/d\ln F$) with the minimum light-crossing time for a source emitting at the Eddington limit (t_{lc}). For a spherical, optically thick source, the minimum light-crossing time is (Cavallo & Rees 1978; Fabian 1979):

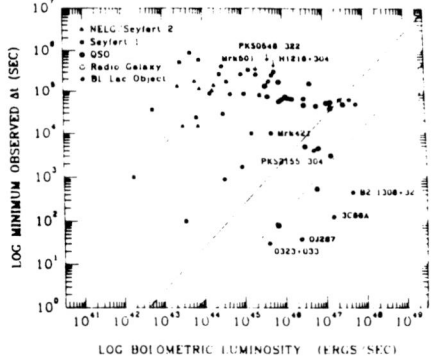

Fig. 2. Plot of most rapid observed X-ray variability doubling times as a function of bolometric luminosity (Bregman 1991). The solid line indicates the light crossing-time corresponding to a Schwarzschild radius of a system radiating a given Eddington luminosity. The dashed line corresponds to the maximum luminosity expected under the most conservative assumptions with asymmetric but unbeamed emission (Abramowicz & Nobili 1982). All of the objects that lie below the line are blazars (BL Lacs or OVV quasars).

$$t_{lc} \approx \frac{3R_s}{c} \approx \frac{6GM_h}{c^3} \approx \frac{L}{4 \times 10^{42} \text{ erg s}^{-1}}.$$

As seen in Figure 2, some blazars, but no Seyfert 1s, were found to vary more rapidly in the X-rays than the minimum permitted light-crossing time. This is evidence that, as in the radio, the X-rays from these blazars are not emitted isotropically, but instead are beamed towards us. The fact that so many Seyfert 1s lie near the solid line supports the standard black hole model, since the only known way to radiate with this high efficiency is through gravitational liberation of the binding energy (Rees 1984). While this overall conclusion is probably still valid, the doubling time depends on the details of the observations (length of observation, sampling rate, and especially signal-to-noise) as well as intrinsic source properties, so it is risky to compare data taken with different telescopes or sampling patterns.

A possibly more robust approach involves examining variability amplitude of sources observed with a single telescope/detector system. Figure 3 shows a clear anti-correlation between long-term ultraviolet variability amplitude and mean source luminosity for the Seyfert 1s. This fits with the predictions of the model that the ultraviolet emission from Seyfert 1s comes from an accretion disk (or some other relatively isotropically emitting source). The more luminous disks would be larger, with longer light-crossing times, and thus would vary more slowly.

There also appears to be a positive correlation between variability and both source luminosity and optical polarization for blazars. A simple explanation is again found in the context of beaming. A more strongly beamed source would have both the apparent luminosity and the degree of polarization boosted, and the apparent time scale would also increase.

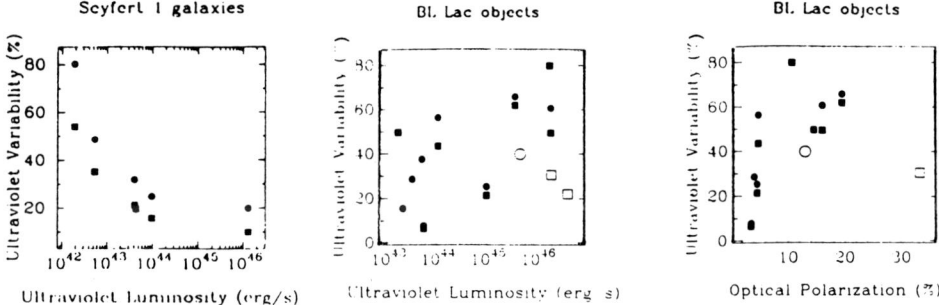

Fig. 3. Long-term ultraviolet variability plotted as a function of luminosity for Seyfert galaxies (left, Edelson, Krolik & Pike 1990), and blazars (center, Edelson 1992). For blazars, variability is also plotted as a function of optical polarization (right). The circles refer to 1400 Å data and the triangles to 2800 Å data. The Seyfert 1 data (on the left) shows a clear anti-correlation, while the blazar data show an apparent positive correlation of degree of variability with both luminosity and optical polarization, especially if one excludes the three open symbols, which refer to poorly observed sources that have a mean time between observations of $\sigma_t < 2$ yr, compared with $\langle \sigma_t \rangle = 4$ yr for the others.

4. Spectral Variability

Spectral variability can bridge the gap between single-band and multi-band variability studies. If the flux is seen to vary while the spectral shape remains constant, this would suggest that the flux in that waveband is dominated by a single component, and that the line-of-sight absorption (if any) does not vary significantly. Likewise, the most natural explanation of significant spectral changes, especially when linked to flux changes, is the interplay of two emission components with different spectral shapes and variability properties. Other possible explanations include variable absorption or a single component with a variable shape.

Figure 4 gives ultraviolet flux–spectral index plots for Seyfert 1s and blazars. The blazars show almost no significant spectral changes, supporting the hypothesis that a single component contributes to their ultraviolet emission. However, the Seyfert 1s show strong spectral changes which are well correlated with flux level. This suggests that there are two components emitting significant ultraviolet flux: a harder, more strongly variable component (the putative accretion disk?) and a softer, less variable one. Other explanations are less tenable: If absorption causes the spectral and flux changes, large changes in reddening indicators such as line ratios and the 2175 Å absorption features would be expected (none are observed), while if it was a single optically thick thermal source with a variable temperature (around 10^5 K, to reproduce the shape of the ultraviolet spectrum), even very large variations would have difficulty producing the observed spectral changes, because $L \propto T^4$ and α_{UV} is measured well below the spectral peak.

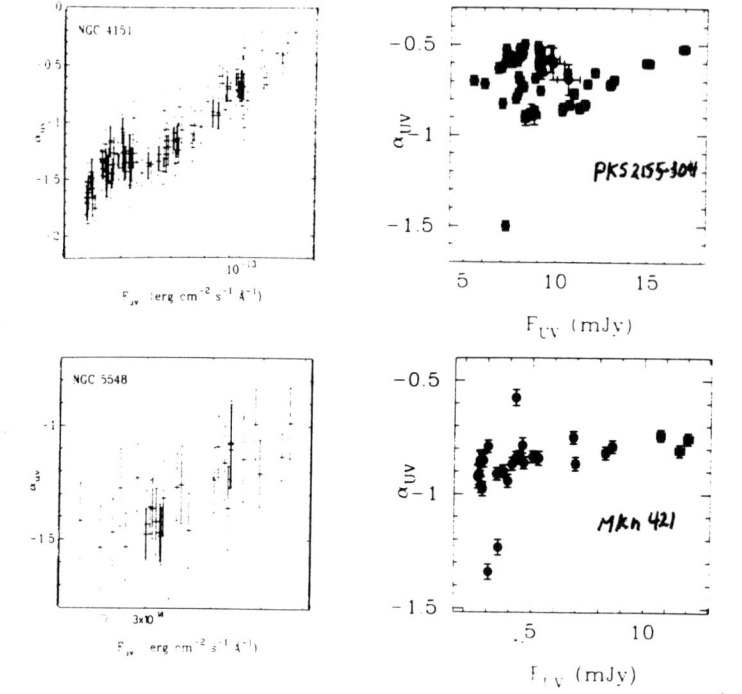

Fig. 4. Ultraviolet continuum spectral index, α_{UV} (defined as $F_\nu \propto \nu^\alpha$, measured between 1400 and 2800 Å) plotted against flux density for the two best observed Seyfert 1s (NGC 4151 and NGC 5548, on the left, from Edelson et al. 1990) and the two best observed blazars (PKS 2155-304 and Mkn 421, right, from Edelson 1992). For the Seyfert 1s, there is a strong correlation in the sense that the spectrum hardens as the source brightens, while for the blazars, there is little spectral variability and no evidence for a correlation between flux and spectral index.

The situation at higher energies is more uncertain, mostly because of the lack of sensitivity of low-resoluiton X-ray and γ-ray telescopes to relatively subtle spectral changes. The 2-10 keV spectra of Seyfert 1s are generally thought to soften as the source gets brighter (e.g., Mushotzky, Pounds & Done 1993), although NGC 7469 shows the opposite behavior (Barr 1986). This softening with increasing flux has also been observed in the soft γ-rays (Paciesas et al. 1993). In the soft X-rays (0.1-1 keV), there is some evidence for variations in the absorbing column, although multiple explanations are possible because of ambiguities introduced when fitting low-resolution X-ray spectra (e.g., intrinsic variability of the power-law slope: Yaqoob & Warwick 1991; variability of the hard tail: Fiore et al. 1992; variable cold absorber: Bond, Matsuoka & Yamanuchi 1993; variable warm absorber: Nandra et al. 1990). Because of its high resolution, the new X-ray satellite *ASCA* will be very helpful in resolving such ambiguities.

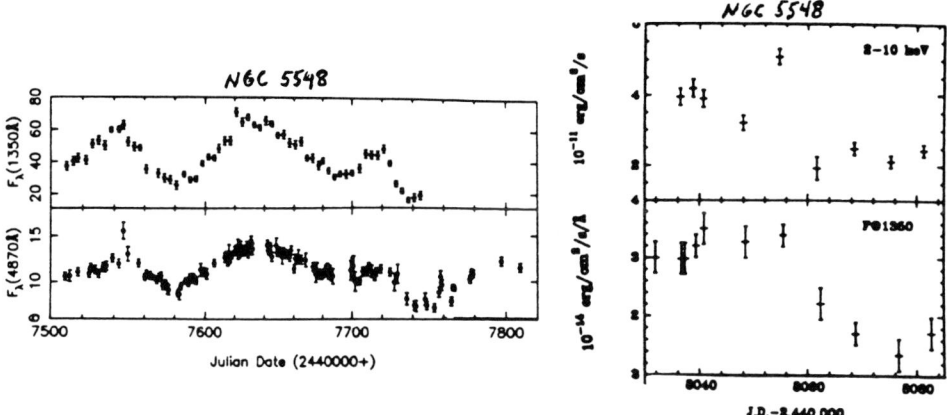

Fig. 5. Coordinated multi-waveband monitoring of the Seyfert 1 NGC 5548. On the left, ultraviolet/optical data show strongly correlated variations. The upper limit on the interband lag is ≲ 4 days. On the right, ultraviolet/X-ray data taken at a different epoch suggest a correlation, although data taken previously do not fit this correlation.

5. Multi-Wavelength Monitoring of NGC 5548

The most constraining recent results resulted from intensive, coordinated monitoring of individual objects over a wide wavelength range. The first of these was the 1989 ultraviolet/optical campaign to monitor the Seyfert 1 NGC 5548 (Clavel et al. 1991; Peterson et al. 1991). Although the goal of this project was to map the structure of the broad emission line region, perhaps the most intriguing result came in the continuum, where it was found that there was no detectable lag between 1300 Å ultraviolet and 5000 Å optical variations, down to a limit of ≲4 days (see Figure 5). In the standard accretion disk model, the 1300 Å radiation is emitted at a mean radius of ~0.1 lt-dy, while the 5000 Å radiation is emitted at ~0.5 lt-dy (Krolik et al. 1991), so the upper limit of ≲ 4 days corresponds to a lower limit of the propagation speed of ≳ $0.1c$. This is much faster than either propagation at the sound speed or the orbital time scale. Similar results have been found for NGC 3783 (Reichert et al. 1993). The only mechanism left by which a signal can propagate at such a high speed is radiative reprocessing (Krolik et al. 1991), a scenario which had not been considered until recently. This potential problem with the accretion disk model has led to a renewed interest in alternative models, but a rigorous test awaits higher time sampling (see §7).

The most promising alternative to the accretion disk model is that the ultraviolet is reprocessed X-ray/γ-ray radiation (e.g., Nandra et al. 1991). This model was supported by observation of an apparent correlation between ultraviolet and X-ray flux variations in NGC 5548 (Clavel et al. 1992; see Figure 5). Perola et al.

(1986) found a similar possible correlation in NGC 4151. Such a correlation would be expected if the X-rays drive the ultraviolet, although a clear test will again require much better sampling. Arguing against this model are: (1) the observed X-ray/γ-ray energy budget is not sufficient to produce the ultraviolet through infrared emission of higher luminosity quasars (although lower luminosity Seyfert 1s like NGC 5548 may not have this problem), and (2) no correlation was seen between X-rays and optical in other Seyfert 1s, most notably NGC 4051 (although a much shorter time scale was sampled; Done et al. 1990).

6. Multi-Wavelength Monitoring of PKS 2155–304

Coordinated X-ray/ultraviolet/optical observations have also strongly constrained emission from the blazar PKS 2155–304 (Edelson et al. 1993). As shown in Figure 6, it showed strong, highly correlated microvariability with the same amplitude in all bands. Careful correlation analysis showed that the X-rays led the ultraviolet by 2–3 hr, with no measurable lag between the ultraviolet and optical bands down to a limit of $\lesssim 1.5$ hr. This important result rules out the hypothesis that the X-rays are produced by SSC scattering of ultraviolet photons, since the X-rays would have to come from a region as large or larger than the ultraviolet, and thus would be expected to track or lag the lower frequencies. Likewise, the claim that the ultraviolet is thermal emission from an accretion disk (Wandel & Urry 1991) can also be rejected, both because a lag of $\gtrsim 50$ hr would be expected between 1400 Å and 5000 Å (by analogy to the argument made in the previous section), while none is seen, and because of the lack of spectral variability (see §4).

By process of elimination, this result supports the hypothesis that the entire optical through soft X-ray continuum arises from direct synchrotron emission in a jet. Broadband spectral fits based on the tapered jet model of Ghisellini, Maraschi & Treves (1985) lead to very short lifetimes for electrons emitting in the X-rays ($t_x \lesssim 0.01$ sec), so the lack of the strong X-ray spectral variability (Brinkmann et al. 1993) means that they would have to be continuously re-accelerated *in situ*. This model also calls for a relatively large distance between the X-ray and optical emitting regions (~ 10 lt-day), which is difficult to reconcile with the observed X-ray/optical lag of 2–3 hr, even after the effects of bulk relativistic motion are accounted for. Instead, this result favors a flattened geometry, such as that which would naturally arise in a shock front in which the X-ray and lower-energy emission are produced in essentially the same place (e.g., Marscher & Gear 1985).

7. Future Plans

These initial large multi-waveband campaigns have been quite successful in eliminating models and constraining AGN physics, but there is clear potential for even greater progress in this rapidly evolving area. The aforementioned studies have taught us about the importance of evenly-spaced, high temporal sampling over

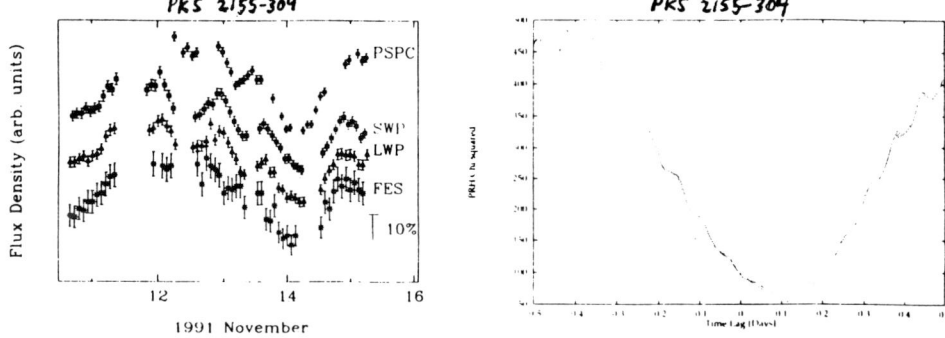

Fig. 6. Intensive monitoring of PKS 2155-304 at X-ray (top), 1400 Å and 2800 Å ultraviolet (middle) and 5000 Å optical (bottom) energies (data from Brinkmann et al. 1993, Urry et al. 1993). The data show clear, correlated microvariability, with essentially no time lag between the ultraviolet and optical bands. The X-ray variations are also strongly correlated, but are ~2-3 hr ahead of the ultraviolet. On the right, correlation analysis using the Press, Rybicki & Hewitt (1992) χ^2 technique shows that this lag is significantly different from zero at the 5σ level.

long time spans, good multi-waveband coordination, and careful source selection. For instance, even and rapid sampling is important because the erratic and aperiodic character of the variations will lead to aliasing if unevenly sampled, and the rapid nature of the variations means that the most valuable information will be lost if the sampling rate (or sensitivity) is too low.

The next round of campaigns has been constructed with these lessons in mind. The AGN Watch consortium (the group responsible for the NGC 5548 and NGC 3783 campaigns) will observe the rapidly variable Seyfert 1 NGC 4151 continuously on 1-10 December 1993, with *IUE*, *Rosat*, *CGRO*, *ASCA* and ground-based telescopes. Because of the fast sampling, multi-waveband coordination, and expected rapid source variability, this campaign should be able to measure ultraviolet lags so short that they would correspond to apparent propagation speeds that would exceed the speed of light, allowing an unambiguous test of the accretion disk model, and correlated ultraviolet/X-ray variations, to test the reprocessing model.

The blazar 0716+714 will also be subjected to continuous, simultaneous *IUE*, *Rosat*, *CGRO*, *ASCA* and ground-based optical, infrared, submillimeter and radio monitoring over a 5.6 day period in March/April 1994. The goal of this campaign is to extend the PKS 2155-304 result to higher and lower frequencies, providing even more constraints on the geometrical relationships between wavebands. We also want to see if this radio-selected blazar behaves like the X-ray selected blazar PKS 2155-304.

Beyond that, we hope to undertake one or two more multi-waveband AGN monitoring campaigns before *IUE* is turned off in September 1994. Terminating this functioning satellite will leave a gaping hole in the ultraviolet (the efficiency of *HST* is too low and its proposal pressure is too high to be of much use for such studies) and will deal a severe blow to the multi-waveband monitoring effort. Thus, this next year will see a flurry of activity in this area which will not be matched for the next few years. Let's hope we make good use of it!

References

Abramowicz, M. & Nobili, L. 1982, Nature, 300, 506
Barvainas, R. 1993, ApJ, 421, 513
Barr, P. 1986, MNRAS, 223, 29p
Brinkmann, W. et al. 1993, ApJ, in preparation
Bregman, J. 1991, Astron. Astrophys. Rev., 2, 125
Bond, I., Matsuoka, M. & Yamanuchi, M. 1993, ApJ, 405, 179
Cavallo, G. & Rees, M. J. 1978, MNRAS, 183, 359
Clavel, J. et al. 1991, ApJ, 366, 64
Clavel, J. et al. 1992, ApJ, 393, 113
Czerny, B. & Elvis, M. 1987, ApJ, 321, 305
Done, C. et al. 1990, MNRAS, 243, 713
Edelson, R., Krolik, J. & Pike, G. 1990, ApJ, 359, 86
Edelson, R. 1992, ApJ, 401, 516
Edelson, R. et al. 1993, ApJ, submitted
Fabian, A. C. 1979, Proc. R. Soc. Lond., 366, 449
Fiore, F. et al. 1992, AA, 262, 37
Ghisellini, G., Maraschi, L. & Treves, A. 1985, AA, 146, 204
Kniffen, D. et al. 1993, ApJ, 411, 133
Krolik, J., Horne, K., Kallman, T., Malkan, M., Edelson, R. & Kriss, G. 1991, ApJ, 371, 541
Laor, A. & Netzer, H. 1989, MNRAS, 238, 897
Madau, P. 1988, ApJ, 327, 116
Madejski, G. et al. 1993, Nature, in press
Maisack, M. et al. 1993, ApJL, 407, L61
Marscher, A. & Gear, W. 1985, ApJ, 298, 114
McHardy, I. & Czerny, B. 1985, Nature, 325, 696
Mushotzky, R., Pounds, K. & Done, C. 1993, Ann. Rev. Astron. Ap., 31, in press
Nandra, P. et al. 1990, MNRAS, 242, 660
Nandra, P. et al. 1991, MNRAS, 248, 760
Paciesas, W. et al. 1993, in *Proceedings of the Compton γ-Ray Observatory Symposium*, ed. N. Gehrels, (St. Louis, MO), in press
Pearson, T. et al. 1981, Nature, 290, 365
Perola, G. C. et al. 1986, ApJ, 306, 508
Peterson, B. et al. 1991, ApJ, 368, 119
Press, W., Rybicki, G., & Hewitt, J. 1992, ApJ, 385, 404
Rees, M. 1984, Ann. Rev. Astron. Ap., 22, 471
Reichert, G. et al. 1993, ApJ, in press
Sun, W.-S. & Malkan, M. A. 1989, ApJ, 346, 68
Tagliaferri, G., Stella, L., Maraschi, L., Treves, A., & Celotti, A. 1991, ApJ, 380, 78
Urry, C. M. et al. 1993, ApJ, 411, 614
Wandel, A. & Petrosian, V. 1988, ApJL, 329, L11
Wandel, A. & Urry, C. M. 1991, ApJ, 367, 78
Yaqoob, T. & Warwick, R. S. 1991, MNRAS, 248, 773

THE X-RAY EMISSION OF QUASARS AS OBSERVED BY *GINGA*

A.J. LAWSON and M.J.L. TURNER
University of Leicester
University Rd.
Leicester LE1 7RH
United Kingdom

Abstract.
We present preliminary results on a spectal analysis of quasars observed by the X-ray observatory *Ginga*. Simple power-law models with fixed Galactic absorbtion provide an adequate description of the spectra for most of the sources in the 2-18 keV band. A small number of sources show evidence for a feature at 6.4 keV (in the source rest frame) due to Fe line emission. Maximum likelihood and Spearman rank tests were used to investigate the relationship between radio loudness and X-ray spectral index in this class of object. These tests showed, respectively, that the mean X-ray spectral index of radio quiet quasars (RQQs) is significantly different from that of flat spectrum radio loud quasars (FRSQs) at the >99% level, and that the dominant relationship with spectral index is radio loudness (not X-ray luminosity or redshift) at >99% significance. This last result has not previously been demonstrated in this band, but agrees with findings in the lower energy *Einstein* band (0.5-3.5 keV)[1]. These results are discussed in the context of current unified models.

Key words: galaxies:active - galaxies:nuclei - quasars:general - radio continuum:galaxies - X-rays:galaxies

1. Method and Results

Ginga observed over 40 quasars ($M_B < -23$) during its 4+ years in orbit. The data from each object were background subtracted using standard techniques[2] and then fitted with a power-law model including hydrogen column absorbtion. There was no requirement in any of the spectra for a column greater than the galactic. Amongst observations with sufficient sensitivity, those of radio quiet objects show evidence for an iron line at 6.4 keV (source frame) with ~ 200 eV equivilent width, but the radio loud objects show no such emission, with EQW <20 eV. Reflection[3] can be fitted to the radio quiet spectra, but is not required statistically. Spectra from 27 objects were used to investigate the relationship between X-ray spectral index and radio loudness. (N.B. Not all the objects observed are used, as some fall below the survey flux limit and others presently have unresolved problems in the background subtraction.)

Previous papers[1, 4, 5] have shown a relationship between X-ray spectral slope and radio loudness in quasars, in the sense that radio loud objects have flatter spectra. There is also some evidence[6, 7] that the slope varies within the radio loud group, with the slope of steep spectrum radio quasars, SRSQs (i.e. those that are lobe dominated) being slightly greater than that of the FRSQs (core dominated objects). Figure 1 shows the plot of index versus radio loudness for this sample (Key: Cross in circle - RQQs ($R_L <1$), Square - SRSQs, Star - FRSQs).

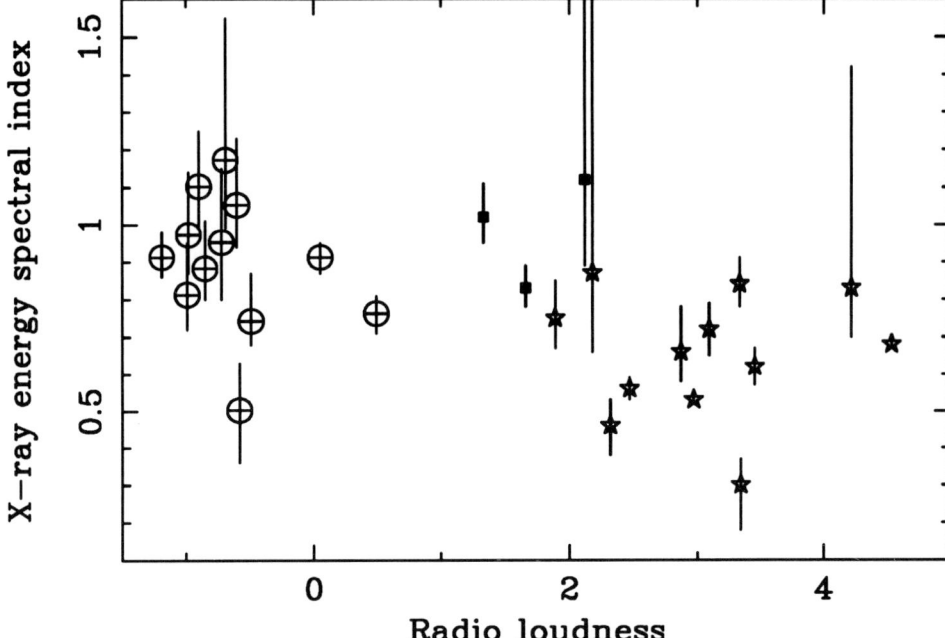

Fig. 1. Plot of X-ray spectral index vs radio loudness for the current *Ginga* sample

Maximum likelihood methods return the following for the mean index ($\bar{\alpha}$) and error on the three groups of interest in this study. Respectivly the results are for the RQQ, FRSQ and SRSQ groups; 0.86±0.03, 0.63±0.07 and 0.95±0.18. These results agree well with those from the previous studies with the $\bar{\alpha}$ of the FRSQs being significantly different from that of RQQs at the >99% level. Also the SRSQ group resembles more the RQQ group than the FRSQ group (the indices of the two radio loud groups are different at 90% significance). Using Partial Spearman tests it is shown, for the first time in the 2-18 keV band, that the dominant relationship with α is that of radio loudness rather than with L_X or z at >99% significance. This agrees with findings[1] in the soft X-ray band.

2. Discussion

2.1. RADIO LOUD OBJECTS

High polarizations, superluminal motion and the EGRET detections of exclusively radio loud objects provides good evidence that relativistic jet motion and beaming are important in these objects and probably dominate the emission at high energies. Recent work[8] has suggested that radio loud objects (FRSQs and SRSQs) and FRII radio galaxies could be the same objects viewed at different angles with respect to the jet, and it has also been shown that jet models are consistent with this orientation effect[9]. Modelling of the jet emission components[10] predicts a slope of $\alpha \sim 0.5$ for the Compton component which would be dominant in the hard X-ray band. Results of our study are consistent with the above findings; an index of ~ 0.6 for the FRSQs, the SRSQs with a different slope to the FRSQs and the lack of a thermal iron emission line possibly due to swamping of the continuum by the doppler boosted component.

2.2. Radio quiet objects

The RQQs, with no instrinsic absorbtion column and 6.4 keV Fe line emission, show similarities in this respect to Seyfert galaxies. However, the spectral slopes of the RQQs are generally steeper from those of the Seyfert 1's[11] at $\sim 99\%$ significance. This difference in spectral index may well be related to luminosity. *ROSAT* studies of AGN[12] have shown absorbtion at a few keV which is thought to be due to ionised material located in the funnel created by the molecular torus. Simulating spectra at the flux levels of a typical RQQ with models with and without a warm absorber show an increase from a slope of ~ 0.7 to ~ 0.9 when the absorber is removed. This neatly corresponds to what is seen in our results, though we do not claim that this is the solution. If luminosity and size scale with mass, the only way to boost the ionization of the material is for the black hole to be accreting nearer the Eddington limit. So RQQs could be Seyfert type objects with a higher effieciency of accretion.

2.3. Concluding remarks

Our results add support to findings of previous studies and agree well with the current beaming and orientation unified models.

References

1. Wilkes B.J., Elvis M., 1987, ApJ, 323, 243
2. Hayashida K. et al., 1989, PASJ, 41, 373
3. Pounds K.A., Nandra K., Stewart G.C., George I.M., Fabian A.C., 1990, Nat, 344, 132
4. Lawson A.J., Turner M.J.L., Williams O.R., Stewart G.C., Saxton R.D., 1992, MNRAS, 259, 743
5. Williams O.R. et al., 1992, ApJ, 389, 187
6. Shastri P., Wilkes B.J., Elvis M., McDowell J., 1993, ApJ, 410, 29
7. Worral D.M., Giommi P., Tananbaum H., Zamorani G., 1987, ApJ, 313, 596
8. Padovani P., Urry C.M., 1992, ApJ, 387, 449
9. Ghisellini G., Padovani P., Celotti A., Maraschi L., 1993, ApJ, 407, 65
10. Ghisellini G., Maraschi L., 1989, ApJ, 340, 181
11. Nandra K., 1991, PhD Thesis, University of Leicester
12. Nandra K., Pounds K.A., 1992, Nat, 359, 215

SOLVING THE MYSTERY OF THE PERIODICITY IN THE SEYFERT GALAXY NGC 6814

Greg M. Madejski[1], Chris Done[1,3], T. Jane Turner[1], Richard F. Mushotzky[1], Peter Serlemitsos[1], Fabrizio Fiore[3], Marek Sikora[4,5], and Mitchell C. Begelman[5]

[1] Lab for High Energy Astrophysics, NASA/Goddard, Greenbelt, MD, USA
[2] Presently at the Physics Dept., Leicester University, Leicester, UK
[3] Center for Astrophysics, Cambridge, MA, USA
[4] Copernicus Astronomical Center, Warsaw, Poland
[5] JILA/University of Colorado, Boulder, CO, USA

ABSTRACT. The reports of periodic X-ray emission from the Seyfert galaxy NGC 6814 have motivated a number of exotic models for the active nucleus. Our ROSAT observation shows that while the nucleus of NGC 6814 is indeed an X-ray emitter, the periodicity is due to another source, most likely a Galactic accreting binary system, ~ 37 arc min away.

1. Introduction

The Seyfert galaxy NGC 6814 has been a subject of much recent study due to its unique variability behavior, being the only AGN so far in which a clear periodicity is seen in the X-ray flux. The existence of the $\sim 12,100$ second period was first detected in the EXOSAT Medium Energy detector (ME; Mittaz and Branduardi-Raymont 1989; Fiore, Massaro, and Barone 1992) and then confirmed in *Ginga* observations (Done *et al.* 1992). A number of interpretations were advanced for the periodicity, all based on orbital motion. These included, among others, gravitational lensing of X-ray emitting hot spots on an accretion disk by the central black hole (Abramowicz *et al.* 1991), or a captured star orbiting the black hole (Syer, Clarke, and Rees 1991; Sikora and Begelman 1992). The combination of very rapid (~ 50 sec) drops in the X-ray flux (Kunieda *et al.* 1990), short period, and large X-ray luminosity observed in NGC 6814 gave strong support for the black hole paradigm in AGN. Clearly, better understanding of this unusual object was needed, and to that end, we observed the field of NGC 6814 with the ROSAT X-ray satellite.

2. Observations and Results

The observation of NGC 6814 was conducted with the ROSAT X-ray telescope Position Sensitive Proportional Counter starting on March 31, 1993. It lasted for 180 ks of running time, and yielded ~ 38 ks of useful data. Figure 1 shows the X-ray image over the 2 degree diameter field of view in the full 0.1–2.4 keV bandpass of the instrument. The observation was centered within a few arc sec of the optical position of NGC 6814 nucleus, which is at RA(2000) = $19^h 42^m 40.4^s$, Dec(2000) = $-10°19'25''$. The positional coincidence of the strong X-ray source within a few arc sec from the the center of the image confirms that the nucleus of NGC 6814 is indeed the X-ray emitter.

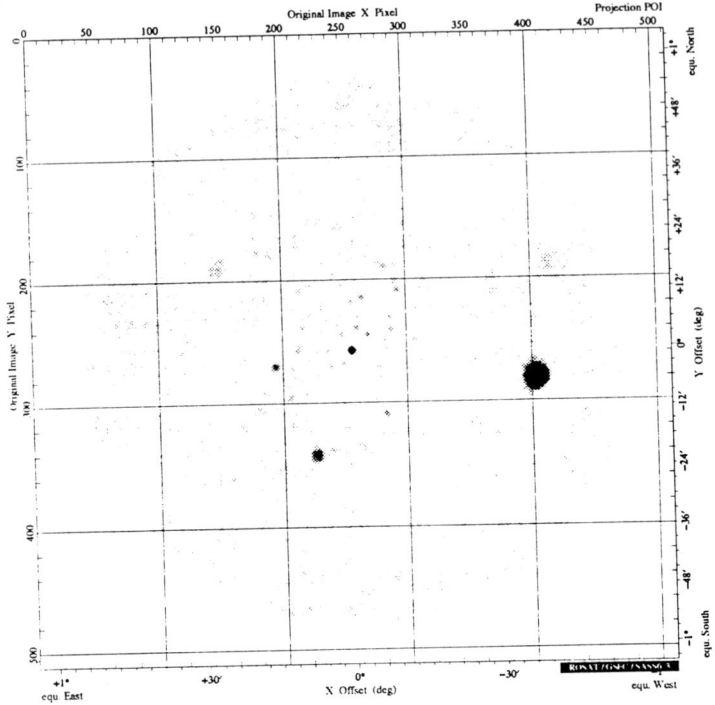

Figure 1. ROSAT PSPC image of the NGC 6814 field. The Seyfert galaxy NGC 6814 is the on-axis source, while the periodic X-ray emitter is 37' off-axis (The image of this second source is smeared by the degraded off-axis point spread function.)

The ROSAT image reveals another bright point source ∼ 37 arc min away, strongly distorted by the off-axis point spread function. The position derived from the ROSAT image is RA(2000) = $19^h40^m13^s$, Dec(2000) = $-10°25'30''$, where we estimate the position error to be ∼ 30 arc sec, primarily because the off-axis image distortion as well as partial obscuration by the PSPC window supporting rib. We extracted light curves for both sources and tested them for periodicity in the 10,000 – 15,000 sec range by repeatedly folding the light curve, and measuring the significance of deviations from a constant by the L-statistic (related to χ^2 but following an F distribution; see Davies 1990; Done et al. 1992). Figures 2a and b show the L statistic for NGC 6814 and the second source; the ∼ 12,000 second periodicity is clearly not associated with the Seyfert galaxy but with the off-axis source. In fact, the best fitting period of 12,142 s (see Fig. 2) is exactly that determined from the most recent *Ginga* observation of the NGC 6814 region (Done et al. 1992). We thus conclude that it is the second, off-axis source that is the origin of periodic X-ray emission.

The spectrum of this second source, derived from background-subtracted data, is well modeled as a black body plus a power law, absorbed by matter with Solar abundances. The fit yields a temperature of $kT_{bb} = 63 \pm 11$ eV and power law energy index of $\alpha = 0.1 \pm 0.25$, absorbed by an equivalent hydrogen column density of $N_H = 0.76 \pm 0.23 \times 10^{21}$ cm^{-2}. The black body contributes $\sim$ 45 % of the 0.2 – 2 keV flux, with the power law contributing the rest. The 0.2 – 2 keV flux averaged over the entire observation is $\sim 6 \times 10^{-12}$ erg cm^{-2} s^{-1}, with $\sim$ 20% uncertainty due to a partial obscuration by the PSPC support ribs. This flat power law index is in fact quite similar to that seen by *Ginga* for the spectrum from the NGC 6814 region (Kunieda *et al.* 1990; Turner *et al.* 1992). We estimate that the 2–10 keV flux, using the extrapolation of the ROSAT spectrum, is only $\sim 1.7 \times 10^{-11}$ ergs cm^{-2} s^{-1}, several times lower than the mean *Ginga* flux. However, this value is quite uncertain, since the overall source spectrum is likely to be more complex than we infer here. The nucleus of NGC 6814, on the other hand, shows a power law spectrum typical of other AGN, with a power law index $\alpha = 0.79(+0.44, -0.33)$, absorbed by $N_H = 1.4(+0.5, -0.7) \times 10^{21}$ cm^{-2}, consistent with the Galactic value. The mean 0.2 – 2 keV flux of NGC 6814 measured in our ROSAT observation is 7.1×10^{-13} erg cm^{-2} s^{-1}.

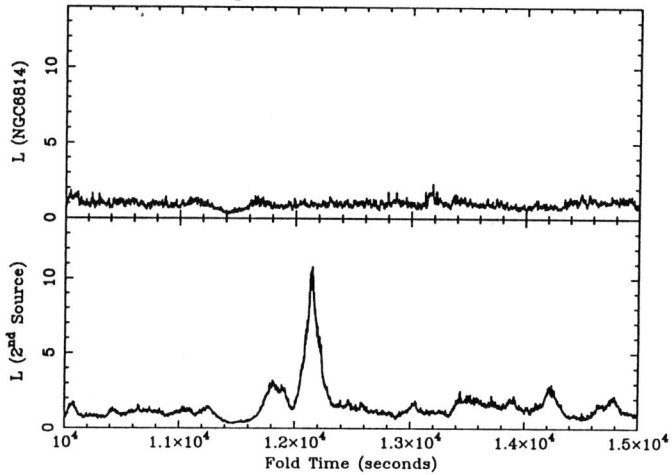

Figure 2. Value of L-statistic tests for periodicity over the range of 10,000 s to 15,000 s for NGC 6814 (a, top) and off-axis source (b, bottom). The 12,142 s period is clearly a property of the off-axis source.

3. Discussion

Our observation shows that the X-ray emission with $\sim$ 12,100 sec periodicity, previously attributed to the Seyfert 1 galaxy NGC 6814, is due to an unrelated object that is located $\sim$ 37 arc minutes from the nucleus of NGC 6814. Why wasn't it discovered earlier? Of course the limited spatial resolution of collimated proportional counter instruments such as HEAO-A2, *Ginga* LAC, and EXOSAT ME precluded separation of X-ray emission from the two sources, especially given the fact that the off-axis source definitely is, and NGC 6814 is likely to be variable. Likewise, the EXOSAT Low Energy telescope was not sensitive enough, and the *Einstein* Observatory IPC had too small of a field of view. A sensitive, large field-of-view imaging instrument such as ROSAT PSPC was necessary for the identification.

Our discovery eliminates a lot of difficulties with modeling of X-ray emission from NGC

6814. The flat ($\alpha \sim 0.4$) *Ginga* X–ray spectrum (Kunieda *et al.* 1990; Turner *et al.* 1992) was quite unusual for an AGN, and $\alpha \sim 0.8$ is more typical. The inferred periodicity required fine-tuning of the source geometry, and the rapid drop of flux observed in the *Ginga* data (Kunieda *et al.* 1990; Done *et al.* 1992) implied an extremely compact source. Finally, the strength and rapid variability of the Fe K line flux (Kunieda *et al.* 1990) required a substantial amount of ionized material very close to the central source, and so a strong gravitational redshift was inferred to reconcile this with the measured line energy (Turner *et al.* 1992). These observational features will now have to be reconciled with another class of an X–ray source.

What is the other source? An optical follow-up (Rosen *et al.* 1993; see also Halpern 1993) indeed shows that indeed, there is a variable object at the ROSAT position, which has the expected $\sim 12,140$ sec period. The optical spectrum of it shows strong He II 4686 line emission, as well as possible cyclotron features, which are strongly suggestive that it is a magnetized CV system in our own Galaxy, consisting of a white dwarf fueled by accretion from a companion star. In fact, the ROSAT X–ray spectrum is typical of such systems (Cordova 1993).

We recognize that the observational data from what was thought to be NGC 6814 was considered a "rosetta stone" for the black hole paradigm for the structure of AGN, but we stress that the inferences from X–ray variability of Seyfert galaxies as a class are unaffected by this finding. These sources often exhibit non-periodic large amplitude variability on time scales of thousands of seconds (cf. McHardy 1989), and the removal of NGC 6814 does not substantially change the strong support that the rapid X–ray variability provides for the black hole hypothesis in AGN (see, e.g., Fabian 1992).

References

Abramowicz, M., Bao, G., Lanza, A., and Zhang, X.-H. 1991, *Astron. Astrophys.*, **245**, 454.
Cordova, F., 1993, in *X–ray Binaries*, ed. W. H. G. Lewin, J. van Paradijs, and E. P. J. van der Heuvel, in press.
Davies, S. R. 1990, *M.N.R.A.S.*, **244**, 93; see also *M.N.R.A.S.*, **251**, 64p.
Done, C., Madejski, G. M., Mushotzky, R. F., Turner, T. J., Koyama, K., and Kunieda, H. 1992, *Ap. J.*, **400**, 138.
Fabian, A. C. 1992, in *Frontiers of X-ray Astronomy*, ed. Y. Tanaka and K. Koyama (Tokyo: Universal Academy Press), p. 603.
Fiore, F., Massaro, E., and Barone, P. 1992, *Astron. Astrophys.*, **261**, 405.
Halpern, J. P. 1993, *Nature*, **365**, 607.
Kunieda, H., Turner, T. J., Awaki, H., Koyama, K., Mushotzky, R. F., and Tsusaka, Y. 1990, *Nature*, **345**, 786.
McHardy, I. 1989, in *Two Topics in X-ray Astronomy*, Proc. 23rd ESLAB Symposium, eds. N. White, J. Hunt, and B. Battrick (Paris: ESA), p. 1111.
Mittaz, J. P. D., and Branduardi-Raymont, G. 1989, *M.N.R.A.S.*, **238**, 1029.
Rosen, S., Done, C., Watson, M., and Madejski, G. M. 1993, *IAU Circ. No.* 5850.
Sikora, M., and Begelman, M. C. 1992, *Nature*, **356**, 224.
Syer, D., Clarke, C. J., and Rees, M. J. 1991, *M.N.R.A.S*, **250**, 505.
Turner, T., Done, C., Mushotzky, R., Madejski, G., and Kunieda, H. 1992, *Ap. J.*, **391**, 102.

MULTIFREQUENCY VARIABILITY OF AGN'S

Continuum variations from the near IR to the X-rays

J. CLAVEL
ESA, ESTEC/SAI Postbus 299
2200-AG Noordwijk, The Netherlands.

1. Introduction

Because they emit copiously over more than 10 decades in frequency, Active Galactic Nuclei (AGN) cannot be understood without the help of multiwavelength observations. On the other hand, variability monitoring has also proven to be invaluable in understanding the continuum and line emission process as well as the geometry of the innermost regions in these objects. Indeed, at the heart of AGN's lies an object which is so compact that the only way to probe its structure is the study of the temporal evolution of its spectrum. The equivalent resolution which can be achieved in this way is of the order of 10 microarcsecs, far beyond the capability of any UV or optical telescope.

The techniques used are fairly straightforward in their principle, but extremely demanding in practice. The holly grail of variability studies are correlated variability between different emission or absorption components. Correlated variability strongly suggests and is often taken as a proof of a causal link between the components. For instance, if emission lines are found to vary together with the continuum, this suggests that the emission line gas is heated and photoionized by the continuum photons. Applying the usual cross-correlation techniques, it is straightforward to measure the the time shift between the two time series and therefore the distance between the corresponding emitting regions. For instance, if the emission line time series is found to be shifted by + 10 days *w.r.t.* that of the continuum, then the gas is likely to lie at a mean distance of the order of 10 lt-d from the continuum source. The variability monitoring techniques even allows us to distinguish between different types of geometry. For that purpose, one needs to reconstruct the Transfer Function (TF) of the correlated time series. The TF describes the response of an emission line to a δ-function continuum burst. Suppose for instance that the gas is concentrated in a disk seen face-on and whose inner and outer radii are respectively r_{min} and r_{max}. It is easy to show then that the TF will be zero everywhere except between $\tau = r_{min}/c$ and $\tau = r_{max}/c$. By contrast, the TF of a thick spherical shell has a flat maximum in the range $0 \leq \tau \leq 2r_{min}/c$ and decreases monotonically to zero at $2r_{max}/c$.

There are formidable logistical obstacles in applying the cross-correlation and TF reconstruction techniques. The fundamental problem is to obtain time series which are long compared to the average time scale of variability τ and with a sampling interval much shorter than τ. In practice, for a moderately luminous Seyfert I galaxy such as NGC 5548, one needs to measure its flux at every 4 days or more during at least 8 months. The amount of observing time necessary to carry-out such a programme rapidly becomes prohibitive. Even if the time is granted, one must still make sure that the observations are scheduled at a regular interval and that there are no gaps in the time series. A couple of weeks of bad weather can easily kill such a project. Moreover, one needs the source to be cooperative in the sense that its variation must be large compared to the typical flux measurement errors. Systematic errors which affect the photometric quality of the data - such as fluctuations in the transparency of the atmosphere or seeing variations which change the relative amount of dilution by stellar light - must also be minimized. These criteria calls for UV monitoring since (1) variations in the space ultraviolet are larger than at other wavelengths (2) one is essentially free from stellar contamination below 3000 Å (3) space observations are unhampered by the earth atmosphere and bad weather. Not surprisingly therefore, the most successful such programmes have been carried-out with the IUE satellite. Of course, the logistical obstacles become an order of magnitude worse when one tries to coordinate observations across different wavebands using different satellites and ground based facilities.

In this review, I will summarize those results obtained from the few multifrequency monitoring campaigns which have been carried-out so far. Because, these data yield information mainly about the continuum emission processes, I will leave aside the results obtained on the emission lines, except where they have a direct relevance to the question of the origin of the continuum. The emission line data have been reviewed recently in a recent and excellent article by Peterson (Peterson, 1993). I will concentrate on "normal" AGN's and shall not discuss "blazars" since these are the subject of another presentation at this conference ((Maraschi, 1993)).

2. UV & Optical variability: constraints on accretion disk models

Because the UV and optical variability data bear on accretion disk models, it is worth reviewing briefly the status of accretion disk models in the context of AGN's.

Since the pioneering works of Lynden-Bell (Lynden-Bell, 1969), Shakura & Sunyaev (Shakura & Sunyaev, 1973) and Shields (Shields, 1978), modeling the Spectral Energy Distribution (SED) of AGN's in terms of thermal emission from an Accretion Disk (AD) surrounding a massive Black-Hole has become a kind of industry. Geometrically thin optically thick AD model are generally successful at fitting the optical & Ultraviolet part of the SED, the so-called "big-bump" (see (Malkan, 1991) and references therein, and more recently (Rokaki et al, 1992)). Leaving aside the fact that this success is by no mean a proof of the existence of thin AD in AGN's, this model is plagued by fundamental problems which have been summarized in

(Courvoisier & Clavel, 1991), namely:
- There are now 4 high redshift QSO's for which the EUV SED is observed to rise sharply up to about 3 Ryd, making the disk supper-Eddington by nearly an order of magnitude (Reimers et al, 1991) (Jakobsen et al, 1993) (Lyons et al, 1993).
- In the LTE approximation at least, thin AD should show a strong discontinuity at the Lyman limit. Such a discontinuity has never been observed with a 15 % stringent upper limit (Antonucci Kinney & Ford, 1989) (Koratkar Kinney & Bohlin, 1992).

There are now several AGN's whose variations have been well sampled contemporaneously at optical ($\sim$ 5200 Å) and ultraviolet ($\sim$ 1400 Å) frequencies. These are NGC 5548 (Clavel et al, 1991) (Peterson et al, 1991), NGC 4151 (Clavel et al, 1990), NGC 3783 (Reichert et al, 1993), (Stirpe et al, 1993), 3C 273 (Courvoisier et al, 1990) and Fairall 9 (Clavel Wamsteker & Glass, 1989). Surprisingly, in all 5 AGN's the various UV and optical continuum band vary in phase with no measurable delay. In Fairall 9, the simultaneity of the variations extends up to 1.2 μ. This is clearly illustrated in Fig.1 & Fig.2. Fig.1 shows the light-curves of various continuum bands and emission lines in NGC 5548 together with the corresponding cross-correlations. The cross-correlations of the optical continuum (upper panel) and H_β emission line (middle panel) with the UV continuum of NGC 3783 are shown in Fig.2. In both sources, it is obvious that, in contrast to the emission lines where a clear lag is present, the optical continuum varies in phase with the UV flux. The upper limits on the delay of the optical flux are quite stringent, e.g. $\leq$ 2 days in NGC 5548 and $\leq$ 1 day in NGC 3783.

Such a finding is in complete contradiction with the thin disk model. In a thin AD, the effective radius of the optically emitting region is about 7 times larger than that of the UV-emitting layer. Since the two regions cannot communicate faster than the sound speed, it is easy to show that the optical should lag the UV by several years. This is true even if the disk is supported by the radiation pressure rather than the gas pressure, the opacity of the disk being so large that the photon diffusion time remains several orders of magnitude than the observed 1-2 days upper limits. In fact, these upper limits suggest that the UV and optical layer are synchronized by an external signal travelling at a speed which is more than 10 % the velocity of light.

3. Correlated UV & X-ray variability: evidence for thermal reprocessing

Further insight into that problem can be gained from simultaneous monitoring of the UV and X-ray flux. Only 4 sources have been the subject of such correlated studies: NGC 5548 (Clavel et al, 1992), NGC 4151 (Perola et al, 1986) and to a lesser extent, Fairall 9 (Morini et al, 1986) and NGC 4593 (Santos-Lleo et al, 1993). As can be seen in Fig.3, the 2–10 keV flux and 1350 Å continuum are clearly

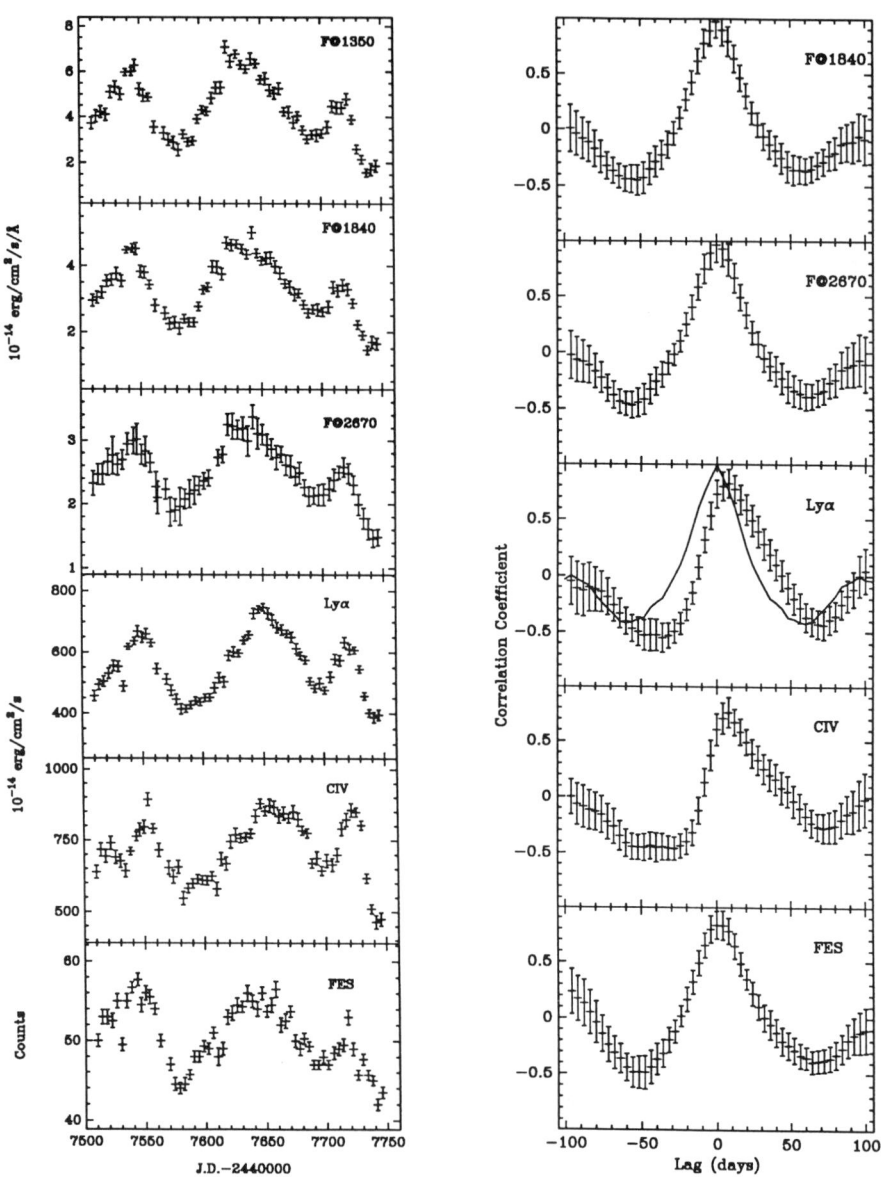

Fig. 1. Left: the light curves of various continuum bands at 1350 Å 1840 Å and 2670 Å) & 5200 Å ("FES") bands and emission lines (Lyα1216, CIV1549) in NGC5548. Right: the corresponding cross-correlations w.r.t. the 1350 Å flux

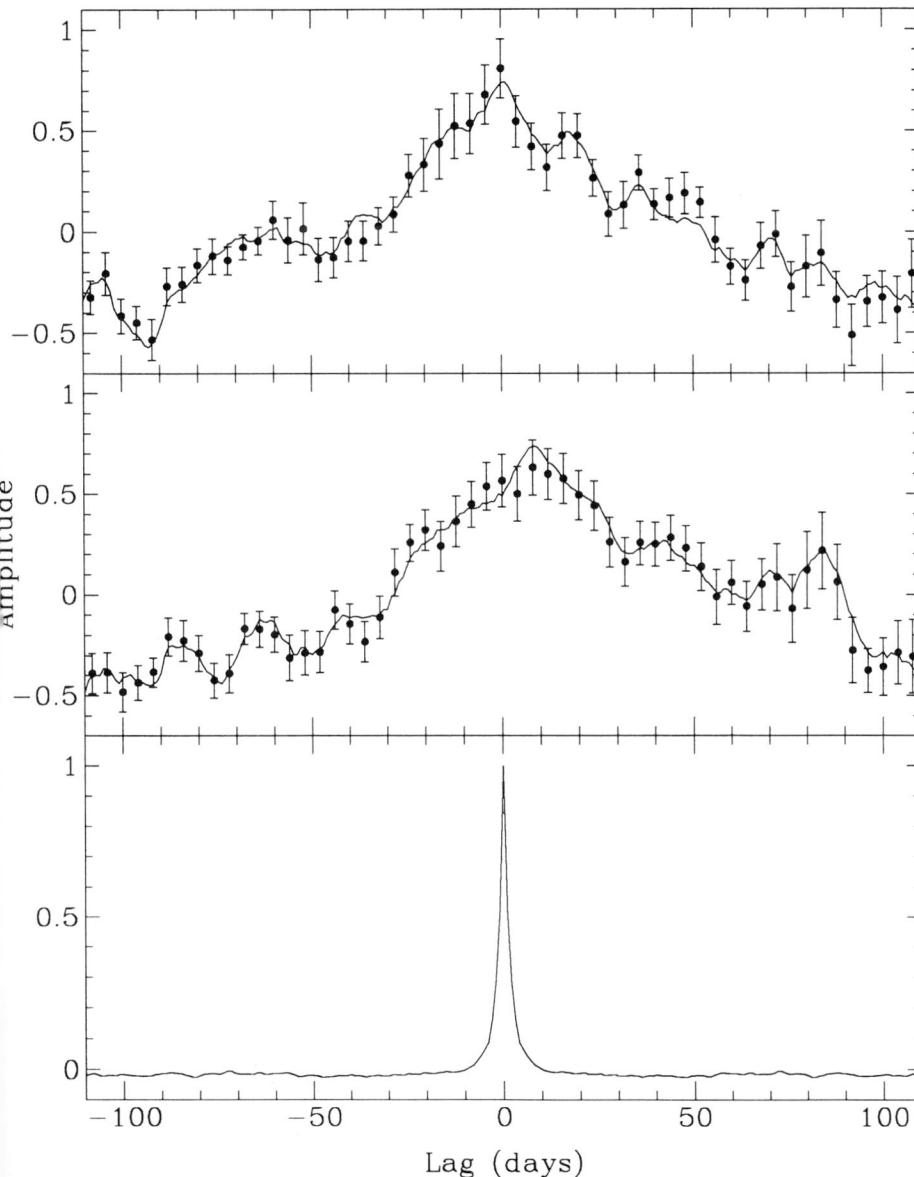

Fig. 2. The cross-correlations of the optical continuum (upper panel) and H_β emission line (middle panel) with the UV continuum in NGC 3783. The bottom panel shows the auto-correlation function of the sampling pattern.

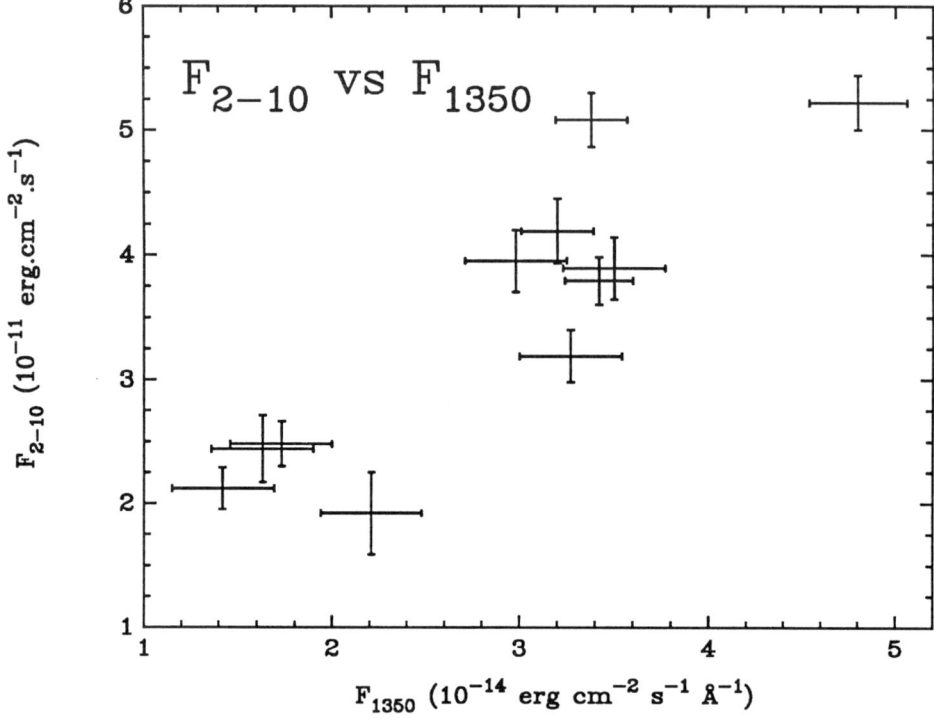

Fig. 3. The 2–10 keV flux as a function of the 1350 Å continuum in NGC 5548

correlated in NGC 5548, and the best-fit regression line goes through the origin, implying that the UV flux varies in direct proportion to the X-rays. Furthermore, a cross-correlation analysis shows that there is no measurable delay ($\Delta t \leq 6$ days) between the 2 bands. A correlation is also present in the 3 other sources.

To understand the meaning of this correlation, it is necessary to put it in the context of other observational results obtained in the X-ray regime. X-ray observations with the GINGA satellite (Pounds et al, 1990), (Nandra et al, 1991a) show that the presence of a strong Fe Kα emission line and a hard X-ray "hump" above 10 keV is a common property of Seyfert galaxies. Both features have been interpreted as evidence for reprocessing of the X-ray spectrum. In this scenario (Georges & Fabian, 1991), the X-rays illuminate a disk of cold gas. Ninety percent of the incident flux is absorbed, but a small fraction is compton scattered back ("reflected") to the observer, thereby creating a broad hump in the spectrum with a maximum around 20 keV. Part of the absorbed flux is re-emitted in the form of Fe Kα emission line photons. The reprocessing model makes one implicit prediction. The 90 % of the X-ray flux that is absorbed generates heat inside the disk. This energy must reappear in the form of thermal emission at frequencies which depend on the effective temperature of the reprocessing medium. Since the latter is of the order of a few 10^5 K (as can be inferred from the energy of the Kα line and iron edge), this is likely to take place in the ultraviolet and extreme ultraviolet region of the spectrum. Furthermore, given the high density of the gas, the absorption and re-emission are essentially instantaneous. Hence, the reprocessing model implicitly predicts that the variations of the X-ray and UV spectrum should be correlated. The correlation does not need to be perfect however. A perfect correlation would imply an implausibly fine tuning, with the light travel time of the X-ray photons matching very closely the characteristic time scale of variability of the X-ray source.

Hence, the correlation of the hard X-rays and UV flux observed in NGC 5548, NGC 4151, NGC 4593 and Fairall 9 can be understood as a consequence of the the X-ray absorption by the disk followed by thermal UV re-emission. Furthermore, the thermal re-emission will also occur at optical wavelength. Hence, the quasi-simultaneity of the UV and optical variability discussed previously can be readily explained with this model. In other words, all available data suggest that the variations of the entire UV-optical spectrum are driven by those of the hard X-rays.

The observed UV/X-ray correlation is far from perfect, as expected in this model. The fastest X-ray variations recorded in NGC 5548 correspond to $\sim$ 30 % fluctuations of the flux in $\sim$ 5 hours (Nandra et al, 1991b). Such rapid variations have never been observed in the ultraviolet, the PDS of the UV light-curve being steep with little or no power on time scales shorter than 1 day (Clavel et al, 1991). In the thermal reprocessing scenario, this implies that the travel time of the X-ray photons is at least of the order of a few times 5 hours, say 1 day. If we combine this lower limit with the observed 6 days upper limit on the delay of the UV flux and translate them into distances, we obtain:

$$2.5\ 10^{15} \leq D \leq 1.5\ 10^{16} \text{cm}$$

where D is the average distance between the X-ray source and that part of the disk which does the bulk of the thermal reprocessing. How does this compare with the dimension expected from the accretion disk in NGC 5548 ? One can combine the BLR effective radius inferred from the delay in the response of the emission lines to the continuum variations together with the mean velocity of the gas measured from the line width to estimate the mass of the black-hole in NGC 5548 (Clavel et al, 1991). Under the assumption that the gas is gravitationally bound, one obtains M = $3.7\ 10^7$ M$_\odot$. This in turns implies that the thermal reprocessing region lies between 225 and 1350 Schwarzschild radii from the X-ray source. This is one to two orders of magnitude larger than the theoretical size of the 1350 Å emitting region on a standard α disk without reprocessing.

An independent way of estimating D comes from the energy distribution itself. Indeed, the fact that the UV varies in strict proportion to the X-rays is consistent with the hypothesis that the entire ultraviolet flux – at least at the time of the monitoring campaigns in 1989 and 1990 – originates from reprocessing of the X-ray photons absorbed in the disk. From the variability time scale, we know that the X-ray source is much more compact than the reprocessing region. The Equivalent Width EW of the Fe Kα line also requires that the disk covers essentially 2 π as seen from the X-ray source. In addition, there is no hint of the strong X-ray absorption at low energy that would be expected if the X-ray source was lying behind or was partially embedded in the reprocessing medium. All these observables calls for a special geometry where the X-ray source lies above the disk, perhaps at the base of the radio jet. The energy distribution of the thermally reprocessed radiation then depends on D, the distance of the X-ray source to the reprocessing medium. This distance can be expressed as a function of the height H of the X-ray source above the disk and the mean effective radius R of the reprocessing ring in the disk, $D = \sqrt{H^2 + R^2}$. For illustrative purpose, let us consider the two extreme cases. In the first case, $H \gg R$, i.e the height of the X-ray source above the disk is much larger than the dimension of the disk itself, at least up to its UV and optical emitting layers. The heating rate per unit surface area of the disk is diluted by the usual D^{-2} factor and it is then easy to show that the effective temperature goes as $T_{eff} \propto R^{-1/2}$ instead of the usual $R^{-3/4}$ dependency in a disk without reprocessing. The emerging spectrum has a frequency spectral index $\alpha = -1$ ($F_\nu \propto \nu^\alpha$). The opposite case is where $H \leq R$. In that case, the above dilution factor is multiplied by a $\cos\theta$ term, where $\theta = \arctan\frac{H}{R}$ is the complement to the angle of incidence of the X-rays onto the disk. One can easily show that in this case, the heating rate recovers its usual $R^{-3/4}$ dependency and the emerging spectrum is the same as that of a disk without external heating i.e. $F_\nu \propto \nu^{1/3}$. The fact that the spectral index of NGC 5548 varies around an average value of -0.7 suggests that the actual situation is intermediate between those two extreme cases, but is probably closer to case (1) than case (2), i.e. $H \geq R$. Detailed calculation of the emergent spectrum from an externally heated disk (Rokaki et al, 1992) show that that the best fit to the NGC 5548 Ultraviolet energy distribution is obtained for 250 $R_s \leq H \leq$ 1000 R_s. It is reassuring that these limits are entirely consistent with the limits on H derived from variability argument.

To make further progress, it will be extremely important to restrict the range of allowed values for H. This can only be achieved with simultaneous X-ray & UV monitoring campaigns that have a much better temporal resolution than previous campaigns. This is precisely the goal of the upcoming observing programme of NGC 4151 which the AGN watch collaboration will undertake in November-December 1993. This AGN will be observed with IUE and CGRO continuously during 10 days, while in the same period ROSAT and ASCA will take repeated measurements of its X and γ ray spectrum. ROSAT will observe every 12 hours. This will hopefully be sufficient to pin down the delay in the response of the UV flux to the X-ray fluctuations and therefore the height of the X-ray source above the disk.

The thermal reprocessing model also seem to solve the problems which plague the thin AD model. X-ray heating from above reduces the temperature gradient in the upper layers of the disk and may therefore weaken or suppress the Lyman Absorption edge. Adding externally supplied energy means that the energy distribution of the "big-bump" can be accounted for with a smaller black-hole mass than in disk models without reprocessing (Malkan, 1991). A smaller central mass allows the disk to be hotter so that its spectrum may extend up to 3 Ryd – as observed in a few high z QSO's – without becoming supper-Eddington. As a matter of fact, if the X-rays are generated outside the accretion flow – in a jet or wind, for instance – their luminosity (and therefore that of the heated disk) is not constrained by the Eddington limit.

The thermal reprocessing model faces a serious energetic problem, however. During the 1989 and 1990 campaigns, there was just enough X-ray flux in NGC 5548 over the 2–100 keV range to power its UV spectrum (Clavel et al, 1992). Since CGRO and SIGMA observations consistently show that the hard X-ray spectrum of Seyfert I galaxies bends steeply above 100 keV (Jourdain et al, 1992); (Cameron et al, 1993); (Maisack et al, 1993), one does not solve the problem by extending the integration range. In 1989 and 1990, the UV spectrum of NGC 5548 was in a medium to low state and the correlation between the X-ray and UV flux holds with the same slope over all time scales from 6 days up to $\sim$ one year. However, on May 21, 1984 when the UV flux reached an historical maximum about 3.5 times higher than the 1990 average, the X-rays as measured by EXOSAT were only 70 % stronger, not enough to account for the energy in the "big-bump". The same situation applies to NGC 4151 (Perola et al, 1986): the X-ray/UV correlation holds with the same slope over time scales from a few days to $\sim$ one year in 1983 and 1984 when the source was moderately bright, but breaks down during a very large UV outburst in May 1979 when the X-ray flux recorded by EINSTEIN was at about the same level as in 1983. In other words, in both NGC 5548 and NGC 4151, the correlation saturates during large outbursts of the ultraviolet flux in such a way that there is a deficiency of X-rays to account for the energy in the "big-bump". This may indicate that the thermal reprocessing model only applies to moderate amplitude (factor of $\sim$ 4) variations of the ultraviolet flux and that another explanation has be found for very large UV bursts. The May 1984 outburst of NGC 5548 was unusual in that it took several months to develop. Such a time scale is compatible with the burst originating from thermal instability in the disk (Clavel et al, 1992).

For QSO's, however, the problem of the energetic becomes overwhelming. This is because statistical studies of large samples show that the X-ray luminosity increases less than linearly with the optical-UV luminosity, i.e. $L_x \propto L_{opt}^{0.8}$ (Avni & Tananbaum, 1986). In other words, Quasars are X-ray deficient so that, in the reprocessing scheme, they lack the amount of high energy radiation necessary to power their prominent ultraviolet "big-bump". Interestingly enough, Quasars do not seem to show much evidence for reprocessing – such as the Fe Kα line or the compton reflection "hump" –in their X-ray spectra either (Williams et al, 1992). It is also worth recalling that the pattern of variability of the UV-optical spectrum of quasars is different from that of the lower luminosity Seyfert I galaxies, the amplitude of their variations being relatively small and the time scales long (Maoz et al, 1993); (Kinney et al, 1991). This suggests that the "big-bump" of QSO's originates from viscous dissipation inside the disk, in contrast to Seyfert's where it is driven by X-ray heating. It is of course not satisfactory to invoke different emission mechanisms in such two so phenomenologically similar classes of objects, the more so because one is now left with no explanation for the absence of a Lyman discontinuity and the super-Eddington SED in quasars. Alternatively, it is possible to solve the energy budget problem if the X-ray emission is not isotropic in such a way that the disk receives more flux than the observer. A partially reflecting screen of hot electrons (i.e. the "warm absorber") could do the job of scattering the X-ray photons back onto the disk. One is then left with the problem of explaining why this screen should be more efficient in high luminosity quasars than in Seyfert's. Another possibility is the completely different model proposed by Ferland where the "big-bump" arises from optically thin free-free emission in a hot plasma heated by X-rays (Ferland Korista & Peterson, 199o) (Barvainis, 1993).

4. Ultraviolet & IR monitoring: mapping the molecular Torus

The ultraviolet (1200–3200 Å) optical and near IR (JHKL) spectrum of the high luminosity Seyfert I galaxy Fairall 9 has been extensively monitored from 1978 up to 1986 (Clavel Wamsteker & Glass, 1989). Dramatic variability was observed, the 1340 Å flux decreasing by a factor $\sim$ 30 in a few years. Though the variations were milder in the optical and in the J Band, they nevertheless tracked closely those of the UV continuum, as already mentioned in the previous section. However, the IR flux at wavelengths longer than 1.2 μ lags behind the Ultraviolet. Moreover, as can be seen from the cross-correlations of Fig.4, the delay increases systematically with wavelength, from 250 days at 1.6 μ, to 385 days at 2.1 μ and 410 days at 3.4 μ. Also, the amplitude of the near IR variations peak at 2.1 μ.

Such a pattern is precisely that expected if the near IR radiation originates from thermal re-emission by dust grains heated by the ultraviolet photons (Clavel Wamsteker & Glass, 1989). The observed delay implies that the dust shell lies at 350 $\pm$ 100 lt-d from the UV source. Using the observed mean luminosity of Fairall 9, one can compute the equilibrium temperature of the grains at such a distance: T_{dust} = 1730 $\pm$ 230 K. This is in excellent agreement with the observed colour temperature of the variable IR component and the fact that it peaks at 2.1 μ. Within the theoretical and observational errors, such a temperature is equal to the sublimation temperature of graphite, the most refractory of all possible dust constituents (Rudy & Puetter, 1982). Detail calculations of a dust model

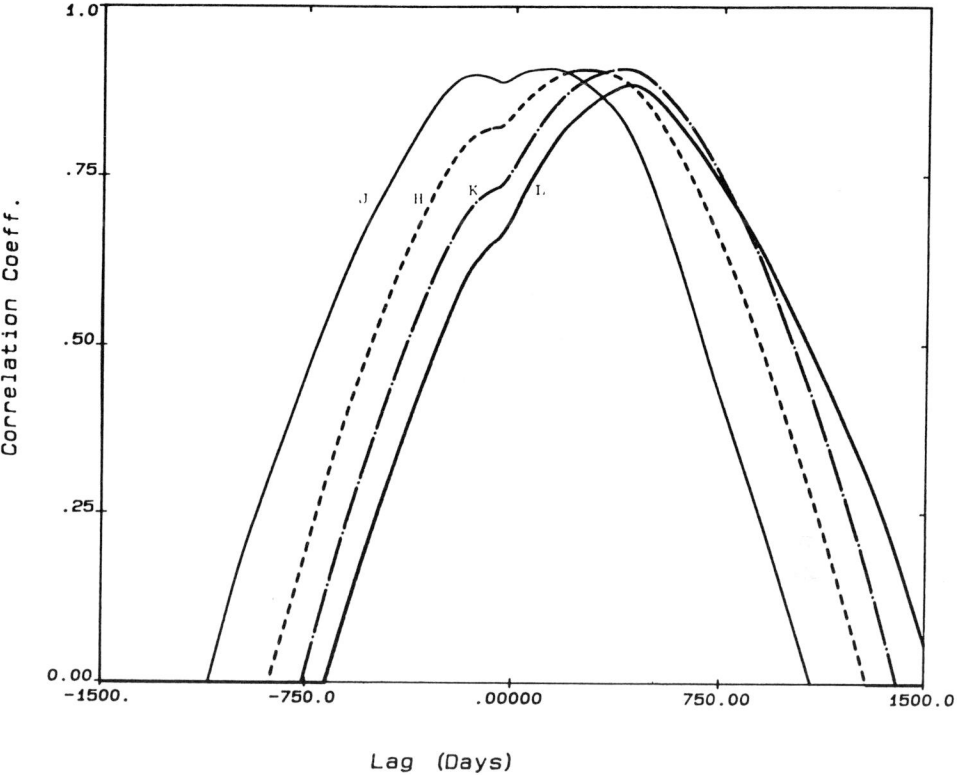

Fig. 4. The cross-correlation of the IR flux in the J H K & L band with the 1338 Å continuum in Fairall 9. The lag clearly increases with increasing wavelength, from 0 at 1.2 μ to 410 days at 3.4 μ.

applicable to the Fairall 9 data confirm and extend the conclusions of Clavel and collaborators (Barvainis, 1992).

The Lyα1216, CIV1549 and MgII2798 emission line intensities in Fairall 9 also varied with a large amplitude (factor $\sim$ 10), and their variations closely tracked those of the UV continuum with a delay of 155 $\pm$ 45 days (Clavel Wamsteker & Glass, 1989). This suggests an effective radius of the order of half a light year for the BLR in Fairall 9, a factor 2 smaller than that of the dust shell. Since the dust shell is outside the BLR but well inside the much more extended Narrow Line Region (NLR), it is tempting to postulate that it lies at the inner edge of the putative molecular Torus which is thought to surround the BLR in AGN's. Moreover, the fact that the dust temperature is equal to that at which graphite grains sublimate further suggests that the inner edge of the Torus is being evaporated by the intense nuclear radiation field, as postulated by theoretical models of the Torus (Krolik & Begelman, 1986). In these models, the evaporation gives rise to an outflowing wind which creates a high latitude screen of hot electrons. This screen acts as a mirror which reflects the nuclear spectrum back to the observer. In Seyfert II's, the Torus is viewed edge-on, the nucleus is obscured and one only sees the reflected nuclear spectrum in polarized light (Antonnuci & Miller, 1985). In Seyfert I's, one has a direct unhindered view of the nucleus and the Torus is otherwise undetectable. Fairall 9 may be the first example of a Seyfert I galaxy for which the Torus leaves a clear imprint on its spectrum. AGN's such NGC 4151 (Bromage et al, 1985) and OI 287 (Antonnuci Kinney & Hurt, 1993), with their blue shifted absorption lines and relatively large polarization, may represent intermediate objects where one sees the nucleus through the base of the wind.

There exist a few other AGN's for which near IR variability data - though not as compelling as those of Fairall 9 - strongly suggest the existence of a dust shell irradiated by the UV source. The best such case is NGC 3783 in which a correlation between the U and K band yields an effective radius of 80 lt-d for the dust shell (Glass, 1992). Note that NGC 3783 has a much lower luminosity than Fairall 9 so that one expects a smaller evaporation radius, as is observed. Another good candidate is GQ Com, a fairly luminous Seyfert I galaxy, in which the IR flux is found to lag behind the UV by $\sim$ 250 days (Sitko, 1991). The last object is the low luminosity Seyfert I galaxy NGC 1566, where the time scale of variability of the near IR flux suggests a radius of about 50 lt-d (Baribaud et al, 1992).

There is currently no direct evidence for the existence of the molecular Torus. Molecular lines and bands have been observed in the spectra of a several AGN's but it is impossible to decide whether they originate from the Torus or from the Interstellar medium (ISM) of the galaxy. A decisive proof would be the detection of variations in the intensity of those molecular features on time scales of a few months to one year. Unlike the ISM, the gas in the Torus is expected to be relatively warm with temperatures in the range 500–1000 K. These rather unusual conditions should favor the emergence of selected spectral features such as the H_2 mid-J rotational lines in the 3–10 μ range or the CO mid-J rotational band at 4.6 μ. Other good candidates are the Polycyclic Aromatic Hydrocarbon (PAH) bands at 3.3, 6.2, 7.7, 8.6 and 11.3 μ. By monitoring the temporal evolution of these features and using the reverberation mapping techniques described here, it may be possible to study in detail the spatial extent and the physical conditions of the gas inside the molecular Torus. Such programmes will be extremely challenging but are within the reach of the upcoming ISO satellite.

References

Antonucci R.R.J. & Miller J.S. 1985: *Ap.J.***297**, 621
Antonucci R.R.J., Kinney A.L. & Ford H.C.: 1989, *Ap.J.***342**, 64.
Antonucci R.R.J. Kinney A.L. & Hurt T. 1993: *Ap.J.***414**, 506.
Avni Y. & Tananbaum H. 1986: *Ap.J.***305**, 83.
Baribaud T. et al 1992: *A & A***256**, 375.
Barvainis R. 1992: *Ap.J.***400**, 502.
Barvainis R. 1993: *Ap.J.***412**, 513.
Bromage G. et al 1985: *M.N.R.A.S.***215**, 1.
Cameron R.A., et al 1993: in *"Compton Gamma Ray Observatory"*, *AIP Proceedings 280*, ed. M. Friedlander, N. Gehrels, & D.J. Macomb (New-York: AIP), 559.
Clavel J., Wamsteker W. & Glass I. 1989: *Ap.J.***337**, 236.
Clavel J., et al 1990: *M.N.R.A.S.***246**, 668.
Clavel J., et al 1991: *Ap.J.***366**, 64.
Clavel J., et al 1990: *Ap.J.***393**, 113.
Courvoisier T.J.-L., et al 1990: *A & A***234**, 73.
Courvoisier T.J.-L. & Clavel J.: 1991, *A & A***248**, 389.
Ferland G.J., Korista K.T. & Peterson B.M. 1990: *Ap.J.***363**, L21
Georges I.M., Fabian A.C. 1990: *M.N.R.A.S.***249**, 352.
Glass I.S. 1992: *M.N.R.A.S.***256**, 23P.
Jakobsen P. et al 1993: *Ap.J.*, in press.
Jourdain E., et al 1992: *A & A.***256**, L38.
Kinney A.L. et al 1991: *Ap.J.Suppl.***75**, 645.
Koratkar A.P., Kinney A.L. & Bohlin R.C.: 1992, *Ap.J.***400**, 435.
Krolik J.H. & Begelman M.C. 1986: *Ap.J.***308**, L55.
Lyons R.W. et al 1993: *Ap.J.*, in press.
Lynden-Bell J: 1969, *Nature***223**, 690.
Maisack M., et al 1993: in *"Compton Gamma Ray Observatory"*, *AIP Proceedings 280*, ed. M. Friedlander, N. Gehrels, & D.J. Macomb (New-York: AIP), 493.
Malkan M.A.: 1991, in *Structure and properties of Accretion Disk*, ed. C. Bertout et al (Paris: Editions Frontieres), p. 165.
Maoz D., et al 1993: *Ap.J.*, in press.
Maraschi L.: 1993, *These Proceedings*.
Morini M., et al 1986: *Ap.J.***307**, 486.
Nandra K., et al 1991a: in *Iron Line Diagnostics in X-ray sources*, ed. A. Treves, G.C. Perola & L. Stella (Berlin: Springer), 177.
Nandra K., et al 1991b: *M.N.R.A.S.***248**, 760.
Perola G.C., et al 1986: *Ap.J.***306**, 508.
Peterson B.M. et al 1991: *Ap.J.***368**, 119.
Peterson B.M.: 1993, *P.A.S.P.***105**, 247.
Pounds K., et al 1990: *Nature***344**, 132.
Reichert G.A., et al 1993: *Ap.J.*, in press.
Reimers D. et al 1989: *A & A***1989**, 218.
Rokaki E., Boisson C.,Collin S.: 1992, *A & A***253**, 57.
Rudy R.J. & Puetter R.C. 1982: *Ap.J.***263**, 43.
Santos-Lleo M., et al 1993: *M.N.R.A.S.*, in press.
Shakura N.I, and Sunyaev R.A.: 1973, *A & A***24**, 337.
Shields G.A.: 1978, *Nature*,**272**, 706.
Sitko M. 1991: in *Variability of Active Galactic Nuclei*, ed. H.R. Miller & P.J. Wiita (Cambridge: Cambridge University Press), p.104
Stirpe G.M., et al 1993: *Ap.J.*, in press.
Williams O.R., et al 1992: *Ap.J.***389**, 157.

VARIABILITY IN THE MILLIMETER AND RADIO DOMAINS

ESKO VALTAOJA
Metsähovi Radio Research Station, Helsinki University of Technology, SF-02540 Kylmälä, Finland

1. Introduction

The radio domain spans only a small fraction of the total electromagnetic spectrum, and is for most active galaxies energetically insignificant in comparison with higher frequencies.However, only in the radio regime can we obtain continuous fluxcurves and be certain that we are not missing anything. We also believe we know where the radiation comes from, and by which mechanisms it is produced. Furthermore, since VLBI generally cannot resolve the smallest core components, multifrequency continuum monitoring remains our only way to study the innermost radio cores of AGN. Finally, continuum monitoring is cheap (although time-consuming) in comparison with most other methods, and therefore we can observe many more sources and with far better time coverage.

The bulk of high radio frequency data comes from two dedicated monitoring programs. At the University of Michigan, 60-200 sources have been observed regulary at 4.8, 8 and 14.5 GHz since 1965 (Aller et al 1985). At the Metsähovi Radio Research Station, some 50-100 sources are being monitored mainly at 22 and 37 GHz, with over 20 000 flux measurements obtained since 1980 (Teräsranta et al. 1992). Both programs are now providing well sampled fluxcurves of the brightest compact northern and equitorial radio sources.

At still higher frequencies observations become increasingly difficult, and there are no similarly dedicated telescopes. We have used the SEST telescope in Chile since 1987 for persistent monitoring of 20-50 mainly equatorial sources at 90 and 230 GHz (Tornikoski et al. 1993). At IRAM, a number of bright quasars are used for pointing checks (Steppe et al. 1992) at 90, 150 and 230 GHz. Finally, the JCMT at Hawaii has been used for mm/sub-mm monitoring of about 20 AGN (Gear et al. 1993). There are also occasional observations with other millimeter telescopes. However, even with combined data from all these the variations at 90 GHz or higher are woefully undersampled.

2. Questions for the radio domain

Radio observations are an essential part of multifrequency studies. But what questions can we attempt to solve using only radio data? We want to know where the radiation comes from. We want to know what mechanisms are producing it, and whether they are similar in all AGN. For the mechanisms, the paradigm to be tested is the shocked relativistic jet scenario. For the differences and similarities we are starting to look increasingly hopefully at unified models of AGN.

2.1. ORIGIN OF THE RADIO FLUX

Essentially all radio flux and variations seem to originate from just three different synchrotron components. Below 10 GHz, most of the radiation comes from the arcsecond-to-mas components also visible in VLBI maps, with only secular changes in flux. Between 10-100 GHz the spectrum becomes dominated by radiation from the inhomogeneous relativistic jet (also called the core jet, the mas jet, or just the core), which appears to have long-term stability (e.g., Valtaoja et al. 1988, Lainela 1993). The flux variations on timescales from a few weeks to a few years are caused by evolving shocks in this relativistic flow. In most VLBI maps the jet and the newest shock appear as a single point source, the VLBI "core".

Is there room for other, as yet undiscovered, major components? The answer seems to be no. The quiescent high-frequency spectra above the turnover associated with the jet are steep and smooth (Valtaoja et al. 1988, Brown et al. 1989; Tornikoski et al. 1993; Gear et al. 1993), indicating that practically all the high frequency flux comes from the optically thin synchrotron jet.

2.2. DATA VS. SHOCK MODELS

The two main models for radio spectra and variations are the numerical code by Hughes et al. (1989), developed to explain the cm-wavelength flux and polarization variations observed in the Michigan monitoring, and the analytical model by Marscher and Gear (1985), which had its impetus in the observed high frequency behavior of 3C 273. While both models assume that the cause of flux variations is the growth and decay of shocks in the jet, the Hughes et al. model is mainly applicable to the cm-wavelength variations, and the MG model for shorter wavelengths.

The simple shock models predict three things. Since the flux variations are assumed to come from a new, and in the first approximation homogeneous, component, the variable component (i.e., the shock) should have a simple, homogeneous synchrotron spectrum with $\alpha_{thin} = 2.5$. The time evolution of the spectral turnover $\nu_m(t)$, $S_m(t)$ is model-dependent, but should show growth and decay stages. Finally, the observed maximum amplitude of the outburst, ΔS and the time delay Δt should both have a definite (although model-dependent) dependence on ν.

2.3. THE SHAPE AND EVOLUTION OF THE SHOCK SPECTRUM

The first prediction has been tested by Valtaoja et al. (1988) and Lainela (1993), who used Metsähovi, Michigan and other data to construct quiescent and outburst epoch cm-to-mm spectra for a sample of sources. Subtracting the quiescent (i.e., the jet) spectra from the outburst spectra they found that within observational accuracy the shock does, indeed, have $\alpha_{thick} \approx 2.5$ and $\alpha_{thin} \approx -0.2$, indicating that the electron energy index $\gamma \approx 1.4$.

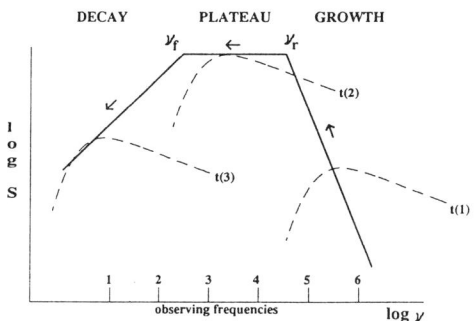

Fig. 1. Evolution of the shock spectrum in the generalized shock model. (Valtaoja et al. 1992a).

It is difficult to follow the time evolution of the turnover. Valtaoja et al. (1988) and Lainela (1993) combined essentially all available radio data in trying to follow the spectral evolution in a number of outbursts. The somewhat disappointing conclusion, also reached by Robson et al. (1993) in the case of the exceptionally well-monitored 3C 273, was that the data agreed with shock models, but were not sufficient to validate them. On the other hand, no clear counterexamples, such as lower frequency variations preceding higher ones, were found.

2.4. FLARE AMPLITUDES AND TIME DELAYS

The maximum amplitudes of the outbursts and the delays between the times of maximum at different frequencies are much easier to observe, since they do not require simultaneous multifrequency data.

We have constructed a simple 'generalized' shock model for comparing the shock framework with observations (Valtaoja et al. 1992a). In this model, depicted in Figure 1, we have a simple homogeneous spectrum evolving in time through parametrized growth, plateau and decay stages. The evolutionary track (S_m, ν_m, t) must of course be obtained from specific models, but even without such models the predicted main features are readily seen. If the observing frequency is lower than the frequency of the plateau, one observes the adiabatic decay stages of the shock, with both ΔS and Δt strongly dependent on ν. If the observing frequency is 'high', the fluxcurves track closely each other, $\Delta t \approx 0$, and ΔS depends only weakly (as $\nu^{-\alpha}$) on frequency. Thus, for each outburst there is a critical observing frequency, below and above which two very different types of dependence are seen (Figure 2a).

Figure 2b shows the results from the large Metsähovi and Michigan data sets. The $\Delta S(\nu)$- and $\Delta t(\nu)$-dependences are in accordance with the shock framework. In particular, from the observed high frequency $\Delta S(\nu)$-dependence we recover the value $\alpha_{\text{thin}} \approx -0.2$, independently derived from the shape of the shock spectrum.

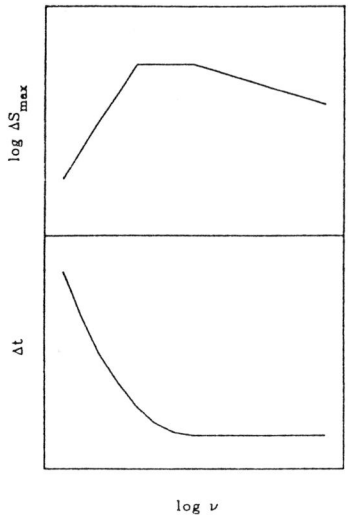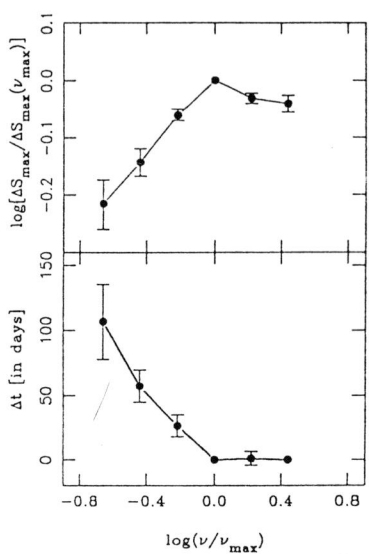

Fig. 2. The predicted and the observed (in a sample of 42 sources) dependence of burst amplitude and time delay on frequency (Lainela 1993).

2.5. CONCLUSIONS FROM THE SPECTRAL AND VARIABILITY ANALYSIS

The shocked jet scenario appears to be capable of explaining all the major radio variations in radio-loud AGN, whether they are radio galaxies, quasars, or BL Lac objects. In our sample of over 50 brightest sources we have found no clear evidence for other mechanisms, such as gravitational microlensing. Alternative mechanisms should stand out either in the detailed behavior of a single source, or in the statistical analysis of samples of sources; we have found neither.

The bad news is that the data are not sufficient to constrain shock models very strongly. It is perhaps not unfair to say that the observations are all *compatible* with the shocked jet scenario, but it would be difficult to single out any observational features which *prove* it to be valid. What is needed, among other things, are fully sampled mm-to-submm multifrequency fluxcurves, mm-polarization data, and improved theoretical models especially for the crucial growth stages of shocks.

3. Variability, spectra and classification

We can attach three separate sets of data to each radio-loud AGN: its classification (or rather, an overabundance of various possible classifications), its spectral characteristics (various spectral indices, turnover frequencies etc.), and its variability characteristics (maximum amplitude of variations, χ^2-related parameters, timescales etc.). Using as complete and large data sets as possible, we may study the relations between these data sets.

An increasing number of such analyses have been published during the last few years (e.g., Valtaoja et al. 1988, Eckart et al. 1989, Ghosh and Gopal-Krishna 1990, Wiren et al 1992, Hughes et al. 1992, Aller et al. 1992, Valtaoja et al 1992b, Lainela and Valtaoja, 1993, Lainela 1993, Gear et al 1993, Teräsranta and Valtaoja 1993). By and large, the conclusions of these studies agree with each other, and conflicting results probably stem from the use of noncomplete samples. I will try to give a short summary of these results before considering their interpretation.

3.1. THE PROPER CLASSIFICATION OF RADIO-LOUD AGN

An impressive, if not outright depressive, number of AGN classifications have been proposed: steep/flat spectrum, VLBI compactness, core dominance, blazar/non-blazar, super/subluminal, OVV/ordinary, ad infinitum. The first task is to find out which divisions are reasonable, in the sense that they are based on intrinsic, persistent properties of the sources, and which divisions are fundamental, in the sense that they do not obviously derive from some other source properties. Much unnecessary confusion has been caused by neglecting to consider these things.

For example, it is common to divide radio-loud AGN into steep- and flat-spectrum sources, usually using decade(s)-old catalogue data of spectral indices around a few GHz and an arbitrary dividing line $\alpha = -0.5$. It would be rather surprising if such a division would properly reflect some fundamental property of the sources, and indeed it does not, despite its widespread and continuous use.

It is not difficult to see why spectral steepness is not a fundamental division. Whatever the frequency range, the spectral index is obviously variable, and a 'steep-spectrum' source may change into a 'flat-spectrum' one, or vice versa, in a few years (e.g., Zhang et al. 1993). The low radio frequency radiation comes to a large extent from old, extended components with at best tenuous connections to the actual core regions. It is true that flat-spectrum sources are more variable than steep-spectrum ones (to take just one example), but this is simply due to the well-known fact that radio galaxies have on the average steeper spectra than quasars, and a flat-spectrum sample thus includes a larger fraction of quasars than a steep-spectrum one. Thus, the proper conclusion is that quasars are different (more variable, flatter spectra) from radio galaxies. The division into flat- and steep-spectrum sources does not add anything except confusion; one might as well mix apples and oranges in a basket, and then try to find out their average taste.

Similar criticism applies to most other groupings; I will mention only one more. If, as several lines of evidence indicate, the intrinsic Lorentz factors the flows in AGN are of the order 5-10, a simple calculation shows that even sources close to the plane of sky are superluminal. However, in such sources the shocks are Doppler deboosted, and probably below the detection threshold of VLBI. Thus, 'superluminal' is not a fundamental property of a source, but rather an observational indication that the shocks are bright enough to be detectable.

What then *are* the relevant classifications? In analyzing our data, we have tried all the above-mentioned (and quite a number of other) classifications. The conclusion, from several independent, large data sets, is the following:

All the statistically significant differences and correlations between the radio properties of AGN are best explained by grouping the sources into radio galaxies, 'ordinary' quasars with low optical polarization (LPQ, $p < 3\%$), and 'blazar-type' quasars with high optical polarization (HPQ, $p > 3\%$). BL Lacs form at best a marginally identifiable class of objects.

One might object that optical polarization is not any more observationally robust than the other indicators I have lambasted. While there is some justification in this criticism (e.g., the optical polarization often is highly variable, and some sources have high polarization produced by scattering), there is the well-known and persistent dichotomy between LPQs and HPQs, and sources once caught exhibiting $p > 3\%$ remain in the HPQ category forever (and usually behave like it).

3.2. INFLUENCE OF PROPER CLASSIFICATION

In order to convince the unrepentant, I give two random examples of how the choice of classification can crucially affect the conclusions. Aller et al. (1992) have studied the spectral properties of the Pearson-Readhead sample of radio galaxies, QSOs and BL Lacs. Using this division, they concluded that BL Lacs are different from quasars. However, by dividing the QSOs into LPQs and HPQs one finds that BL Lacs are different from ordinary quasars but similar to HPQs, leaving open the question whether they form a separate class of objects.

Gear et al. 1993) compare the mm-spectra of BL Lacs and flat-spectrum quasars, finding that they form two different classes. Again, using the LPQ/HPQ division and repeating their analysis, one finds the proper conclusion to be that BL Lacs are different from ordinary quasars, with the question of association between BL Lacs and HPQs left open.

3.3. PROPERTIES OF RADIO GALAXIES, LPQs, HPQs AND BL LACS

Taking a grand average of the published analyses, the radio properties of these classes I have claimed to be fundamental can be shortly summarized as follows. For details, the reader is referred to the original papers listed in the beginning of this section; some examples are shown in Figures 3 and 4.

Highly polarized quasars have flatter spectra, are more variable, have shorter characteristic timescales and higher brightness temperatures than *low polarization quasars* (ordinary quasars). For all quasars, the spectral flatness and the amount of variability are positively correlated. *Radio galaxies* have the steepest spectra, are the least variable, have the longest characteristic timescales and the lowest brightness temperatures associated with outbursts.

BL Lacs possibly have flatter spectra than quasars, but as a class they are not, repeat *not*, more variable than quasars, nor do they have shorter timescales or higher brightness temperatures than quasars. Many BL Lacs are among the *least* active and variable of all bright compact radio sources. In general, BL Lacs do not

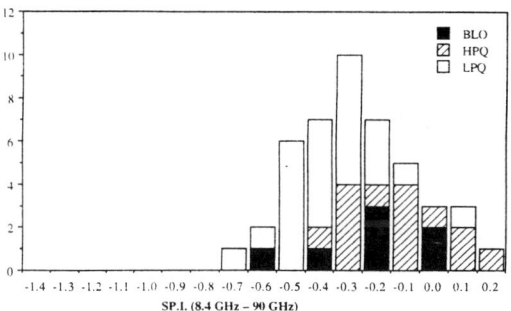

Fig. 3. Cm-to-mm spectral index in a complete sample of AGN (Tornikoski et al. 1993).

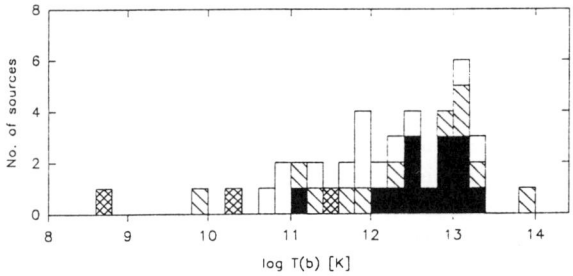

Fig. 4. The highest observed brightness temperature in a sample of HPQs (black), LPQs (white), BL Lacs (hatched) and radio galaxies (crossharched) (Teräsranta and Valtaoja 1993).

form a class clearly identifiable by their radio properties.

4. Unified models?

Is there a simple way to explain these results? Since one does know that the AGN will have different orientations relative to the line of sight, a natural way to proceed is to consider whether at least a part of the observations can be explained if orientation is assumed to be the decisive, fundamental underlying property. Of course, such an approach is highly simplistic - we do also know that sources have different intrinsic luminosities, different ages, different environments, etc. - but one may hope that in large enough samples these other factors will average out.

4.1. Spectra, variability, and orientation

From the basic premise of radiation coming from a directed relativistic outflow and shocks immediately follow a number of testable predictions. As the viewing angle decreases, the Doppler boosting increases, the apparent luminosity increases, and the timescales are compressed, with the net result that the amplitude and the rapidity of variations increases.

With simple qualitative models (Eckart et al. 1989; Valtaoja et al. 1992a) the unified model predictions can be made more explicit. If the low-frequency radiation comes mainly from the unboosted extended component(s) and the high-frequency radiation from the boosted jet and shocks, the observed source parameters will depend on the boosting factor D (i.e., the viewing angle) roughly as follows: the last turnover (=jet turnover) as D, the timescales as D^{-1}, the brightness temperatures as D^3 and so on. The low-to-high radio frequency spectral index and the different variability indices will also depend on D in a more model-specific manner.

If we now, following Barthel (1989), identify the sources with largest viewing angles (least boosted) with radio galaxies, the intermediate sources as ordinary low polarization quasars, and the sources with smallest viewing angles as 'blazar'-type HPQs, both the differences between these classes and the correlations between, e.g., spectral flatness and variability, are explained very satisfactorily.

The fact that optical polarization is a good tracer of orientation is also easy to understand if we assume that a quasar core is surrounded by a dust torus (analogous to Seyferts), which hides the highly polarized optical core from view if the viewing angle is larger than the critical angle separating HPQs from LPQs. Although the optical polarization of a HPQ may occasionally drop to low values (when the core flux is small), the geometry of the source does not change and the classification based on polarization is robust.

We may even use the observed differences between LPQs and HPQs to estimate the relative amounts of Doppler boosting in these two classes of sources. Tornikoski et al. (1993) found that the last spectral turnover for LPQs occurs around 20 GHz and for HPQs around 40-90 GHz. This then gives $D_H/D_L \approx 2-4.5$. Lainela (1993) used structure function analysis to derive characteristic timescales of 3.8-4 years for LPQs and 1.3-1.7 years for HPQs, giving $D_H/D_L \approx 2-3$; a reanalysis of the Michigan cm-data (Hughes et al. 1992) gives a value of ≈ 4. Finally, Teräsranta and Valtaoja (1993) determined the brightness temperatures of outbursts (Figure 4), finding median values of $5.4\,10^{12}$ K for HPQs and $8.1\,10^{11}$ K for LPQs, consistent with $D_H/D_L \approx 2$. Thus, four idependent estimates indicate that HPQs have, on the average, 2-4 times more Doppler boosting than LPQs.

4.2. Viewing angles of radio sources

Different amounts of Doppler boosting can result either from different viewing angles or different intrinsic Lorentz factors; the correct alternative cannot be deduced from continuum data. But by combining continuum with VLBI we can do much better, and actually derive the viewing angles and Lorentz factors of individual sources from observed (v/c) and $D \propto T_b^{1/3}$ (Teräsranta and Valtaoja 1993).

If we make a plot of (v/c) vs. log T (Figure 5a), unification schemes predict that the parent population will be found towards the lower left corner and the daughter population towards the upper right corner, with a similar range of Γ but smaller θ. In Barthel's unification scheme, the progression should be radio galaxies

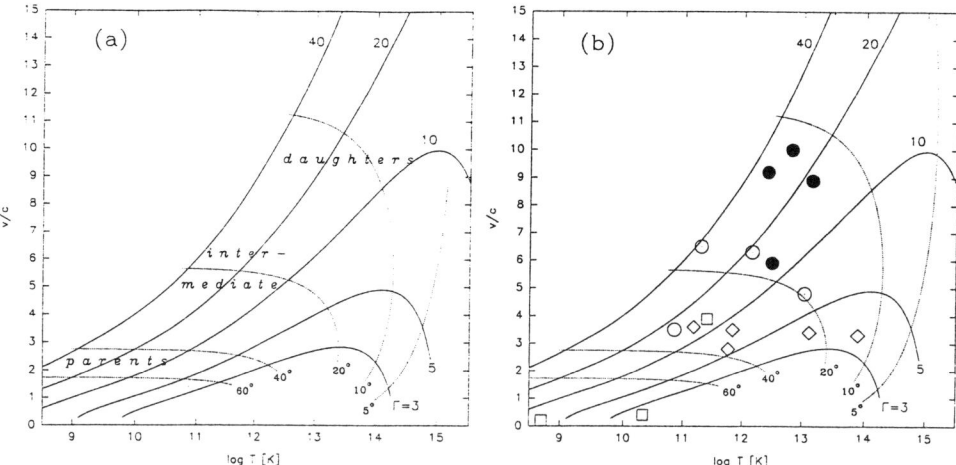

Fig. 5. (a) The relationship between the observable quantities $(v/c, T_b)$ and the intrinsic source parameters (Γ, θ). (b) The observed distribution (Teräsranta and Valtaoja 1993). Filled circles, HPQs; open circles, LPQs; squares, radio galaxies; diamonds, BL Lacs.

- LPQs - HPQs. If the BL Lacs come from a different parent population, they should occupy a different (Γ, θ)-range.

In figure 5b are plotted all the sources with sufficient VLBI and continuum data from Teräsranta and Valtaoja (1993). The statistics are small, but the distribution is as predicted. The estimated values are in good accordance with those derived from number counts, luminosity functions or other independent means. We find that a source appears as a HPQ if the viewing angle $\theta \leq 18°$, as a LPQ if $18° \leq \theta \leq 35°$, and as a radio galaxy if $\theta \geq 35°$; for all, the Lorentz factors range between 10 and 35 (assuming $H_o = 100$km/s/Mpc, $q_o = 1/2$)

BL Lacs clearly occupy a different regime; they have $\theta \leq 40°$ and $4 \leq \Gamma \leq 12$. The large viewing angles we derive for several BL Lacs are in accordance with the lack of rapid variability in most BL Lacs, and also with values derived from detailed model fittings. However, they do not agree with the small average viewing angles derived from luminosity function fits, where the assumption is that BL Lacs are beamed FR I radio galaxies.

5. Conclusions

The main conclusions derived from the analysis of high radio frequency spectra and variations are the following.

1) The framework of evolving shocks in a relativistic synchrotron jet is sufficient

to explain the observations, without the need to invoke alternative explanations for either the spectral shape or the variations (with the possible exception of intraday variability, not discussed here). However, at present the observations provide only weak constraints for the shock models.

2) The unification of radio galaxies (FR II), ordinary low polarization quasars and higly polarized quasars is in accordance with, and almost certainly required by the data. In particular, the observed brightness temperature distributions of radio outbursts are very hard to explain without invoking orientation as the decisive factor differentiating radio galaxies, LPQs and HPQs.

3) The unification of radio galaxies (FR I) and BL Lacs does not find much support in the radio data. Several lines of evidence indicate that BL Lacs are not especially closely beamed. We favor the explanation that the 'complete samples' of BL Lacs used in various analyses are mixtures of two different populations, the nearby 'true' BL Lacs (with FR I parents) and the distant 'BL Lacs', which really are the extreme end of the HPQ population (FR II parents).

References

Aller, H.D., et al. 1985, *ApJS* 59, 513
Aller, M.F., et al. 1992, *ApJ* 399, 16
Barthel, P.D. 1989, *ApJ* 336, 606
Brown, L.M.J., et al. 1989, *ApJ* 340, 129
Eckart, A., et al. 1989, *MNRAS* 239, 381
Gear, W.K., et al. 1993, it MNRAS, in press
Ghosh, T., Gopal-Krishna 1990, *A&A* 230, 297
Hughes, P.A., et al. 1989, *ApJ* 341, 54
Hughes, P.A., et al. 1992, *ApJ* 396, 469
Lainela, M. 1993, *A&A*, in press
Lainela, M., Valtaoja, E. 1993, *ApJ*, in press
Marscher, A., Gear, W.K. 1985, *ApJ* 298, 114
Robson, E.I., et al. 1993, it MNRAS 262, 249
Steppe, H., et al. 1992 *A&AS* 96, 441
Teräsranta, H., Valtaoja, E. 1993, *A&A*, in press
Teräsranta, H., et al. 1992, *A&AS* 94, 121
Tornikoski, M., et al. 1993, *AJ* 105, 1680
Valtaoja, E., et al. 1988, *A&A* 203, 1
Valtaoja, E., et al. 1992a, *A&A* 254, 71
Valtaoja, E., et al. 1992b, *A&A* 254, 80
Wiren, S., et al. 1992, *AJ* 104, 1009
Zhang, Y.F., et al., 1993, *ApJ*, in press

CONTEMPORANEOUS MULTIWAVEBAND OBSERVATIONS OF BLAZARS

ALAN P. MARSCHER, STEVEN D. BLOOM and YUN FEI ZHANG*
Department of Astronomy, Boston University, 725 Commonwealth Ave., Boston, MA 02215-1401, USA

and

WALTER K. GEAR
Royal Observatory, Blackford Hill, Edinburgh EH9 3HJ, Scotland, UK

October 28, 1993

Abstract. We have observed a number of blazars at wavebands ranging from radio to γ-ray. We find bright γ-ray emission to be associated with strong synchrotron flares observed at lower frequencies. The X-ray flux and entire radio spectrum of 4C 39.25 have each increased in strength by 30% over a 2-year period, in agreement with the prediction of the bent relativistic jet model.

Key words: quasars, BL Lac objects, γ rays, multifrequency observations

1. Introduction

While much has already been learned about the nonthermal emitting regions in radio-loud, highly variable quasars and BL Lac objects (collectively referred to as "blazars"), we are still ignorant concerning many of the most important aspects of these sources. Multiwaveband observations of flares in blazars currently hold the greatest promise for exploring the region where the relativistic plasma is generated and the jet formed, focused, and accelerated. The timescales of variability can be as short as a few days, which indicates that the flares originate deep within the jet, perhaps near its base. By measuring time delays of variations in brightness as a function of frequency, it is potentially possible to determine the geometry of the inner jet as well as infer the steepness of gradients in magnetic field, relativistic electron density, and bulk Lorentz factor of the jet flow (e.g., Marscher 1993).

The most exciting recent development is the detection of strong, highly variable γ-ray emission from about 40 blazars (see Kurfess et al. and Thompson et al., these proceedings). The apparent γ-ray luminosity is typically 1–2 orders of magnitude greater than that observed at other wavebands. Therefore, unless the γ-ray emission is more highly beamed than that at lower frequencies, most of the nonthermal energy emerges as very high energy photons during the γ-ray high states. Variability on timescales as short as a few days is observed, which indicates that the γ rays are produced in the inner jet. Understanding the γ-ray emission mechanism is therefore crucial to the exploration of the energetics and structure of the innermost regions of blazars.

* Also Center for Astrophysics, 60 Garden St., Cambridge, MA 02138, USA

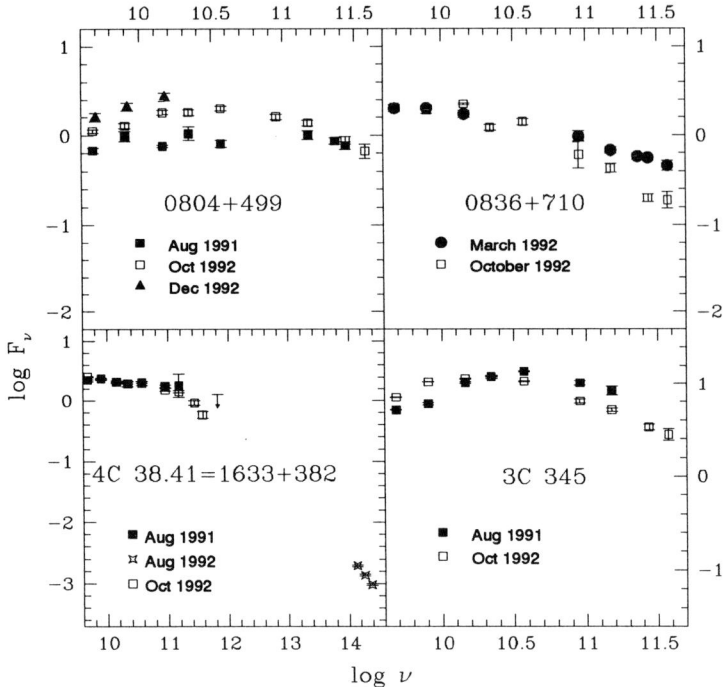

Fig. 1. Simultaneous radio to submillimeter spectra of four blazars. Epochs of measured strong γ-ray emission are: 0804+499 and 0836+710 — Jan & Mar 1992; 4C 38.41 (note different frequency scale) and 3C 345 — Sept 1991. From Bloom et al. (1994).

2. Relation between γ-ray and Lower Frequency Emission

We are in the midst of studies involving contemporaneous multiwaveband observations of blazars, consisting of VLBI, radio–submm, IR (for a few sources), X-ray, and γ-ray observations, in many case at multiple epochs. Instruments used include the VLBA and VLBI network, the U. Michigan and Metsähovi radio antennas, IRAM, the JCMT, UKIRT, ROSAT, and CGRO (mainly the EGRET detector). There are many collaborators, the most important for the work discussed here being H. D. and M. F. Aller, H. Teräsranta, E. Valtaoja, and S. D. Hunter.

Early results for γ-ray bright quasars are shown in Fig. 1. In all cases except 4C 38.41 (for which the data are rather sparse), *periods of high γ-ray flux are contemporaneous with enhanced levels of synchrotron emission* (see also Reich et al. 1994).

Nearly simultaneous multiwaveband spectra of a small number of sources show that the X-ray luminosity is not midway (on a logarithmic scale) between the γ-ray and infrared luminosities. Because of this, the spectra are inconsistent with

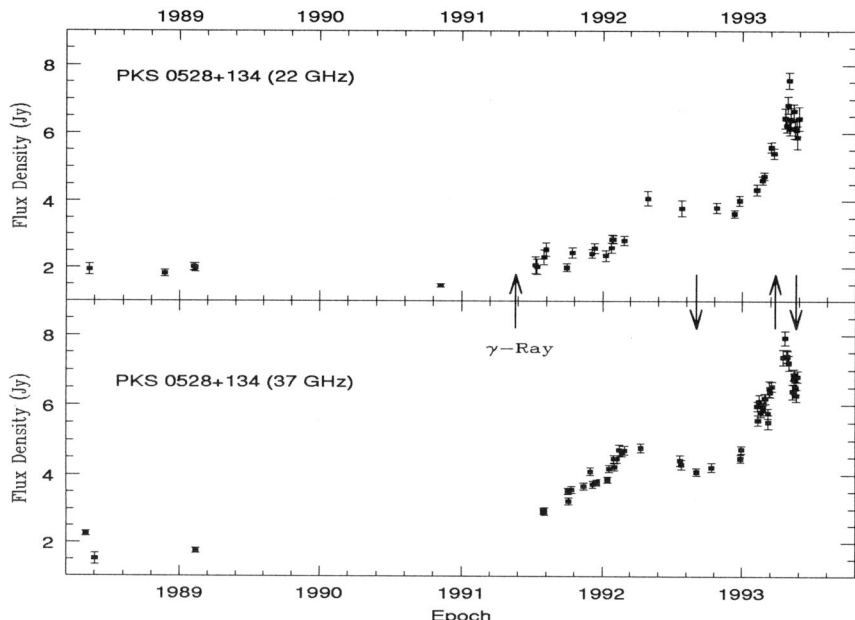

Fig. 2. 22 and 37 GHz light curves of PKS 0528+134. Epochs of CGRO γ-ray observations are indicated by the arrows; up-arrows: very strong detections; down arrows: upper limits to flux, ~ 10 times lower than the strongest detection. From Zhang *et al.* (1994).

second-order (but *are* consistent with first-order) self-Compton emission as the origin of the γ rays (see Bloom and Marscher 1993).

The multiple CGRO observations of PKS 0528+134 provide an opportunity to compare γ-ray and radio emission, since the light curve has been well sampled since mid-1991 at 4.8, 8.0, 14.5, 22, and 37 GHz (Zhang *et al.*, 1994). As shown in Fig. 2, γ-ray high states (Hunter *et al.* 1993; Sreekumar *et al.* 1993) occurred near the beginning of a major high-frequency radio flare in mid-1991, which later propagated to lower frequencies as the outburst became progressively less opaque. The very high γ-ray flux of March 1993 preceded the peak of the strongest 37 GHz flux observed thus far by 25 days, whereas the factor of ~ 10 lower γ-ray flux observed in late May 1993 (Nolan *et al.*, 1993) was measured during the declining phase of the 37 GHz outburst. The source was opaque at this frequency (the turnover frequency increased from 7 to 60 GHz, indicating that the core had become very active after a quiescent period), hence the 37 GHz peak was probably time delayed relative to the optically thin maximum at $\nu \gtrsim 100$ GHz. Nevertheless, these observations show that lags between γ-ray and synchrotron flares can be measured given sufficient time coverage with mm, submm, IR, and optical telescopes during and after (and before, if possible) the CGRO pointings.

It may be that all γ-ray flares are associated with synchrotron outbursts. The converse is not true, however. In the quasar 4C 39.25, the site of the synchrotron flare is a knot in the jet rather than the VLBI core region (Marscher et al. 1991; Alberdi et al. 1993). Despite a flux density that has risen to over 10 Jy at radio wavelengths, no γ rays were detected by CGRO in September–October 1993 (Mattox, private communication). We tentatively conclude that the synchrotron outburst must be in the core for there to be a γ-ray flare. The above authors model the radio behavior of 4C 39.25 in terms of a bent relativistic jet, with the fractional time rate of increase of flux density being the same at all observed radio frequencies above 4.8 GHz. We find that this extends to the X-ray emission, as predicted by the bent jet model if the X-rays arise from synchrotron self-Compton emission: both the X-ray and radio flux densities increased by 30% between April 1991 and April 1993.

3. Conclusion

The best way to establish and study the connections between γ-ray and X-ray emission and lower frequency synchrotron emission is through simultaneous multi-waveband monitoring. In general, past programs have not had sufficient time coverage to fulfill the promise of this technique. Still, much has been learned. More intensive monitoring is planned during the next few years, which should result in major advances in our knowledge of the physics of blazars.

4. Acknowledgments

This work was supported in part by NSF grant AST-9116525 and by NASA grants NAG5-1637, NAG5-1943, and NAG5-1566 (ROSAT and CGRO Guest Investigator Programs).

References

Alberdi, A., Marcaide, J.M., Marscher, A.P., Zhang, Y.F., Elósegui, P., Gómez, J.L., and Shaffer, D.B. 1993, *Astrophysical Journal*, **402**, 160–172
Bloom, S.D., et al., *Astronomical Journal*, submitted
Bloom, S.D., and Marscher, A.P. 1993, in *Compton Gamma-Ray Observatory, AIP Conference Proceedings*, **280**, ed. M. Friedlander, N. Gehrels, and D.J. Macomb (New York: American Institute of Physics), pp. 578–582
Hunter, S.D., et al. 1993, *Astrophysical Journal*, **409**, 134–138
Marscher, A.P. 1993, in *Astrophysical Jets, STScI Symposium Series*, **6**, ed. D. Burgarella, M. Livio, and C. O'Dea (Cambridge University Press), in press
Marscher, A.P., Zhang, Y.F., Shaffer, D.B., Aller, H.D., and Aller, M.F. 1991, *Astrophysical Journal*, **371**, 491–500
Nolan, P.L., et al. 1993, *IAU Circular* no. 5802
Reich, W., et al. 1993, *Astronomy and Astrophysics*, **273**, 65–78
Sreekumar, P., et al. 1993, *IAU Circular* no. 5753
Zhang, Y.F., Marscher, A.P., Aller, H.D., Aller, M.F., Teräsranta, H., and Valtaoja, E. 1994, *Astrophysical Journal*, submitted

MULTIWAVELENGTH EXPERIMENTS ON AGN:
REVERBERATION EXPERIMENTS

W. WAMSTEKER and R. VIO
ESA IUE Observatory, P.O. Box 50727, 28080, Madrid, Spain

Abstract. The nature of the successful reverberation campaigns on variable active galaxies is summarized. A general summary of the completed campaigns is given. A discussion of the results of the first order (CCF) analysis is given and some comments are made on the directions needed to obtain more physcally meaningful solutions to the inversion problem presented by such observations of the sources of activity in the centers of active galaxies

Key words: Seyfert I; Ultraviolet ; Multiwavelength; Broad-line-region

1. Introduction

Since the original suggestion that the line emission in active galaxies (AGN) is caused by photo-ionization and reverberation will permit to probe the Broad-Line Region (BLR) structure, many observations have been made in the optical, but especially in the UV with the IUE (International Ultraviolet Explorer), towards that end. Many early attempts were defined to try to match the *fast variations* in Seyfert I galaxies discovered in the X-Ray domain. The original concept that the continuum radiation was correlated over a large range of energies from IR to X-Rays, motivated this approach. This basic assumption was of course dramatically biased by the very incomplete sampling of the Fourier domain of AGN variation (most X-Ray observations consisted of single observations of a duration of the order of hours) as well as the lack of simultaneous multiwavelength observations. Only in a few cases a different and more systematic approach to these problems was taken such as the long term observational studies by Lloyd (1984) and the long term AGN monitoring at Florida State.

Only after some 7 years of extensive AGN studies with IUE, with the availability of the archival data of IUE, and the recognition that archival studies in astronomy are a respectable activity, was it realized that a major concerted effort of the AGN community would be needed to obtain sufficient observing time to address the variability of, especially Seyfert I galaxies, in a profitable way. A compilation of the results for all 522 active galaxies observed before 1990 with IUE, together with relevant multiwavelength information, is given in the IUE–ULDA Guide No.4 (Courvoisier and Paltani, 1992). Examples illustrating the directions in which progress could be made with the use of IUE archival data as well as dedicated observations with IUE are given in the UV and IR study of F-9 (Clavel, Wamsteker and Glass, 1989) and the UV and optical study of NGC 5548 (Wamsteker *et al.*, 1990).

During a workshop held in Segovia under the auspices of the *ESA IUE Obser-*

vatory in 1987, a group of AGN researchers agreed to share their resources and attempt to obtain enough observing time both on public and private observatories, to do what could be considered to be a "Reverberation Experiment" in Astrophysics. Under the leadership of a small group, the "*Central Committee*" (grown somewhat in recent years, but still less than 10 people) a call was issued to the world-wide community and a loosely bound group of scientists, the "*AGN Watch*" (some 100 AGN researchers), constituted itself with the purpose to support, through their individual specialized contributions, in many different ways the first reverberation experiment. This was done on the intermediate Seyfert I galaxy NGC 5548 in 1989 with an extensive IUE observing run of observations every four days. In parallel with this, the *AGN Watch* made also a worldwide optical campaign which was performed equally successful (Peterson et al., 1991). The original plans by the *AGN Watch* were drawn up on the basis of the well documented of variability history of NGC 5548, especially in the UV. The results of this first IUE campaign have been published by (Clavel et al., 1991). It is important to recognize the cooperation of the staff and management of both IUE Observatories, since without their support this first innovative program would not have been completed as successfully as it was.

2. The reverberation Experiments

The reasons why the initiative for the reverberation experiments was closely associated with the IUE Space Observatory are:
- The wavelength range accessible with IUE (115.2 nm$< \lambda <$320 nm, (half of this range in a single exposure) covers important photo ionization diagnostics at different ionization levels, such as Ly α λ 121.6 , CIV λ154.9, He II λ164.0, CIII] λ190.9 and Mg II λ279.8 and is close to the ionizing continuum.
- Variability amplitudes had been found to be larger in the UV than in other wavelength domains.
- The operations mode of IUE would easily allow the introduction of the complex scheduling requirements associated with a regular spacing and frequent return to a specific target.
- The fact that IUE is a facility available to astronomers from all countries made a worldwide coordinated effort easier.
- The IUE orbit and ground control mode allowed uninterrupted scheduling without additional constraints.
- The use of a space observatory removed the problems associated with the terrestrial weather.
- The IUE users had already extensive experience with multiwavelength observations using different observatories and could this way obtain considerable added value through coordinated observations from the X-Rays to the Infrared.

Even before these systematic studies were started the first firm evidence of delays between the variations in the continuum of an active galaxy and the delays

TABLE I
Results of Major AGN Reverberation Experiments (unit: Days)

Mean sample interval	NGC 4151 3 Days		NGC 3783 3 Days		NGC 5548 4 Days		F-9 60 Days	
	ΔT	FWHM	ΔT	FWHM	ΔT	FWHM	ΔT	FWHM
$F\ \lambda 184$	1.5	22	0	14.2	0	38	0	63
$F\ \lambda 270$			-1.4	18.5	0	40		
F Opt.	3.5	25			4	42	6	86
$He\ II\ \lambda 164.0$			0.5	11.3	10	40		
$CIV\ \lambda 154.9$	5.0	26	5.5	18.6	16	36	200	80
$Ly\ \alpha\ \lambda 121.6$			3.8	15.6	12	34	115	70
$Mg\ II\ \lambda 279.8$	5.0	23	8.5	14.7	~ 50	72	160	90
$H\ \beta\ \lambda 486.1$			7.5	40.7	18	50		

between the continuum were found for the luminous galaxy Fairall 9, which after its dramatic drop in brightness seen in the UV (Clavel, Wamsteker and Glass, 1989) and X-Rays (Morini et al., 1986), was followed on a semi-regular basis and cross correlation showed significant differences in the times of UV continuum minimum and the times of minimum emission line flux (see table I). The following objects have been subjected to systematic variability monitoring studies:

3C273: *UV* (Ulrich et al., 1993), *Optical, IR, Radio*

NGC 5548: *X-Ray* (Clavel et al., 1992), *UV* (Clavel et al., 1991), *Optical* (Peterson et al., 1991)

NGC 4151: *X-Rays* (Perola et al., 1986), *UV* (Clavel et al., 1990)

NGC 3783: *UV* (Reichert et al., 1994), *Optical/IR* (Stirpe et al., 1994).

Pks 2155-304 (Edelson et al., 1994),: *X-Ray* (Brinkmann et al., 1994), *UV* (Urry et al., 1993), *Optical/IR/Radio* (Courvoisier et al., 1994).

Since for 3C273 the emission lines do not vary significantly with the UV continuum (Ulrich et al., 1993) and (O'Brien et al., 1989), and for the BL Lac object Pks 2155-304 (Urry et al., 1993) and (Edelson et al., 1994) the absence of emission lines, precludes the application of reverberation studies, even though the results for these objects are very important for the analysis of the continuum variations in themselves.

3. Results

The main results from these campaigns on NGC 4151, NGC 3783 and NGC 5548 are given in table I, together with the earlier results on F-9. Table I gives the time delay (Δ T) measured from the centroid of the cross correlation function (CCF) of the short wavelengths UV continuum with the parameter given in the first column.

The optical continuum is measured either with the IUE Fine- Error-Sensor or as a spectroscopic flux measured near 550 nm. The second parameter in table I (FWHM) indicates the Full- Width-Half-Maximum of the respective CCF. This does not give the accuracy of the delays, but is rather a measure of the correlation length in the lightcurves if the sampling is sufficiently small with respect to the variability timescale. For the case of NGC 5548, also a shorter campaign was made in coordination between IUE and GINGA. It was found that the 2-10 keV X-Rays were within the resolution of the sampling simultaneous, i.e. the delay between the X-Ray continuum and the UV continuum was less than 4 days (Clavel et al., 1992).

The results in Table I show clearly that reverberation experiments were extremely successful and represent a powerful tool to study Active Galaxies. The delays derived are in general much shorter than would be expected from classical central photoionization theory and also accretion rate changes are imcompatible with the variations shownfor the energy distribution in the "big blue bump".On the other hand, it has become obvious that more sophisticated tools than the CCF alone must be applied to interpret such data properly. The various methods to perform lightcurve inversion are discussed in detail elsewhere in these proceedings by Krolik. It has also become clear that for various well studied objects e.g. NGC4151 (Clavel et al., 1990) and F-9 (Clavel, Wamsteker and Glass, 1989) the emission lines vary rather differently at different velocity ranges. These specific velocity limited variations could make that, although the parameters derived through CCF and MEMECHO methods are of the right order of magnitude for the BLR, a diferent approach might be needed for the analysis of such data sets to acquire a satisfatory understanding of the nature of the BLR. The UV reverberation studies have supplied a completely new tool to study active galaxies, where access to the Ultraviolet wavelength range is vital.

References

Brinkmann, W., et al., 1994, ApJ, in press
Clavel, J., Wamsteker,W., Glass,I.S., 1989,ApJ, 337, 236
Clavel, J., et al., 1990, MNRAS, 246, 668
Clavel, J., et al., 1991, ApJ, 366, 64
Clavel, J., et al., 1992, ApJ, 393, 113
Courvoisier, T.J.-L., et al., 1994, ApJ, in preparation
Courvoisier, T.J.-L. and Paltani, S., 1992, IUE-ULDA Access Guide # 4, ESA SP 1153
Edelson, R., et al., 1994, ApJ, in press
Lloyd, C., 1984, MNRAS, 209 697
Morini, M., et al., 1986, ApJ, 307, 486
O'Brien, P.T., Zheng, W., Wilson,R., 1989, MNRAS, 240,741
Perola, G. C., et al., 1986, ApJ, 306, 508
Peterson, B.M., et al., 1991, ApJ,368, 119
Reichert, G.A., et al., 1994,ApJ, in press
Stirpe,G.M., et al., Astron. Astrophys.,1994, in press
Ulrich,M.-H., Courvoisier,T.J.-L., Wamsteker, W., 1993, ApJ, 411, 125
Urry, C.M., et al., 1993, ApJ, 411, 614
Wamsteker, W., et al., 1990, ApJ, 354, 446

THE STRUCTURE OF THE BROAD EMISSION LINE REGION AS SHOWN BY VARIABILITY MONITORING

JULIAN H. KROLIK
Johns Hopkins University

Abstract.
The various methods used to infer the physical conditions and location of the material responsible for broad line emission in AGN are reviewed. Recent efforts have focussed on reverberation mapping, whose basic concepts and experimental constraints are discussed. A new method for analyzing the results of monitoring experiments, regularized linear inversion, is presented. This method is then applied to published data from the 1989 IUE campaign on NGC 5548, and the results found contrasted with those obtained by the previous standard method, maximum entropy.

1. Introduction

Studies of the broad emission lines in active galactic nuclei have engaged very large numbers of astronomers ever since the discovery of AGN. Precisely because there is such a large volume of work in this field, it is very easy to get lost in the details. For this reason, I will begin by summarizing why this work is felt to be valuable, and will try to highlight what our principal goals are.

The first and lowest level goal is to understand the physical conditions in the source(s) of these emission lines, and to determine their principal production mechanisms. There has been, I believe, substantial progress toward accomplishing this task. A related problem is to locate the line emission region with respect to the other structures in AGN. While there has been much recent work in this area (whose discussion will form the bulk of this review), this problem is by no means completely solved.

The next step up in sophistication is to understand the dynamics of the line-emitting material. At present, while we have a measure of the volume-integrated velocity distribution function in the form of line profiles, we do not know whether the *local* velocity distribution has any special spatial orientation with respect to the center of the AGN, *i.e.*, we don't know whether the matter is falling in, streaming out, or following stable bound orbits. Nor do we know which forces are responsible for accelerating the matter. Whatever these are, there must be some very special constraints built into the hydrodynamics because the observed 1-d velocity dispersion (typically several thousand km/s) corresponds to Mach numbers of several hundred with respect to the sound speed inside the line-emitting gas.

An area about which we would certainly like to learn, but about which even less is known is the "natural history" or life-cycle of the emission line gas. While there have been many speculations voiced about the origin and ultimate fate of the emission line gas, none has achieved any wide acceptance. To give a sense of the vast range of uncertainty in this area, I'll simply list a sampling of these suggestions: thermal instabilities in an accretion flow (Krolik 1988); winds from red giant stars in a dense stellar cluster (Scoville and Norman 1988; Kazanas 1989); or the surface of the accretion disk in which many of us would like to be able to believe (Dumont and Collin-Souffrin 1990).

Finally, if we are fortunate, we will be able to study AGN emission lines not just to learn about their own origin and character, but as a device to learn about other elements of the AGN system. It has long been a hope in this field that once we understand well the properties of the emission line region itself, we will be able to use the emission lines as "plasma diagnostics" of surrounding regions. If the line-emitting matter is in pressure balance with ambient, less-observable gas (or even if it is not in pressure balance, if there is a calculable relation between the two pressures), then knowledge of the physical conditions in the emission line region carries over into knowledge of those in the hot gas. If in addition there is dynamical coupling between emission line gas and nearby unseen gas (through drag, for example, or viscosity in an accretion disk), then the emission lines tell us about the ambient gas's motions. Reliable results in this area still remains elusive.

2. Techniques for Reaching These Goals

A variety of methods have been invented for accomplishing these objectives. The simplest—and oldest—way to find the physical conditions prevailing in emission region is to construct a single-zone photoionization model, an idea that goes back almost twenty-five years to Bahcall and Kozlovsky (1969), and, in its first detailed realization, to work by Davidson (1972) and MacAlpine (1973). The fundamental idea behind this method is to assume that the emission lines are powered by photoionization. While this idea was initially controversial, the success of these models in crudely reproducing the observed relative line strengths, and, more recently, the excellent correlation (at a delay) found between fluctuations in the continuum and fluctuations in the lines (*e.g.* Clavel *et al.* 1991; Peterson *et al.* 1991; Reichert *et al.* 1993), very strongly support its basic assumption.

To carry out this method, one varies three free parameters: the gas pressure (or density), its thickness (either in length or column density), and the ionization parameter at its exposed edge (defined variously as the ratio of ionizing photon density to gas density, or ionizing photon energy density to gas pressure (this version is denoted Ξ), or in one of several other almost equivalent forms). Using known atomic physics data, one then computes the local ionization equilibrium, local thermal balance, and local excited state population balance in selected atoms and ions for gas subjected to a continuum of a specified spectral shape. It is, of course, a weakness of this method that while we know a great deal about AGN spectra at energies of several to 10 eV, and also 0.5 to 20 keV, we have very little direct knowledge of their spectra in the range between these energies, which is exactly the range of most interest for photoionization. From the computed local physical conditions, one then calculates the total photon emission from the gas, and compares the relative strengths of the emission lines to the values observed. By varying the free parameters, one eventually finds values which more or less reproduce the observed line strengths. In order to get the best fit to the greatest amount of line flux using only a single zone, one usually finds pressures $p \sim 0.01 - 0.1$ dyne cm^{-2}, column densities $N \sim 10^{22} - 10^{23}$ cm^{-2}, and ionization parameters (defined in the pressure ratio form) $\Xi \lesssim 1$ (*e.g.* Kwan and Krolik 1981; Krolik and Kallman 1988; Rees, Ferland, and Netzer 1990). Strikingly, because the relative line strengths vary comparatively little from object to object, these parameters are crudely similar for all AGN. They may be combined with the luminosity in any particular object to estimate the mean distance of the emission line region from the source of the continuum, assuming free-streaming propagation in between:

$$\langle r \rangle \simeq \left[\frac{L_{ion}}{4\pi cp\Xi}\right]^{1/2} \sim 100 L_{ion,44}^{1/2} \text{lt-d}, \qquad (1)$$

where L_{ion} is the luminosity in the ionizing band.

However, this method has a number of unsatisfactory points. First of all, while the quality of fit to the line strengths one obtains with a single zone model is surprisingly good (there is no *a priori* reason why a single zone model should come anywhere close), it is by no means perfect, and because there is no clear statistical measure of the quality of fit, one can never be quite sure how well determined the mean parameters are, or whether the quality of the fit is as good as could be expected from data of a given signal/noise. In addition, there are a number of hidden assumptions making it somewhat model-dependent. For example, nearly all these calculations assume slab geometry. In addition, while assuming solar abundances gets one close to the observed line strengths, there is the possibility that the abundances are actually different. Unfortunately, it is very difficult to find an unambiguous signature of any particular abundance deviation (Davidson 1975; Kwan and Krolik 1981; Hamann and Ferland 1992). The final disability of this procedure lies at its very heart: we certainly do not expect that the distribution in physical conditions is a perfect delta function, and we would like to know about the spread in conditions, if possible correlating that spread with location and/or kinematics.

The next simplest procedure is to construct photoionization models as a function of line of sight velocity, *i.e.* with respect to the line profiles (Kallman *et al.* 1993). By doing so, one effectively projects the line emissivities onto a set of nested surfaces, the surfaces of constant line of sight velocity. The mean physical conditions inferred for each model then correspond to the conditions averaged over each of these surfaces. Unfortunately, this technique suffers from two nasty illnesses: we do not know the shapes of these isovelocity surfaces *a priori*, so we do not know where these inferred conditions apply; and worse, we do not even know if these surfaces exist. To the degree that the line emitting gas moves like a fluid, its velocity is well-defined at each point; however, if it moves collisionlessly (for example, if the gas is broken up into a large number of small clouds, or if the gas resides on the surfaces of stars), its velocity distribution at any given point could be quite broad. When that is the case, the isovelocity surfaces become thick and start to overlap. Interpreting a profile-based photoionization model in such a situation is virtually hopeless.

3. Reverberation Mapping

Variability in AGN permits the application of a quite different method to this problem, the method of "reverberation mapping". While first proposed in concept many years ago (Bahcall, Kozlovsky, and Salpeter 1972), and first developed mathematically more than a decade ago (Blandford and McKee 1982), it was first attempted only in the last few years (Clavel et al. 1991; Maoz et al. 1991). The reason for this long delay, as we shall see, has to do with the very large quantity of data required, and the special planning which must go into the collection of this data if they are to be useful. Before discussing these issues, I digress to outline the basic idea behind the method.

Careful monitoring of AGN continuum emission (particularly for those of lower luminosity) shows that they are often quite variable in the optical and ultra violet. If the ionizing continuum follows the same light curve (as is suggested by the excellent correlation at zero lag between the optical and non-ionizing UV bands: Krolik et al. 1991), then the emission line response from any small volume within the broad emission line region should vary along with the ionizing continuum. Because it takes a finite time for the gas to adjust its equilibrium to the new value of the continuum flux, in principle there can be a small local lag; however, at the densities indicated by single-zone photoionization modelling, the local equilibration time is very short, perhaps ~ 1 minute. By contrast, as we have already estimated, the light travel time across the region is far longer, at least days and possibly many weeks even in low luminosity AGN. Consequently, the line light curve we measure should be a delayed, and smoothed, replica of the continuum light curve. Because we understand (or think we do) how emission line gas responds to changes in the ionizing continuum, we can use monitoring data for both the continuum and the emission lines to invert this argument and give us information about the internal geometry of the emission line region. That is to say, we can project out a map of the emission line region, where the surfaces of projection are the surfaces of constant light travel-time with respect to us, defined by

$$r = c\tau(1 - \cos\theta), \qquad (2)$$

where the polar axis with respect to which θ is defined is the direction towards us.

There are (at least) two major difficulties with this method: First, it requires a very large quantity of data, and the data must be obtained in the right way. To see just how much, and what the right way is, we write the relation between the line light curve and the continuum light curve as

$$\delta F_l(t) = \int d\tau \, \Psi(\tau) \delta F_c(t-\tau), \tag{3}$$

where $\delta F_{l,c}$ are the fluctuations in the line and the continuum with respect to their mean values, and the response function $\Psi(\tau)$ is the marginal emissivity of the line with respect to changes in the continuum flux, averaged over the surface with delay τ. Assuming a linear relation between continuum fluctuations and line fluctuations is only a good approximation when $\delta F_l/\langle F_l \rangle \ll 1$, but this is commonly the case. Equation 3 is a convolution relation, and so has the formal solution

$$\Psi(\tau) = \int df \, e^{-2\pi i f \tau} \frac{\hat{\delta F_l}(f)}{\hat{\delta F_c}(f)}, \tag{4}$$

where the symbol $\hat{X}$ denotes the Fourier transform of X. Thus, if the response function has structure on the scale τ_o, for us to measure it the AGN must have significant variability on that timescale, and we must sample that variability on a timescale shorter than $\tau_o/2$. In addition, to obtain a statistically reliable measure of that variability, we should extend our observations for times considerably longer than τ_o. To do this well, and to be sensitive to a reasonably wide range of possible timescales, requires many many observations. Simulations, and experimental experience, show that $\simeq 50$ observations is a bare minimum. Moreover, to obtain the least biased picture of the variability components on different timescales, the observations should preferably be evenly-spaced. As several recent campaigns have shown (Clavel et al. 1991; Peterson et al. 1991; Reichert et al. 1993) it is possible to do this, but a very large amount of labor is required.

The second difficulty is actually inverting the convolution equation. Despite the existence of the formal solution just shown, this is not so easy. Unfortunately, in most cases the Fourier solution cannot be used. The problem is that most AGN fluctuation power spectra are fairly "red", *i.e.* most of the power is found at low frequencies. As a result, measurement error, which has a white noise spectrum, overpowers the signal at high frequencies. At the same time, though, this lost high frequency information is necessary if we are to obtain maximal resolution in the response function.

Another possible approach is to discretize the convolution according to the actual sampling. The integral equation then has the appearance of a conventional linear equation, which one might think should be directly soluble. But this, too, does not work because the smoothing produced by the convolution creates a large ambiguity in the possible solutions (*i.e.* many different kinds of smoothing all create the same final result). This ambiguity is expressed mathematically by the fact that the matrix representing the integral equation kernel always has at least a few eigenvalues many orders of magnitude smaller than the typical value of elements in the matrix. If there is any noise in the data with projection onto the corresponding eigenvectors, the implicit matrix inversion of the linear solution multiplies these noise components by numbers of order the inverse of the very small eigenvalues; these inverses are, of course, very large. In practise, this noise amplification is completely catastrophic.

Hitherto, this problem has been solved by a variety of model-fitting techniques, most prominently maximum entropy (Krolik et al. 1991; Maoz et al. 1991). While model-fitting is easy to make stable, it, of course, always entails some level of model-dependence in the answer. In the maximum entropy version, it is particularly hard to trace the impact of particular model assumptions on the solution. Maximum entropy also has the further drawback of requiring positive values of Ψ because the nonlinear function it maximizes to select otherwise equally acceptable solutions is undefined for negative arguments. While most lines respond positively to continuum fluctuations in most circumstances, this is not true in general (Gaskell and Sparke 1986; Sparke 1993), and it would be desirable to test for negative responses.

4. A New Inversion Method: Regularized Linear Inversion

Fortunately, there is a direct inversion method, called regularized linear inversion, which does not suffer from these defects, and is also computationally very efficient (Press et al. 1992). Christine Done and I have recently shown how to apply this method to the reverberation mapping problem (Done and Krolik 1993), and I summarize our results in the remainder of this review. The heart of this method is the recognition that in order to break the degeneracy of the inversion one must inject some *a priori* information. This is done by simultaneously minimizing both the deviation of the solution from the measured data and the deviation of the solution from one's *a priori* assumption. Obviously, this method entails model-dependence; its beauty is in the clear dependence of the solution on these model assumptions through a single tunable parameter.

More quantitatively, if we wished solely to find the solution which best reproduced the observed data, we would minimize the quantity

$$\chi^2 = (C \cdot \Psi - \mathbf{L})^2, \tag{5}$$

where $\mathbf{L}$ is the list of observed line flux fluctuations normalized by their uncertainty

$$L_i = \delta F_l(t_i)/\epsilon_i, \tag{6}$$

Ψ is the list of values of the response function at lags τ_j, and the matrix C is the discretized kernel of the integral equation similarly normalized

$$C_{ij} = \delta F_c(t_i - t_j)/\epsilon_i. \tag{7}$$

On the other hand, if our *a priori* condition is to look for the "smoothest" possible solution, this means that we are searching for solutions with small derivatives, and these can be represented by a differencing operator R. Thus, the total quantity to be minimized is the sum $\chi^2 + \lambda \boldsymbol{\Psi} \cdot R^T \cdot R \cdot \boldsymbol{\Psi}$. The balance between satisfying the data and satisfying our prejudices is expressed by the tunable parameter λ. Because both quantities to minimize are quadratic forms in Ψ, the solution is found by solving a simple linear equation

$$\left(C^T \cdot C + \lambda R^T \cdot R\right) \cdot \boldsymbol{\Psi} = C^T \cdot \mathbf{L}. \tag{8}$$

Even though the solution of either minimization separately would be disastrously unstable because of the small eigenvalue problem, their simultaneous solution is quite stable because the probability that the eigenvectors of the two pieces coincide is extremely small. This method also possesses the virtues of very clear error propagation, and directly testable model-dependence through manipulation of λ.

5. Application to Real Data

There have now been four major monitoring programs with sampling good enough to attempt a solution for the response function: a purely ground-based program on NGC 4151 (Maoz et al. 1991); one on NGC 5548 combining IUE data with coordinated observations from the ground (Clavel et al. 1991; Peterson et al. 1991); another IUE plus ground program on NGC 3783 (Reichert et al. 1993; Stirpe et al. 1993); and, most recently, a return to NGC 5548 using HST, IUE, and more ground observations (Korista et al. 1994). Of these, the one most suitable for immediate analysis with linear regularization is the 1989 IUE campaign on NGC 5548. As can be seen from the forms just presented, the method works best with evenly-sampled data, so the ground-based NGC 4151 data is eliminated.

The HST data is not yet completely reduced (see Peterson's article in this volume for a preliminary look at the light curves), so it is eliminated. Finally, the NGC 3783 campaign was the victim of bad fortune: the utility of any monitoring data clearly depends on the ratio of real variance on the relevant timescales to noise variance, and this Seyfert galaxy happened not to vary terribly much on the right timescales.

To place this program in context, I will just remind you that NGC 5548 is a nearby type 1 Seyfert galaxy whose monochromatic luminosity νF_ν in the UV is $1.6 \times 10^{43} h^{-2}$ erg s^{-1}. Interestingly, *no* single-zone photoionization model provides a good match to its line strengths: at least two zones are required, one at $\simeq 10 h^{-1}$ lt-d distance, the other at $\sim 200 h^{-1}$ lt-d (Krolik et al. 1991), When maximum entropy analysis is applied to the seven UV line light curves (Krolik et al. 1991), seemingly robust solutions can be found for five of the lines, HeII 1640, NV 1240, Lα, CIV 1549, SiIV 1400, and CIII] 1909. The other two lines, MgII 2800 and OI 1304, did not vary enough relative to the measurement error. In the maximum entropy solution, the response functions of all but CIII] 1909 peak either at or very near zero lag, and trail off towards higher lags. The maximum lag at which significant response is found increases more or less with declining ionization level, from HeII 1640, whose response function is small beyond $\simeq$ 15d, to SiIV 1400, which retains significant response out to 30d or more. However, in evaluating these results, it is important to bear in mind that the criterion of acceptable χ^2 was defined with respect to all line light curves simultaneously, so substantial deviations in individual line light curves could be, and are, present.

The picture as seen by linear regularization is surprisingly different. Taking the solutions one line at a time, one finds that only three lines yield solutions with acceptable χ^2: Lα, CIII] 1909, and SiIV 1400. For all three of these, the reduced χ^2 is $\simeq$ 1.5. Even with $\lambda = 0$, there is *no* solution yielding acceptable χ^2 for the other lines. The reason for this discrepancy between the two methods is partly that the χ^2 for some of the individual light curves found by maximum entropy is, in fact, rather large, and partly that the maximum entropy solution used treated the continuum light curve as a set of free parameters constrained, but not fixed, by the observations. This extra freedom makes the number of free parameters actually rather larger than the number of unknown variables.

In the case of Lα, the response function derived by regularized linear inversion has very small, and possibly negative, amplitude at zero lag, rising to a peak at around 8d. Beyond 16d the response amplitude is fairly small. In contrast, the maximum entropy solution peaks at zero, and falls smoothly to low levels beyond 20d.

The regularized inversion solution for CIII] 1909 is strikingly different from the maximum entropy solution. The new method finds a response amplitude which is large and negative at zero lag, but which rises sharply with increasing lag. It passes through zero around 4d, and peaks about 10d, where it rolls off gently. The maximum entropy solution for this same data was positive everywhere (by assumption), but had a clean peak between 20 and 30d.

The response function found by regularized linear inversion for SiIV 1400 qualitatively resembles that of CIII] 1909, but it is possible that an acceptable reduced χ^2 is achieved not because the deviation from the observed light curve is small, but because the error bars are very large. A similar comment applies to the maximum entropy solution for this line.

What are we to make of the lines for which no acceptable solution is found ? This class is best exemplified by CIV 1549, the line with the best effective signal/noise of all the lines. The solution for this line produces a light curve which tracks the observed light curve fairly well up until its final positive excursion. At that point, while the line flux increases sharply, the continuum immediately before has only a weak maximum. Consequently, it is very hard for *any* time-steady linear response function with a width shorter than the interval between major continuum fluctuations to reproduce the line light curve. In the maximum entropy solution, this problem was cured by introducing significant response at a lag of 180d. Because the stretch of data constraining the solution at such large lags is relatively short, we cut off the linear inversion response function at a maximum lag of 80d. Thus, use of the linear inversion technique brings clearly into focus that the simplest model of line response fails to explain a significant element in the light curve. Just what is needed to fix it is unclear. Possibly response at very large lag, as suggested by the maximum entropy solution, is the correct answer. Possibly the response function changes on a timescale shorter than the duration of the monitoring, as suggested by Netzer and Maoz (1990). Possibly the line is responding to a continuum component varying in a way which is not simply proportional to the near-UV continuum. At this stage we do not have enough guidance from the data to choose the correct explanation.

6. Conclusions

Much effort has been expended over a very long period of time to try to understand the physical conditions and location of the broad emission line gas in AGN. The basic idea of photoionization is amply confirmed, but many details remain to be worked out.

The most active portion of this field at the moment is the application of variability monitoring techniques to the inference of the geometrical structure of the broad emission line region. A number of monitoring campaigns have generated very impressive data sets, but their interpretation is still problematic. The dramatic contrast between the response functions obtained by maximum entropy and regularized linear inversion, operating on identical data, demonstrates that even with these very large data sets, we are not in a position to unambiguously determine the response functions. In a sense, maximum entropy provides the "prettiest" possible solution, the one which is found by looking for that continuum light curve which, within the constraints of the measurement, allows the solution to very closely track the observed line light curve. On the other hand, regularized linear inversion, by only considering the "most likely" continuum light curves, presents the most conservative and stringent test of the simple time-steady linear response model.

Homing in on the correct response function is not, of course, the end of our job. Even with that in hand, it will be necessary to construct photoionization models whose values of $\partial F_l/\partial F_c$ averaged over the isodelay surfaces best match the response functions. Velocity-resolved reverberation mapping (as we hope to obtain from the recent HST campaign) will add another layer of complexity to the problem; however, if carried out successfully, it will also carry us to a much deeper level of understanding. In particular, it will only be with the conclusion of that program that we can answer some of the basic question raised at the beginning of this paper, such as the direction of flow within the broad line region.

7. References

Bahcall, J.N. and Kozlovsky, B.-Z. 1969, Astrophysical Journal 155, 1077.
Bahcall, J.N., Kozlovsky, B.-Z., and Salpeter, E.E., 1972, Astrophysical Jour-

nal**171**, 467.
Blandford, R.D. and McKee, C.F., 1982, Astrophysical Journal**255**, 419.
Clavel, J. et al. 1991, Astrophysical Journal**366**, 64
Davidson, K. 1972, Astrophysical Journal**171**, 213.
Davidson, K. 1975, Astrophysical Journal**195**, 285
Done, C. and Krolik, J.H. 1993 in preparation
Dumont, A.M. and Collin-Souffrin, S. 1990, Astronomy and Astrophysics**229**, 313
Gaskell, C.M. and Sparke, L.S. 1986, Astrophysical Journal
Hamann, F. and Ferland, G. 1992, Astrophysical Journal**391**, L53
Kallman, T.R., Wilkes, B., Krolik, J.H., and Green, R.F. 1993, Astrophysical Journal**40 3**, 45
Kazanas, D., 1989, Astrophysical Journal**347**, 74
Korista, K. et al. 1994 in preparation
Krolik, J.H., 1988, Astrophysical Journal**325**, 148
Krolik, J.H. and Kallman, T.R. 1988, Astrophysical Journal**324**, 714
Krolik, J.H., Horne, K., Kallman, T.R., Malkan, M.A., Edelson, R.A., and Kriss, G.A. 1991, Astrophysical Journal**371**, 541
Kwan, J.Y. 1984, Astrophysical Journal**283**, 70
Kwan, J.Y. and Krolik, J.H. 1981, Astrophysical Journal**250**, 478.
MacAlpine, G.M. 1972, Astrophysical Journal**175**, 11.
Maoz, D. et al. 1991, Astrophysical Journal**367**, 493.
Netzer, H. and Maoz, D. 1990, Astrophysical Journal, Letters to the Editor**365**, L5.
Press, W.H., Teukolsky, S.A., Vetterling, W.T., and Flannery, B.P. 1992, it Numerical Recipes, 2d ed., (Cambridge University Press: Cambridge)
Peterson, B.M. et al. 1991, Astrophysical Journal**368**, 119
Rees, M.J., Netzer, H., and Ferland, G.A. 1989, Astrophysical Journal**347**, 640.
Reichert, G. et al. 1993, Astrophysical Journalin press
Scoville, N.Z. and Norman, C.A. 1988, Astrophysical Journal**332**, 163
Sparke, L. 1993, Astrophysical Journal**404**, 570
Stirpe, G. et al. 1993 in preparation

THE AMPLITUDE–TIME LAG RELATION FOR EMISSION-LINE FLARES OF AGN

IVAN I. SHEVCHENKO
Institute of Theoretical Astronomy, Russian Academy of Sciences
Nab. Kutuzova 10, St.Petersburg 191187, Russia
E-mail: iis@iipah.spb.su

Abstract. The amplitude–time lag ("ΔA–Δt") relation is considered in order to describe behaviour of the emission-line spectrum of an active galactic nucleus during a separate active event. Here ΔA, called the amplitude, is the maximum relative increment of the flux in a line, and Δt is the time lag between the maximum of the ionizing continuum flare and the maximum of the flare in a line. As suggested by Shevchenko (1988), the construction and analysis of such relations can be used to discriminate between broad-line region models. Comparison of theoretical "ΔA–Δt" relations with the observed one composed by data for flares in various lines during a separate active event, is proved to be a useful tool for investigating the geometry of a broad-line region, for studies of the form of phase functions of a typical line-emitting cloud in various lines, as well as for clearing up the duration and amplitude of the initial flare in the ionizing continuum. The advantage of this method is that it utilizes the most general observed characteristics of the emission-line flares and nevertheless provides basic information on the allowed BLR models before the detailed modelling of emission-line light curves is performed.

1. Introduction

Since the discovery of rapid emission-line variability of AGN in emission lines (Lyutyj and Cherepashchuk, 1971; Cherepashchuk and Lyutyj, 1973), this phenomenon has been extensively studied (cf. review by Peterson, 1993). It has been usually analyzed in the framework of the "reverberation mapping" approach, introduced by Blandford and McKee (1982). This technique requires detailed information on emission-line light curves. A supplementary approach consists in an attempt of interpretation of basic characteristics of emission-line variations such as time lags and amplitudes (Shevchenko, 1984, 1985a, 1988). Here the interrelation between these characteristics is discussed.

2. The model of a BLR and the phase functions

Hereafter the standard picture of a broad-line region is adopted, according to which the BLR is an aggregate of a large number of line-emitting clouds surrounding the central source of ionizing radiation; the line emission of an individual cloud responds to continuum variations instantaneously as compared to the light-crossing time of the BLR (cf. e.g. the review by Peterson, 1993). The duration of the ionizing flare is assumed to be much less than that of the accompanying line variation, i.e. the whole analysis is performed in a flare approximation. The dependence of the luminosity of a BLR cloud in a line upon the incident ionizing flux is represented by the power law $L \propto F^s$ ($s \geq 0$); the index of this power law is referred to as "the parameter s".

Let us specify the form of phase functions defining how the flux in a line received by a distant observer from an individual cloud depends on the cloud phase angle. These functions are important because the anisotropy of line emission, upon which a typical cloud emits mainly from the side illuminated by the ionizing source, represents a necessary condition for a time lag to occur in case of a spherically symmetric BLR, if one adopts the flare approximation (Shevchenko, 1984). Hereafter the time lag of the maximum of a line variation is implied. Assume that the surface of the cloud emits in a line orthotropically, and the dependence of the line luminosity of the surface element on the incident ionizing flux is described by the power law $l \propto f^s$. First let the cloud represent a flat "pancake" orthogonal to the direction to the ionizing source. Then the phase function has the form

$$j(\theta) \propto \cos\theta + |\cos\theta|, \qquad (1)$$

where θ is the angle between the ionizing source and the observer as seen from the cloud. Consider next the case of a spherical cloud under the same assumptions. A close approximation to the phase function of a sphere, according to Shevchenko (1985b), is as follows

$$j_s(\theta) \propto (1 + \cos\theta)\left(1 + \frac{s}{2}\cos\theta\right), \qquad (2)$$

valid on the interval $0 \leq s \leq 2$.

3. The time lag–parameter s relation

The simplest possible model of cloud distribution is the homogeneous one; it turns out to be very illuminating. Define $\alpha = \sigma n \equiv \text{const}$, where σ is the mean section of a typical cloud in the plane orthogonal to the ionizing source direction, and n is the number of clouds in a unit volume. The BLR radius in the homogeneous model is nothing but the radius of the illuminated domain, confined by cloud shadowing. This radius R is of order α^{-1}, hereafter we adopt $R \equiv \alpha^{-1}$. The general covering factor formally equals to unity, but due to large velocities of clouds the line photons may freely escape from the BLR. Formally this model is equivalent to a non-homogeneous one with no cloud shadowing (zero covering factor), but with an exponential cut-off in radial distribution of clouds.

According to Shevchenko (1985a, 1988), if there is a central cavity of radius R_0 in the homogeneous cloud aggregate, the time lags are described by the relation

$$\Delta t = \begin{cases} W(1-s)/\alpha c, & \text{if } 0 \leq s \leq 1 - 2\alpha R_0/W \\ 2R_0/c, & \text{if } s \geq 1 - 2\alpha R_0/W, \end{cases} \qquad (3)$$

where W is a constant, which in case of phase functions (1) or (2) has corresponding values $W = 3.19$ or $W = 2$; c is the speed of light.

Define a mean harmonic radius of the BLR in a line as $\langle R \rangle_h = \langle r^{-1} \rangle^{-1}$, where luminosity of a cloud in the line $L \propto r^{-2s}$ corrected for the local covering factor is used as a weight function. In the homogeneous model

$$\langle R \rangle_h = 2(1-s)/\alpha = 2(1-s)R, \tag{4}$$

where $0 \leq s \leq 1$ (Shevchenko, 1985a). Then $\Delta t = W \langle R \rangle_h / 2c$, i.e. the time lag is directly proportional to the mean harmonic size. Differences in time lags are often taken as an evidence for a "stratification" of the BLR in multiple regions emitting in different lines. Eq. (4) clearly demonstrates that this stratification is irrelevant to matter distribution.

4. The amplitude–time lag relation

Define the amplitude of a flare in a line as the maximum relative increment of the observed flux: $\Delta A = \Delta F_{\max}/F$. In case of the model with a central cavity of radius R_0, if one adopts phase function (2), the ΔA dependence on s is as follows

$$\Delta A = \begin{cases} \frac{12(1-s)^{2(1-s)} e^{s-1}}{(s+6)\Gamma(3-2s,\chi)} c(s)((1+\Delta A_c)^s - 1) \, d\tau, & \text{if } 0 \leq s \leq 1-\chi \\ \frac{3}{(s+6)(s-1)(2s-1)} \left(\chi(2-s-\chi) + \frac{\chi^{3-2s} e^{-\chi}}{\Gamma(3-2s,\chi)} \right. \\ \left. \times \left(\chi - s + \frac{2(s^2-1)}{\chi} \right) \right) ((1+\Delta A_c)^s - 1) \, d\tau, & \text{if } 1-\chi \leq s < 1-d\tau, \end{cases} \tag{5}$$

where $\Delta A_c = \Delta F_c/F_c$ is the fractional amplitude of the flare in the ionizing continuum, $d\tau = \alpha c \, dt/2 = c \, dt/2R$ is the duration of the continuum flare measured in light-crossing times of the BLR; $0.95 < c(s) \leq 1$ for $0 \leq s \leq 1$; $\chi = \alpha R_0 = R_0/R$ is the radius of the cavity measured in units of the characteristic radius of the BLR. On the reason that the duration of the ionizing flare fixes the time resolution of time lags, Eq. (5) is applicable if $s < 1 - d\tau$.

Eqs (3) and (5) parametrically define the amplitude–time lag dependence. It is generally in accordance with observational data, presented by Cherepashchuk and Lyutyj (1973), Antonucci and Cohen (1983), Ulrich et al. (1984), Peterson (1993), as the amplitudes in lines are predicted to be much smaller than that in the continuum, and are expected to decrease with the time lag value.

For realistic values of $d\tau > 0.1$ (accordingly $s < 0.9$) the homogeneous model and the model with a cavity for even high values of χ (say, equal to 0.5) differ very slightly, by some percents, in the predicted amplitudes. Therefore one can hardly expect that any information on the amplitudes solely can say much about radial structure of the BLR. The "ΔA–Δt" diagram is much more informative in this respect. According to Eq. (3), if the cavity exists, the lags smaller than some value constant for all lines emitted anisotropically cannot be observed, and therefore the diagram is strongly sensitive to deviations in radial structure.

The ratio of amplitudes predicted for phase functions (1) and (2) in the homogeneous model, according to Shevchenko (1988), is confined between 0.72 and 0.88

for allowed values of s, i.e. the amplitudes are slightly affected by the choice of the phase function. The time lags are again better indicators, as they differ by 1.6 times (see Eq. (3)). Therefore the "ΔA–Δt" dependence can still be used to discern between various forms of phase functions.

The amplitude in a line for the BLR in a general standard model (described in Section (2)) in the flare approximation can be represented as

$$\Delta A = f(s) \, \Delta A_{qs} \, c \, dt/D, \qquad (6)$$

where function $f(s)$ is shaped by the BLR geometrical structure as well as the form of cloud phase functions, and normally depends on s weakly and is of order of unity; $\Delta A_{qs} = (1+\Delta A_c)^s - 1$ is the amplitude corresponding to the quasistationary state, for which the continuum variations are much longer in duration than the light-crossing time of the BLR; D is the diameter of the BLR. Thus the amplitudes in lines are determined mainly by the amplitude and duration of the flare in the continuum and by the size of the BLR, and therefore can be used to estimate them.

The amplitudes of line variations in the flare approximation are less than in the quasistationary case by factor $c\,dt/D$. This marked difference indicates that the values of the parameter s cannot be derived by means of application of the quasistationary law to rapid events, whether the procedure of displacement of a line light curve on the value of the time lag is utilized or not. Application of this law is self-consistent on the time scale of slow variability, i.e., according to Lyutyj (1977), on the time scale of years.

5. Conclusions

Consideration of the amplitude–time lag dependence may often be helpful on the reason that, along the amplitude axis, it is strongly sensitive to the parameters of the continuum flare as well as to the size of the BLR, and, along the time lag axis, it is affected by the BLR geometrical structure and the form of phase functions of the BLR clouds. Therefore this diagram can provide basic information on the allowed BLR models before the detailed modelling of light curves is accomplished.

References

Antonucci, R. R. J. and Cohen, R. D.: 1983, *Astrophys. J.* **271**, 564.
Blandford, R. D. and McKee, C. F.: 1982, *Astrophys. J.* **255**, 419.
Cherepashchuk, A. M. and Lyutyj, V. M.: 1973, *Astrophys. Letters* **13**, 165.
Lyutyj, V. M.: 1977, *Astron. Zh.* **54**, 1153. (*Sov. Astron.* **21**, 655).
Lyutyj, V. M. and Cherepashchuk, A. M.: 1971, *Astron. Tsirk.* **633**, 3.
Peterson, B. M.: 1993, *Publ. Astron. Soc. Pacific* **105**, 247.
Shevchenko, I. I.: 1984, *Pis'ma Astron. Zh.* **10**, 896 (*Sov. Astron. Lett.* **10**, 377).
Shevchenko, I. I.: 1985a, *Pis'ma Astron. Zh.* **11**, 83 (*Sov. Astron. Lett.* **11**, 35).
Shevchenko, I. I.: 1985b, *Pis'ma Astron. Zh.* **11**, 432 (*Sov. Astron. Lett.* **11**, 178).
Shevchenko, I. I.: 1988, *Astrofizika* **28**, 59 (*Astrophysics* **28**, 35).
Ulrich, M.-H., et al.: 1984, *Monthly Notices Roy. Astron. Soc.* **206**, 221.

INTENSIVE SPECTROSCOPIC MONITORING OF NGC 5548 WITH HST AND IUE

B.M. PETERSON
Department of Astronomy, The Ohio State University,
Columbus, Ohio, USA

and

K.T. KORISTA
Space Telescope Science Institute,
Baltimore, Maryland, USA

Abstract. We present preliminary results on a combined *HST/IUE*/ground-based monitoring campaign on the Seyfert 1 galaxy NGC 5548 undertaken by the International AGN Watch in order to answer questions that require both high temporal resolution (one day) and high signal-to-noise ratios. Our preliminary conclusions are (1) the ultraviolet and optical continuum variations are simultaneous to within a day (2) the He II λ1640 variations lag behind the continuum by about 1.7 days, and (3) the velocity field of the C IV-emitting region is not characterized by bulk by radial motion, but the higher velocity gas seems to originate closer to the continuum source than the lower velocity gas.

Key words: Spectroscopy, variability, multiwavelength observations

1. Introduction

In 1988 – 89, a large consortium known as the International AGN Watch (Alloin et al. 1993) carried out a program of monitoring continuum and emission-line variability in the Seyfert 1 galaxy NGC 5548 with *IUE* and ground-based telescopes (Clavel et al. 1991; Peterson et al. 1991). This highly successful effort (summarized by Peterson 1993) showed (1) that the UV and optical continua vary in phase to within approximately 2 ± 2 days, (2) that the broad emission lines respond to the continuum variations on short time scales (days to weeks) and (3) that the highest ionization lines (e.g., He II λ1640) respond more rapidly than the low ionization lines (e.g., Hβ), providing clear evidence for radial ionization stratification of the broad-line region (BLR). A second campaign by the International AGN Watch on the southern hemisphere Seyfert galaxy NGC 3783, discussed in these proceedings in papers by Reichert and by Stirpe, produced results that are similar to those obtained on NGC 5548.

The surprisingly short response times for the emission lines and the apparent simultaneity of the UV/optical variations leaves some important questions unresolved, specifically:

1. Is there a phase difference between the UV and optical continuum variations? Whether or not the variations in the different wavebands are truly simultaneous can provide a fundamental constraint for models of the continuum emission.

2. What is the response time of the most rapidly varying high-ionization lines? The rapid variability of the highest ionization lines indicates that there are ionization fronts within a few light days of the continuum source.
3. What is the velocity field of the BLR? Determination of the velocity field provides some of the strongest possible constraints on the origin of the BLR and physical conditions within a few light days of the central source, and indeed might also lead to an unambiguous determination of the mass of the central object. The results from the original campaign are ambiguous, with Clavel (1991) arguing for random cloud motions with higher velocities close to the central source, and Crenshaw & Blackwell (1990) arguing for gravitational infall.

Resolution of these important questions requires not only better temporal sampling than was achieved in the original campaign (4 days), but higher signal-to-noise ratios as well.

2. The 1993 Monitoring Campaign

In order to address these key issues, an intensive monitoring program was undertaken with *HST*, *IUE*, and ground-based telescopes. The *IUE* observations were made once every two days between 1993 March 16 and May 27. During the 39-day period from April 19 to May 27, *HST* FOS spectra of the UV spectrum between 1150 Å and 2330 Å were obtained once per day at nearly regular intervals. Ground-based observations have been obtained throughout the entire observing season and are continuing at the time of this conference.

All of the scheduled observations were made as planned. At the present time, both the *IUE* and *HST* spectra have been reduced and closely compared. Various calibration difficulties have been encountered with approximately one quarter of the FOS spectra, and the calibration is currently being refined. We have performed preliminary analysis on most of the optical data as well. The results presented here are based solely on the preliminary versions of the reduced *HST* spectra and optical spectra; only one FOS spectrum has been excluded in this first look at the results.

The light curves for the UV and optical continua and for the He II and C IV emission lines are shown in Fig. 1, along with the cross-correlation functions (CCFs) obtained by cross correlating each light curve with the UV continuum light curve. The UV/optical continuum CCF shows that to within our currently estimated uncertainties the two continuum bands vary simultaneously. The peak of the continuum/He II CCF is at about 1.7 ± 0.5 days. The peak C IV response is at a longer delay, around 7 – 10 days, which is consistent with the result of the first campaign. The C IV line can be arbitrarily divided into four velocity ranges, a blue and red line core extending from line center to $\pm 3000 \, \text{km s}^{-1}$, and wings at larger relative velocities. The red wing is probably unreliable because of contamination by He II $\lambda 1640$. However, blue core/red core CCF shows that they are perfectly in phase, which argues strongly against a predominantly radial velocity field, at least

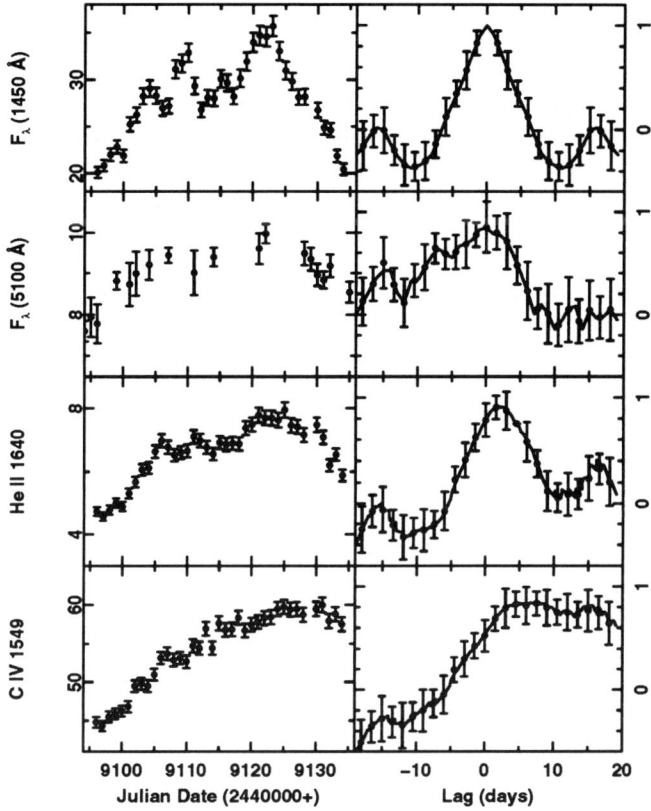

Fig. 1. The left column shows the light curves for the UV continuum, optical continuum, and two emission lines from the *HST* FOS data and contemporaneous ground-based data. The right column shows the CCFs (both interpolation and DCF methods, as described by White & Peterson 1993) produced by cross-correlating the light curve to the immediate left with the UV continuum light curve shown in the top panel (the top right panel is the continuum autocorrelation function). For the UV/optical continuum CCF, the interpolation was done *only* in the UV continuum.

out to $3000\,\mathrm{km\,s^{-1}}$. The blue wing/blue core CCF reveals that the *wing leads the core* by about 2 days, i.e., there is evidence that the higher velocity material lies closer to the central source.

3. Conclusions

Preliminary analysis of the *HST* spectra gives us great confidence that the data will allow us to address each of the questions we posed in a definitive way. We find that the UV and optical continua vary simultaneously. The He II λ1640 emission line responds to continuum variations on a time scale of about 1.7 days. A simple attempt to examine the C IV λ1549 emission line for radial velocity-dependent variations indicates that the higher velocity material lies closer to the continuum source than does the lower-velocity material. We find no evidence at this stage for large-scale radial motions in the BLR. The results presented here are very preliminary, and may change with (a) completion of the calibration of the FOS data, (b) integration of the *IUE* results, (c) completion of the optical data base analysis.

Acknowledgements

The success of this program is attributable in large part to the great cooperation of the 89 AGN Watch co-investigators for this project, the dozens of astronomers who contributed ground-based support, and the tremendous support and assistance we received from the directors and staff members at STScI and at the Goddard and Vilspa *IUE* observatories. The space-based portion of this project was supported by NASA through grants NAG5-1824 and STScI grant GO-3484.01-91A. We acknowledge support for the ground-based observations at OSU through NSF grant AST-9117086.

References

Alloin, D., Clavel, J., Peterson, B.M., Reichert, G.A., & Stirpe, G.M. 1993, in *Frontiers of Space and Ground-Based Astronomy*, ed. W. Wamsteker, M.S. Longair, & Y. Kondo, Kluwer Acad. Pub., in press

Clavel, J., in *Variability of Active Galactic Nuclei*, ed. H.R. Miller & P.J. Wiita, Cambridge Univ. Press, p. 301

Clavel, J., et al. 1991, ApJ, 366, 64

Crenshaw, D.M., & Blackwell, J.H., Jr. 1990, ApJL, 358, L37

Peterson, B.M. 1993, PASP, 105, 247

Peterson, B.M., et al. 1991, ApJ, 368, 119

White, R.J., & Peterson, B.M. 1993, in preparation

AGN VARIABILITY AND VLBI

L. B. BÅÅTH
Onsala Space Observatory
S-439 92 Onsala
Sweden

25 November 1993

Abstract. This contribution discusses the connection between variability in radio and optical with structural variations observed with VLBI. Structural changes do not have to start in the core, and intensity variations may be caused by components in the jet outside the core. The scenario is probably more complicated than present day theories assume.

Key words: Active Galactic Nuclei, Quasars, Very Long Baseline Interferometry

1. Introduction

The activity in Active Galactic Nuclei is manifested in the grand radio jets we can observe in some source, e.g. Cyg A, and in the intensity variations over the whole electromagnetic spectrum. The intensity variations have been discussed previously at this symposium (Bregman), but let me remind you that the very first detection of such variablity was in CTA102 (Sholomitskii 1965). At that time VLBI was not available and the lower limit of the angular size of the source was believed to be 0".01. The size scale was calculated from the intensity variations by Sholomitskii to be ≈ 0.1 pc, resulting in a distance to the CTA102 of ≈ 2 Mpc. Later it was shown that CTA102 has a redshift of $z=1.037$, but theories developed since then have clearly shown that beaming would allow a seemingly contradiction of the Causality theorem.

The purpose of VLBI is to study the structural changes within the fine-scale strucure of radiojets. The purpose of this talk will be to discuss the connection between such structural changes and the intensity variations:
- do large intensity changes result in structural changes?
- do structural changes result in changes in intensity?

As a corollary we can also raise the following questions:
- do all intensity variations originate in the core?
- do all structural changes originate from the core?

2. Where is our point of reference?

In order to discuss structural changes in an object it is necessary to define a point of reference with regard to the rest of the universe. A fundamental question for our further discussion is thus where the socalled "core" is. This is in VLBI usually assumed to be the component at the end of the milliarcsecond jet which has a flat

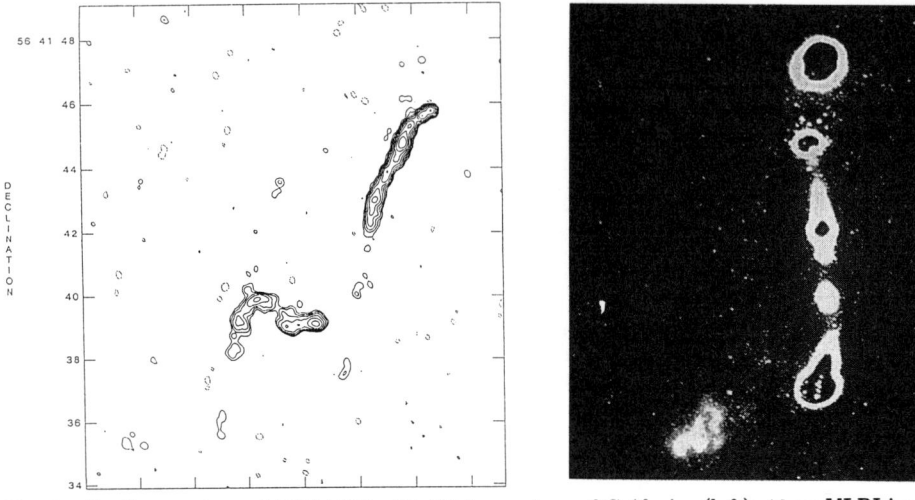

Fig. 1. Radio structure of 1857+566 with VLA, courtesy of C.Akujor (left); 18cm VLBI image of 1422+202(right)

spectrum between 5–10 GHz. It has been shown, however, that this component is *not* at the same position on the sky at all frequencies. Marcaide et al. (1985) showed e.g. that the core of 1038+525A has a positionoffset on the sky of about 1 mas between 2 and 8 GHz. Therefore we may well assume that what we state to be the core is simply the first visible part of the jet, and that the structural motions we measure are differential motions between components within the jet.

There are further indications for this from mmVLBI. The spectrum of 3C273 can be divided up into several parts, with the quiet radiospectrum broken down into a cm-wavelength part, coming from the jet emission and seen with VLBI at cm wavelengths, leaving a radiocomponent which peaks at around 150-200GHz and which is observed with VLBI at 3mm wavelength (Bååth 1992). This may well be the real "radiocore" which is practically invisible at cm wavelengths.

3. Radio structure

A significant part of the discussion about the nature of extragalactic radiosources is based on comparisons of the brightness of the large-scale structure, usually being the outer radio lobes, relative the brightness of the compact radio core. Such comparisons all assume that the radio jet is straight. Figure 1a shows one of many examples of a definitely curved radio jet, and we can from such images deduce that most radiojets are curved in three dimensions on some scale. This means that Doppler boosting effects may differ significantly between various parts of the jet and any comparison between lobes and core is meaningless if it does not take differential Doppler boosting into effect.

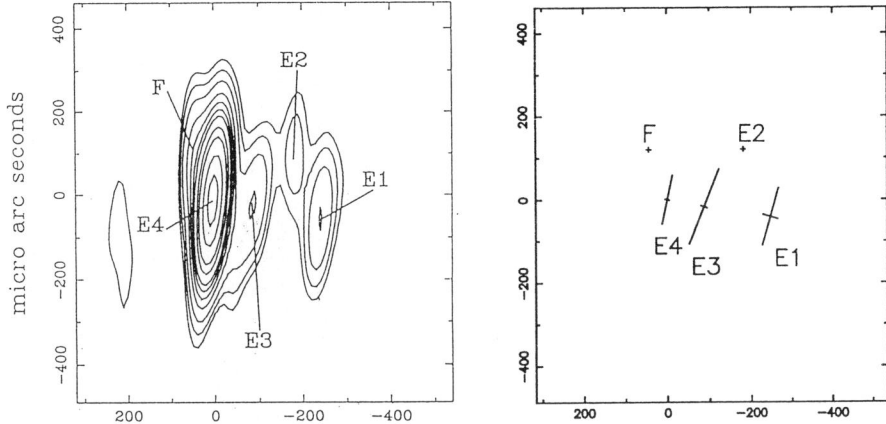

Fig. 2. Radio structure of 3C273 observed with λ3mm VLBI: hybrid map (left), major and minor axes of deconvolved Gaussian fit (right)

Figure 1b shows two images of the radio structure of the quasar 1422+202. The large scale structure is from the VLA, while the overlayed fine scale structure was observed with VLBI at λ18cm. The VLBI observations were made with phase-referencing technique and the two images can therefore be correctly overlayed on the sky. This comparison of images show that the radiojet contains fine scale structure all the way out into the southern radiolobe. and that the radio jet is curved when studied on the milliarcsecond scale level, even though it looks straight with the lower resolution of the VLA. This is another indication that differential Doppler boosting over the radio jet may cause significant misinterpretation on any lobe/core ratio.

When discussing structure, and especially when we compare structure observed with different instrument, it is important to realize the vast different in scale we may observe. Images observed with the VLA can have resolution about 3", with cm wavelength VLBI about 3–1 mas, while mmVLBI may reach 40 μas: ranging over five magnitudes in resolution! A truely amazing dynamic, resulting in a very powerful arsenal of observing instruments. The physics over such a large variation in size scales may differ though, and we may not be able to apply the same discussion on the structure of the lobes as we do on the core.

4. Intensity variations vs. structural variations

One of the best examples of a close relationship between flux outbursts and birth of new VLBI components is BLLac. Mutel et al. (1990) have shown that a series of outbursts in BLLac can be identified with new components emerging from the core. There are a number of other examples of this phenomenon. We managed in 1988 to observe 3C273 (Bååth et al. 1991) with λ3mm VLBI only about 60 days after a major outburst. The resulting hybrid map was dominated by a very bright component which could be identified with the outburst by subtracting the quiescent spectrum from the current radio spectrum, leaving a synchrotron spectrum with a λ3mm flux density that fitted the VLBI component very well. A fit of gaussian components showed that the deconvolved component (Fig. 2) was thin along the jet and extended across the jet at a ratio of about 10:1, in good agreement with the burst-model by Marscher and Gear (1985).

The BLLac objects are of course of special interest for this discussion since one of their defining characteristics is the high degree of variability. The object 1749+701 (Bååth and Zhang 1992) is especially interesting here as a good counter-example to BLLac. We did find this source to be superluminal, with components moving from the core with apparent superluminal speed. Knowing this speed it is then possible to deduce the probable time of birth of the component. We could *not* find any significant increase in flux density at any time close to this, but in all cases we did find that the birth seemed to be associated with a flip in polarization angle. This is also in accordance with the thin shock model, and the lack of a flux outburst can be explained if the geometry in this source is such that the Doppler boosting effect will not produce any enhancement of the radioation from the matter moving behind the shock front. There can be several explanations for this:

– The geometries for maximum Doppler boosting and maximal apparent speed are different.
– The structural speed does not have to be the same as the speed of the flow.

The latter point is a valid possibility, especially if the pattern, or structural, speed is caused by phase motion as suggested by Hardee (1989). Hardee suggests that the components are formed at the interception point between two modes of a reflective Kelvin-Helmholtz instability within a flow of plasma through a galactic/intergalactic medium of a different density. We have found (Rantakyrö et al. 1993) that the jet in 3C345 is confined up to a point about 20 mas from the core where it starts to expand in a seemingly adiabatical way. Our explanation to this is that this point corresponds to where we expect the jet to enter the narrow line region and thus enter a region of less density where it can expand more freely. Therefore we do suspect that the density of the surrounding medium will indeed have an effect on the jet expansion and its internal structure.

The lesson we can draw from this is that VLBI components do not necessarily have to be born at a time of a flux outburst, but there will probably have been a change in polarization angle at that time.

5. Do all changes originate from the core?

0735+178 is a good example of that flux outbursts do not have to originate in the core (Baath, Zhang and Chu 1991). Our VLBI studies of this source showed for many years two components which were *not* moving relative each other, but with a third component moving between the two. We could from this motion predict

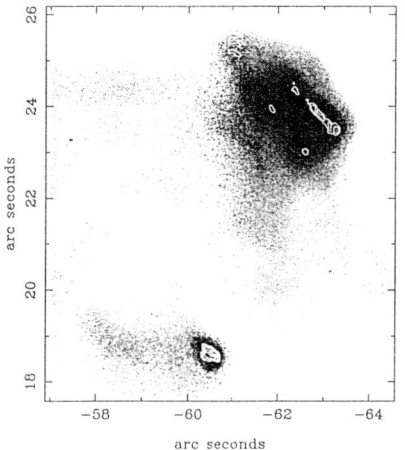 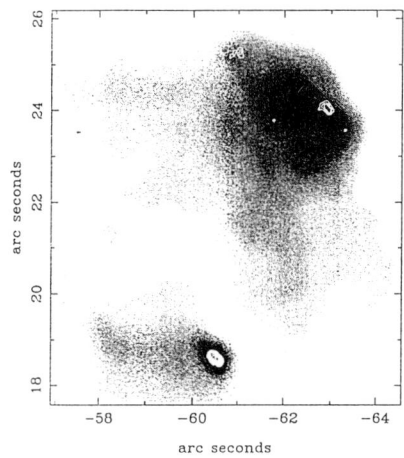

Fig. 3. The northern lobe of Cygnus A observed with the VLA (gray-scale) and MERLIN in 1983 (left) and 1993 (right)

when the moving component would reach the stable one, which took place at around 1990. At this time we observed the source with VLBI at $\lambda 1.3$ cm, and clearly the *outer*, stable, component showed up as a large, but bright, component, even though it had been invisible some years earlier at this wavelength. At the same time, the flux density of the source had steadily increased for some years, and the moving component had slowed down its apparent motion. We interpreted this as a source scenario where the jat starts at an angle of about 10° to the line of sight and then curves so that at the point of the outer, stable, component the flow is directed directly towards the observer. This change in viewing angle would explain both the slowing down of the component, since the geometry would be less favourable, and the flux increase. Such phenomenon could very well happen in any source with a curved jet, and therefore conclusions from statistics may not be very trustworthy as long as we do not know the physics and the detailed structure of the plasma flow within the jet.

For many years we have neglected to look for any structural motions within the radio lobes of large radio sources, e.g. CygA. The reason for this neglect is mainly caused by that programme committees usually prefer more safe projects, and the general impression has been that no motion should be seen at sucg large distance from the core. MERLIN gave us time for studying Cygnus A at two epochs 10 years apart, 1983 and 1993. The source is too large for conventional mapping techinques, and the observations were made in spectral line mode, where the observing band was divided up into a number of narrow frequency channels. This procedure was first tested on a 5C12-field (Okopi and Bååth 1991) and then used to map the full area covering Cygnus A with the full resolution of MERLIN at 6cm. The resulting image contains about 4000 pixel^2, and is using the VLA position of the core as the phase reference point on the sky. Our two epochs maps are shown in Figure 3, overlayed on a gray scale VLA map at λ6cm by Perley and collaborators (e.g. Dreher, Carilli, and Perley 1987). The VLA map was observed only a few days in time from our MERLIN observations in 1983, and there is indeed a very good correspondence between features. Our image from 1993 clearly indicates structural changes. The primary hotspot remains at the same place, but the north-west thin elongated feature has moved outwards. The apparent motion is about 70c, which indeed is superluminal. I do not, however, want to claim that this features started from the core and has moved all that way at a speed very close to the speed of light. A more reasonable suggestion is that this feature is a coherent structure caused by the turbulence which must be very important in the lobe, as well as within the jet itself. Such coherent structure is mainly a phase motion and can move in all three dimensions. It is therefore not very surprising if some of such structures will move in a direction which is very favourable to create apparent superluminal motion. I want to stress here that if such motion can happen in the lobes, then it can also happen in the jet which is believed to be highly turbulent. It would be very interesting to see some jet simulations where also turbulence is allowed to be an important factor.

Finally I would like to remind you that structural changes and intensity variations may arise in the medium between the source and us. Such changes may be created by gravitational lensing, especially by micro-lensing caused by star or starclusters in an intermediate galaxy. An effect which should be more common is refractive interstellar scattering (e.g. Rickett 1986), especially at long wavelengths.

References

Bååth,L.B., Padin,S., Woody,D., Rogers, A.E.E., Wright, M.C.H., Zensus,A., Kus, A.J., Backer, D.C., Booth, R.S., Carlstrom, J.E., Dickman, R.L., Emerson, D.T., Hirabayashi, H., Hodges, M.W., Inoue, M., Moran, J.M., Morimoto, M., Payne, J., Plambeck, R.L., Predmore, C.R., and Rönnanäng, B.: 1991, *Astron.Astroph.*, **241**, L1

Bååth,L.B., Zhang,F.J., and Chu,H.S.:1991, *Astron.Astroph.*, **250**, 50

Bååth,L.B.: 1992, *Variability of Blazars, eds. Valtaoja,E. and Valtonen,M. (Cambridge Univ. Press, Cambridge)*, 229

Bååth,L.B. and Zhang,F.J.: 1992, *Astron.Astroph.*,**262**, 1

Dreher,J.W., Carilli,C.L., and Perley,R.A.: 1987, *Astroph.J.*, **316**, 611

Hardee,P.E.: 1989, *Parsec-scale Radio Jets, eds. Zensus,J.A. and Pearson,T.J. (Cambridge Univ. Press, Cambridge)*, 266

Marscher,A. and Gear,W.: 1985, *Astrophys.J.*, **298**, 114

Mutel,R.L., Phillips,R.B., Bumai Su, and Bucciferro,R.R.: 1990, *Astroph.J.*, **352**, 81

Okopi,J. and Bååth,L.B.: 1991, *Radio Interferometry: Theory, Techniques, and Applications, eds. Cornwell,T.J. and Perley,R.A. (Astron. Soc. Pac. Conf. Series)*, 253

Ricket,B.J.: 1986, *Astroph.J.*, **307**, 564

mm-VLBI: Jets in the Vicinity of Galaxy-Cores

T.P.Krichbaum[1], K.J.Standke[1], D.A.Graham[1], A.Witzel[1], C.J.Schalinski[2], J.A.Zensus[3]
[1] Max-Planck-Institut für Radioastronomie (MPIfR), Bonn, Germany
[2] Institut de Radio Astronomie Millimetrique (IRAM), Grenoble, France
[3] National Radio Astronomy Observatory (NRAO), Socorro, N.M., USA

Abstract. Millimeter-VLBI provides an angular resolution of up to a few tens of microarcseconds and allows imaging of compact radio sources, self-absorbed at longer wavelengths, with unsurpassed angular resolution. At 43 GHz the participation of the VLBA and the 30 m-MRT at Pico Veleta (e.g. Krichbaum et al., 1993 a&b), and at 86 GHz the addition of the 100 m-RT at Effelsberg and the 30 m-MRT (Schalinski et al., 1993, and this volume) have improved the imaging capabilities of mm-VLBI observations.
Results: The increased sensitivity of mm-VLBI observations allows the investigation of fainter objects, previously not accessible. As one example we show in Fig.1 the first detection of the compact radio source Sgr A* in the Galactic Center with VLBI at 43 GHz in May 1992 (Krichbaum et al., 1993d) and at 86 GHz in April 1993 (Krichbaum et al., 1994). In both observations the size of Sgr A* appeared to be larger than its expected scattering size, indicative of intrinsic source structure showing up at mm-wavelengths. Future monitoring with mm-VLBI is necessary to search for (not unexpected) structural variability.
Monitoring of AGN with mm-VLBI reveals in all cases observed in sufficient detail jet curvatures of increasing amplitude towards the self-absorbed VLBI-cores (e.g. in 1803+784: Krichbaum, 1990, OJ 287: Krichbaum et al., 1993c), and sub- or superluminal motion along 'quasi-helically' bent trajectories (e.g. 3C 84: Krichbaum et al., 1993b; 3C 273: Krichbaum et al., 1993c), which differ sometimes for adjacent jet components (e.g. 3C 345: Krichbaum & Witzel, 1992, Krichbaum et al., 1992&1993a). In 3C 84, 3C 273 and 3C 345 the apparent velocity of jet components varies systematically along the jet axis, in 4C 39.25 (Alberdi et al., 1993) a moving component decelerates and brightens, all of this suggesting differential Doppler boosting and motion along three-dimensionally curved trajectories. In 3C 345 the complex kinematics of C4 and C5 (Zensus, this volume) has been geometrically modeled by motion along a helical path on the surface of a conical jet (Qian et al., 1992, Steffen et al., 1993, and this volume; see also Camenzind, this volume). As a new example, the oscillations of the inner jet and its velocity variations $\beta_{app}(r)$ are shown for the BL Lac object 1803+784 in Fig. 2 (see the maps in: Krichbaum et al., 1993b). The frequent occurence of 'quasi-sinusoidal' bends in the inner jets of very different classes of AGN (QSO's, BL Lac's, Seyfert's) suggests that this effect is common in a large fraction of AGN and that the underlying jet-physical process may be fundamental for the understanding of the creation of jets.

Acknowledgements: *We thank the staff of the observatories for their help. Special thanks for their help are due to S. Britzen, A.-M. Gontier, A. Greve, C. Naundorf, H. Steppe, R. Wegner, and W. Steffen. The work of T.P.K. was supported by the German BMFT-Verbundforschung.*

References

Alberdi, A., Krichbaum, T.P., Marcaide, J.M., et al., A&A, 271, 93.
Krichbaum, T.P., 1990, in: *Parsec-scale radio jets*, ed. J. A. Zensus and T. J. Pearson (Cambridge University Press), p. 83.
Krichbaum, T.P., and Witzel, A., 1992, in: *Variability of Blazars*, ed. E. Valtaoja and M. Valtonen (Cambridge University Press), p. 205.
Krichbaum, T.P., Witzel, A., Graham, D.A., Zensus, J.A., 1992, in: *Physics of Active Galactic Nuclei*, ed. W.J. Duschl and S.J. Wagner (Springer, Heidelberg), p. 574.
Krichbaum, T.P., Witzel, A., Graham, D.A., et al., 1993a, A&A, 275, 375.
Krichbaum, T.P., Witzel, A., Graham, D.A., et al., 1993b, in: *Sub Arcsecond Radio Astronomy*, ed. R.J. Davis and R.S. Booth, Cambridge University Press, p. 181.
Krichbaum, T.P., Witzel, A., Graham, D.A., 1993c, in: *Jets in Extragalactic Radio Sources*, ed. H.-J. Röser and K. Meisenheimer (Springer, Heidelberg), p. 71.

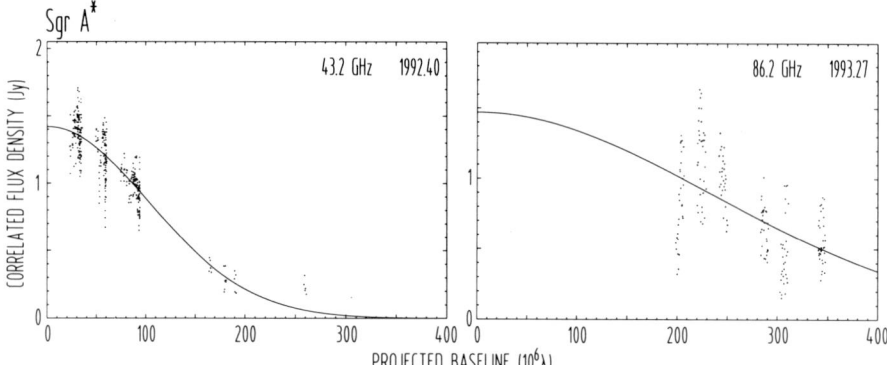

Fig. 1. Correlated flux density of Sgr A* plotted versus projected uv-distance (left: at 43 GHz, right: at 86 GHz). The solid line represents a circular Gaussian component fit to the data with flux density and size (FWHM) of $S_{43\,GHz} = (1.42 \pm 0.10)\,\text{Jy}$, $\theta_{43\,GHz} = (0.75 \pm 0.08)\,\text{mas}$, respectively $S_{86\,GHz} = (1.47 \pm 0.75)\,\text{Jy}$, $\theta_{86\,GHz} = (0.33 \pm 0.14)\,\text{mas}$. The corresponding brightness temperatures are $T_B(43\,GHz) = 1.7 \cdot 10^9$ K, and $T_B(86\,GHz) = 2.2 \cdot 10^9$ K. The scattering sizes extrapolated from VLBI observations at $\nu \leq 22$ GHz are $\theta_{43\,GHz}^{scat} = 0.53 \pm 0.02$ mas and $\theta_{86\,GHz}^{scat} = 0.13 \pm 0.01$ mas, both smaller than the source sizes given above (Krichbaum et al., 1993d&1994). Note that at $r_0 = 8.5$ kpc an angle of 0.1 mas corresponds to $1.3 \cdot 10^{13}$ cm = 0.9 AU.

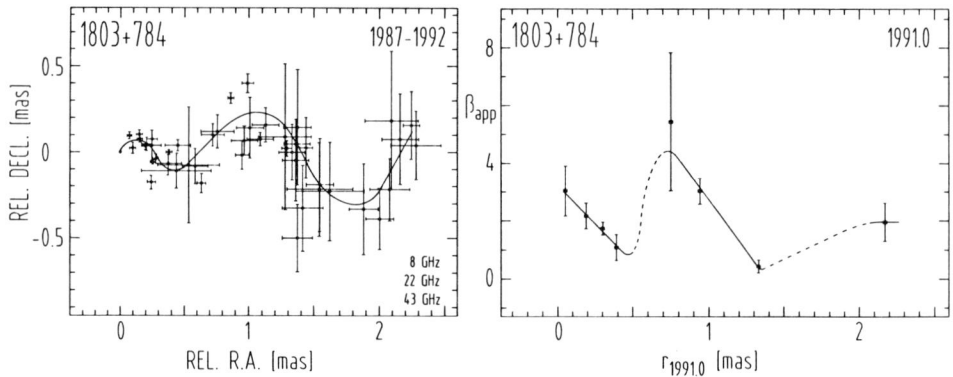

Fig. 2. Left: Relative positions of the VLBI components of the inner jet of 1803+784, obtained between 1987-1992 at 8, 22, and 43 GHz with respect to the stationary assumed VLBI-core. Data at 8 GHz are from Britzen et al., this volume. Right: Apparent velocity β_{app} ($z = 0.864$, $H_0 = 100$ km s^{-1} Mpc^{-1}, $q_0 = 0.5$) of the jet components plotted versus core-separation at epoch 1991.0. The oscillations of the 'mean jet-axis' and the systematic variations of $\beta_{app}(r)$ along the jet strongly indicate motion along a three-dimensionally bent path, e.g. a helically bent jet.

Krichbaum, T.P., Zensus, J.A., Witzel, A., et al., 1993d, A&A, 274, L37.
Krichbaum, T.P., Schalinski, C.J., Witzel, A., et al., 1994, in: *The Nuclei of Normal Galaxies: Lessons from the Galactic Center*, ed. R. Genzel (Kluwer, Dordrecht), in press.
Qian, S.J., Witzel, A., Krichbaum, T.P., et al., 1992, *Chin. Astron. Astrophys*, 16/2, 137.
Schalinski, C.J., Greve, A., Grewing, M., et al., 1993, in: *Sub Arcsecond Radio Astronomy*, ed. R.J. Davis and R.S. Booth, Cambridge University Press, p. 184.
Steffen, W., Krichbaum, T.P., Witzel, A., Zensus, J.A., in: *Sub Arcsecond Radio Astronomy*, ed. R.J. Davis and R.S. Booth, Cambridge University Press, p. 363.

High-resolution radio observations of nearby galaxies

D.J. Saikia
Tata Institute of Fundamental Research, NCRA, Post Bag 3, Ganeshkhind, Pune 411 007, India
and
A. Pedlar
Nuffield Radio Astronomy Laboratories, Jodrell Bank, Macclesfield, Cheshire SK11 9DL, UK

The central regions of many nearby active galaxies are often heavily obscured by dust and gas and are visible at only the far infrared (FIR) and radio wavelengths. High-resolution radio observations provide an invaluable tool for clarifying the dominant power source in an active galaxy which could be due to either an intense burst of star formation or an active galactic nucleus (AGN). An AGN is normally assumed to be powered by a supermassive black hole with an accretion disk, where the black hole forms as the end product of stellar evolution. There have been suggestions linking both these forms of activity to galaxy interactions, but the conditions that might lead to the formation of an AGN and how it is fuelled are not well understood.

The observed FIR emission in these galaxies is largely due to dust which absorbs the radiation from the population of hot young stars and re-radiates it at infrared wavelengths. On the other hand, the extended sources of non-thermal radio emission are due to acceleration of particles by supernovae and the subsequent evolution of the supernova remnants (SNRs), while the compact radio sources could be due to young SNRs, compact HII regions or AGN. Several diagnostic tests are however required to clarify the nature of these compact radio sources. The brightness temperature of a compact source produced by a starburst is less than about 10^5K and its radio spectrum becomes flat when the brightness temperature significantly exceeds the electron temperature of about 10^4K. On the other hand, non-thermal sources associated with an AGN have flat radio spectra when synchrotron self-absorption becomes important at brightness temperatures exceeding about 10^{10}K. The radio structures can also provide a useful diagnostic to identify whether there is an AGN in the nuclear region of a galaxy. For example, well-collimated radio jets such as the ones seen in Mkn3 (Kukula et al 1993) which traverse outwards from a flat-spectrum nuclear component (Ghosh et al., in preparation) must be due to an AGN since starbursts cannot produce such structures. The radio brightness distribution of a starburst galaxy is intimately related to the distribution of massive stars, but could be broadened by the diffusion of cosmic-ray particles due to supernova-driven winds as well as the expansion and evolution of old remnants. Another useful diagnostic for identifying whether an AGN is the dominant source of power is to look for significant departures from the very tight FIR-radio luminosity correlation which spans several decades of luminosity (see

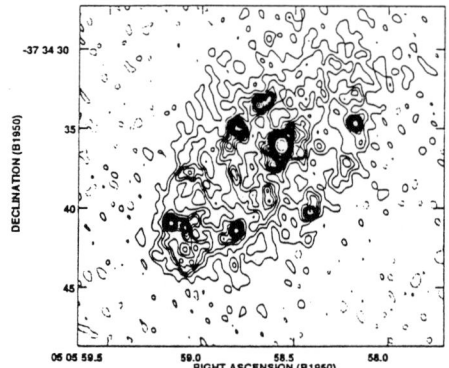

Fig. 1. Radio image of the starburst galaxy NGC1808 at 8.4 GHz

Condon 1992 for a review).

Radio images of two of the archetypal starburst galaxies NGC253 and NGC1808 observed with the VLA at 8.4 GHz (Collison et al., in preparation), show the compact radio components in the nuclear region. Comparison of these images with earlier observations of NGC253 (Antonucci & Ulvestad 1988) and NGC1808 (Saikia et al. 1990) at 5 GHz suggests that most of the components have spectral indices $\alpha \geq 0.4$ (S$\propto \nu^{-\alpha}$), suggesting that they are SNRs occurring in regions of high star formation. In order to identify a larger sample of such galaxies and also those which might be harbouring an AGN, and understand the conditions which could lead to the formation of an AGN, we started a project to study the properties of a sample of galaxies with morphologically peculiar nuclei (Saikia et al., in preparation). These galaxies, which have been christened Sérsic-Pastoriza or S-P galaxies include galaxies with diffuse and amorphous nuclei in addition to the well-known hot-spot systems. They reflect a broad spectrum of properties with some of them also having a Seyfert nucleus. A radio survey of a sample of about 50 such galaxies with an angular resolution of about an arcsec shows that while almost all the ones with prominent hot-spots are detected as radio sources, those with weak nuclei and no well-defined hot-spots at optical wavelengths are rarely detected. Compact radio components have been seen in most of the galaxies detected as radio sources. Using the peak brightness of the components and assuming their sizes to be similar to the resolution of the observations, the brightness temperatures are invariably less than about 10^5K, and hence cannot be identified unambiguously with AGN. The steep radio spectra of the individual peaks of emission for which we could determine spectral indices suggest that most of them are likely to be SNRs. The distributions of the luminosities of the compact components observed with the highest resolution at, say, 8GHz, show that they are somewhat more luminous than those in M82 (Muxlow et al. 1993), where the median value is about 10^{18} W Hz^{-1}. This is significantly higher than for SNRs in the LMC and in our Galaxy by about an order of magnitude. For the more distant galaxies, only the most luminous ones

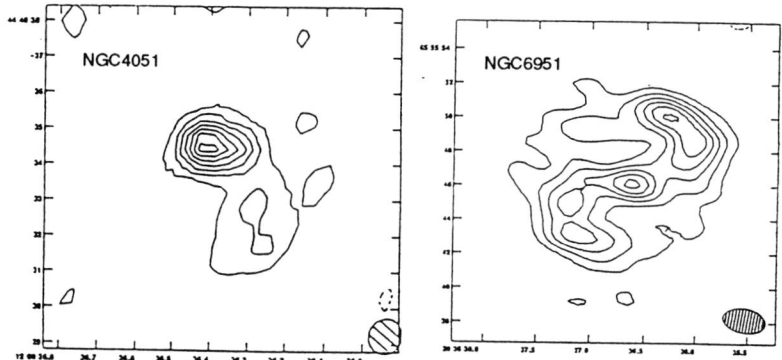

Fig. 2. Radio images of the Sérsic-Pastoriza galaxies NGC4051 and NGC6951

are detected with the weaker ones being either below the detection limit or confused by the extended emission. The luminosites of these putative SNRs are, however, smaller than the compact components in the nuclear regions of Seyfert galaxies (Kukula et al., in preparation). Of the four most luminous components seen in our sample of S-P galaxies, NGC2196 has a compact flat-spectrum component with $\alpha \sim 0.1$, NGC613 has a weak nuclear component (Hummel & Jörsäter 1992) while NGC2782 and NGC3504 have dominant compact components surrounded by more extended diffuse emission. We also find evidence of weak radio rings in the galaxies NGC1530, NGC2997 and NGC3351, in addition to the well-known ones in NGC613 (Hummel & Jörsäter 1992) and NGC1365 (Sandqvist et al., in preparation). The formation of rings could be due to a bar which causes these structures to form at the Inner Lindblad Resonance. The radio continuum emission appears to trace a spiral pattern in the nuclear regions of NGC5430 and NGC6951.

In the nearby galaxy M82, MERLIN observations have revealed details of the structure of individual SNRs. For example, at 5GHz a majority of the 23 detected SNRs show evidence of shell or partial shell structures. Their surface brightness–diameter relation is consistent with that seen for Galactic and LMC SNRs with the ones in M82 extending the relation to smaller diameters and higher values of surface brightness. The plot of the cumulative numbers of SNRs less than the diameter D is linear for the 23 detected remnants whose sizes range from about 0.5 to 4.5 pc suggesting that they are in a state of free expansion. The flux density of the SNRs is inversely proportional to its diameter, which is inconsistent with simple adiabatic losses in a synchroton emitting source (Muxlow et al. 1993).

High-resolution radio observations at different frequencies have played an important role in identifying possible AGN and establishing the close alignment of the radio structures and the cones of emission seen at optical wavelenghts (e.g. Pedlar et al. 1993 for NGC4151; Hummel & Saikia 1991 for NGC4388). The orientation of the cones relative to the plane of the galaxy, however, varies over a wide range being nearly orthogonal in NGC4388 to being closer to the plane in

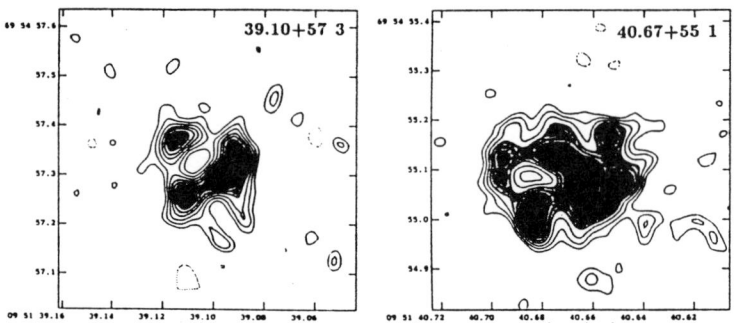

Fig. 3. MERLIN images at 5GHz of two of the remnants in M82

NGC4151.

In the somewhat more distant low-luminosity Fanaroff-Riley class I radio galaxies the radio axes tend to be along the optical minor axes of the parent elliptical galaxies. However, this minor axis trend appears to be almost entirely due to sources with prominent cores whose contribution to the total flux density at, say 2.7 GHz, is greater than about 10 per cent (Kapahi & Saikia 1982). For those with weaker cores, there does not seem to be a significant trend. The tendency for sources with stronger cores to be aligned with the minor axes may be related to the efficiency of fuelling in the nucleus depending on the relative orientation of the accretion disk, stellar and gas rotation axes and the parent galaxy. High-resolution radio observations with the VLA A-array at λ20, 6, 2 and 1.3cm of about 30 galaxies selected from the earlier sample with roughly equal number of sources with weak and strong cores confirm the earlier result and show no significant difference in the spectra of the cores (Saikia et al., in preparation). In these high-resolution observations, only the bases of the jets are often detected and they are usually very asymmetric if not completely one-sided, and tend to occur in sources with stronger cores. If the relative strength of the core is due to relativistic beaming, this would suggest that the observed asymmetry of the jets close to the nucleus in these low-luminosity Fanaroff-Riley class I galaxies is also due to relativistic beaming.

References
Antonucci R.R.J., Ulvestad J.S. 1988, ApJ, 330, L97
Condon J.J., 1992, ARA&A, 30, 575
Hummel E., Jörsäter S., 1992, A&A, 261, 85
Hummel E., Saikia D.J., 1991, A&A, 249, 43
Kapahi V.K., Saikia D.J., 1982, JA&A, 3, 165
Kukula M.J. et al., 1993, MNRAS, in press
Muxlow T.W.B. et al., 1993, MNRAS, in press
Pedlar A. et al., 1993, MNRAS, in press
Saikia D.J. et al., 1990, MNRAS, 245, 397

CORRELATED X-RAY AND MILLIMETRE VARIABILITY OF THE BLAZAR 3C273

I M M^C HARDY*, I PAPADAKIS*, C M LEACH*, E I ROBSON[♯], W JUNOR[†],
R STAUBERT[♭], H STEPPE[♮] and W K GEAR[‡]
* Department of Physics, The University, Southampton SO9 5NH, UK.
[♯] Joint Astronomy Center, Hilo, Hawaii;
[†] National Radio Astronomy Observatory, Soccoro, NM;
[‡] Royal Observatory Edinburgh, UK;
[♭] Astronomisches Institut, Universitat Tubingen, Germany;
[♮] Instituto de Radioastonomia Milimetrica, Granada, Spain.

Abstract. We present the results of X-ray and millimetre monitoring of the blazar 3C273 at 1-2 day intervals over the period 12 December 1992 to 24 January 1993. No large flares are seen in this period but variations in both wavebands of $\sim 30\%$ on few day timescales are apparent. The ROSAT PSPC X-ray spectrum consists of 2 power-law components with the harder component dominating above 0.5 keV. There is very little correlation between the variability of the soft and hard components. The soft component does not correlate with the millimetre variations, but the hard component correlates reasonably well and leads the millimetre variations by about 10 days. These results show that the hard X-ray component cannot be a simple extrapolation of the millimetre/IR synchrotron component but may be explained as a self-Compton component in a shocked jet.

Key words: Blazars, X-rays, millimetre observations

1. Introduction

It is now generally accepted that most of the emission from blazars such as 3C273 arises in a relativistic jet oriented relatively close to our line of sight, but there are many possible models for such jets. Observations of correlated X-ray and millimetre variability can, in principle, provide one of the best diagnostics for distinguishing between alternative models however previous studies have produced no firm conclusions because of poor temporal coverage (eg Courvoisier et al. 1990). Rapid variations (days - weeks) do occur in the millimetre and infrared bands (eg Robson et al. 1993) and in jet models (eg Marscher and Gear 1985) X-ray/millimetre lags of this order are expected and so we monitored 3C273 on 1-2 day intervals in the X-ray and millimetre bands for over a month to search for any such lags.

2. The X-ray Observations

3C273 was observed 14 times by ROSAT with the PSPC detector and once with the HRI detector in the period 12 December 1992 - 10 January 1993. The observations were of duration $\sim$ 2ksec and the average PSPC count rate was ~ 10 s^{-1}. For each PSPC observation it was possible to measure the spectrum in the region 0.1-2.4 keV. It proved impossible to fit the spectra adequately with a simple two component model such as a power-law plus absorbing column. The best fits are provided by a model consisting of two power-laws plus an absorbing column. When the absorbing column is allowed to float free it reaches a value below the galactic column and so it was fixed at the galactic value. The energy index of one of the power-laws was fixed at -0.5, the value found by EXOSAT and GINGA in the 2–10 keV band. The energy index of the second power-law was then derived to be ~ -1.7. This second, softer, component dominates below ~ 0.5 keV.

We derived lightcurves in various energy bands, 0.1-0.5 keV (soft), 0.5-2.4 keV (hard), 1.5-2.4 keV (very hard) and 0.1-2.4 keV (total - figure 1). No huge flares are seen but all lightcurves show variations of amplitude $\sim 20\%$ on $\sim$daily timescales. These observations show, contrary to some previous suggestions, that 3C273 does show short timescale X-ray variability; it is just that the timescales of variability are somewhat longer than in the lower luminosity Seyfert galaxies.

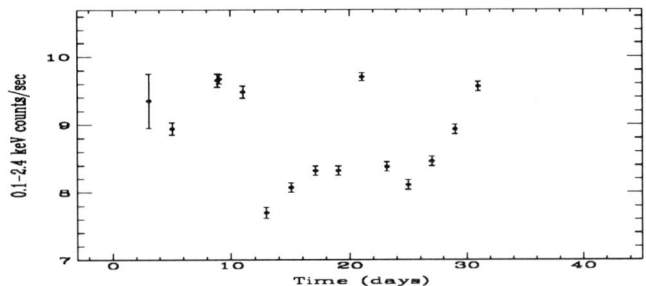

Fig. 1. 0.1-2.4 keV X-ray lightcurve. Day 0 = 0 hrs UT on 10 December 1992

The existence of two components in the PSPC band is already known (Staubert 1992) but here we are able to study their relative variability (figure 2). There is no obvious correlation; the two spectral components vary essentially independently.

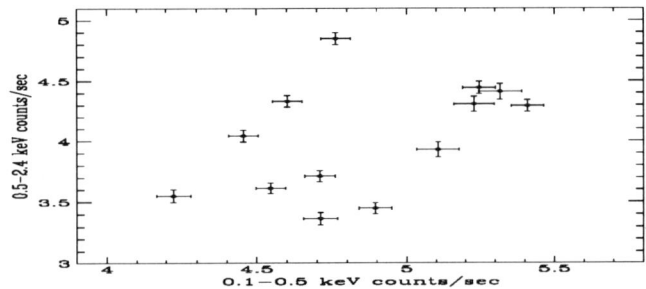

Fig. 2. Soft (0.1-0.5 keV) versus hard (0.5-2.4 keV) count rate.

3. The Millimetre Observations

Observations of 3C273 at 2, 1.1, 0.8 and 0.45mm were scheduled with the JCMT on Mauna Kea almost every night during the period 12 December 1992 - 24 January 1993. Unfortunately much of the programme was lost because of very poor weather conditions. Fortunately observations at 230 GHz were made during part of the lost period at IRAM and a small number of observations were made at 230 GHz by the NRAO 12m telescope on Kitt Peak. The millimetre spectrum of 3C273 was optically thin at this time and so, using the average spectral index (~ -0.7) we have extrapolated our 230 GHz fluxes to 273 GHz (1.1mm – figure 3). Variability of comparable amplitude to the X-ray variability, and on similar timescales, is apparent in the 1.1mm lightcurve. We also obtained some 104 GHz observations from SEST. Extrapolation to 1.1mm gives fluxes consistent with the other 1.1mm observations but, as the spectral extrapolation is quite substantial we do not, at least for the present, include these in our 1.1mm lightcurve.

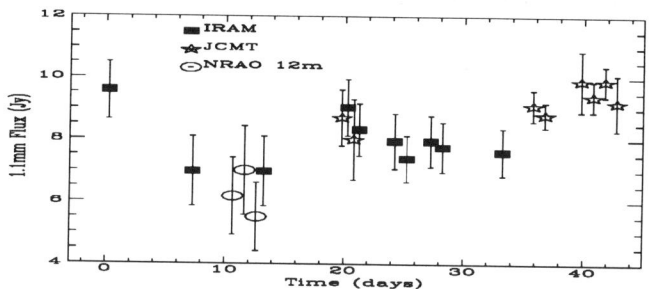

Fig. 3. 1.1mm lightcurve of 3C273. Day 0 = 0hrs UT on 10 December 1992

4. Cross-correlation of the X-ray and millimetre lightcurves

We separately cross-correlated the 1.1mm lightcurve with the various X-ray lightcurves. We used the discrete cross-correlation method of Edelson and Krolik (1988). There is a weak correlation with the 0.1-2.4 keV lightcurve with the X-rays possibly leading the millimetre variations by ~ 10 days. There is no correlation with the 0.1-0.5 keV X-rays. The best correlation is with the hardest (1.5-2.4 keV) X-rays (figure 4). The correlation is significant at about 3σ.

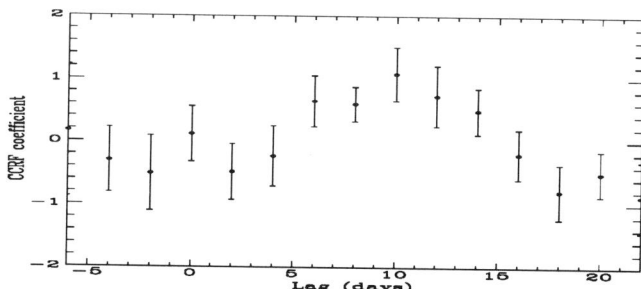

Fig. 4. Cross correlation of 1.5–2.4 keV X-ray and 1.1mm lightcurves.

5. Discussion

We have seen that the ROSAT pspc spectrum of 3C273 is described well by two power-law components, both of which vary rapidly and with large amplitude. The soft component fits broadly onto an extrapolation of the general millimetre/ infrared/optical continuum, which is generally assumed to be synchrotron emission, but does not correlate strongly with it. More work is required to determine the origin of the soft X-ray component.

The hard X-rays do correlate reasonably well with the synchtrotron component, but are not in phase. They are certainly not an extrapolation of the millimetre synchrotron component but may be produced by inverse compton emission associated with the synchrotron component. As there is a lag between the X-ray and millimetre bands the emission region cannot be optically thin and homogeneous and so a possible explanation is provided by inhomogeneous jet models (eg Marscher and Gear 1985). This model was derived to explain a flare of factor a few in the radio emission from 3C273 by means of a strong shock, but may well be applicable to the smaller amplitude variations here where disturbances propogating down the beam may encounter smaller scale weather. In such models we expect X-ray synchrotron self-compton emission to precede the synchrotron emission by $\sim$weeks and so our observations are consistent with this model. A possible problem is that the millimetre spectral index (-0.7) is slightly steeper than the hard X-ray spectral index (-0.5). In simple SSC models these indices should should be the same. It is not yet clear whether this discrepancy is a result of measurement error, or whether a real physical effect remains to be explained.

Models not involving shocks (eg Maraschi *et al.* 1992; Melia and Konigl 1989) have greater difficulty explaining both the X-ray/millimetre lag and the hard X-ray spectral index but cannot be absolutely ruled out at this stage.

References

Courvoisier, T. *et al.* 1990. A & A., 234, 73.
Edelson, R.A. & Krolick, J.H., 1988. ApJ, 333, 646.
Maraschi, L., Ghisellini, G., & Celotti, A., 1992. ApJ, 397, L5.
Marscher, A.P. and Gear, W.K., 1985. ApJ., 298, 114.
Melia, F. & Konigl, A., 1989. ApJ 340, 162.
Robson, E.I. *et al.* 1993. Preprint.
Staubert, R., 1992. AIT Preprint 1/92.

HIGH S/N ROSAT–PSPC OBSERVATIONS OF THE QUASAR PG 1116+215: POWER LAW SHAPE OF THE SOFT X-RAY SPECTRUM

M.-H. ULRICH
*European Southern Observatory, Karl-Schwarzschild-Str. 2,
D–85748 Garching, Germany*
mulrich@eso.org

and

S. MOLENDI
*Max-Planck-Institut für Extraterrestrische Physik,
Giessenbachstr. 1, D–85748 Garching, Germany*
sil@mpe-garching.mpg.de

1. Observations and data extraction

1.1. THE QUASAR PG 1116+215

The quasar PG 1116+215 (Schmidt and Green, 1983) was selected because it is an intrinsically bright low redshift quasar. With $z = 0.177$ the soft X-ray emission is not significantly redshifted outside the ROSAT band. Its apparent magnitude $m_v = 15.04$ (Véron and Véron, 1989) corresponds to $M_v = -25.0$ for $H_0 = 50$ km s^{-1} Mpc^{-1} and $q_0 = 0$. An additional selection criterion was low N_H. Radio observations give $N_H = 1.44 \times 10^{20}$ cm^{-2} (Elvis, Lockman and Wilkes, 1989).

1.2. ROSAT POINTED OBSERVATIONS

PG 1116+215 was observed at four different epochs with the ROSAT-PSPC: one observation during the ROSAT All Sky Survey and three pointed observations (Table 1). The first two observations were analyzed with the "old" calibration of the response matrix, which was the default one until December 1992. The last two observations were analyzed with the calibration online at present (July 1993). During the first pointed observation the quasar was at the center of the field and the spacecraft wobble turned off. An unknown fraction of the incident flux was intercepted by the wire structure. The analysis of the counts (within a circle of 180 arcsec radius and not corrected for background) versus time shows that the counts in bins of 400 seconds remained constant within ±15%. These fluctuations are attributed to the effect of the wire structure and small drifts in the satellite attitude. Because there were no large flux variations during the pointed observation, we estimate the uncertainty on the flux to be less than 10%. No count rate variations larger than 10% are present in the light curves of the 2 other pointed observations.

For the spectral analysis of the pointed observations, the counts were extracted in a circle 125 arcsec in radius centered on the source centroid. Only the signal between channels 11 and 210 was kept in the spectral analysis. The background was determined from a circle 250 arc sec in radius centered 600 arc sec west of the quasar in a region devoid of obvious background sources. The background subtracted spectrum was corrected for dead time and mirror vignetting, divided into bins with S/N $\gtrsim$ 10 in each bin (Table 1) and compared to trial models.

1.3. ROSAT ALL SKY SURVEY

PG 1116+215 was observed during the ROSAT All Sky Survey in the period 24–27 November 1990. The source counts were extracted from a circle 400 arcsec in diameter (362 counts). The nearby background was estimated from a circle of the same radius with its center at the same ecliptic latitude as the source (51 counts). A time analysis of the count rate was performed on the source counts before background subtraction following the method of Schaeidt, Hasinger and Trümper (1992) and Molendi, Maccacaro and Schaeidt (1993). There is no clear evidence for variation. χ^2_{red} is 1.2 consistent with the source being constant.

2. Search for spectral variations through hardness ratio analysis

We made a first investigation of the spectral shape of the PSPC spectra through hardness ratio analysis. In addition to the standard ROSAT hardness ratio, HR, (HR = (H−S)/(H+S) where H and S are the counts in the channel intervals 11–41 and 52–200 respectively), we used two hardness ratios, one in the soft band, HR_S, the other in the hard band, HR_H, defined as follows: HR_S = (B−A)/(B+A), HR_H = (D−C)/(D+C) with A, B, C and D being the counts in the following channel intervals: A 11–41, B 52–90, C 91–150 and D 151–210. No difference in the value of the hardness ratios is detected except for a hardening of the harder part of the spectrum in December 1992, valid at the 1.5σ level.

This is suggestive of and consistent with the emergence in the PSPC range of a hard component as is commonly seen in the EXOSAT and Ginga spectra of Seyfert galaxies and radio quiet quasars (Pounds et al., 1990); (Williams et al., 1992). It is also interesting to note the similarity of the spectral shape in June 1991 and May 1993 whereas the flux differs by a factor 1.5 between the two dates.

3. Power law models of the PSPC spectra

The PSPC spectra at the four epochs of observations were fitted with a power law and photoelectric absorption. A good fit is obtained at each epoch using the galactic N_{H}. Furthermore the energy index has the same value of $\Gamma \sim 2.7$ (Table 2). We note here that most of the counts are in the soft part of the PSPC spectra. The relatively large value of HR_H of December 1992 is thus diluted by the contribution of the counts in the soft part of the spectrum to the power law fit. This explains why Γ has the same value in December 1992 as at the other epochs in spite of a different value of HR_H; the reduced χ^2 is as expected larger in December 1992.

4. Power law fitting of the PSPC + IUE spectra

IUE spectra of PG 1116+215 were obtained near the time of the PSPC observations of June 1991 and April 1993 (Table 1). In our analysis of the PSPC+IUE spectra we used the same algorithms as we used for fitting the PSPC spectra alone. We present here the results with a power law model.

We ran a fitting procedure to the PSPC+IUE spectra of June 1991 and April / May 1993. The result is that a power law with a photon index $\Gamma = 2.57$, which is close to but statistically different from the index of the power law which fits the PSPC spectra, meets the UV spectrum at its short wavelength end (Table 2a,b). The slope $\Gamma = 2.57$ is less steep than the slope of the PSPC spectra alone but steeper than the slope of the UV spectra.

The IUE spectra after keeping only the continuum windows with the best S/N and free of emission lines have a harder slope than the UV / soft X-ray spectra: $\Gamma = 1.83$ for the SWP+LWP of June 1991, and $\Gamma = 2.16$ for the SWP of April 1993.

5. Conclusions of the power law fits of the PSPC spectra

The conclusion of the fitting exercise, and the main result of this work (Ulrich and Molendi, 1993) is that the high S/N PSPC spectra of PG 1116+215 are well represented by power laws. The photon index ($\Gamma \sim 2.7$) is the same for count rates differing by a factor 1.5. Comptonization of the UV photons by energetic electrons could re-shape the exponential cut off of the inner disk component into a power law similar to the one which is observed. Similarly, excellent single power law fits (+ galactic N_H) have been found for the high S/N spectra of other bright low z quasars such as GQ Comae and PG 0157+001 (in preparation) suggesting again thermal comptonization of the UV photons.

Acknowledgements

G. Hasinger is thanked for useful discussion on the calibration of the PSPC response matrix. S.M. acknowledges financial support from MPE.

References

Elvis, M., Lockman, F.J. and Wilkes, B.J.: 1989, *Astronomical Journal* **97**, 777
Molendi, S., Maccacaro, T. and Schaeidt, S.: 1993, *Astronomy and Astrophysics*, in press
Pounds, K.A., Nandra, K., Stewart, G.C., George, I.M. and Fabian, A.C.: 1990, *Nature* **344**, 132
Schaeidt, S., Hasinger, G. and Trümper, J.: 1992, MPE Report **335**, 191
Schmidt, M. and Green, R.F.: 1983, *Astrophysical Journal* **269**, 352

Ulrich, M.-H. and Molendi, S.: 1993, *Astronomy and Astrophysics*, accepted
Véron, M.-P. and Véron, P.: 1989, ESO Scientific Report No. 7
Williams, O.R., et al.: 1992, *Astrophysical Journal* **389**, 157

TABLE 1
ROSAT and IUE Observations.

ROSAT–PSPC

	Date	Effective exposure (seconds)	Count rate (second^{-1})	Number of spectral bins
Survey	24–27 November 1990	—	1.060	7
A01	29–30 May 1991	24585	1.029	84
A03(1)	20 December 1992	6109	0.993	42
A03(2)	25 May – 1 June 1993	7874	1.560	60

IUE

Date and images		Flux (10^{-14} erg cm^{-2} s^{-1} Å^{-1})
2 June 1991	SWP 41744	2.40 at 1344 Å
17 June 1991	LWP 20619	1.24 at 2640 Å
26 April 1993	SWP 47550	4.30 at 1344 Å

TABLE 2
Results of power law fits[1].

a) Power law fits for the PSPC spectra

Date	Γ	A	$\chi^2/d.o.f. = \chi^2_{red}$
November 1990	2.54	1.52 10^{-3}	3.7/5 = 0.74
June 1991	$2.71^{2.73}_{2.70}$	1.36 10^{-3}	105/82 = 1.28
December 1992	$2.65^{2.69}_{2.62}$	1.42 10^{-3}	63.7/40 = 1.59
May 1993	$2.67^{2.69}_{2.64}$	2.21 10^{-3}	44.4/58 = 0.76

b) Power law fits of the PSPC+IUE spectra

Date	Γ	A	
June 1991	2.58	1.53 10^{-3}	
May 1993	2.56	2.48 10^{-3}	

(1) All models with $N_H = 1.44 \times 10^{20}$ cm^{-2}; normalization A in photons cm^{-2} s^{-1} keV^{-1} at 1 keV. Quoted intervals are 90% confidence intervals ($\Delta\chi^2 = 2.71$) on the one interesting parameter (Γ).

Correlations between Emission Components

EMISSION COMPONENTS IN SEYFERT GALAXIES AND QUASARS

THIERRY J.-L. COURVOISIER
Observatoire de Genève
CH-1290 Sauverny
Switzerland

Abstract. The continuum emission components of Seyfert galaxies and quasars are reviewed with a particular emphasis on the correlations observed between these components. It is shown that the blue bump emission which is observed in Seyfert galaxies and quasars and not in BL Lac type objects is not hidden in the latter objects by a strong beaming of the jet emission. The shape of the blue bump is described and some consequences of these observations for the interpretation of this component are given. The relationship between the X-ray emission of quasars and the radio beaming properties of the object should be re-visited when the shape of the soft excess emission can be better established.

1. Introduction

The emission of Active Galactic Nuclei (AGN) is a mix of a large number of components of very diverse physical origin. The same component (meaning the same physical process) is observed in objects belonging to different classes of AGN. For example, synchrotron emission is observed in BL Lac objects and in radio loud QSOs and emission from warm dust is found in Seyfert galaxies as well as in quasars.

It is expected that different mixes of the components can explain the different AGN classes that are observed. One typical example for this is the difference between Seyfert galaxies of type 1 and those of type 2. Assuming that Seyfert galaxies of both types are intrinsically very similar Antonucci and Miller (1985) explained the observed differences between the two classes with the presence of a thick torus that lies in our line of sight to Seyfert 2 nuclei and thus absorbs the light emitted by the central regions (ionising continuum source and broad line region). The torus is expected to exist also in Seyfert 1 galaxies. In these objects it lies, however, outside of the line of sight to the continuum source. This model explains the difference between the two types of Seyfert galaxies by taking one of the emission component from the Seyfert 1 galaxies away. Although this model is very successful attention should be brought to the paper presented by Z. Tsvetanov in this volume (Tsvetanov et al. 1994). This paper shows that, against the expectations, HST images seem to imply that in the Seyfert 1 galaxy NGC 4151, our line of sight lies outside of the ionisation cone. According to the above picture this would imply that NGC 4151 should be observed as a Seyfert 2 galaxy rather than as a Seyfert 1.

In several unification models (e.g. Barthel 1989 but also above) the different mix of components is due to geometrical effects. Viewing an object from one direction or from another will cause a different classification, even for intrinsically very similar objects.

One emission component, the radio emission, cannot be easily hidden by interstellar matter. So extended isotropic radio emission must be a genuine component existing in only a fraction of the AGN. We will show in section 2 that there are other components that are also genuine to the class of AGN in which they are observed and cannot be due solely to geometrical considerations.

The tools we have to study the relationships between the different components in AGN are repeated observations spanning the electro-magnetic spectrum. These observations are difficult to obtain, because they imply the use of a large set of ground based and space born instruments. As a consequence, few objects have been observed well enough to be subjected to this analysis. Most of these are quoted in one place or another of the present proceedings.

2. Radio emission and the blue bump

In radio loud objects the radio emission is due to synchrotron emission. This is established from the polarisation of the emission and its variability properties (see for example the lecture notes of the 20th advanced course of the Swiss Society for Astrophysics and Astronomy, Springer 1990).

Similar models are used to explain the emission in radio loud quasars and in BL Lac objects. These models are based on the synchrotron emission of electrons moving with relativistic bulk motions at different angles to the line of sight (see the reviews by E. Valtaoja and L. Bååth, this volume).

One could wonder whether the weakness of emission lines and the absence of a blue bump in BL Lac objects is intrinsic or whether it is due to the fact that synchrotron emission is strongly beamed towards the observer in these objects and thus apparently dominates the other components. If this was the case, BL Lac objects seen from another direction than down the axis of the jet should have a significant blue bump and emission lines. Two arguments can be brought against this picture and show that the blue bump is intrinsic to the quasars and not present in BL Lac objects.

It can be shown (Courvoisier 1992) that the synchrotron emission of the radio loud quasar 3C 273 should be enhanced by a factor $\sim$ 1000 in order to be at a level similar to that of the ultraviolet flux at 1000Å. An object like 3C 273 (and at a similar red-shift) but in which beaming would be such that the blue bump and the emission lines are lost in the beamed continuum would have a radio flux of 40 Jy or more. There are no BL Lac whose radio flux is anywhere near this value. In general BL Lac object are fainter than bright QSOs (Véron-Cetty and Véron 1991).

The previous argument is based on a single object which may not be appropriate for a general argument. Using several methods Ghisellini et al. (1993) have, however, shown that the beaming factor in BL Lac objects is in average less than that of radio loud quasars. Thus if BL Lac objects had intrinsically a blue bump of the same relative strength (compared say to the radio emission measured at 90 degrees from the jet axis) as the one of quasars it would in average be apparently stronger than that of quasars (compared to the radio emission). This is in strong contrast with the fact that BL Lac objects have weak or no blue bump and emission lines.

I conclude from these two arguments that the blue bump is an intrinsic property of quasars and that it is weak or absent in BL Lac objects. The intrinsic weakness of the blue bump implies that BL Lac objects have a weak ionising flux compared to objects which have in addition to the synchrotron emission a strong blue bump that peaks in the far ultraviolet domain. The weakness of the ionising flux in BL Lac objects may explain in turn the absence (or at least the relative weakness) of the emission lines in these objects.

A further consequence of these arguments is that the blue bump may not be taken as a primary indicator for the ultimate source of energy in AGN.

The existence of intrinsic differences between radio loud quasars on one side and BL Lac objects on the other side argues in favour of having two unification schemes, one for each class. This has been suggested by Browne and Jackson (1992) who considered one such scheme for the high luminosity objects and one for the low luminosity objects. The arguments presented here tend to indicate that a strong ionising continuum is a characteristic of the high luminosity objects. It is also worth noting that weak radio galaxies, like BL Lac objects have weak or no emission lines (Woltjer 1990).

3. The blue bump

3.1. THE NATURE OF THE BLUE BUMP

The large excess of continuum radiation over the extrapolation of the infrared continuum energy distribution, the blue bump, has very often been associated with accretion disks as originally suggested by Shields (1978). This association has met difficulties in recent years (Courvoisier and Clavel 1991). The main problem is that the observed properties of the UV and optical continuum variability do not match those expected from accretion disks.

These difficulties lead us to re-examine the origin of the blue bump. We have analysed the observations taken simultaneously with IUE and ROSAT during the ROSAT all sky survey (RIASS program) for a small sample of objects (Walter et al. 1994 and Orr et al. 1994). These observations span the blue bump up to the soft excess X-ray emission. It is striking that the continuum energy distributions have their maximum and their minimum (in the soft X-rays) at energies that are in a small range (within a factor of a few), irrespective of the luminosity of the objects which differ by factors larger than 10^4. This result can be seen both in a model independent way by measuring flux ratios in the soft X-rays and ultraviolet domains and by modelling the blue bump with a power law and a cut-off. In this latter case, the energy of the cut-off is related to the slope of the ultraviolet power law in such a way that steeper power laws have higher energy cut-offs than flatter power laws, showing clearly that the soft excess is closely related to the blue bump.

These observations imply that the shape of the blue bump is independent of the luminosity of the object and of the strength of the blue bump relative to the other components. Walter and Fink (1993) studied a large sample of objects for which the ultraviolet and X-ray observations are not simultaneous. They find essentially the same results as those described here.

Standard accretion disk models require considerable fine tuning in order to explain these results. Comptonised disks can account for the spectral properties of individual objects. However, they too need fine tuning to explain the results described above. The importance of the Comptonised flux relative to the underlying disk flux needs be such that the position of the maximum flux and that of the minimum remain independent of the luminosity of the object.

Optically thin emission has been much discussed at this conference to explain the blue bump (Barvainis 1994 and references therein) as an alternative to the optically thick disk emission. It is possible to account for the soft X-ray emission of quasars using optically thin emission (see e.g. Bühler Courvoisier and Staubert 1994). Figure 1 shows, however, that in the temperature range most often considered the emission is dominated by the line emission of the plasma rather than by the continuum flux, an effect that needs be taken into account. It is interesting to note that the range of temperatures in which the line emission, and therefore the plasma emission, is strongest corresponds to the temperatures deduced from the modelling of the blue bump.

3.2. ULTRAVIOLET VARIABILITY

Important clues on the physical nature of the blue bump emission is expected from the study of its variability. Several papers in these proceedings deal with the variability of objects (see review by J. Clavel in this volume).

In order to probe whether the ultraviolet variability of AGN depend on their type we analysed all the data obtained with IUE on AGN up to 1988 (Paltani and Courvoisier 1994). We expected to find marked differences in the variability properties of different classes of AGN because the ultraviolet flux in BL Lac type objects and in some quasars is thought to be due to synchrotron emission, whereas the blue bump of quasars and Seyfert galaxies was thought to be of thermal origin.

Our study considered all low dispersion IUE spectra of AGN (Courvoisier and Paltani 1992). Sixty seven objects in this sample were observed 8 times or more and had data of adequate quality, these objects were the object of our variability study. This sample is not complete in any sense of the word except that it contains all available data.

For each object we constructed the following measure of the variability as a

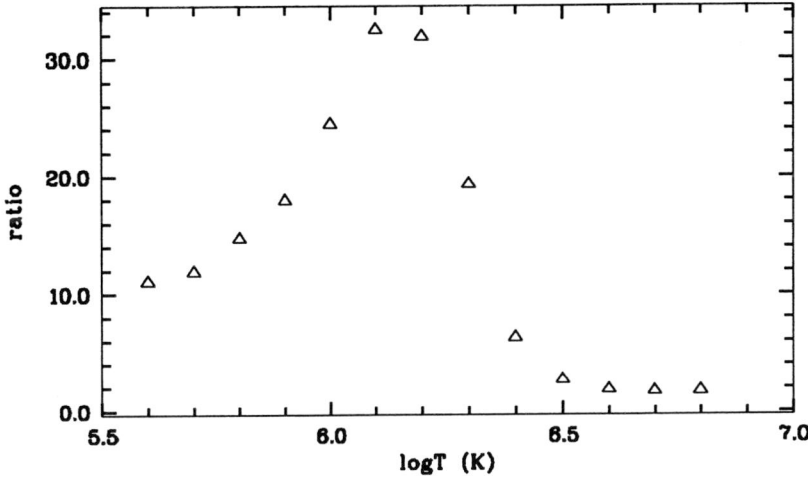

Fig. 1. Ratio of line to continuum flux in a 1.3Å to 300Å band as a function of the temperature for an optically thin plasma. The plasma model is from Mewe et al. 1986

function of wavelength:

$$s(\lambda) = \frac{[\frac{1}{N-1}\sum[f_{50,i}(\lambda) - \bar{f}_{50}(\lambda)]^2]^{1/2}}{\bar{f}_{50}(\lambda)},$$

where $f_{50,i}(\lambda)$ is the flux at epoch i averaged over a 50Å bin and $\bar{f}_{50}(\lambda)$ is the same quantity averaged over all epochs.

The wavelength dependence of $s(\lambda)$ was then parameterised using the variability at 2000Å: $\sigma_{f,2000}$ and the gradient of $s(\lambda)$ as a function of λ: $\Delta_\lambda(\sigma_f)$. It was found that 80% of the objects vary by more than 20% at 2000Å using the above measure and that in general the variability increases when the wavelength decreases. Figure 2 gives the distribution of both parameters for the 67 objects considered.

Analysis of this parameterisation of the variability as a function of various properties of the AGN led to the following findings:
- The dispersion of both parameters when correlated with the luminosity of the objects, their flux, the type of object or their spectral slope is large. We conclude that none of these properties are closely related to the observed characteristics of the variability.
- There is a small increase of variability as the luminosity decreases. This trend is much less than expected for models in which the variability is due to independent events of similar luminosity (Paltani and Courvoisier 1994).
- We could not find a significant difference between the mean of the 2000Å variability and its wavelength gradient for the different types of objects (fig. 3).

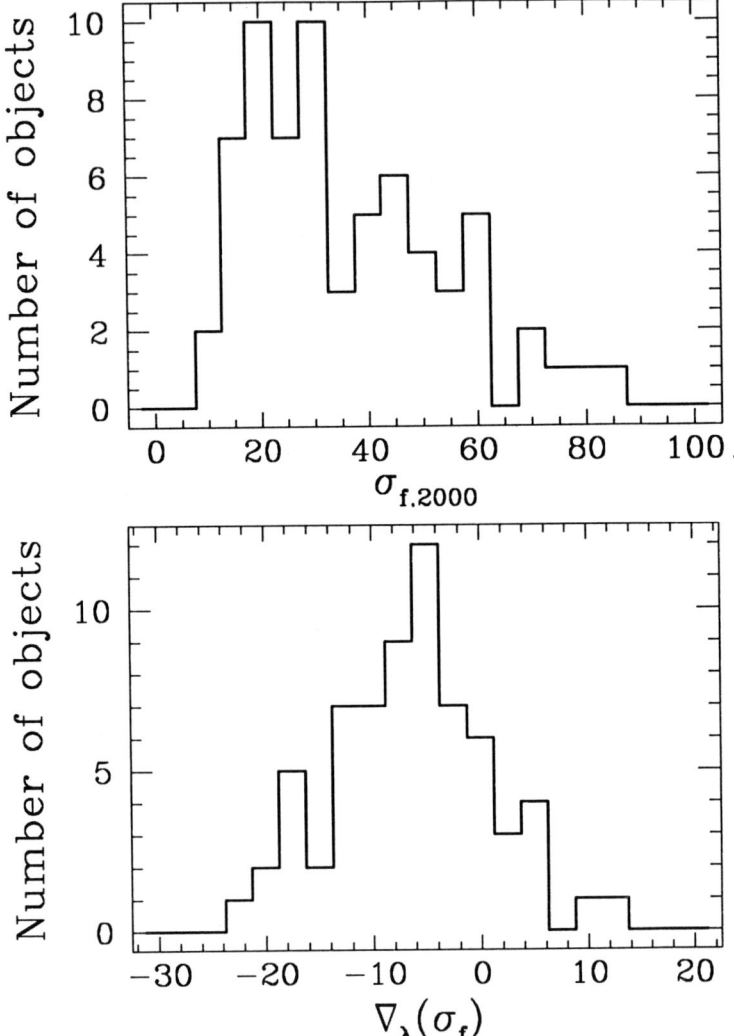

Fig. 2. Distribution of the variability at 2000Å and of the gradient of variability for 67 AGN

A more extensive study of the ultraviolet variability of AGN including the IUE data up to the end of 1992 is in preparation.

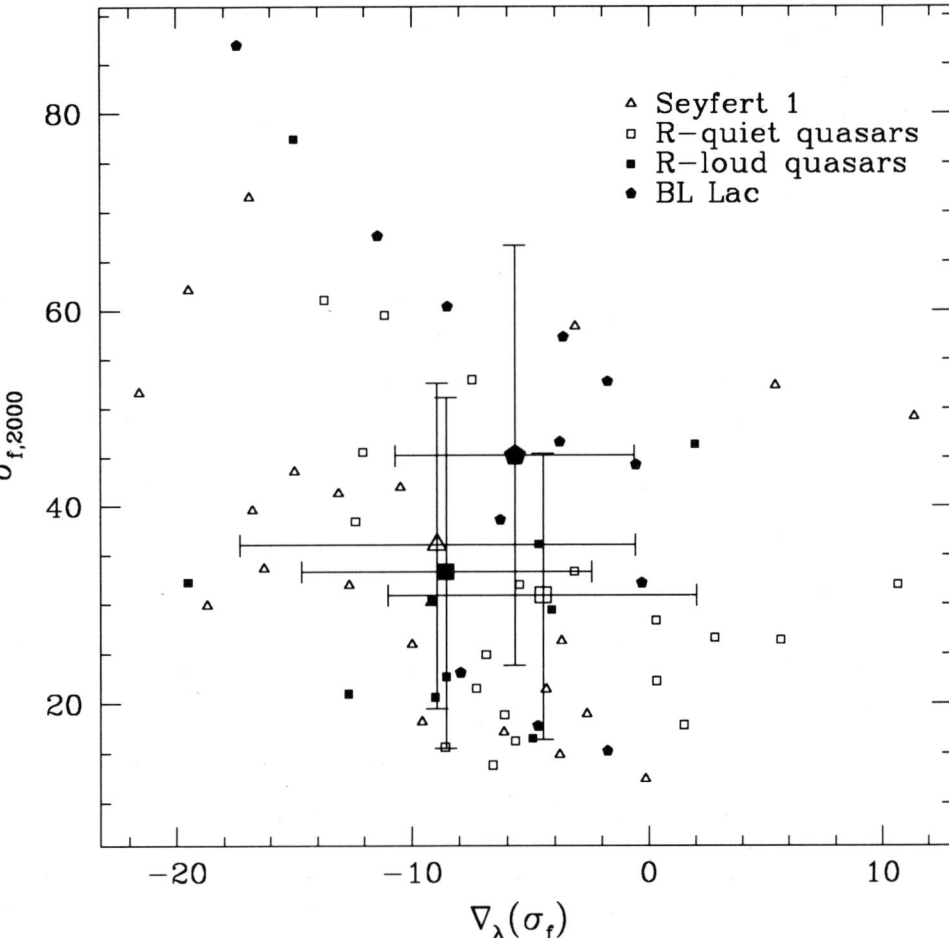

Fig. 3. Variability at 2000Å and gradient of the variability for 67 AGN. The data include all IUE observations up to 1988. The different symbols correspond to different classes of AGN. The average of each class of AGN is indicated together with the dispersion of the subsample. No significant differences can be measured between the different classes of AGN.

4. Blue bump and radio emission flux correlations

A very significant correlation is observed between the light curves of the blue bump and the radio domain for the bright quasar 3C 273 (Courvoisier et al. 1990). Recent data confirm this correlation (fig. 4). The correlation is observed when the blue bump is sampled through the ultraviolet emission, but also when it is sampled with optical data. The lag between the two light curves indicates that the blue

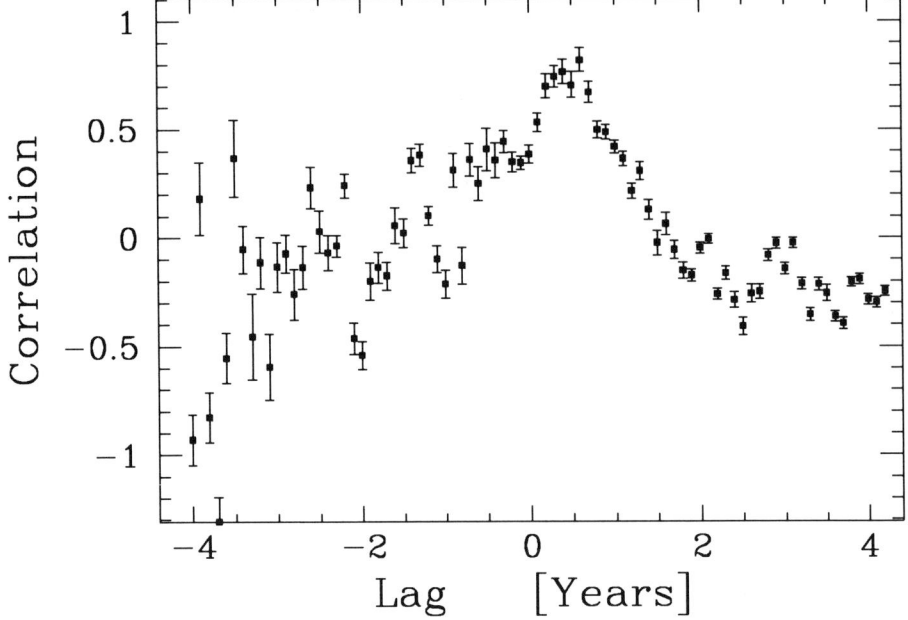

Fig. 4. Cross correlation between the V and the 22 GHz light curves of 3C 273.

bump flux leads the radio emission by approximately 6 months. It is difficult to quote a formal uncertainty on this lag.

Paltani et al. (1994) used this result together with the findings of VLBI repeated observations (Zensus et al. 1990) to estimate the distance of the 22GHz emitting region from the blue bump under the hypothesis that the observed lag is due to the time needed for a central perturbation (possibly a shock) signalled by the ultraviolet variability to move down the jet at the speed of the VLBI components. This distance was found to be $6.5/h^2 l.y.$, where $H_0 = 100 h km/(s \cdot Mpc)$.

Tornikoski et al. (1994) also studied the possible existence of correlations between the blue bump emission and the radio emission of AGN. They find a significant correlation between the two components in a number of objects. This shows that flux correlations between the ultraviolet and radio parts of the spectrum may well be a characteristic of AGN rather than an exception.

5. X-ray emission

High energy emission from AGN is reviewed in several papers in these proceedings. I therefore limit this section to a few remarks relating the X-ray emission to other components.

X-ray emission apparently results from several components: the soft excess which may be the high energy tail of the blue bump (see above), a steep isotropic power law, a flatter beamed power law (e.g. Jackson, Browne and Warwick 1993 and references therein) and signatures of the reflection of an "intrinsic" power law by cold material (Pounds, Nandra and Stewart 1992).

Jackson et al. (1993 and this volume) attempted to correlate the X-ray slopes in the ROSAT and EINSTEIN energy domains with the radio properties of the AGN. Their aim was to measure how the hard power law depends on the beaming properties of the objects. Their study could not confirm such a link. Bühler, Courvoisier and Staubert (1994), however, show that this difficulty is related to the assumed shape of the soft excess. Jackson et al. (1993) describe the soft excess with a power law which extends to high energies. Modelling the soft excess with a thermal model (or equivalently with a power law with cut-off) leads to very different results for the normalisation of the other power laws used to model the X-ray emission. Since the energy resolution of ROSAT is insufficient to uniquely characterise the shape of the soft excess, the data available cannot be used to discard a model in which part of the X-ray flux is beamed.

6. Conclusions

The emission of AGN and in particular of Seyfert galaxies and quasars results from many different components. These include line emitting gas (Broad Emission Line Region and Narrow Emission Line Region), a synchrotron emitting jet, emission from dust with a large range of temperatures, accretion disk (with or without a Comptonising corona), optically thin emission, soft X-ray excess, X-ray power laws, a reflection hump, Comptonised components, and emission resulting from pair processes.

These components are described by a large number of parameters to take into account the physical conditions of the emitting gas or plasma but also the geometrical arrangement of the emission region and its bulk velocity with respect to the observer. It is therefore quite natural that we can produce reasonable fits to the observed multi-wavelength spectra of AGN, even if not all components are present in each object. The wealth of parameters is also the cause of the lack of predictive power that many models of AGN have shown.

Although this state of affairs is not very satisfactory, significant progress has been obtained in recent years with the identification of several of the emission components and a much better description of those not firmly identified. In particular, it seems that the long wavelength emission of radio loud and radio quiet AGN is well described with the models of beamed synchrotron emission and dust emission, even if the details of the jet models are not well established yet.

In contrast, several of the discussions during the meeting showed that the nature of the blue bump emission is not understood, nor is that of the X-ray emitting components. The question as to why some AGN are radio loud whereas the majority is radio quiet is also still unanswered.

Acknowledgements

It is a pleasure to thank A. Orr, P. Bühler and S. Paltani for many discussions while preparing this paper and for the preparation of the figures.

References

Antonucci R. and Miller J., 1985, Ap.J. 297, 621
Barvainis R. 1994, this volume
Browne I.W.A. and Jackson N., 1992, in Physics of Active Galactic Nuclei, Eds Duschl W.J. and Wagner S.J. Springer, p. 618
Bühler P., Courvoisier T.J.-L. and Staubert R., 1994, submitted
Courvoisier T.J.-L., 1992, in Variability of Blazars, Eds Valtaoja E. and Valtonen M., p. 399
Courvoisier T.J.-L. et al., 1990, A&A 234, 73
Courvoisier T.J.-L. and Clavel J., 1991, A&A 248, 389.
Courvoisier T.J.-L. and Paltani S., 1992, ESA SP 1153 volumes A and B
Ghisellini G., Padovani P., Celotti A. and Maraschi L., 1993, Ap.J. in press
Jackson N., Browne I.W.A. and Warwick R.S., 1993, A+A 274, 79
Mewe R. et al., 1986, A&A Suppl. Ser. 65, 511
Orr A. et al., 1994, this volume
Paltani S. and Courvoisier T.J.-L., 1994, this volume.
Paltani S., Courvoisier T.J.-L., Bouchet P., Robson E.I., Staubert R., Tornikoski M., Ulrich M.-H., Valtaoja E. and Wamsteker W., 1993, in "The multi wavelength approach to gamma ray astronomy", submitted to Ap.J.Supp.
Pounds K.A., Nandra K. and Stewart G.C., 1992, in Physics of Active Galactic Nuclei, Eds Duschl W.J. and Wagner S.J. Springer, p. 32
Shields G.A., 1978, Nature 272, 706
Tornikoski M. et al., 1994, this volume
Véron-Cetty M.-P. and Véron P., 1991, ESO Scientific Report No. 10
Walter R. and Fink H.H., 1993, A+A 274, 105
Walter R., Orr A., Courvoisier T.J.-L., Fink H.H., Makino F., Otani C., and Wamsteker W., 1994, submitted.
Woltjer L., 1990, in Active Galactic Nuclei, Eds Courvoisier T.J.-L. and Mayor M. Springer p. 1.
Zensus A.J., Unwin S.C., Cohen M.H. and Biretta J.A., 1990, Astron.J. 100, 1777
Tsvetanov Z. et al., 1994, this volume

X-RAY EMISSION AND REFLECTION IN AGNS

HAGAI NETZER
School of Physics and Astronomy and the Wise Observatory
Tel Aviv University

Abstract. X-ray spectra of many AGNs are dominated by strong absorption-like features at low (0.5-1 keV) energies. Recent progress in modeling AGNs indicate that pure X-ray absorption is unlikely to be observed and the contribution of emission and reflection must be significant. Several such models are shown and compared with what is likely to be observed by two X-ray experiments, *ROSAT* and *ASCA*. It is argued that X-ray colors may be useful diagnostic tools in analyzing such spectra.

1. Introduction

Many low luminosity AGNs show evidence of X-ray absorption at around 1 keV (e.g. Reichert et al. 1985) and recent observations (e.g. Weaver et al. 1993) hint to the presence of X-ray emission lines at around 1-2 keV. The absorption has been interpreted as due to neutral ("cold") intrinsic absorber with partial covering (e.g. Reichert, Mushotzky & Holt, 1986) or highly ionized ("warm") intrinsic absorber (e.g. Halpern, 1984; Krolik & Kallman, 1984, Yaqoob, Warwick & Pounds 1989). The observed strength of the features have been used to estimate the column density and level of ionization of the absorbing gas (the "warm absorber" case) or the covering factor of the cold material.

A recent work by Netzer (1993) discusses the physical conditions in the gas thought to give rise to the observed X-ray absorption, and shows that emission and reflection by the highly ionized material is important in modifying the shape and the depth of the absorption features. Such effects are likely to show up in medium and high resolution X-ray spectra.

This short contribution gives several examples of recently calculated X-ray spectra, demonstrates the importance of X-ray emission and reflection, and suggests new ways of analyzing data by means of X-ray colors.

2. X-ray Absorption, Emission and Reflection

The strength of X-ray absorption features depend on the column density, level of ionization and the fractional obscuration of the source. X-ray emission and reflection are important at all situations and at most energies. The relative strength and equivalent width of such components is a function of the ionization parameter and the covering fraction.

The appearance of the spectrum depend on the origin of the X-ray reprocessing material. If these are clouds on the line of sight, then strong absorption features are likely to show up. If this is a warm material near the X-ray source, such as a thin accretion disk, the spectrum is dominated by emission and reflection. Most

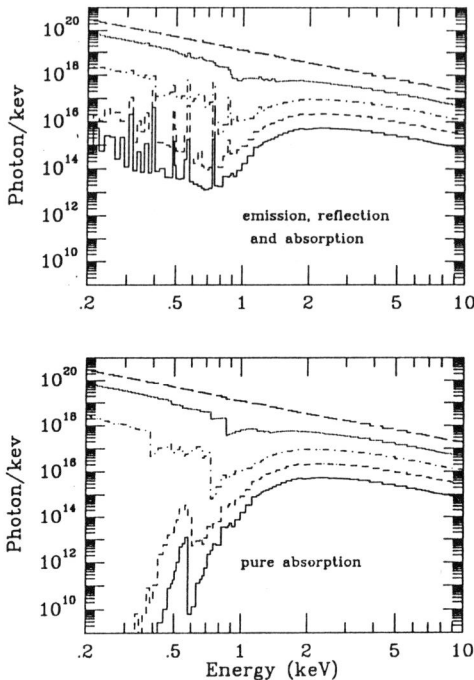

Fig. 1. Calculated X-ray spectra for an absorbing column of $10^{22.5}$ cm^{-2} and a range of ionization parameters (lowest curve is smallest ionization parameter). Top: Full spectrum (absorption, emission and reflection for a covering factor of 0.5). Bottom: The same but absorption only.

previous studies have neglected X-ray emission and low energy reflection by ionized gas. As shown by Netzer (1993), and below, this is not justified and can result in a wrong interpretation.

Typical reflection by moderate column density gas, at 0.5-3 keV, is of order 1% of the incident continuum flux for a covering factor of 0.5. At lower energies, and for larger column density material (e.g. the surface of an accretion disk) it can be much larger, close to 100%. This sets a limit to the depth of the observed X-ray absorption features (clouds on the line of sight) and gives the impression of a soft excess (the ionized accretion disk case)

Recent calculations that I have performed demonstrate this case. Fig. 1 shows an example of a $10^{22.5}$ cm^{-2} column density cloud, on the line of sight to a central X-ray continuum source, for a range of possible ionization parameters. It demonstrates the very large difference between the *pure absorption* case (bottom part) and the more realistic case where emission and reflection are included. Clearly visible are emission lines and edges and the rising continuum, at low energies, resulting from X-ray reflection (this effect was not discussed in detail by Lightman

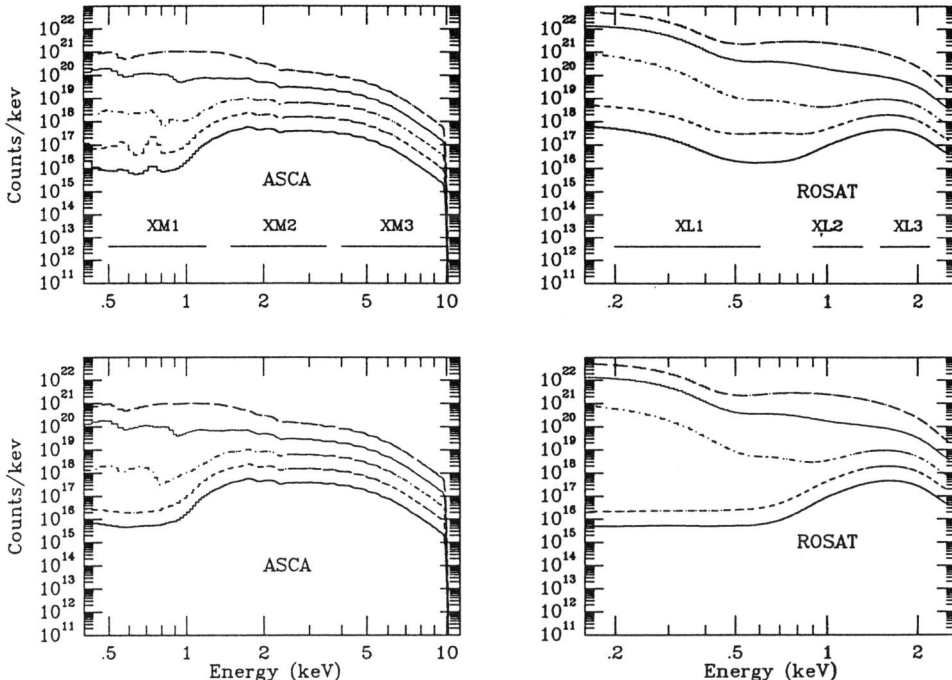

Fig. 2. The X-ray spectra of Fig. 1 as will be seen by *ROSAT* PSPC and *ASCA* SIS. Note that *ASCA* has little capability in distinguishing pure absorption from the combined spectrum. *ROSAT* with its lower energy sensitivity, is much better in this respect. The X-ray colors mentioned below are marked on the top panels.

and White, 1988, who emphasized reflection by neutral gas).

The appearance of observed X-ray spectra depend on the experiment resolution and sensitivity. Thus the *ROSAT* PSPC has very poor resolution, small energy range (0.1-2.5 keV) and high sensitivity at low energies. This makes it difficuly to detect X-ray features. The recently launched *ASCA*, with its much improved resolution and range (0.4-10 keV), is potentially better, but the limited sensitivity at low energies, and partial charge in the CCD detectors, make it difficult to distinguish pure absorption cases from others. This is illustrated in Fig. 2 that shows the predicted theoretical spectra after convolving with the response matrices of the above two experiments. The shape of the spectrum at low energies is enough to separate the two cases of Fig. 1, using *ROSAT* (PSPC), but not in the *ASCA* (SIS) case.

3. X-ray Colors

The idea of broad band data, and their ratios (*colors*), extensively used in the analysis of optical and infrared observations, can be adopted in the X-ray range. This method suffers from the poor spectral resolution but has the advantage of improved S/N and of uncovering the overall energy distribution of the source. Recently we (Netzer, Turner & George, in praparation) have studied this idea by investigating several possible X-ray colors and their relations to the spectrum. Such bands must be specified for every experiment and our best choice for *ROSAT* and *ASCA* are shown in figure 2. We find this method very useful in separating AGNs into groups according to their continuum properties. We also find that the X-ray colors, combined with time variability data, can distinguish those AGNs with intrinsic continuum shape variations from those where the spectral changes are due to changes in the level of ionization of the absorbing material. We expect that the method will be very useful in analyzing the newly obtained *ASCA* data.

Acknowledgements

It is a pleasure to acknowledge much help from members of the Laboratory for High Energy Astrophysics at Nasa/Goddard and a senior NRC research fellowship. The comments and contributions of my collaborators, Jane Turner and Ian George, helped a lot in preparing this paper.

References

Halpern, J.P, 1984, ApJ, 281, 90.
Krolik, J.H., and Kallamn, T.R., 1984, ApJ, 286, 366
Lightman, A.P., and White, T.R., 1988, ApJ, 335, 57
Netzer, H., 1993, ApJ, 411, 594.
Reichert, G.A., Mushotzky, R.F., Petre, R., and Holt, S.S, 1985, Ap.J. **296**, 69.
Reichert, G.A., Mushotzky, R.F., and Holt, S.S, 1986, Ap.J. **303**, 87.
Weaver, K.A., *et al.*, 1993, ApJ (submitted)
Yaqoob, T., Warwick, R.S., and Pounds, K.A., 1989, MNRAS, 236, 153

THE SOFT X-RAY EXCESS IN QUASARS AND DEEP X-RAY SURVEYS

AJIT KEMBHAVI
Inter University Centre for Astronomy and Astrophysics
Ganeshkhind, Pune 411 007
India

and

A. C. FABIAN
Institute of Astronomy
Madingley Road, Cambridge CB3 0HA
England

Abstract. We examine the surface density of quasars expected in the *ROSAT* PSPC surveys, using quasar optical luminosity functions and model continuum spectra including a soft X-ray excess. We show that the surface density is a sensitive function of the characteristic energy of the soft excess, and the absorption within the quasar as well as the interstellar medium. Comparison of the predicted surface density with the results of various surveys allows useful constraints to be put on the nature of the soft excess.

Key words: *ROSAT, Quasars, Soft X-ray Excess*

1. Introduction

It is now well established that in the continuum spectrum of quasars and AGN, there is excess emission in the soft X-ray region with $\leq 0.5\,\mathrm{keV}$, over the extrapolation of the power-law spectrum which is observed in the $\sim 1 - 10\,\mathrm{keV}$ region. This excess has been observed directly as residuals from power-law fits to the spectrum of 3C273 (Turner *et al.* 1990) and a few other quasars and Active Galactic Nuclei (AGN). Indirect evidence for the ubiquitous presence of the soft excess is also available from spectral fits to *EINSTEIN* IPC and *EXOSAT* data. Many objects in these samples show X-ray measured coloumn density N_H lower than the galactic line-of-sight-value (Wilkes and Elvis 1987, Turner *et al.* 1989), an obvious explanation for which is the presence of excess soft X-ray emission. The energy resolution presently available at soft X-ray energies is however too poor for the spectral shape to be determined, or even to constrain the parameters of a given model.

The optical luminosity function of quasars is now well determined to a reshift $z \simeq 3$, and given a continuum spectrum which extends from the optical to the X-ray region, this can be used to predict the number of quasars per unit area of the sky expected to appear in an X-ray survey. We will show that the predicted surface density, as a function of the flux in the soft X-ray region, depends sensitively on the assumed soft X-ray excess. Comparison with the results of recent deep X-ray surveys (Hasinger *et al.* 1992, Shanks *et al* 1989) then allows constraints to be put on parameters which define the excess, and to see their relationship with other factors which determine the broad-band shape of the continuum spectrum.

2. The Continuum Spectrum

In the optical and $\sim 1-10\,\text{keV}$ X-ray regions, the continuum spectrum is reasonably well represented by a simple power law, with spectral indices $\alpha_{op} \simeq 0.5$ and $\alpha_{xh} \simeq 0.7$ respectively. In order to keep the number of free parameters to the minimum, we represent the soft excess by a simple exponential with characterstic energy E_{xs}, and express the continuum by

$$L(E) = B_{op} E^{-\alpha_{op}} exp(-E/E_{xs}) + B_{xh} E^{-\alpha_{xh}}, \qquad (1)$$

where B_{op} is determined given the optical luminosity, and B_{xh} using the spectral index α_{ox} which connects the emission at 2500Å with that at $2\,\text{keV}$. This spectrum clearly reduces to a power-law in the optical and $\sim 1-10\,\text{keV}$ regions as required, and has excess emission at $E \simeq E_{xs}$. The excess can also be represented by a Planckian, but in this case there is an extra parameter.

3. The Quasar Optical Luminosity Function

Large complete samples of quasars, selected in ultra-violet excess (UVX) or multi-colour surveys, have become available over the last several years, and using these it has been established that the optical luminosity distribution of quasars may be be represented as a power-law which is relatively flat at low luminosities and steep at high luminosites; the characteristic luminosity at which the change in slope occurs increases with redshift, leading to pure luminosity evolution. We will use the luminosity function of Boyle et al. (1991), which is based on a UVX sample, augmented by quasars found using other techniques for $2.2 < z < 2.9$. The sample is complete to $m_B < 21$ in the redshift range $0.3 < z < 2.9$, and the derived luminosity function fits the observed surface density very well except at the bright apparent magniude. The form of the function is

$$\Phi(M_B, z) = \frac{\Phi^*}{10^{0.4[M_B - M_B(z)](\alpha+1)} + 10^{0.4[M_B - M_B(z)](\beta+1)}}, \qquad (2)$$

with $M_B(z) = M_B^* - 2.5 k_L \log(1+z)$. The values of the constants are $\alpha = -3.8$, $\beta = -1.6$, $M_B^* = -22.6$, $k_L = 3.45$ and $\Phi^* = 6.5 \times 10^{-7}$. A similar function has been proposed by Warren et al. (1993) at higher redshifts. It follows from their luminosity function that the space density of quasars decreses rapidly beyond $z \simeq 3$, so that their contribution to the surface density of the X-ray sources is not considerable.

4. Quasar Surface Density in the X-ray Band

The PSPC on *ROSAT* has a large collecting area and extremely low intrinsic background, and is especially sensitive at soft X-ray energies, which makes it ideally suited to detect a soft X-ray excess. The number of counts per second produced in some energy band of the PSPC by a quasar is determined by the form of the luminosity as a function of energy given in Equation (1), its redshift and the effective area of the detector as a function of energy. The surface density of quasars as a function of the count rate can be found by : (1) finding the number of quasars in small ranges of M_B and z from their luminosity function, (2) finding their PSPC count rate and (3) repeating the procedure over the range of M_B and z required.

The results of such a calculation are summarized in the Figure below, where we have shown the integral surface density of quasars as a function of the count rate in the $(0.15 - 2\,\text{keV})$ band of the PSPC.

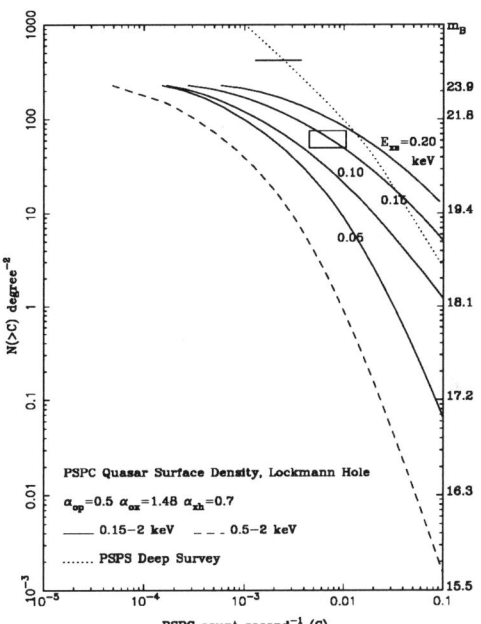

The numbers on the right edge of the box indicate the apparent blue magnitude at which the integral surface density of quasars shown on the left edge is reached. The surface density depends most sensitively on the characteristic energy E_{xs} of the soft excess and the hydrogen coloumn density N_H. The ROSAT Deep Survey (Hasinger *et al.*, 1993) was carried out in the region of the *Lockmann Hole* with $N_H = 5.7 \times 10^{19}\,\text{cm}^{-2}$, which is the minimum value known in the sky. Qusars as a class are not known to have significant intrinsic N_H, and therefore we choose N_H as in the Deep Survey. Results of the Deep Survey have been reported in the $(0.5-2\,\text{keV})$, energy band, which misses the soft X-ray photons. The translation of count rate from this band to ours depends on the shape of the spectrum and redshift, and for the range of these parameters being considered, 1 PSPC count in the $(0.5-2\,\text{keV})$ band corrsponds to 10 ± 0.04 counts in the $(0.15-2\,\text{keV})$ band. The horizontal line in the diagram shows the surface density of $443\,\text{deg}^{-2}$ of all X-ray sources at the limit of the Deep Survey observed by Hasinger et al. , translated to our band. The dotted line shows the integral surface density of X-ray sources in the Deep survey as a function of the count rate, and it is clear that quasars contribute only a fraction of the sources, especially at the faintest limits obtained through fluctuation analysis. A direct constraint on the soft excess is provided by the results of Shanks et al. (1991) who carried out an optical identification programme to obtain a surface density of 63 ± 12 quasars deg^{-2} at the limiting $0.5-2\,\text{keV}$ flux of $\sim 10^{-14}\,\text{erg}\,\text{cm}^{-2}\,\text{s}^{-1}$. The observed surface density, together with the 1σ error-bar translates in our band to the small rectangle shown in the figure. It is clear that soft X-ray emission with $E_{xs} > 0.15\,\text{keV}$ would produce an excess of quasars, while emission with $E_{xs} < 0.05\,\text{keV}$ would lead to a deficit relative to this limit. The allowed range of E_{xs} depends upon the value of α_{ox} with the bound $E_{xs} < 0.1\,\text{keV}$ for $\alpha_{ox} = 1.38$, $E_{xs} < 0.13\,\text{keV}$ for $\alpha_{ox} = 1.48$ and $E_{xs} < 0.15\,\text{keV}$ for $\alpha_{ox} = 1.58$. If the excess is represented by a black-body spectrum, the temperature which leads to a surface density consistent with observation is $kT \simeq 40\,\text{keV}$, but this value depends upon the total energy present in the black-body radiation.

At the flux limit of the Deep Survey, quasars contribute only a fraction of all the sources. Since the surface density of quasars at this level is close to the total surface density of quasars, it is clear that other populations will be making an increasing contribution at the faint levels. We have shown (Kembhavi and Fabian, 1993) that Seyfert galaxies contribute at most several hundred sources per square degree, so that the remaining sources, most of which are extragalactic (Hasinger et al., 1992), should be liners, starburst galaxies or normal galaxies.

Acknowledgements AK wishes to thank Professor D. Lynden-Bell, Director, Institute of astronomy, Cambridge for funding a visit during which much of this work was done. ACF acknowledges the Royal Society for financial support.

References

Boyle, B.J. et al., 1991. In Crampton, D., ed., *ASP Conference Series* **21**.
Hasinger. G., et al., 1993. Preprint 238, *Max-Planck-Institut für Extraterrestrische Physik*.
Kembhavi, A. K. and Fabian A. C. , 1993. Preprint.
Shanks, T., et al., 1991. *Nature*, **353**, 315.
Turner, M.J.L. and Pounds, K.A., 1989. *MNRAS*, **240**, 833.
Turner, M.J.L. et al., 1989. *MNRAS*, **244**, 310.
Wilkes, B.J. and Elvis, M., 1987. *Astrophys. J.*, **323**, 243.

ON THE BROAD BAND ENERGY DISTRIBUTION OF BLAZARS

Laura Maraschi
Dipartimento di Fisica, Università di Milano, via Celoria 16, 20133 Milano, Italy

Gabriele Ghisellini
Osservatorio di Torino, Strada Osservatorio 20, 10025 Torino, Italy

and

Annalisa Celotti
Institute of Astronomy, Madingley Road, Cambridge CB3 0HA, United Kingdom

Abstract. The broad band energy distributions of blazars are revisited with particular emphasis on the sources detected in γ-rays by the Compton Observatory (GRO). The observed distributions can be broken down into two main components, corresponding to two broad peaks in the νF_ν representation. The first occurs in the FIR–optical range, the second in the MeV–GeV region. In the case of MKN 421, which may be representative of X-ray selected BL Lacs, the first peak is shifted to higher frequency ($\simeq 10^{16}$ Hz) and the γ-ray spectrum extends to TeV energies. There is general agreement that the first spectral component is due to synchrotron radiation from a relativistic jet, although some problems remain in deriving the spectrum and location of the emitting relativistic electrons. The second component, which in most objects extends from the X-ray to the γ-ray range, can be naturally interpreted as inverse Compton scattering by the same electrons producing the synchrotron photons, either on the synchrotron photons themselves (SSC) or on photons external to the jet. It is argued that multifrequency studies of these sources including γ-rays will allow to test Inverse Compton models and to distinguish between different sources of photons.

Key words: Active Galactic Nuclei: blazars - Radiation Mechanisms: Synchrotron, Inverse Compton - UV - X-rays - γ-rays

1 Introduction

A blazar usually exhibits a compact radio core with flat or inverted spectrum, rapid variability at all wavelengths and significant optical polarization. There are two main subgroups within blazars: objects characterized by the absence or weakness of emission lines (EW$\leq$ 5 Å) are called BL Lacs while objects exhibiting typical quasar lines are designated as flat spectrum (FSQ) or core dominated (CDQ) or optically violently variable (OVV) or highly polarized (HPQ) quasars depending on which of the above properties is present and/or needs to be emphasized.

Of all Active Galactic Nuclei, blazars have the widest electromagnetic spectrum, extending from the radio to the γ-ray range. There is presently general agreement that the blazar continuum is due to non-thermal radiation emitted by plasma moving at relativistic speed (relativistic jet) in a direction close to the line of sight, as originally proposed by Blandford & Rees (1978). Synchrotron radiation can account for the emission up to the UV and, in some cases, X-ray band. Several models attribute the γ-ray emission to Inverse Compton scattering of relativistic electrons off various sources of soft photons (see e.g. Sikora 1993), but alternative mechanisms have also been proposed (see e.g. Mannheim 1993, Ghisellini 1993)

A critical parameter in estimating the physical conditions in these objects is the Doppler or beaming factor $\delta = (\Gamma - \sqrt{\Gamma^2 - 1}\cos\theta)^{-1}$, where Γ is the bulk Lorentz factor and θ is the angle between the velocity and the line of sight. The observed power is related to that emitted in the rest frame of the jet by $L_{obs} = L_{int}\delta^p$, where $p = 4$ for a moving blob and $p = 3$ for a continuous jet of plasma.

Blazars as a class are characterized by bulk Lorentz factors of order 10, as directly shown by the superluminal knots and derived by a simple application of the SSC theory to the radio core (see Ghisellini et al. 1993). It is interesting to recall that the very condition of transparency to γ-rays, together with the short time scale variability, also allows to derive a lower limit to the beaming factor (Maraschi, Ghisellini & Celotti, 1992).

On average BL Lacs have similar Lorentz factors as CDQs, but lower beaming factors, implying somewhat larger average viewing angles (Madau et al. 1987, Ghisellini et al. 1993). Thus the idea that BL Lacs have emission lines comparable to those of CDQs, which are invisible because of a highly enhanced continuum, can be dismissed. We then conclude that the "thermal" power (including emission lines and the UV bump) in BL Lacs is intrinsically weak. Exceptions are possible, and some 'hidden' CDQs may be present among high redshift BL Lacs.

Here we summarize recent observational results on the broad band energy distributions of BL Lacs and CDQs, most notably the discovery of γ-ray emission, but also observations in the IR–optical–UV and X-ray bands, which improve our knowledge of the overall continuum shape (§2). In §3 we discuss synchrotron models for the continuum up to the UV and in some cases X-ray band. An interesting constraint (§4) is imposed by the observed correlation and delay of few hours between the UV and X-ray light curves of the BL Lac object PKS 2155-304. This object, together with MKN 421 and other X-ray bright BL Lacs may represent a subclass of BL Lacs with extreme values of the X-ray to radio flux ratio.

In §5 we discuss inverse Compton models for the production of γ-rays, where the scattering electrons are the same that produce the synchrotron continuum in the jet and the scattered photons may be either the synchrotron photons within the jet or ambient photons produced by other mechanisms outside the jet. We examine possible ways of discriminating between different models by studying the correlation of variability of γ-rays with other frequency bands.

2 Observed Energy Distributions

Systematic studies of the Broad Band Energy Distributions (BBED) of blazars were started by Impey and Neugebauer (1988) and followed by Ghisellini et al. (1986) and Landau et al. (1986). These authors found that the spectra from the radio to the UV band were "curved" in the sense that the "average slope" increased with frequency.

However, recent accurate simultaneous measurements in the IR–optical–UV

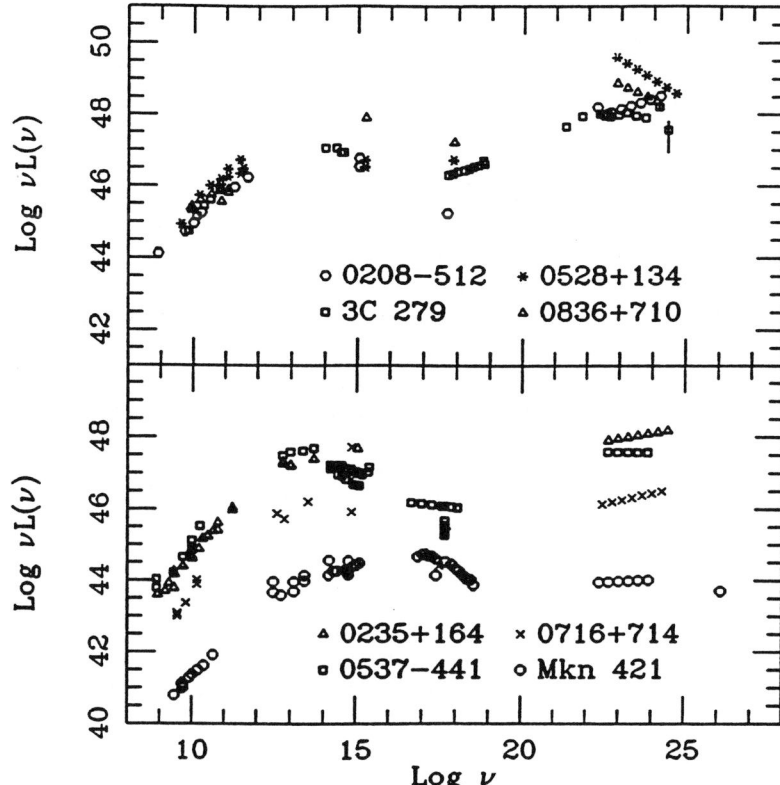

Fig. 1. The broad band energy distributions of 4 HPQs (upper panel) and 4 BL Lacs (lower panel) are plotted using the $\nu L(\nu)$ representation. The data are collected from the literature, and in general are not simultaneous. For 3C 279 see Hartmann et al. (1992) and Maraschi et al. (1992) and references therein; for PKS 0208-512 see Bertsch et al. (1993). For the other sources, radio data are from Kuhr et al. (1981), Wiren et al. (1992), Wall & Peacock (1985), Ledden & O'Dell (1985), Valtoja et al. (1992); far IR data from Impey & Neugebauer (1988); IR, optical and UV data from Maraschi et al. (1986), Ghisellini et al. (1986) and references therein, Tanzi et al. (1989), Kuhr et al. (1981); X-ray data from Worral & Wilkes (1990), Treves et al. (1993); γ-ray data from Hunter et al. (1993 a,b) Thompson et al. (1993 a,b) Lin et al. (1992), Punch et al. (1992).

bands (e.g. Falomo et al., Edelson, these proceedings) show that the continua of several objects are well described by a single power law in this extended range, contrary to the notion of continuous curvature. Giommi et al. (these proceedings), using archival data on BL Lacs, show that the BBED from the mm to the X-ray band can be described by two power laws with a break. Different objects are characterized mainly by a different break energy, while the spectral shape below and above the break is similar for all objects. Thus a new picture seems to emerge in which in individual objects changes of slope occur in relatively narrow frequency ranges. This is not inconsistent with the earlier results, if the increasing slope in the "average" spectrum is due to an increasing fraction of objects in the sample having break energy below the considered band.

Ghisellini at al. (1986) noted that the optical–UV continuum of X-ray selected BL Lacs was flatter than that of radio selected ones. This has been recently confirmed on a larger sample (25 radio selected, 8 X-ray selected) by Pian and Treves (1993). Adopting the above point of view, the flatter spectra may be due to a spectral break in X-ray selected objects occuring at a very high frequency.

The BBED up to the γ-ray range of 4 HPQs and 4 BL Lac objects are shown in Fig. 1. The chosen objects represent all BL Lacs, and representative cases among the 11 CDQs, for which the γ-ray flux and α_γ were available (Hartman et al. 1993). The BBED of the remaining 7 CDQs closely resemble those plotted in Fig. 1.

We note some remarkable features: the BBED of all objects show a relative maximum between the radio and X-ray bands. For HPQs the maximum should fall between the mm and IR bands (10^{12-14} Hz), a spectral region where only for few objects IRAS data are available.

For three of the BL Lacs the maximum falls in the same frequency interval as for HPQs (note that all the four BL Lacs have IRAS data.) However for MKN 421 the maximum occurs at a much higher frequency, close to 10^{16} Hz. In this case the X-ray spectrum is very steep and appears to be an extension of the optical–UV continuum. This seems to be a general property of the X-ray emission of radio weak, X-ray selected BL Lacs, which have an average spectral index of $<\alpha_X> = 1.4 \pm 0.1$ (Sambruna et al. 1993). Here and in the following α is always defined as $F_\nu \propto \nu^{-\alpha}$. On the contrary the X-ray spectra of HPQs are generally flat with $<\alpha_X> \sim 0.5$ while those of radio-strong BL Lacs appear intermediate, with $<\alpha_X> \sim 1$ (Worral & Wilkes, 1990)

Joining the X-ray flux with the γ-ray flux, a rising slope [in $\nu F(\nu)$] is obtained in all cases except for MKN 421. This indicates a new spectral component which we attribute to the inverse Compton process. For the four BL Lacs $\alpha_\gamma \leq 1$ indicating that the maximum power of this component is in or beyond the GeV range. In the case of MKN 421 the γ-ray emission extends up to TeV energies.

Of the 15 quasars with known α_γ, 10 have $\alpha_\gamma \geq 1$, indicating that the maximum power is emitted in or below the GeV range. The four cases reported in Fig. 1 have been chosen so as to include a flat and a steep case.

Another indication from Fig. 1 is that the ratio between the γ-ray luminosity and the maximum luminosity in the 10^9–10^{17} Hz range tends to be ≥ 1 for quasars, while it is ≤ 1 for BL Lacs. However, since for most objects the measurements in other bands are not simultaneous with the γ-ray observations and the objects are extremely variable, the ratio L_γ/L_{IROUVX} is at present highly uncertain.

In summary the BBED of blazars from radio to γ-ray frequencies may be described as two smooth components: the first extends from 10^9 to 10^{17} Hz, with a broad maximum in the power per decade between 10^{12} and 10^{14} Hz for CDQs and radio selected BL Lacs, and at 10^{15-16} Hz for X-ray selected BL Lacs. The second extends from 10^{17} to 10^{25} Hz, peaking often in the 10^{22-24} Hz range but sometimes even beyond. The first will be called here the synchrotron component, the second will be called the inverse Compton component. Some theoretical problems relevant in the two domains are discussed below.

It is worth stressing that the present body of data on the BBED of blazars, though impressive, is still very incomplete. In particular simultaneous measurements in different frequency bands are needed.

3 The synchrotron component

The success of models explaining the evolution of radio flares, the significant polarization often observed in the optical and the spectral "continuity" leave little doubt that the emission mechanism from the radio to the UV is synchrotron radiation. It is also clear that the flat spectra observed in the radio ($\alpha_R \simeq 0$) are due to the superposition of synchrotron self-absorption turnovers of different emission regions (i.e. the jet is inhomogeneous), higher frequencies corresponding to more compact regions. Early models attempted a continuous "fluid" description of the inner jet to reproduce the overall radio to X-ray spectrum (e.g. Königl 1989), but models involving one or more discrete emission regions, probably associated with shocks, are also viable (e.g. Marscher 1993).

Above about 10^{12} Hz the emission becomes transparent and the spectrum steepens to $\alpha_{IR} \simeq 1$. Since only few objects have been detected with IRAS, the initial slope of the transparent synchrotron spectrum is poorly known. At higher frequencies, usually in the optical–UV band or in the X-ray band for the X-ray selected BL Lacs, the spectrum steepens to $\alpha \geq 1$, perhaps through a localized change in slope (break), rather than in the continuous fashion previously believed.

A steepening in the spectrum of non thermal electrons can be interpreted as due to radiation losses. In the case of continuous injection the break occurs at the energy for which the radiative life time equals the escape time of the particles from the emission region. Assuming the minimum escape time R/c and energy losses due to only the local energy density of radiation, it can be easily shown that the break occurs at $\gamma_b \sim 1/\ell$ where ℓ is the compactness parameter defined by $\ell = \sigma_T L/(R m_e c^3)$. This compactness refers to the frame of the emitting plasma, and is related to the observed quantities through $\ell = \ell_{obs}/\delta^{p+1}$, if R is derived from an observed variability time scale, $R = c t_{obs} \delta$. Sources with $\delta = 10$, $L_{obs} = 10^{46}$ erg/s and $t_{obs} = 1$ day should have $\ell = 10^{-3}$ (for $p = 4$), and therefore $\gamma_b \sim 10^3$. This corresponds to a synchrotron break frequency $\nu_b \sim 3 \times 10^{13} B$ Hz (where B is the magnetic field).

An alternative explored by Ghisellini, Maraschi & Treves (1985, hereinafter GMT) and Ghisellini & Maraschi (1989) is to interpret the steepening as a convolution of high energy cut-offs in emission regions of decreasing size with increasing energy. This choice was motivated by the short ($\simeq$ hours) timescales observed in the X-ray band. In this model the radiative loss timescales can be much shorter than R/c, which does not need to be the same at all frequencies.

More realistically, one or more shocks may be present in the inner regions of the jet. At the shock front particles should be accelerated and diffuse downstream, those of higher energy filling a smaller volume due to the shorter life time (see Bregman 1985). This case is intermediate between the two above, in that at all energies above the break the particle life time and the size of the emitting region are related. Bregman (1985) obtained an almost exponential steepening (larger than observed), but a more realistic treatment of particle acceleration and diffusion may yield better agreement with the present data.

4 The case of PKS 2155-304

Intensive multiwavelength observations of PKS 2155-304 have provided two major results: 1) the variability in the optical-UV and in X-rays is highly correlated on timescales of hours; 2) the UV light curve is delayed by $\simeq$ 3 hours with respect to X-rays, as measured in the ROSAT band (see Edelson these proceedings). The simultaneous overall spectrum of PKS 2155-304 obtained from these observations is shown in Fig. 2. The γ-ray upper limit is derived converting the limiting sensitivity of EGRET of 10^{-7} ph/(cm^2 sec) above 100 MeV.

These results show unambiguously that the X-ray and UV emitting regions are not coincident, in agreement with the GMT model. However the measured lag is small compared to the size of the UV emitting region, if the latter is related to the observed month long doubling timescale. Unfortunately the X-ray observations spanned only a four day period, due to the wisdom of allocation committees, and we have no information concerning the correlation of X-rays with the larger amplitude UV variation. Considering UV and X-ray observations in the higher GINGA band (Sembay et al. 1993), the correlation appears less tight, but no delay can be determined due to the sparse sampling.

In the context of emission from relativistic jets, studies of the variability behaviour of the high frequency emission have been carried out by Celotti, Maraschi & Treves (1991) and Marscher, Gear & Travis (1992).

The former consider a perturbation (shock) moving through an otherwise stationary jet, described by the model of GMT, causing the emission from different regions to vary at different times. Emission below the break frequency, i.e. in the IR-optical-UV bands, is produced in the same volume (see §3). Therefore it is expected to vary simultaneously, without spectral changes, if the perturbation affects the particle density but not their spectrum. Emission above the break (X-rays in the case of PKS 2155-304) derives from a smaller region, closer to the nucleus. Therefore it is affected first by a perturbation moving outward. The typical cooling time of the high energy electrons in the latter region can be shorter than the

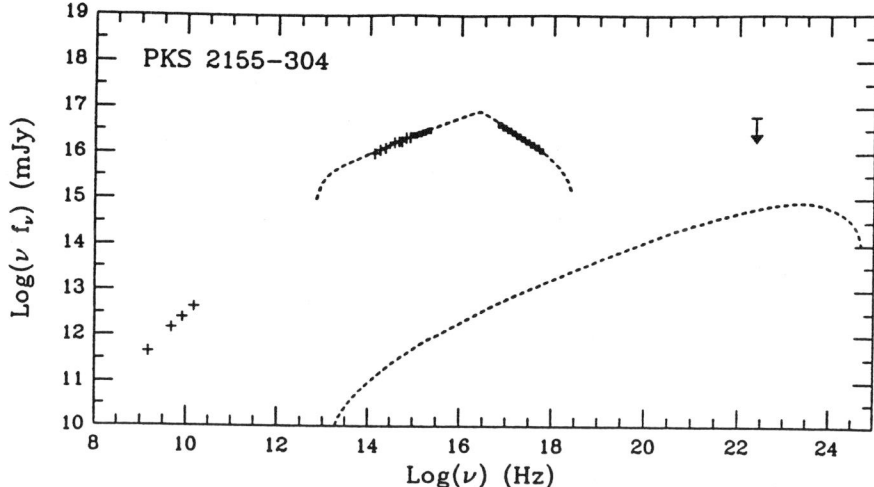

Fig. 2. The BBED as measured simultaneously on Nov. 13 1992. The arrow represents a generic EGRET upper limit. The dashed lines represent synchrotron and Inverse Compton emission from a relativistic jet described by the GMT model, with $\delta = 10$, $R_0 = 3\ 10^{13}$ cm, $R_{max} = 3\ 10^{16}$ cm, $B_0 = 3\ 10^3$ G, $B(R_{max}) = 150$ G.

size of the region and continuous acceleration is required.

Marscher et al. (1992) propose that the whole emission is produced by electrons accelerated at a shock at large distance from the nucleus allowing for lower energy densities and consequently longer cooling timescales, of the order of the observed variability timescales. In the case of PKS 2155−304 this implies a magnetic field of the order of ∼ 0.25 G. The emission region has a sheet–like geometry determined by the radiative cooling length of particles accelerated in a shock front. X-rays would be closer to the front and the UV emission further "downstream". If the downstream region is on the observer side (reverse shock), the geometry is similar to the previous model. In the opposite case (direct shock) the Doppler effect would act so as to lengthen the observed lag, which would be a severe difficulty.

The tight limits on any observed spectral variability between the UV and X-ray bands could also pose a problem to the above models. In fact in both cases energy dependent variability implies spectral changes. However the ROSAT band is very close to the spectral break above which spectral variability is expected and the intensity amplitude of the observed variations is also small.

Specific computations are required to verify whether any of these models can quantitatively reproduce the observed light curves. As already mentioned the basic problem is to account for a large IR–UV emitting region and a short UV–X–ray time lag (and in fact both the models assume that the UV emitting region is larger than the X-ray one). It is important to stress that relativistic corrections depend not only on the velocity but also on the geometry of the emitting regions and on the origin/nature of the variations. For instance in the case of a propagating wave the simple Doppler correction for a time interval is $\propto \Gamma^2$. Therefore it is possible to ease the difficulty mentioned above if different relativistic corrections have to be applied to the variability and lag timescales.

5 The inverse Compton component: external vs. internal photons

Since blazars emit high frequency synchrotron radiation from regions of high photon density, it is inevitable that some γ–rays be produced by first order Compton scattering of the relativistic electrons off the synchrotron photons. This process is called "self-Compton" (SC). Using the inhomgeneous jet model by GMT, we were able to reproduce the observed BBED of the HPQ 3C 279 up to γ–ray energies (Maraschi, Ghisellini & Celotti 1992). We further applied the same model to the BL Lac object MKN 421 and to PKS 0537–441, which shows intermediate properties between the two classes (Maraschi, Ghisellini & Boccasile 1993). The required Γ's were between 5 and 7 and the radiation energy density was larger than the magnetic energy density, reflecting the ratio of SC and synchrotron luminosities.

It has been recently pointed out that the same relativistic electrons could also give rise to γ–rays by upscattering radiation external to the jet (EC). This could be due to photons coming directly from the accretion disk (Dermer, Schlickeiser & Mastichiadis, 1992), or by these very photons after isotropization by scattering in surrounding material, or by photons directly emitted by the broad line region (Sikora, Begelman & Rees, 1993). The latter two components (to which we will refer cumulatively as BLR), being isotropic within a region of radius R_{blr}, are seen enhanced in the comoving frame, due to relativistic aberration and blueshift, while the first is severely dimmed except for small distances from the disk. In the comoving frame, the ratio between the energy density of external BLR radiation U_{blr} and the magnetic energy density U_B is (omitting factors of order unity)

$$\frac{U_{blr}}{U_B} \sim \frac{L_{45}}{B^2} \frac{a}{0.1} \left(\frac{\Gamma}{10}\right)^2 \left(\frac{10^{18} \text{cm}}{R_{blr}}\right)^2 \tag{1}$$

where a is the luminosity fraction scattered or emitted in the BLR, $R_{blr} > R_\gamma$, and R_γ is the location of the γ–ray emitting region. Assuming that the energy density of synchrotron photons U_S is comparable to U_B, eq. (1) shows that U_{blr} can be dominant provided that: 1) Γ is relatively large; 2) the 'BLR' exists; 3) the γ–ray emitting region is inside the 'BLR'.

In Ghisellini & Maraschi (1993) we have reconsidered the cases of 3C 279, PKS 0537–441 and MKN 421, suggesting that the EC process may contribute significantly to the observed γ–rays in the first two cases, but probably not in MKN 421 and other BL Lacs with *intrinsically* weak lines.

In general, the relative luminosities of the three processes, synchrotron (SY), Self Compton (SC), and inverse Compton off external photons (EC), are determined by the relative values of the energy densities U_B, U_S and U_{blr}:

$$L_S \propto N U_B; \qquad L_{SC} \propto N U_S; \qquad L_{EC} \propto N U_{blr} \qquad (2)$$

where N is the total number of relativistic particles, and $U_S \propto L_S$.

A first criterion to distinguish observationally between the SC and EC mechanisms is based on the spectral shape of the two components. In fact the SC spectrum is fixed if the SY spectrum is known. Since the external photons are expected to have a narrower spectrum and a higher average energy than the SY ones, for a given SY spectrum, L_{EC} is more peaked toward the highest energies, while L_{SC} is broader. However, due to the wide bands where observations are lacking (FIR, hard X–ray soft γ–ray), a purely spectral discrimination though possible in principle seems hard in the near future.

Variability may be a more powerful test of SC or EC models. In fact we know that γ–rays are strongly and rapidly variable. 3C 279 was found to vary by a factor two in few days and by larger factors over few months (Kniffen et al. 1993). External radiation is not expected to vary by large factors on these time scales. The first prediction of both Compton models is that L_S and L_{SC} or L_{EC} should vary in a correlated fashion.

From eq. (2) we see that if variations are due to N, $\Delta L_{EC} \propto \Delta L_S$, while $\Delta L_{SC} \propto \Delta L_S^2$. Thus by measuring the variability in the ranges close to the maximum power output we can hope to test Compton models and distinguish between different sources of soft photons.

A less naive approach may consider a feed back of the large energy losses on the particle distribution function. In fact the estimated cooling times are short in most cases. Assuming that relativistic particles are continuously injected in the emitting region, the distribution function can be derived by means of a continuity equation. If the injected power is P_i, the steady particle density N (at energies where cooling dominates over escape and adiabatic losses) is

$$N \propto \frac{P_i}{U_B + U_S + U_{blr}} \qquad (3)$$

Note that, for the energies we are interested in, only the first order scattering is important, the second order being cut off by the Klein Nishina regime. This is taken into account approximately by neglecting U_{SC} and U_{EC} in eq. (3).

Assume for the moment that U_{blr} is negligible. For sufficiently small P_i, we will have $U_B > U_S$, so that synchrotron radiation dominates the radiation losses. Combining eq. (2) with eq. (3) one can easily find that $L_S \propto P_i$ and therefore $L_{SC} \propto P_i^2$. Viceversa, at sufficiently large P_i, we will have $U_B < U_S$. In this case SC dominates the bolometric output, so that $L_{SC} \propto P_i$ and $L_S \propto P_i^{1/2}$.

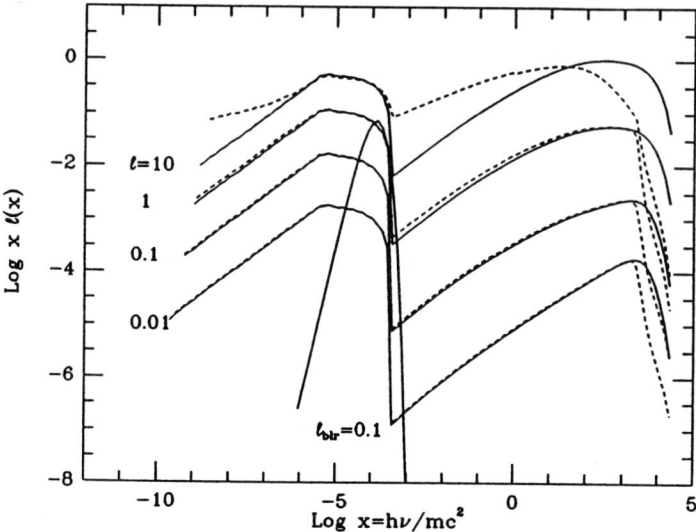

Fig. 3. *Continuous lines:* "Self–consistent" Synchrotron and inverse Compton emission spectra from relativistic electrons injected with given compactness ℓ into a region with $B = 10$ G, $R = 3\ 10^{16}$ cm. Electrons cool by SY, Self Compton (SC) and External Compton (EC) on UV photons of compactness $\ell_{blr} = 0.1$ (shown in the fig. as the narrow peaked distribution). For small injected ℓ, the main cooling process is EC so that the high energy spectrum reflects the electron spectrum. For large ℓ, the high energy radiation is mainly produced by the SC mechanism, resulting in a smoother high energy spectrum. *Dashed lines:* Same cases, but including pair production and radiation by the created pairs. The highest energy photons are absorbed by photon–photon collisions even for very small ℓ, due to the presence of the UV photon distribution. For $\ell > 10$ pair effects modify the entire spectrum. Note that the SY luminosity $\propto \ell$ for small compactnesses, and becomes $\propto \ell^{1/2}$ for large ℓ.

Let us now consider U_{blr}. Since it does not depend on P_i, we have for L_{EC} the same behaviour as L_S.

Hence, in all cases, when changing the injected power in relativistic electrons the SC luminosity should change more than the S one, while the EC and S emission should change by the same amount.

Therefore we have a simple and potentially powerful way of checking if the origin of the γ–rays in blazars are due to SC or EC radiation. Of course this is simple in theory, but it may prove difficult in reality, because:

i) the comparison discussed above refers to the entire (frequency integrated) S, SC and EC luminosities, while we observe in restricted energy bands;

ii) the emission region may not be homogeneous, with multiple components;

iii) we have assumed a constant bulk Lorentz factor and a constant energy density for the external radiation. If Γ changes, instead of P_i, then U_{blr} changes also, possibly inducing a variation of the γ–ray emission larger than the synchrotron one even when the external radiation is dominant. In this respect estimates of Γ

at different epochs, from the superluminal motion, are extremely valuable.

To illustrate the above points we have computed SY, SC and EC spectra with the following assumptions: relativistic electrons are injected continuously in a source of radius $R = 3 \; 10^{16} cm$, $B = 10 \; G$, with spectrum $Q(\gamma) \propto \gamma^{-2}$ between $\gamma_{min} = 3 \times 10^3$ and $\gamma_{max} = 3 \times 10^4$. The injected power corresponds to an injected compactness ℓ, while the external radiation energy density (as seen in the comoving frame) corresponds to $\ell_{blr} = 0.1$. The spectra are computed with an iterative method, which allows the effects of photon–photon absorption and pair cascades to be taken into account. The computed spectra are shown in Fig. 3 for varying ℓ and fixed B and ℓ_{blr}.

6 Conclusions

The BBED of blazars can be decomposed in two smooth broad components, the first extending from the radio to the UV and in some cases X-ray spectral regions, the second from the X-ray up to the γ-ray bands. The relation between the two is a strong constraint for theoretical models in which the first component is interpreted as synchrotron and the second as Compton emission from the same relativistic electrons. A complete and simultaneous coverage of this enormous frequency range is desirable, but not easy to obtain in the short term. However, a very important quantity for an understanding of the physics of the emission region, is the luminosity ratio of the two components. Thus simultaneous observations in frequency bands close to the peaks in the power per decade, i.e. IR–opt–UV and γ-ray observations are essential.

Depending on the soft photons available for upscattering (either synchrotron photons or photons external to the jet), the relation between the low frequency and high frequency spectrum is expected to be different, with regard to both spectral shape and correlated variability. More theoretical work is needed in the area of time dependent models to explore a range of possibilities.

The long life–time expected for the Compton Gamma Ray Observatory gives us a tremendous opportunity to carry out systematic studies of correlated variability which may give us direct insight on the energy input and radiation losses of high energy particles in relativistic jets.

7 References

Bertsch, D.L. et al., 1993, ApJ, 405, L21
Blandford, R.D. & Rees, M.J., 1978, in "Pittsburgh Conference on BL Lac Objects", ed. A.N. Wolfe (Pittsburgh University Press), p. 328
Bregman, J.N., 1985, ApJ, 288, 32
Celotti, A., Maraschi, L. & Treves, A., 1991, ApJ, 337, 403
Dermer, C.D., Schlickeiser, R. & Mastichiadis, A., 1992, AA, 256, L27
Ghisellini, G., 1993, proc. of the Cospar Symposium (Washington), in press.

Ghisellini, G. & Maraschi, L., 1989, ApJ 340, 181
Ghisellini, G. & Maraschi, L., 1993, Proc. 2nd Compton Symp., Maryland, in press
Ghisellini, G., Maraschi, L. & Treves, A., 1985, A.A., 146, 204 (GMT)
Ghisellini, G., Maraschi, L., Tanzi, E. & Treves, A., 1986, ApJ, 310, 317
Ghisellini, G., Padovani, P., Celotti, A. & Maraschi, L., 1993, ApJ, 407, 65
Hartmann, R.C. et al. 1993, Proc. 2nd Compton Symp., Maryland, in press.
Hartman, R.C. et al., 1992, ApJ, 385, L1.
Hunter, S.D. et al., 1993a, ApJ, 409, 134.
Hunter, S.D. et al., 1993b, AA, 272, 59
Impey, C.D. & Neugebauer, G., 1988, AJ, 95, 307
Kniffen, D.A. et al. 1993, ApJ, 411, 133
Königl, A., 1989, in "BL Lac Objects", eds. L. Maraschi, T. Maccacaro and M-H. Ulrich (Springer-Verlag), p. 321
Kühr, H., Witzel, A., Pauliny-Toth, I.I.K & Nauber, U., 1981, AA (Suppl), 45, 367.
Landau, R. et al., 1986, ApJ, 308, 78
Lin, Y.C., et al., 1992, ApJ, 401, L61
Madau, P., Ghisellini, G. & Persic, M., 1987, MNRAS, 224, 257
Mannheim, K., 1993, AA, 269, 67
Maraschi, L., Ghisellini, G. & Boccasile, A. 1993, Proc. Physics of AGNs, Cambrige, in press
Maraschi, L., Ghisellini, G. & Celotti, A. 1992, ApJ, 397, L5
Maraschi, L., Ghisellini, G., Tanzi, E.G., & Treves, A., 1986, ApJ, 310, 325.
Marscher, A.P., Gear, W.K. & Travis, J.P. 1992, in "Variability of blazars", eds. E. Valtaoja and M. Valtonen (Cambridge Univ. Press), p. 85
Marscher, A.P., 1993, in Astrophys. Jets, eds. Burgarella, D., Livio, M. & O'Dea C., Cambridge University Press, in press.
Pian, E. & Treves, A., 1993, ApJ, 416, 130
Punch, M. et al., 1992, Nature, 358, 477
Sambruna, et al., 1993, ApJ, submitted.
Sikora, M., Begelman, M.C. & Rees, M.J., 1993, ApJ, in press
Sikora, M., 1993, ApJ Suppl. in press (proceed. of the 142 IAU Symp.)
Tanzi, E. et al., 1989, in BL Lac Objects, Lecture Notes in Physics, 334, p.171, eds. Maraschi, L., Maccacaro, T. & Ulrich M.H., Springler-Verlag
Thompson, D.J., et al., 1993a, ApJ, 410, 87
Thompson, D.J., et al., 1993b, ApJ, 415, L13
Treves, A. et al., 1993, ApJ, 406, 447
Valtoja, E., Lahteenmaki A. & Terasranta H., 1992, preprint
Wall, J.V., & Peacock, J.A., 1985, MNRAS, 216, 173.
Wiren S., Valtoja E., Terasranta H. & Kotilainen, 1992, preprint
Worral, D.M. & Wilkes, B.J., 1990, ApJ, 360, 396

MODELING BENT RADIO JETS:
ANOTHER PROBE OF THE CENTRAL ENGINE

P.A. HUGHES, M.F. ALLER and H.D. ALLER
Astronomy Department, University of Michigan

Abstract. We present a model for curved parsec-scale jets that includes most of the key radiation physics and relativity, but which is simple enough to allow an exploration of diverse geometries and flow parameters.

1. Introduction

Recent VLBI observations have shown that parsec-scale jets commonly exhibit bent, sometimes complex, and/or changing, morphology. Although enhanced by line-of-sight projection, there must be a nonlinearity of the flow. The strong sensitivity of the total and polarized fluxes to flow speed and orientation (through Doppler frequency shift, Doppler boost, aberration and time delays) means that modeling *must* allow for such flow nonlinearity and *may* in fact elucidate the flow geometry. Once determined, the flow geometry should give a strong indication of the origin of nonlinearity, *e.g.*, precession of the central engine; an ambient pressure gradient; an ambient density gradient, or 'clouds'; back action of a cocoon flow on the jet; Kelvin-Helmholtz instability – and hence probe the nature of the flow *and the ambient conditions* on parsec and sub-parsec scales. This article addresses a method of modeling nonlinear flows.

Fitting multi-frequency, multi-epoch total and polarized flux data calls for detailed radiation transfer calculations (Gómez, Alberdi & Marcaide 1993; Hughes, Aller & Aller 1989; Jones 1988; van der Walt 1993). However, this procedure is too computationally intensive to admit extensive exploration of parameter space. Something simpler – ideally, interactive – but quantitative is needed (Marscher & Zhang 1991; Steffen 1992). We describe a model that includes most of the key radiation physics and relativity, albeit in a simple way, and provides on-the-fly maps and light curves. We illustrate the usage of the model by interpreting the observations of 4C 39.25. The model enables us to reproduce/understand the *general* features of this source: namely the stationary and propagating components seen on VLBI maps and the derived $\beta_{\rm app}$; and the profiles and percentage polarization exhibited by single dish data.

2. The Model

The geometry is an arbitrary locus on which are 'threaded' circularly-cylindrical pills of increasing radius and diminishing field and particle energy density.

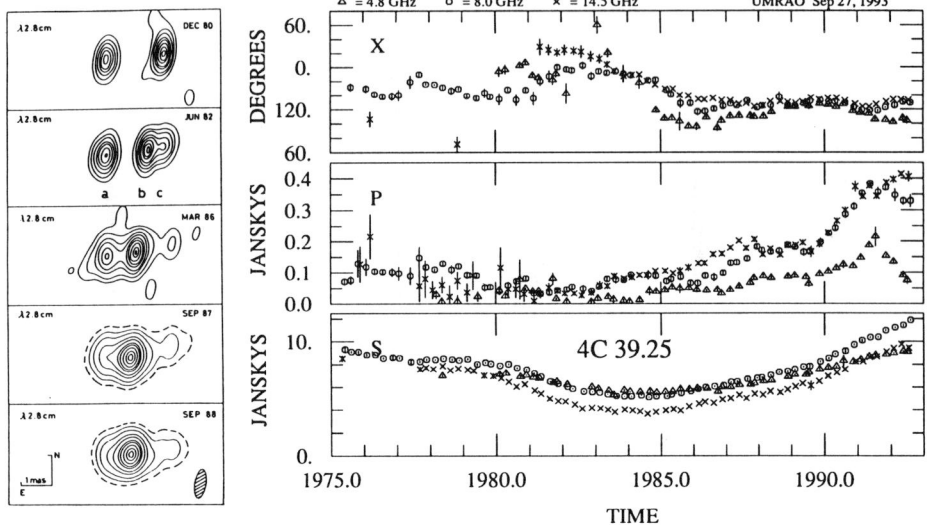

Fig. 1. Maps and light curves for 4C 39.25. The panels from bottom to top are: total flux, polarized flux and PA of the electric vector. Each point is a tri-monthly average.

The total and polarized fluxes of *each* pill are calculated assuming a homogeneous synchrotron plasma, and these are converted to intensities by 'distributing' the emission over the projected area of the pill.

A 'superluminal' component may be generated by specifying a strength and extent of shocked jet. The jump conditions are used to determine the state of the shocked flow, and the polarized emission is associated with a compressed tangled magnetic field.

For an arbitrarily oriented observer, the spectral shift, Doppler boost, aberration (which in part determines the degree of observed polarization) and time delay with respect to the jet 'base' are computed for each pill. The emissivity in both quiescent and shocked parts of the flow takes account of these effects.

For each pill a set of 'weights' is computed, associated with the degree of overlap of juxtaposed pills. Maps and light curves are made using the information contained in these weights and a measure of the opacity of each pill, using complex polarization to correctly sum the individual polarized fluxes.

The calculations are 'packaged' to provide an interactive display, wherein a source may be rotated, parameters (*e.g.*, the rate of decline of the emissivity along the jet, the 'observing' frequency, the quiescent flow Lorentz factor and the direction of the shock) changed, and shocks propagated with on-the-fly map and light curve generation (see Fig. 3).

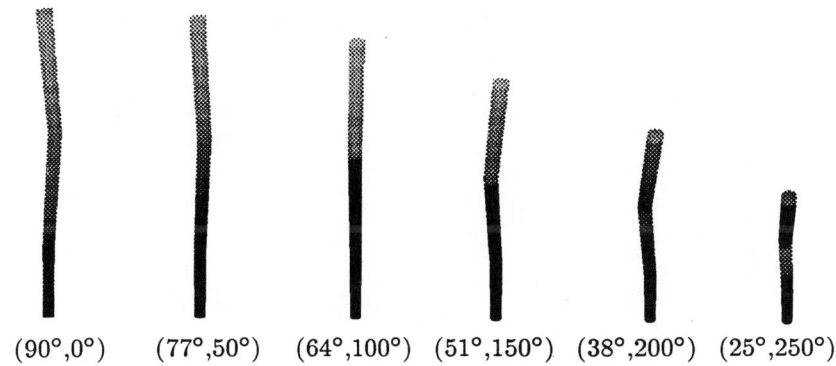

(90°,0°)   (77°,50°)   (64°,100°)   (51°,150°)   (38°,200°)   (25°,250°)

Fig. 2. The adopted jet morphology at various polar and azimuthal angles to the line of sight.

3. Results and Conclusions

Fig. 1. shows multi-epoch VLBI maps of 4C 39.25 (Marcaide *et al.* 1990) and the UMRAO monitoring data for the same period.

Fig. 2. shows the adopted jet morphology, the first panel being in the plane of the sky, with subsequent panels rotated as indicated by the labeling. The last panel – in which intensity has been rescaled to account for the Doppler boost – shows two stationary components despite the very slight nonlinearity of the flow. The top panel of Fig. 3 shows a display at intermediate epoch during the propagation of a shock, exhibiting a geometry with two stationary and one propagating component, the latter with apparent speed close to that inferred from VLBI maps. The bottom panel shows the corresponding light curves, which mimic those of Fig. 1.

The model was built to provide guidance for more detailed study, but in fact illustrates the great sensitivity of maps and light curves to orientation and flow parameters, allowing these to be well-defined with minimal effort. The model can give insight into the flow conditions: its interactive nature facilitated the exploration of parameter space, showing that opacity cannot play a significant role in determining the apparent structure of 4C 39.25 – supported by observed spectral properties. The model highlights the fact that modulation of a single outburst by flow curvature may lead to a rich light curve structure (particularly for the polarized flux) wherein (as noted by Gabuzda, private communication) there is not necessarily a one-to-one correspondence between well-defined single dish 'events' and the appearance of *new* components on VLBI maps.

This work was supported in part by grant AST 9120224 from the National Science Foundation

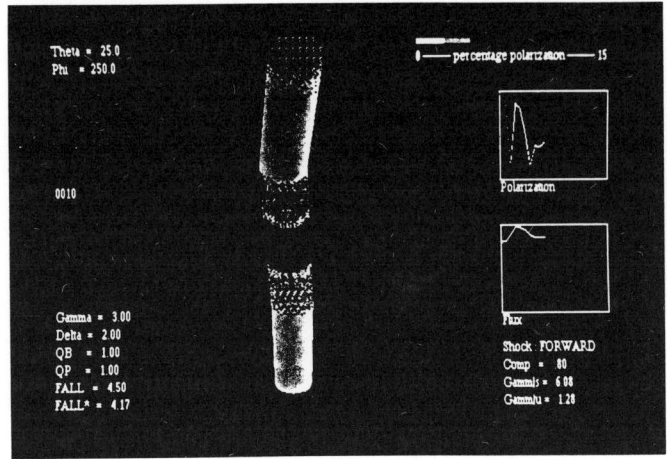

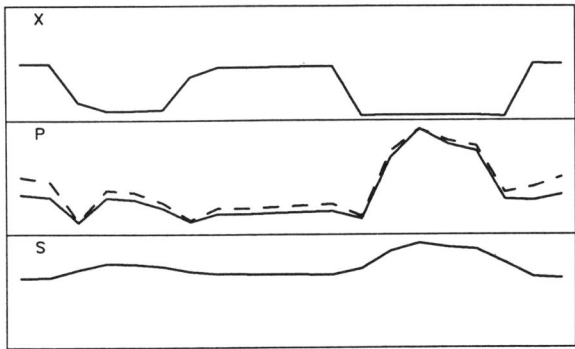

Fig. 3. Shock evolution. In the top panel the propagating component appears dark in this monochrome reproduction. The time evolution of total flux (S), polarized flux (P; percentage polarization is shown as a dashed line and reaches $\sim 5\%$) and PA of the electric vector (X) are shown in the bottom panel, and agree with those seen in Fig. 1. The early rising and late falling portions of the model light curves are presumably not seen in the data because of confusion from other outbursts.

References

Gómez, J. L., Alberdi, A., & Marcaide, J. M. 1993, *Astr. Ap.*, in press.
Hughes, P. A., Aller, H. D., & Aller, M. F. 1989, *Ap. J.*, **341**, 68.
Jones, T. W. 1988, *Ap. J.*, **332**, 678.
Marcaide, J. M., et al. 1990, In *Parsec Scale Radio Jets*, ed. Zensus, J. A. & Pearson, T. J. (Cambridge: Cambridge University Press), p. 59.
Marscher, A. P., Zhang, Y. F., Shaffer, D. B., Aller, H. D., & Aller, M. F. 1991, *Ap. J.*, **371**, 491.
Steffen, W. 1992. Diplomarbeit, Max-Planck-Institut für Radioastronomie, Bonn.
van der Walt, D. J. 1993, *Ap. J.*, **409**, 126.

AGN Physics and Models

MODELS FOR VARIABILITY IN AGNS

MARTIN J. REES
Institute of Astronomy
Madingley Road
Cambridge, CB3 0HA, U.K.

1. Introduction

In this talk I shall address three different processes relevant to continuum variability in AGNs. The first two refer to the physical conditions in the regions responsible for the non-thermal emission, and the implications of high brightness temperatures. The third is the distinctive type of flare that results when a star is tidally disrupted by a massive black hole; this process, which merits much further study, it likely to be specially important as a diagnostic of physical conditions in low-luminosity nearby nuclei.

2. How high can jet Lorentz factors be?

It has for been evident for several years that radio jets, at least on the scales probed by VLBI, display bulk flows with Lorentz factors Γ of up to 10. The evidence comes from the brightness temperature limit on any incoherent synchrotron source which avoids a synchrotron self-Compton catastrophe, and directly from VLBI observations of superluminal motions. Gamma-ray data from the EGRET experiment on the Compton Gamma-Ray Observatory offer independent indications of highly relativistic outflow, as has been discussed by several speakers at this meeting. Some sources have been detected as strong gamma-ray sources which vary on timescales as short as days. If the sources were actually as small as a few light days, then no gamma rays would escape (irrespective of the emission mechanism), because the photon density would be so high that pair production would occur. This process can be avoided by highly relativistic outflow, which allows a larger source size, and raises the photon energy threshold for pair productions in the observer's frame. If the gamma rays are relativistically beamed rather than being isotropic, the implied jet luminosity need be no more than 10^{46} ergs/s; on the other hand, isotropic emission would imply uncomfortably high powers of several times 10^{48} ergs/s.

Although the evidence requires $\Gamma \simeq 10$ (and a jet with power Lj up to a few times 10^{46} ergs/s) it is interesting to ask whether Γ could be still higher? The data do not obviously exclude this, though there are well-known theoretical constraints: in particular, Compton drag near the base of the jet is important, especially if the energy is mainly carried by electron-positron pairs (Rees 1984, Phinney 1985). The issue is now more open, however, because of the realisation that magnetic fields probably play a dominant role, especially near the base of the jets. The wound-up field would collimate the jet. Moreover Li, Begelman and Chiueh (1992) have shown that, in Poynting-dominated jets, the bulk flow is gradually accelerated despite Compton drag. The drag enhances the energy stored in the field, and this energy can be given back to the jet (as in a decompressing spring). There is therefore no reason for feeling too inhibited about envisaging higher Γ. After all, we are quite comfortable with the idea that the relativistic wind from the Crab pulsar may have $\Gamma \simeq 10^4$.

Neither superluminal motions nor gamma-rays yet force us to Lorentz factors much above 10. But there is now another phenomenon that may do – intraday radio variability (IDV). Intensity changes within a single day, with more than 20 per cent amplitude, have been detected in about a quarter of all compact extragalactic sources. (Quirrenbach et al 1989, Witzel 1992). The claimed correlation with variations in the optical band would preclude interpretations in terms of interstellar scintillations, and imply that they were intrinsic to the source. If the source dimensions were limited to ct, these rapid variations would indicate vastly higher brightness temperatures than are compatible with a synchrotron source. In 0917+624 (Qian et al 1991) the "formal" brightness temperature is 2.10^{18} K. To bring this down below the limit set by the Compton catastrophe would require Γ to be at least 100; the relevant emitting region would then be at a distance of $\sim 10^4$ light days, i.e. several parsecs, from the base of the jet. Is this possible?

The general consequences of postulating higher Γ can be readily envisioned simply by considering how various key timescales depend on Γ and on Lj. Let us consider a jet of fixed luminosity and cone angle, and focus attention on a particular point along its length. Relativistic electrons will cool due to inverse Compton scattering of ambient radiation – radiation from the nucleus itself, and from the gas that emits the broad emission lines. The radiation energy density is $\sim L/r^2c$. As viewed in the frame of the jet the energy density is enhanced by a factor $f\Gamma^2$.

The factor f, equal to 1 for isotropic radiation, depends on the angular distribution of the radiation. Scattering of ambient radiation becomes increasingly important for high Γ. The maximum magnetic field is that for which Poynting flux carries all the jet luminosity Lj. When Γ is larger, straightforward Lorentz transformation tells us that the maximum magnetic field in the comoving frame becomes weaker for a given $L_j (\propto \Gamma^{-1})$. Synchrotron lifetimes then become longer, and synchrotron emission consequently becomes very much less competitive with Compton scattering, for which the characteristic lifetime *decreases* for higher Γ. Moreover synchrotron losses also become slow compared to adiabatic cooling: the timescale for adiabatic expansion at a fixed location along the jet, measured in the comoving frame, goes as Γ^{-1} because of the ordinary time dilation. (A long synchrotron lifetime also guarantees that synchrotron self-Compton emission would be unimportant if a substantial part of the jet energy were transported as Poynting flux)

To produce an apparent synchrotron luminosity $\sim 10^{44}$ erg/s might require only 10^{-4} efficiency if $Lj \simeq 10^{46}$ erg/s and the beaming solid angle were (say) 10^{-2}. But for sufficiently high Γ, the synchrotron efficiency, given by $\min[t_{exp}, t_{Comp}]/t_{sync}$, would fall below even such a low value as this. There is therefore an upper limit to Γ, set by the requirement that the synchrotron efficiency is not too low, and that the high-energy output should not exceed the variable radio luminosity by an unacceptable factor. (These two contributions would be beamed and doppler-boosted in an essentially similar way). The presently-claimed IDV data turn out to be marginally consistent with a synchrotron hypothesis, with Γ in the range 100-200. This reduces the motivation for invoking coherent processes (eg Benford 1992). However the speed of these variations is right at the limit: any radio variations that were substantially more rapid could not be accommodated without invoking radiation processes that permit higher brightness temperatures.

Note that the above arguments set an upper limit to Γ specifically for a jet (or a part of a jet) that emits synchrotron radiation above some threshold efficiency. They do not rule out far higher Γ in a jet that is simply a conduit for energy, or which is detectable only via inverse Compton scattering of ambient radiation. It is indeed quite possible that Poynting flux can be carried by very high-Γ jets. Moreover it is possible that in some jets (eg that of M 87) the material near the axis has $\Gamma \gg 100$, the observed synchrotron radiation coming from a violently-sheared boundary sheath that moves more slowly (though perhaps still relativistically).

3. Induced (or stimulated) Compton scattering

Whatever radiation mechanism the IDV sources turn out to involve, they offer a further motive for considering a process which is important in high-surface-brightness sources – induced compton scattering. This process, originally known as the "Dirac-Kapitza effect", has been known for a long time, and causes apparent changes in the spectrum and structure of a source. Its effects are inherently non-linear, and therefore tricky to calculate, but it may well be significant in compact radio sources. Induced scattering dominates spontaneous scattering whenever the occupation number n exceeds unity. At first sight, one would have thought that it should consequently be colossally important. However, if there are two beams with respective occupation numbers n_1 and n_2, induced scatterings from 1 to 2 ($n_1 \times n_2$) are compensated by those in the reverse direction ($n_2 \times n_1$). The cancellation would be exact were it not for the effects of electron recoil, which amounts to a fraction $h\nu/m_e c^2$ of the photon momentum. A necessary condition for induced scattering to be significant is therefore that the photon occupation numbers exceed $m_e c^2/h\nu$; equivalently, a brightness temperatures above $m_e c^2/k$ is required.

The brightness temperatures of many compact radio sources are directly observed by VLBI to be as high as 10^{12}K (this is, of course, well known to be close to the limit allowed by a self-absorbed incoherent synchrotron source). Induced scattering effects could therefore be significant in these source components even if the Thomson depth in or around them were only $\sim (m_e c^2/kT_b) \simeq 10^{-2}$.

The consequences of induced scattering are frequency-dependent, and depend on the radiation intensity and spectrum in non-linear ways. The simplest effect, first discussed by Sunyaev (1970) and Levich (1972), is a "Bose-condensation" of radiation towards lower frequencies. If induced scattering is too important, the radiation is, in effect, completely absorbed; but for induced optical depths of order unity the low-frequency brightness temperature can actually be enhanced above the self-absorption limit (though generally only by a modest factor), and steep gradients can be created in the spectrum. An ultra-compact high brightness core surrounded by tenuous thermal plasma may consequently produce a spectrum with a second component, due to induced scattering, peaking at a lower frequency than the intrinsic spectrum from the core.

When spatial structure can be resolved, the apparent brightness of each part of the source can be modified by induced scattering; moreover these modulations change in a complicated way when the source varies. If such effects were important, they would confuse the interpretation of variability and superluminal motions (Wilson 1982, Coppi, Blandford and Rees 1993). The firmest indications of induced scattering would be strong spectral variations in source structure and large frequency-dependent linear polarization.

There is no compelling evidence that induced Compton effects have been observed in any compact radio source. This already sets non-trivial upper limits on the density of thermal plasma within the central few parsecs of active galaxies. Future observations with VLBI arrays will have greater dynamic range; detection of such effects would offer a new probe for physical conditions in AGNs. Blandford (1993) has pointed out that stimulated Raman scattering off collective plasma waves, which has similar observational characteristics, may be even more important.

Induced scattering could be catastrophic in coherent sources. For instance, suppose that the region at the base of the jets, where the field strengths may be $\sim 10^4$ G and the gyrofrequency at centimetre wavelengths, emitted coherent cyclotron radiation. (This is not an absurd hypothesis, and would certainly remove the problem of explaining intraday variability.). But the brightness temperatures would then be $\gtrsim 10^{20}$ K, and induced scattering would quench the spectrum unless τ_{es} were below 10^{-8}. Since τ_{es} is characteristically of order unity, this would seem to require a rather special geometry.

4. Flares from tidally-disrupted stars, especially in low-luminosity nuclei

Tidal disruption of stars was investigated in the 1970s as a possible fuelling mechanism for AGNs. However, it became clear (eg Frank 1978) that this process was unlikely to be significant in high-luminosity objects – to get a high enough disruption rate, the stellar concentration would have to be so extreme that stellar collisions would supply even more fuel. Tidal disruption is a relatively more significant process in low-luminosity AGNs. Indeed it is of greatest interest in quiescent galactic nuclei. If one accepts that massive black holes lurk in the centres of many (maybe even most) galaxies, then gas accretion must, in these systems, be suppressed, and it is not difficult to think of how this might come about. On the other hand, the occasional deflection of a star onto a near-radial orbit, bringing it within the Roche radius of the massive hole, seems unavoidable. It is therefore of interest to explore the consequences of such events: these may be manifested as occasional flares in quiescent galaxies, thereby allowing us to test the hypothesis that massive black holes indeed lurk in such places.

Evidence is strengthening, from studies of the central stellar velocities, for dark central mass concentrations in the nuclei of several nearby galaxies (see Kormendy 1993 for a recent review). For example, the centre of M 31, even less active than that of our own Galaxy, may contain a black hole of $\sim 3.10^7$ $M_\odot$. The quiescence implies that accretion of gas is proceeding very slowly, and/or is very inefficient radiatively. However, there would be occasional tidal disruption of stars that diffused onto nearly radial orbits. The rate of such events depends on the parameters of the central star cluster. It is therefore somewhat uncertain, but is a "clean" problem involving stellar dynamics rather than gas dynamics. Simple estimates based on models for the stellar distribution in the innermost few parsecs suggest a rate of around 10^{-4} per year. (Now that the HST is providing more detailed information about the central stellar distribution, it would be worthwhile making better estimates)

If this capture rate is translated into a time-averaged accretion rate, it could potentially yield a luminosity higher than is seen from the nucleus of M31. This is not a contradiction, provided that the debris is swallowed or ejected in a flare which lasts only a small fraction of the interval between successive events. What would be the "light curve" and the spectrum of such a flare, and the likely energy output in kinetic form? These depend on the complicated question of how the material, falling in an unsteady, non-axisymmetric fashion, gets accreted, and on the effects of radiation pressure during phases when the dissipation rate is "super-Eddington".

4.1. THE FATE OF THE DEBRIS

Earlier investigations of this phenomenon (eg Rees, 1988, Evans and Kochanek 1989, Canizzo et al 1990) have led to the following inferences:

(i) The actual disruption process is inconspicuous. Tidal forces induce violent shocks, which more than double the star's internal temperature so that its self-gravity cannot prevent it from flying apart. However, the radiation thereby generated is trapped within the star and would be attenuated by adiabatic cooling before being able to escape. (The situation is similar to a supernova explosion, which would not appear bright if there were only a "prompt" energy input at the time of the explosion itself.),

(ii) The debris in the most tightly-bound orbit reaches apocentre and falls back after a few months. This material acquired, during the tidal disruption, a velocity deficit of order $v_* \simeq (Gm_*/r_*)^{\frac{1}{2}}$ relative to the star's centre of mass, and consequently an energy deficit larger than the star's binding energy by the ratio of the orbital speed at pericentre to v_*.

(iii) At much later times the infall rate declines in proportion to $t^{-5/3}$ as more loosely bound material returns to the hole. About half the debris moves on orbits bound to the hole; the rest is on hyperbolic orbits, which escape with speeds up to $\sim 10^4$ km/s.

What is much less obvious is what happens to the debris after it completes its first orbit? Does it quickly circularise? And how quickly and efficiently is radiation emitted when it accretes onto the hole?.

A tidally disrupted star, as it moves away from the hole, develops into an elongated banana-shaped structure, the part on the most tightly-bound orbit (the first to return to the hole) being at one end (Laguna et al 1993, Kochanek, 1993, Lee and Monaghan 1993). If the debris were moving in a Newtonian "1/r" potential, then in principle it could wind up into an elliptical spiral, where successive turns would eventually merge to make an elliptical accretion disc. Such a disc could survive for many orbits if its viscosity (or α-parameter) were low enough (cf Gurzadyan and Ozernoi 1980, Syer and Clarke 1992). It would not then yield a spectacular flare; on the other hand, the decay timescale could be so slow that one would expect to see evidence of a residual disc in most galactic nuclei containing black holes

But the possibility that the debris could neatly wind up into an elliptical disc is precluded by relativistic precession. Consider first the case of a Schwarzschild hole. Orbits then remain in a plane, but each pericentre passage precesses the orbital axis through an angle $\theta_p \simeq 3\,(r_g/r_T)$, where $r_g = GM_h/c^2$ is the hole's gravitational radius. This means that there can be high-speed collisions between bits of debris that have completed different numbers of pericentre passages. This "self-intersection" process begins soon after the most tightly bound debris begins its second orbit: this material can run into inward-falling material returning towards the hole for the first time. Kochanek (1993) and Lee and Monaghan (1993) have tracked the debris to the stage when this encounter starts.

The relative velocity v_{rel} between the outgoing and incoming material is high – enormously supersonic with respect to its internal sound speed. All the debris has essentially the same specific angular momentum around the hole. If the encounter were dissipative, then the coalesced material would have an orbit that was much tighter (and more nearly circular). In fact, even though the post-shock densities are high and the radiative cooling timescales short, the shock is actually nearly elastic. This is because, owing to the high density, the electron-scattering optical depth is large compared with c/v_{rel}; before it can escape, the resultant radiation therefore gets adiabatically cooled, converting the dissipated energy into expansion of the shocked material, which sprays out in a range of directions with speed $\sim v_{rel}$. The shocked material would have a broad spread in angular momentum relative to the hole: it would not be confined narrowly to the original orbital plane, and a substantial fraction would be counter-rotating with respect to the original orbit. This material would be reshocked before it had completed more than one further orbit, and would circularise within a time shorter than the period of the original elliptical orbit.

There is a second self-intersection point near pericentre, when material returning for the first time meets other material on a more tightly bound orbit returning for the second time. The angle between the two streams is then $\sim \theta_p$. This is a relatively small angle; but since the velocities are largest at pericentre the energy dissipated can still be significant.

A further complication arises if (more realistically) the hole is described by a Kerr rather than Schwarzschild metric. The orbital angular momentum of the captured star will generally be mis-aligned with the hole's spin. Owing to Lense-Thirring precession, the debris will not then remain in a plane. This can in some cases prevent self-intersections; on the other hand, the relative velocity of the collisions can be increased. Overall, this complication provides a further parameter which would render these tidal-capture events highly non-standardised in their outcomes.

After being rendered axisymmetric, the bound debris could, without redistribution of its angular momentum, settle into a torus whose density maximum would be at a radius $\sim 2r_T$, corresponding to an orbital period of only a few hours. If there were no viscosity to redistribute angular momentum, it would cool into a thin ring. However, it is more likely that the "alpha-viscosity" is high enough to maintain it as a thick torus, and allow accretion into the hole – all that this requires is $\alpha \gtrsim 10^{-3}$. [There is, moreover, a special reason why, in this situation, a very low viscosity is unlikely: the Bardeen-Petterson effect would twist the disc around a Kerr hole, and there would be an extra transfer of angular momentum associated with this non-axisymmetric effect.]

4.2. THE "LIGHT CURVE"

These processes are too complex and poorly modelled to permit any confident inferences. However, I believe one can with fair confidence argue that the luminosity is determined by the rate at which debris is "raining down" onto the hole after completing one orbit – in other words, there is not a substantial timelag between the "first return" of the debris and its accretion or ejection. When the infall rate exceeds what is needed to produce the Eddington luminosity for high efficiency, L will be of order L_{Ed}. However, it is harder to assess whether the surplus mass is then expelled by radiation pressure, or swallowed with low efficiency. The location of the effective photosphere, and hence also the spectrum and bolometric correction, depend on the answer to this question, and therefore remain uncertain. The radiation would be predominantly thermal, with a temperature of order 10^5K; however the energy dissipated by the shocks that occur during the circularisation process would provide an extension into the X-ray band.

As illustrative examples, suppose a solar-type star passes just within the tidal radius r and is tidally disrupted by a hole of mass M_h, for the cases $M_h = 10^6 M$ and $M = 10^8$ M$_\odot$. The orbital period for the most tightly bound debris depends on $M_h^{\frac{1}{2}}$, and is respectively 2 months and 2 years in these two cases. In the first case, the infall rate is "super-Eddington" for about a year. We would therefore expect a luminosity to stay at a "plateau" level (though perhaps with a changing spectrum) throughout that period, and then to fade as $t^{-\frac{5}{3}}$. In the higher-mass (10^8 M$_\odot$) case, the accretion rate never gets high enough to generate L_{Ed}, so the luminosity would smoothly rise and fall with a timescale of 1-2 years, without there being necessarily any radiation-driven outflow.

The events would not have standardised properties. They would depend on the impact parameter: in particular, for passages much closer than r_T the timescale is shorter. They also depend on the type of star captured, and the spin of the hole. Observed effects may also depend on orientation, the amount of reprocessing, and the effects of directionality, etc.

Supernova-type searches with $\gtrsim 10^4$ galaxy-years of exposure should either detect instances of this phenomenon, or else place limits on its nature. Such evidence would also help to test whether every galaxy has been through an AGN phase, leaving a 'fossil' black hole at its centre (cf Haehnelt 1993). The possibility of such a bonus should be an added motivation for groups engaged in such searches.

4.3. STARS IN RELATIVISTIC ORBITS?

A separate question, related somewhat to the issue of tidal disruption, is whether a star can be captured into a tightly bound orbit around a massive black hole without being destroyed. Orbits just outside the tidal radius would be close enough to the hole to manifest interesting relativistic effects, such as have been computed in detail by Karas and Vokrouhlicky (1993a,b).

It is readily shown (eg Rees 1988) that a star cannot reach such an orbit by the kind of "tidal capture" process that can create close binary star systems. This is because the binding energy of the final orbit is far higher when the companion is a supermassive hole than when it is also of stellar mass, and this energy cannot be dissipated by the star without destroying it: a star whose orbit brings it within (say) $3r_T$ of a massive black hole may not be destroyed on first passage, but if it is then on a bound elliptical orbit, it will surely get puffed up and disrupted before the orbit has circularised. However, Syer, Clarke and Rees (1991) pointed out a different mechanism. A star's orbit can be "ground down" by successive impacts on a disc (or any other resisting medium): the orbital energy lost does not, in this case, have to be radiated by the star. Other constraints on the survival of stars in the hostile environment around massive black holes – tidal dissipation when the orbit is eccentric, irradiation by ambient radiation, etc – are explored by Podsiadlowski and Rees (1993) (See also King and Done (1993)). Such stars would not be directly observable, though they might cause quasiperiodic modulation of the AGN emission. Such a phenomenon could offer important information on strong-field gravity, as well as on the geometry of the continuum-emitting region in AGNs; it is therefore well worth searching for.

5. Concluding comments

The three topics I have outlined here all need (and offer timely opportunities for) theoretical clarification and development. The collimation and generation of jets is a challenging problem in relativistic MHD; the variability stems probably from internal shocks quite far out in the jet, which are triggered either by variability at the base, or by interaction with "obstacles". Radiative transfer when induced scattering must be taken into account (section 3) is sufficiently complicated that even the simplest "test problems" await a proper treatment. The phenomenon of tidal disruption, discussed in section 4, offers a specially daunting computational challenge – it involves relativistic, non-axisymmetric unsteady gas dynamics and radiative transfer, with an exceptionally large dynamic range.

I am grateful to Mitch Begelman, Roger Blandford, Paolo Coppi and Marek Sikora for collaboration on topics briefly summarised in this talk.

6. References

Begelman, M.C., Sikora, M. and Rees, M.J. 1993 MNRAS submitted
Benford, G. 1992, Astrophys. J. (Lett.), 391, L59 Blandford, R.D. 1993 in preparation.
Canizzo, J.K., Lee, H.M. and Goodman, J 1990 Astrophys. J. 351, 38
Coppi, P., Blandford, R.D. and Rees, M.J. 1993 MNRAS 262, 603.
Evans, C.R. and Kochanek C.S. 1989 Astrophys. J. (Lett) 346, L13.
Frank, J. 1978, MNRAS 184, 87
Gurzadyan, V.G. and Ozernoi, L.M. 1980 Astron. Astrophys. 86, 315.
Haehnelt, M. 1993, these proceedings.
Karas, V., and Vokrouhlicky, D., 1993a MNRAS in press
Karas, V., and Vokrouhlicky, D., 1993b Astrophys. J in press
Khokhlov, A., Novikov, I.D. and Pethick, C.J. 1993 MNRAS in press
King, A.R. and Done, C. 1993 MNRAS 264, 388
Kochanek, C.S., 1993 Astrophys. J. in press
Kormendy, J. 1993 in "Testing the AGN Paradigm" ed Holt, S (AIP Conference Proceedings)
Laguna, P., Miller, W.A., Zurek, W.H., and Davies, M.B. 1993 in "Testing the AGN Paradigm" ed S.Holt (AIP Conference Proceedings)
Lee, H.M. and Monaghan, J.J. 1993 in "Nuclei of Nearby Galaxies" ed Genzel, R and Harris, J. (Kluwer, in press).

Levich, E.V. 1972 Sov. Phys. JETP 34, 59.
Li, Z-Y, Begelman, M.C. and Chiueh, T., 1992. Astrophys. J. 384, 567.
Phinney, E.S. 1985 in "Astrophysics of Active Galaxies and Quasi-Stellar Objects" ed J. Miller p453 (University Science Books, California)
Podsiadlowski, P. and Rees, M.J. 1993 in preparation.
Quirrenbach, A. et al 1989 Astr. Astrophys. 226, L1
Qian, S.J. et al 1991 Astr. Astrophys. 241, 15
Rees, M.J. 1984 in "Very Long Baseline Interferometry" eds R. Fanti et al p.207 (Reidel, Holland).
Rees, M.J. 1988 Nature 333, 523
Sunyaev, R.A. 1970 Astrophys. Lett 7, 19.
Syer, D and Clarke, C.J. 1992 MNRAS 255, 92 (errata in 260, 463)
Syer, D., Clarke, C.J. and Rees, M.J. 1991 MNRAS 250, 505.
Wilson, D.B. 1982 MNRAS 200, 881
Witzel, A. 1992 in "Physics of Active Galactic Nuclei", eds Wagner, S., and Duschl, W. (Springer-Verlag, Berlin)

MAGNETIZED ACCRETION DISKS DRIVING JETS

G. PELLETIER, J. FERREIRA and F. ROSSO
Laboratoire d'Astrophysique de l'Observatoire de Grenoble
Domaine Universitaire. BP 53X. 38 041 Grenoble. France

1. Framework

In this brief communication, we present some progress in the investigation of a most promising model that was designed to combine ejection with accretion. In this model, a bipolar configuration of opened magnetic field lines that thread the accretion disk, allows the extraction of angular momentum, the acceleration of matter up to super Alfvénic velocities and the self collimation of the jet. However, important issues have remained unsolved. First, a systematic method for solving the jet MHD equations with their critical surfaces was lacking. Second, the capability of accretion disks to generate super Alfvénic jets was unknown. Third, the back-reaction of the ejection on the accretion disk dynamics and its energetics remained to be done. Solving these three points led us to draw some noteworthy consequences for the understanding of AGNs.

2. Jets from quasi Keplerian disks

Solving for the jet MHD equations is very difficult either analytically or numerically (Lovelace et al. 1987), because of the existence of unknown critical surfaces, mostly Alfvén (A) surface and fast magnetosonic (FM) surface, and of unknown surfaces where the PDEs change their type (from elliptical to hyperbolic and vice versa). These difficulties were rounded (Rosso and Pelletier 1993) by using a functional, whose extremalization with respect to the local Alfvén Mach number m and to the flux function ψ leads to the equations of Bernoulli and Grad-Shafranov. This functional reads:

$$\int (\frac{1}{2}\frac{m^2-1}{r^2}(\nabla \psi)^2 + U(m^2, \psi; r, z))r dr dz$$

So the solution of the problem was obtained by extremalizing the functional under the conditions for the flow to pass through the critical surface and to satisfy regularity conditions, and also to match a boundary condition at the accretion disk surface. Because of the dominent Keplerian motion, some self-similarity properties of the accretion disk hold on a wide radius interval. This implies the same similarity to the outflow, which is exactly admitted by the MHD equations for the jet. These classes of self-similar solutions contains Blandford and Payne (1982) and Konigl (1989) solutions as particular solutions. We obtained a flow diagram on a given magnetic surface that revealed that the Alfvén points are bifurcation points where three types of flow can be discriminate: a sub-A flow, a super-A but sub-FM flow, and a super-FM flow.

Three noteworthy results were obtained. First, as we considered jets carrying a poloidal current, we confirmed the result, obtained by Heyvaerts and Norman (1989), that the jet undergoes cylindrical self-collimation after having passed through the critical surfaces. A return current inside the jet itself would cancel the self- collimation (Lovelace et al. 1991). Second, whatever the enthalpy in the disk corona, significant bending is required to get a jet that crosses the A and FM surfaces. We know from Blandford and Payne (1982) that a cold outflow from a disk can become a super-A jet provided that the field lines bend more than 30^0 from axis at the surface of the disk. However a wind can be launched from the disk even if the field lines are straight, provided that a corona give it enough enthalpy. But our results show the requirement of an important bending to get a super-FM jet. Third, despite the magnetic surface widening above the accretion disk, the centrifugal acceleration does not lead to asymptotic velocities much larger than the Keplerian velocity at the foot of the field lines because it turns out that the A-surface is close to the accretion disk. We have explored a wide range of parameters and we did not find any situation where the A-surface is remote. So jets from quasi keplerian accretion disks are more likely subrelativistic. We think that the only possibility to launch relativistic jets is in the vicinity of a fastly rotating black hole (Blandford and Znajek 1977, Camenzind 1986) and probably in the form of relativistic flow of electrons and positrons (Baskin et al. 1991). However it is not yet sure that the Compton drag can be overcome. Alternately, a model of relativistic ejection of e^+e^- pairs that starts between 50 and $100 r_G$ from the black hole, a distance where Compton drag is weakened, was proposed (Henri and Pelletier 1991). This ejection, that is powered and channeled by the subrelativistic disk outflow, takes the form of a continuous or sporadic relativistic beam and likely explains the high energy spectrum of radio quasars (Henri et al.1992, Henri and Markowith 1993).

Is it possible for an accretion disk to display such bended open field lines?

3. Accretion disks revisited

R. Pudritz (1986) proposed that an MHD jet could extract the angular momentum of the accretion disk and thus makes the turbulent viscosity useless. In this scheme, the jet controls the accretion process and properties of such a jet was derived (Pelletier and Pudritz 1992). Indeed the most powerfull jets are those that extract most of the angular momentum and thus modify significantly the accretion disk properties. Our first attempts (Ferreira and Pelletier 1993a) to solve completely the MHD equations for a magnetized accretion disk with no viscosity succeeded to give a solution where the MHD wind extract completely the angular momentum, but with bipolar straight field lines. In this particular configuration, all the gravitational power is stored in the magnetic field, the disk emission being just a secondary process. Thus, the luminosity is not a measure of mass flux and Eddington estimates are merely fiducial numbers. We expected that the magnetic surfaces would widen at a remote distance, so that the wind could become super

Alfvénic; but our previously mentioned results about jets indicate that this is not possible. New attempts (Ferreira and Pelletier 1993b), including a weak viscous effect, led to successful solutions with a significant bending of the field lines. An interesting class of solutions was studied with more details because all the involved physical effects can be put together on a wide interval of radii of the disk. This is for a flux distribution on the midplane such that $\psi \propto r^{-3/4}$. Hence we solved the complete problem, including the vertical structure, with i) angular momentum extraction by both magnetic and viscous torques, ii) heating by both Joule and viscous dissipation, iii) vertical equilibrium by both magnetic tension and gravitation compression. Even for a magnetic torque 50 times larger than the viscous torque, we obtained jet solutions with field lines bended of 50^0 from axis, provided than the magnetic Reynolds number be sufficiently larger than unity (of order 10). Indeed the more the magnetic Reynolds number $\mathcal{R}_m$, the more the field lines bending. In the evaluation of $\mathcal{R}_m$, defined by

$$\mathcal{R}_m \equiv \frac{r u_r}{\nu_m},$$

we need an anomalous magnetic diffusivity ν_m such that $\nu_m = \alpha_m V_A h$, in order to have a stationary flow of matter crossing the magnetic surfaces with a thin disk configuration. This is required by steady MHD equations; moreover the number α_m must be of order unity to quench resistive instabilities, because it is the ratio of the Alfvén time across a disk of thickness h over the resistive time for the same scale. Increasing $\mathcal{R}_m$ by reducing α_m leads to convection instabilities by producing superadiabatic gradients and resistive instabilities. Inside the disk the resistive instability is expected to produce magnetic islands by field line reconnections. But this occurs under conditions where the internal pressure plays an important role. This is not the case in solar corona nor in tokomaks; and so the dynamics of reconnections in high β plasmas is not really known. However we could expect that the reconnections occuring in the disk corona where the magnetic configuration is almost force free are similar to those of the solar corona. In fact, the microturbulence generated along the neutral lines of reconnections can bring the amount of anomalous resistivity needed, and moreover the electric fields along the neutral lines are likely to generate the relativistic electrons responsible for the high energy spectra.

The magnetic field reroutes part of the stream lines in the accretion disk, even in the optically thick region; they turn to more than 120^0 and tends asymptotically towards the field line outside the disk, achieving an almost ideal MHD flow, at few disk thicknesses, with a magnetic Reynolds number of order 10^3. All these classes of self-similar solutions for the disk henforce to the same similarity for the jet solutions, because they correspond to exact similarity properties of the MHD equations. So the matching between the accretion disk and the ideal MHD jet is realised, at least for some parameter range. We got these results even under conditions where the magnetic pressure is smaller than the radiation pressure.

4. Synthesis for AGNs

Our view of AGNs exhibiting jets can be summarized as follows. Three types of situations can be discriminated, likely due to the conditions of magnetization of the accretion disk.

i) In AGNs with FR I radiojets, the jets seem clearly non relativistic and their production by a quasi Keplerian accretion disk is very likely. However, some relativistic outburts can occur (in Bl Lac, for instance).

ii) In AGNs with FR II radiojets, large scale structures and hot spot parameters do not dictate relativistic motions (Roland et al. 1988) (Meisenheimer et al. 1989). Relativistic velocities are only clearly suggested by the superluminal motions of VLBI knots. We consider that the accretion disk can power most of the ejection in the form of a mildly or sub relativistic MHD flow. The relativistic ejection would come from a beam (Sol et al. 1988), either from a fastly rotating black hole if Compton drag is really overcome, either from pair creation at the edge (50 to 100 r_G) of the "magnetized cauldron" (Henri and Pelletier 1991).

iii) In AGNs of Seyfert galaxies, the outflow from the accretion disk does not pass through the critical surfaces, but some possibility of an aborted jet exists; which is convenient to explain the high energy spectrum.

Indeed, in the three cases, we propose that a magnetic cauldron made by the MHD outflow, heated by Alfvén turbulence, contains a plasma of relativistic pairs at a decentred location, illuminating the disk with hard X-rays and interacting with the hot absorbing medium of the subrelativistic outflow (See Ferreira et al.. 1994). The results we obtained in the framework of a quasi Keplerian disk driving MHD jets are encouraging to understand the main energetics of the accretion-ejection process and deserve deeper physical investigations, especially regarding instabilities and transport phenomena.

References

Beskin V.S., Istomin Y.A., Pariev V.I. (1991), "Extragalactic Radio Sources", proc. 7th IAP meeting, Paris 1991.
Blandford R.D. and Payne D.G. (1982), MNRAS, 199, 883
Blandford R.D. and Znajek D.G. (1977), MNRAS, 179, 433
Camenzind M. (1986), A. and A., 156, 137
Ferreira J. and Pelletier G. (1993a), A. and A., 276, 625 and A. and A., 276, 637
Ferreira J. and Pelletier G. (1993b), A. and A.,(submitted)
Ferreira J., Henri G., Markowith A., Pelletier G., Rosso F. (1994) A. and A. (in preparation)
Henri G. and Pelletier G. (1991), ApJ Letter, 383, L7
Henri G., Pelletier G., Roland J. (1992), ApJ Letter, 404, L41
Henri G. and Markowith A. (1993), proceeding of the conference
Heyvaerts J. and Norman C. (1989), ApJ, 347, 1055
Konigl A. (1989), ApJ, 342, 208
Lovelace R.V.E., Wang J.C.L., Sulkanen M.E. (1987), ApJ, 315, 504
Meisenheimer K., Roser H.J., Hiltner P.R., Yates M.G., Longair M.S., Chini R., Perley R.A. (1989), A. and A., 219, 63
Pelletier G., Pudritz R.E. (1992), ApJ, 394, 117
Pudritz R.E. (1986), ApJ, 301, 571
Roland J., Pelletier G., Muxlow T.W.B. (1988), A. and A., 207,16
Rosso F. and Pelletier G. (1993), A. and A., (submitted)
Sol H., Pelletier G., Asséo E. (1989), MNRAS, 237, 411

VARIABILITY AND SLIM ACCRETION DISKS

F.H. WALLINDER
NORDITA, Blegdamsvej 17, DK-2100, Denmark

Abstract. Many compact galactic and extra-galactic sources have a luminosity in the slim accretion disk regime, i.e. $L/L_E \sim 1$, where L_E is the Eddington luminosity. The correspondingly large accretion rates diminish the relevance of variability interpretations based on the thin disk model. This paper explores the possible connection between variability in AGN and local instabilities in slim disks, and points out some relevant areas of future research.

Key words: Black holes, accretion disks, variability

1. Introduction

Accretion disks surrounding massive central objects are probably the main power source in compact sources, such as AGN. Models of such objects often employ the standard thin disk model (Shakura & Sunyaev, 1973, 1976), in spite of the well-known fact that luminosities comparable to the Eddington one imply a break-down of the thin disk approximation. Thin disks have also unphysical singularities at the inner edge, due to an oversimplified treatment of the inner region. In contrast, slim accretion disk models have an accretion rate $\dot{m} = L/L_E \sim 1$, where $L_E = 10^{46} M_8$ erg s^{-1} is the Eddington luminosity and M_8 the central mass in units of $10^8 M_\odot$. The treatment of the inner region take into account transonic radial motion, non-Keplerian rotation, advection of heat into the presumed black hole as well as radial pressure and velocity gradients, none of which are included in the standard model (e.g., Abramowicz et al. 1989; Wallinder 1991a,b; Chen & Taam 1993). Another advantage is the limited number of input parameters, comprising $(\alpha, \dot{m}, m)$, where α is the standard viscosity parameter and m the central mass in solar units. If AGN variability is due to instabilities in slim disks, information about these parameters may be deduced. Combining for instance the inferred central mass with a size estimate would provide additional evidence of black holes in galactic nuclei.

The main aim of this paper is to study the local stability of slim disks as a function of radius. This will provide information about where the various instabilities operate and what time scales they have.

2. The radial profile of the luminosity

The issue of where most of the luminosity is generated is essential, since the most noticeable variability should come from the most luminous region of the disk. If one plots radial profiles of flux against radius, one usually finds that the flux has a maximum at $x = r/r_g \sim 5$, where $r_g = 10^{-5} M_8$ pc is the Schwarzschild radius. Many authors therefore assume that most of the variability arises close to $x \sim 5$.

However, the emitting *area* around this radius is rather small compared to the one for the whole disk. What seems more relevant is the radial dependence of the integrated flux. Fig. 1a shows the obtained luminosity profile for the emission from both sides of the disk. The examples shown are for two accretion rates, $\dot{m} = 0.01$ and 6.3. Here and henceforth, the parameters $(\alpha, m) = (10^{-3}, 10^7)$ were adopted. As can be seen, the luminosity rises quickly but approaches a constant value only when $x \gtrsim 10^3$.

In Fig. 1b it has been assumed that all the luminosity is produced within $x \lesssim 10^3$, so that the scaled luminosity $L/L_{\max} = 1$ at $x = 10^3$. Apparently ~ 80 % of the radiation comes from the region $x \lesssim 10^2$, whereas only about 30 % arises within $x \lesssim 10$. This argues against the notion of a "preferred" radius around $x \sim 5$ where most of the variability is generated. The combined effect from many radii may be a better interpretation.

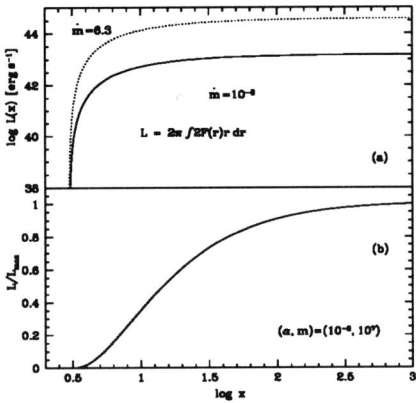

Fig. 1. (a): Integrated flux profiles from slim disks, i.e. the luminosity emitted within the radius $x = r/r_g$, where r_g is the Schwarzschild radius. The luminosity approaches a constant value only when $x \gtrsim 10^3$. (b): The same luminosity profiles, but scaled in terms of $L_{\max} \equiv L(10^3)$. Due to the almost parallel curves in (a), values virtually coincide. The fraction of the total luminosity which arises within $x \lesssim 10$ and $\lesssim 10^2$ is ~ 30 and ~ 80 %, respectively.

3. Method of stability analysis

The radial structure of the disk is governed by the conservation laws of mass, momenta and energy. Together with e.g. an equation of state they form of complete set of equations. The time-independent equations are solved numerically, in order to obtain the slim disk equilibria. The resulting solutions are fed into the coefficients of a fourth-order dispersion relation, which is obtained from the linearly perturbed time-dependent equations. The analysis is local, so that boundary conditions are avoided. The dispersion relation is solved numerically, yielding growth rates for the thermal/viscous and the two acoustic modes in various parts of parameter space.

4. Results

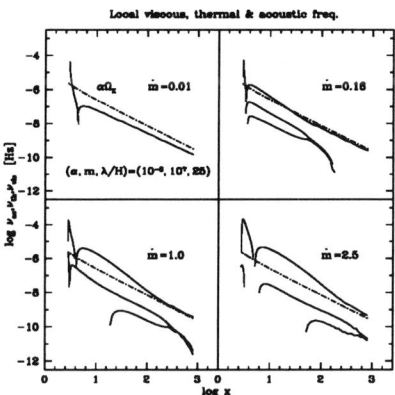

Fig. 2. Local instability frequencies against radius. The dash-dot curve corresponds to the frequency $\alpha\Omega_K$. The modes (solid curves) can be distinguished through their different growth rates, which increase in the order viscous-thermal-acoustic. However, in the top/left fig. the accretion rate is so low that no thermal/viscous instabilities are present. The only unstable mode is the outward propagating acoustic one, and the inward propagating one in the innermost region. In the top/right fig. the thermal/viscous instabilities turn on at about $x \sim 10^2$. Advection cooling results in a stable inner region, whose extent increases with $\dot{m}$. The time scales vary from $\sim 10^2$ yrs in the outer disk to fractions of days in the inner.

Fig. 2 shows thermal, viscous and acoustic instability frequencies as function of radius. For $\dot{m} = 10^{-2}$, only the outward propagating acoustic mode is unstable over a large fraction of the disk, whereas the inward propagating acoustic mode becomes highly unstable in the innermost region. When $\dot{m} \sim 0.1$, the viscous and thermal instabilities turn on at around $x \sim 10^2$, but in general they have much lower growth rates than the acoustic modes. This implies that the advection cooling in the inner region stabilizes the viscous mode first, then the thermal and lastly the acoustic modes. It can also be noted that the time scales adhere to the standard $(\alpha\Omega_K)^{-1}$ one only at specific combinations of the input parameters, i.e. approximating instability time scales with the latter value may be inappropriate in general. Another thing to note is the huge span of time scales across the disk, from hundreds of years in the outer part to fractions of days in the inner one.

5. Conclusions

Despite the rapid increase of the luminosity profile, observed variability should come from an extended region (say $x \lesssim 10^2$) instead of a narrow annulus around

$x \sim 5$. The selection of a few radii due to wave trapping inside the epicyclic barrier at $x \sim 4$ (Okazaki et al. 1987; Nowak & Wagoner 1991) may be irrelevant, since only a minor fraction of the total luminosity arises within $x \lesssim 10$. Evaluation of variability time scales may be hazardous without knowledge of detailed stability properties. As the examples above show, the whole range from totally stable to partially unstable or fully unstable disks is possible. The result depends partly on which type of instability is considered, partly on the position in parameter space. The range of time scales is rather large, so it should come as no surprise that variability power spectra often show no significant peaks.

6. Future work

A growing number of researchers have realized that slim accretion disks constitute a good framework in which additional physics may be included. Thus, a number of projects are under development at present. One will attempt to include an optically thin plasma near the inner region, in order to produce the observed hard X-rays. Another will attempt to solve the time-dependent slim disk equations. The combination of these will allow computation of theoretical power spectra, which convolved with the detector response of whatever X-ray satellite will produce results which can be compared with observations in detail.

As pointed out by Lin et al. (1990), a realistic accretion disk has probably a sound speed which decreases in the z-direction. It follows that a wave front which starts out perpendicular to the disk plane will be refracted into the corona. If a sufficient amount of energy can be transferred to the modes, the result may be acoustic coronal heating. Thus, information about the acoustic instability frequencies may be propagated into the X-ray producing region.

Acknowledgements

Financial support from the Swedish Natural Science Research Council is gratefully acknowledged.

References

Abramowicz M.A., Szuszkiewicz E., Wallinder F.H.: 1989, proc. *Theory of Accretion Disks*, eds. F. Meyer et al., (Kluwer), p. 141
Chen X. and Taam R.E.: 1993, ApJ **412**, 254
Lin D.N.C., Papaloizou J.C.B., Savonije G.J.: 1990, ApJ **364**, 326
Nowak M.A. and Wagoner R.V.: 1991, ApJ **378**, 656
Okazaki A.T., Kato S., Fukue J.: 1987, PASJ **39**, 457
Shakura N.I. and Sunyaev R.A.: 1973, A&A **24**, 337
Shakura N.I. and Sunyaev R.A.: 1976, MNRAS **175**, 613
Wallinder F.H.: 1991a, A&A **249**, 107
Wallinder F.H.: 1991b, MNRAS **253**, 184

STRUCTURE AND EMISSION OF PARSEC–SCALE JETS IN QUASARS

MAX CAMENZIND
Landessternwarte Königstuhl, D-69117 Heidelberg, Germany

Abstract. The moving knots observed in VLBI of compact quasars are due to off–axis substructures in the parsec–scale jets which are dragged along by the underlying plasma flow. This is suggested by detailed MHD models for the structure of relativistic jets on the parsec–scale that are based on exact solutions of the nonlinear Grad-Schlüter-Shafranov equation in the asymptotic domain. These jet models are characterized by a current–carrying core and a current–free envelope. The corresponding core–radius $R_c = u_p R_L$ is related to the poloidal jet velocity u_p and the light cylinder radius R_L generated by the rotation of the magnetic surfaces. The Poynting flux, which has an inhomogeneous distribution in the jet, provides an off–axis energy reservoir that could be tapped and converted into particle acceleration by non–axisymmetric instabilities in the core–jet structure. Evidence is presented that quasi–periodic synchrotron emission (in the IR–optical) and inverse Compton emission (in X– and γ–rays) from freshly accelerated electrons are due to a lighthouse effect of the rotating knots.

1. Introduction

Rapid optical and radio activity has now been observed during outbursts of a number of radio–loud quasars and BL Lac objects. There is some evidence that gamma–emission, detected by EGRET at energies of a few hundred MeV, occurs simultaneously with this optical activity. Two years ago, we proposed for the first time that this rapid variability could be explained in terms of emission from plasma blobs which move relativistically within the parsec–scale jets of these objects (Camenzind & Krockenberger 1992a,b). Due to the non–vanishing angular momentum of the plasma in MHD jets, the rotation of the plasma blob leads to a lighthouse effect for the beamed emission.

In the meantime, this idea has been tested for the well–documented optical outburst of the quasar 3C 345 beginning in 1991 and ending in 1993 (Schramm et al. 1993). This model can explain the appearance of equidistant peaks in the emission and the simultaneous increase in the radio lightcurve. Both features, VLBI knots and emission activity appear therefore related to the underlying structure of the parsec–scale jets. VLBI maps taken at successive epochs also show quite convincingly a correlation between flux outbursts seen in the far–infrared – optical region of the spectrum and the birth of a new superluminal knot in the parsec–scale jet. In this sense, detailed monitoring of compact radio sources provides very powerful tests for the structure of the parsec–scale jets of these sources.

2. Formation of Relativistic Jets in Quasars

In the following we discuss the structure of relativistic jets in quasars and BL Lac objects following from the relativistic magnetic jet model developed in the last years (Camenzind 1986; 1987; 1990; 1992; 1993a; Appl & Camenzind 1993a,b). This is the only model for extragalactic jets that can account for the observed relativistic motion and the tight collimation required by VLBI measurements on the milliarcsecond level. According to this picture, jets are nothing else than magnetically confined plasma flows supported by currents of the order of 10^{18} Ampères. These axisymmetric jet channels are the natural carriers for the knots seen in superluminal motion. Non–axisymmetric perturbations in the jet channel are dragged along by the underlying plasma and the induced rotation is responsible for the observed flaring activity.

Relativistic jets are formed under quite unique conditions (Camenzind, 1990; 1993a). Relativistic speeds cannot be attained in pure hydrodynamic acceleration, magnetic fields are a necessary ingredient for jet formation. Dynamo action in the inner accretion disk creates a rotating magnetosphere which is filled with disk plasma. Along open flux surfaces, the plasma is accelerated to relativistic speeds and collimated outside the light cylinder $R_L = c/\Omega_K(R_F)$ for footpoint radii R_F near the inner last stable orbit at $r_{\rm ms}$ around a rapidly rotating Kerr black hole

$$R_L = c/\Omega^F \simeq 10\, GM_H/c^2 \simeq 1.5 \times 10^{15}\,{\rm cm}\, M_H/10^9\, M_\odot\,. \qquad (1)$$

As in the Newtonian case, axisymmetric MHD flows can be described in terms of several invariants. The rotating field itself forms axisymmetric flux surfaces which are either collimated by the toroidal magnetic field or external pressures (Appl & Camenzind 1993b). Since the host galaxies of quasars are expected to be giant ellipticals – as demonstrated by 3C 273 – these galaxies dispose of large cores with core radii of a few hundred parsecs. This entire region can be filled up with molecular gas forming in this way a molecular torus (Camenzind 1993b). Steep pressure walls around the rotational axis will collimate the outflowing disk winds into jets with extremely small opening angles on the parsec–scale.

Disk plasma streams along these collimated magnetic surfaces in such a way that the total energy and total angular momentum are conserved. The toroidal magnetic field in the flow also follows from conservation laws. It turns out that the magnetization σ_D provided by the central accretion disk around a rapidly rotating black hole is the only parameter that is essential for the question of the outflow speed (Camenzind 1993a). Relativistic speeds can only be obtained for $\sigma_D > 1$ with outflow speeds $\gamma_j \leq \sigma_D$. When the magnetization of the disk is low, $\sigma_D < 1$, only nonrelativistic speeds will result, $\beta_j \simeq \sigma_D^{1/3}$.

In the innermost part of a galaxy, these newly created plasma flows are then probably collimated by the external pressure of the molecular torus. Jets would otherwise expand outside the light cylinder until their self–collimation radius were achieved (Appl & Camenzind 1993b).

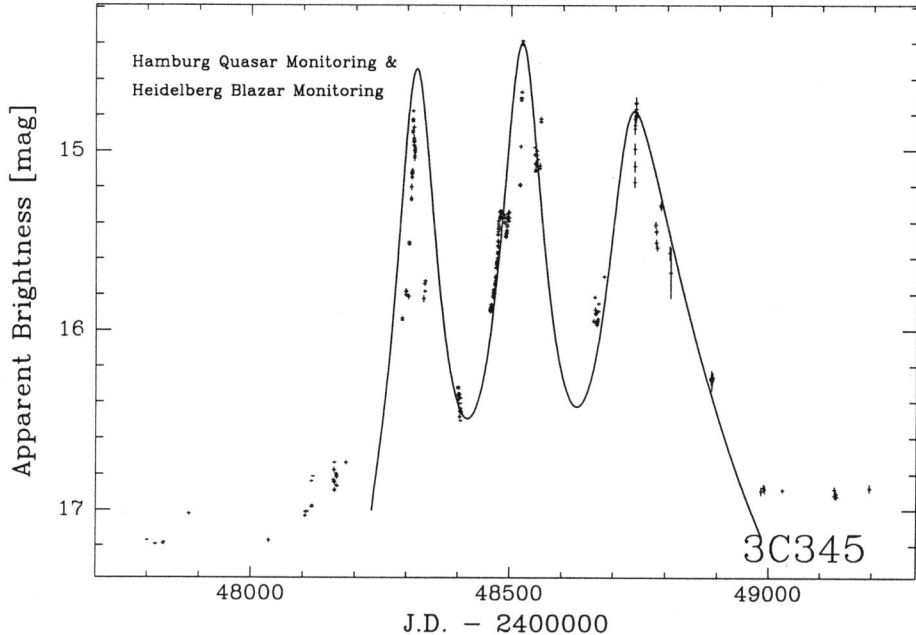

Fig. 1. A theoretical optical light curve (solid line) for 3C 345 based on the lighthouse effect of a relativistic magnetized jet (Dreissigacker & Camenzind 1994). The data points are from Schramm et al. (1993). A knot is moving with bulk Lorentz factor $\gamma_j = 16$ at a radius $R_0 = 16 R_L \simeq 0.3$ pc for a central black hole $M_H \simeq 4 \times 10^{10} M_\odot$. The inclination angle between the line of sight and the jet axis is $0.95°$. Equidistant peaks are produced by a completely collimated jet, which however opens up at a distance of $\simeq 50$ pc from the center resulting in a sudden decay of the optical synchrotron emission.

3. Knot Propagation and Knot Emission

The observed variability of radio–loud quasars in the non–thermal emission is interpreted in terms of substructures moving relativistically through the underlying jet channel. These structures need not be necessarily shocks, they could be created by instabilities (Appl & Camenzind 1992) or plasmoids ejected along the magnetic surfaces. In the initial phase, the relativistic electrons in these structures are cooling by incoherent synchrotron emission and inverse Compton losses due to scattering of the electrons on infrared and UV–photons. Outbursts driven by such structures have been well documented for 3C 273, 3C 345 and many other sources. In the case of 3C 273, the cycle time between various outbursts is of the order of two years, for 3C 345 somewhat longer. In the case of BL Lac objects, the cycle time is of the order of months or weeks.

The motion beyond the light cylinder occurs in such a way that the specific angular momentum stays constant. The rotation of such a bubble around the central axis of the jet produces therefore a lighthouse effect which mimics a time–dependence of the received flux (Camenzind & Krockenberger 1992a,b). The flux emitted by the bubble is quasi–periodic. The observed period appears drastically shortened due to the projection effect

$$P_{\text{obs}} = P_{\text{jet}} (1+z)(1-\beta_j \cos \Theta) \simeq (1+z) \frac{P_{\text{jet}}}{2\gamma_j^2} \qquad (2)$$

This reduction in the time–scale is fundamental for the understanding of short time–scales observed in quasar emission. For knots moving with $\gamma_j = 20$, all time–scales appear shortened by a factor of 800 —

In Fig. 1 we show the optical light curve for 3C 345 (solid line) superposed onto the optical data for the last outburst starting in 1991 (Schramm et al. 1993). The first three peaks are practically equidistant in time requiring an initial intrinsic opening of the jet of $0.05°$. The decay of the third peak and the absence of further peaks is due to a larger opening of the jet beyond a projected scale of 0.15 mas, corresponding to $\simeq 50$ parsecs away from the central source.

The detailed structure of the lightcurve (showing various precursors and shoulders) cannot be explained by this simple model. This time–structure in the lightcurve might signal the presence of a corresponding spatial substructure in these non-axisymmetric knots. Such structures are expected to be formed, if the electrons accelerated to relativistic energies in these knots are due to magnetic reconnection. It is interesting that the amount of electrons needed to explain the observed synchrotron fluxes is of the order of the density found in the jet matter ($\simeq 10^5$ cm^{-3}). This would require that most of the available electrons are accelerated to relativistic energies in the knot. There are no pairs necessary to explain the synchrotron emission.

References

Appl, S., Camenzind, M.: 1992, *Astron. Astrophys.* **256**, 354
Appl, S., Camenzind, M.: 1993a, *Astron. Astrophys.* **270**, 71
Appl, S., Camenzind, M.: 1993b, *Astron. Astrophys.* **274**, 699
Camenzind, M.: 1986, *Astron. Astrophys.* **162**, 32
Camenzind, M.: 1987, *Astron. Astrophys.* **184**, 341
Camenzind, M.: 1990, in G. Klare, ed., *Rev. Mod. Astron.* **3**, 234 (Springer–Verlag, Heidelberg)
Camenzind, M.: 1992, in M. Locher, ed., *Symp. Interface of Astrophys. with Nuclear and Particle Phys.*, p. 251 (PSI, Villigen)
Camenzind, M.: 1993a, in H.–J. Röser, K. Meisenheimer, eds., *Lecture Notes in Physics* **421**, 109
Camenzind, M.: 1993b, *The Parsec–Scale Structure of Relativistic Jets in Quasars and the Origin of Gamma–Emission*, preprint (Heidelberg)
Camenzind, M., Krockenberger, M.: 1992a, in W.J. Duschl, S.J. Wagner, eds., *Physics of Active Galactic Nuclei*, 551 (Springer–Verlag, Heidelberg)
Camenzind, M., Krockenberger, M.: 1992b, *Astron. Astrophys.* **255**, 59
Dreissigacker, O., Camenzind, M.: 1994, in prep.
Schramm, K.–J., Borgeest, U., Camenzind, M., et al.: 1993, *Astron. Astrophys.* **278**, 391

EMISSION OF ACCRETION DISKS

B. CZERNY*
Copernicus Astronomical Center, Bartycka 18, 00-716 Warsaw, Poland

Abstract. Accretion disks surrounding massive black hole is an attractive scenario of nuclear activity. A number of arguments support it although there is no unquestionable proof of the existence of accretion disks in active galactic nuclei. Meaningful comparison of the disk model prediction with the data can only be made if emission of accretion disks is calculated taking into account the existence of optically thin parts responsible for the emission of x-ray radiation. Nonlocal reprocessing phenomena have to be also included. Since we have no real understanding of the viscous processes operating in accretion disks some ad hoc parameterization of these processes has to be used and its applicability should be checked by broad band comparison of predictions for continuum emission and spectral features with available data.

Key words: accretion disks, thermal emission, optical/uv spectra, x-ray spectra

1. Introduction

The nature of nuclear activity in galaxies remains a mystery although a huge amount of data have already been accumulated over the energy range from radio to gamma. This is the result of unquestionable complexity of the phenomenon.

It is widely believed that the ultimate source of energy is accretion onto supermassive black holes, as suggested by Zeldovich & Novikov (1964) and Salpeter (1964). The clear advantage of such a scenario (Rees 1984) is the compactness of the activity center, its efficiency and stability, and the possibility of multiple reactivation of the nucleus by the fresh supply of matter. However, attempts are made to explain all the nuclear activity through starburst phenomenon (e.g. Terlevich et. al. 1992).

The overall character of accretion is not so clearly specified as in the case of x-ray binaries (e.g. Frank, King and Raine 1985). The accreting gas in active galactic nuclei most probably comes from the inner ~10 kpc losing its angular momentum through gravitational interaction (Schlosman, Begelman & Frank 1990) but an inflow more resembling the cooling flow may also be a possibility (see Heckman 1992).

The overall view of an active nucleus is determined not only by the energy release close to the center of activity but also by the circumnuclear processes: collimation and acceleration of jets and modification of outgoing radiation by scattering and absorption by dust and ionized medium surrounding an nucleus at 0.01 – 100 pc. Because of anisotropic character of these phenomena, the result strongly depends on the inclination angle. Unification schemes suggested recently (see Barthel 1992), although not perfect (Antonucci, this volume), help to disantangle the central phenomenon from these secondary effects.

* Also Observatoire de Paris-Meudon

The AGN spectra and observed variability clearly indicate the coexistence of hot and cold gas within the inner ~1 pc (see Mushotzky, Done & Pounds 1993). However, this is possible within two basically different scenarios. According to the first one, most of the gravitational energy is released within hot accreting medium in a form of x-ray emission, and the cold blobs coexist with the hot phase (as a result of thermal instabilities and confinement by magnetic field) reprocessing a fraction of the primary radiation. According to the second one, accreting gas is relatively cold, disk like and optically thick, and only in some (innermost ?) part of the flow, due to thermal instabilities, the temperature rises and a fraction of energy is released in a form of x-rays.

Some other, more complex, possibilities can be imagined as well (e.g. Fabian & George 1990) but they combine the properties of the two cases described above. In this review I discuss the second scenario, with particular emphasis on the observational verification of the models through the prediction of their emission spectra.

2. Disk accretion

Accretion disks surrounding massive black holes were suggested as a plausible scenario of nuclear activity by Lynden-Bell (1969). The early papers of Shakura and Sunyaev (1973) and Novikov and Thorne (1973) formed a theoretical basis (see Pringle 1981). The contribution of the disk to the AGN spectrum was expected in the uv band and it was identified with observed blue bump component by Shields (1978). However, if we want to explain the overall activity of a nucleus we have to go beyond those first simple models. Particularly, we should include:
— possible radial and vertical stratification of the disk,
— global effects of reprocessing of radiation emitted by one part of the disk by the other parts,
— time evolution.

2.1. Arguments in favor of disk accretion

There are no proofs of the existence of accretion disks at the active nuclei. However, a number of arguments have been broadly discussed in a number of review articles (e.g. Collin-Souffrin 1992). To summarize them briefly, there are two basic arguments. The first is the observational one: clearly axial - not the spherical! - symmetry of the nucleus, as adopted in all unification schemes (see Antonucci, this volume). A number of phenomena indicates that: (i) presence of jets and anisotropic radio emission (ii) dust/molecular torus (iii) ionization cone/ENLR (iv) x-ray reflected component (v) double profile component in emission lines. The second is the theoretical one: the disk accretion is highly efficient (5 % for non-rotating black hole and up to 42 % for a rotating one) whilst the efficiency of spherical accretion is usually much lower (and difficult to estimate).

There are also some arguments against accretion disks (e.g. Antonucci 1992) but these arguments are related to particular disk models and cannot be dealt with before establishing the vertical disk structure (Malkan 1992).

2.2. EXPECTATIONS FROM THE MODEL

We do not expect to explain the blazar component at this stage of modelling as the formation of jet is a very complex problem in itself. Although there are attempts to describe this phenomenon in the context of accretion disks (e.g. Romanova & Lovelas, this volume) they necessarily involve assumptions about the global properties of magnetic field within the disk and outside. In this review I concentrate on AGN which are both radio-quiet and unabsorbed by circumnuclear dust.

2.2.1. Optical/uv continuum

Large samples of AGN (e.g. Neugebauer et. al. 1987, Sanders et. al. 1989) clearly show the presence of big blue bump extending from optical frequencies upwards. The universal position of the minimum at $\log\nu \sim 14.5$ (or $\lambda \sim 1\mu m$) results from the evaporation of dust grains at the temperature about 1500 K so the dust contribution to the spectrum does not extend beyond near-ir (see Barvainis 1992). High frequency extension of the big bump is not well constrained. In (unobscured) Seyfert galaxies and low redshift quasars the decline of the spectrum is frequently observed in far-uv (Edelson and Malkan 1986, O'Brien, Gondalekhar and Wilson 1988) whilst in many high redshift quasars no cut-off is observed even up to $\log\nu \sim 15.7$ ($\lambda \sim 650$ Å) (e.g. Bechtold et al. 1984, Reimers et al. 1992). The soft x-ray excesses present in 50% - 90% of Seyferts and quasars (Wilkes and Elvis 1987, Walter and Fink 1993) do not necessarily mean that the big blue bump extends into soft x-rays. In contrary, the universal shape of most of the excesses (Walter and Fink 1993) are better explained as spectral features due to reprocessing of x-rays by weakly ionized gas, e.g. accretion disk (Czerny et al., this volume). This spectral component is expected to come from optically thick (in a sense of effective optical depth) parts of accretion disk.

2.2.2. X-ray continuum

Recent data from Compton Observatory indicate that the x-ray continuum in radio-quiet objects does not extend beyond ~ 100 keV and it is predominantly thermal (see Zdziarski, this volume). This conclusion is based both on the positive detection for a few sources as well as non-detection of a number of Seyfert galaxies. This spectral component is expected from optically thin parts of accretion disk, mostly cooled by Compton scattering.

2.2.3. Spectral features

High resolution of x-ray data allows to study such features as absorption edges, emission lines (particularly K_α iron line) and recombination continua as well as Compton reflection. Significant part of these features may (or should) form as the result of reprocessing of x-ray flux by the optically thick cool parts of accretion disks. Also some uv and optical low ionization emission lines may (and should) come from the disk (e.g. Dumont & Collin-Souffrin 1990)

3. Keplerian stationary local black body model

This simplest model has been widely used to model the AGN spectra, starting from Malkan (1983). The underlying assumptions (see Pringle 1981, Frank, King and Raine 1985) reduce the problem of disk structure so significantly that the flux generation at a given disk radius r is given by a simple formula and the disk spectrum is fully described by the mass of the black hole M, accretion rate $\dot{M}$, and the inclination angle i. The applicability of the model is constrained by the critical value of the accretion rate dependent on the efficiency η of accretion: $\dot{M}_{crit} = L_{Edd}/c^2/\eta$, as 'slim disks' effects become important when the luminosity of the disk is comparable or larger than the Eddington luminosity L_{Edd} (e.g. Abramowicz et al. 1988).

Assumptions of stationarity and local emmissivity given by black body cause that the description of the spectrum neither contains nor requires any information about the disk structure.

First applications of this model were encouraging as the model gives a spectrum peaking in the uv range on νF_ν diagram. This is, however, not surprising, as any thermal emission at a fraction of the Eddington luminosity expected for a $10^8 M_\odot$ black hole would give the same range of temperatures (Rees 1984). Also, closer examination reveals a number of problems: (i) the energy index in the optical/uv slope ($F_\nu \sim \nu^{-\alpha}$) predicted by the model (asymptotically approaching -1/3 at long wavelengths) is too small; mean values are ~ 0.2 for quasars (Neugebauer et al. 1987) and ~ 0.5 for Seyferts (Malkan 1988) (ii) x-ray emission is not predicted for by the model. Although a plausible scenario explains x-ray emission as originating in shocks developing in the gas outflowing along the symmetry axis, i.e. above the accretion disk (Henri & Pelletier 1991), this x-ray emission is not accounted for energetically.

4. Transfer of radiation and of angular momentum

Any departure from local black body assumption requires:
- description of opacities,
- description of angular momentum transfer and the energy dissipation.

Inclusion of bound-free and bound-bound opacities is essential, particularly in the outer part of the disk (Hure, Collin and Pineau des Forêts, this volume). Some progress has been made along this line (e.g. Kolykhalov and Sunyaev 1984, Czerny and Elvis 1987, Ross, Fabian and Mineshige 1993) but still more improvements are required.

As for the angular momentum transfer, we have no real understanding of its mechanism. From the practical point of view, there are three possibilities:(i) angular momentum is transported locally and rather uniformly within disk (ii) the transport and dissipation is still local but significantly enhanced either close to the disk surface or to the equatorial plane (iii) the transport is truly global, e.g. through global magnetic field or spiral waves.

5. Alpha viscosity disks

Simple parameterization of the angular momentum transfer was introduced by Shakura and Sunyaev (1973). Angular momentum is transported by some anomalous viscosity and the viscous torque τ is proportional to the pressure. Two cases are widely discussed: torque proportional to the gas pressure and torque proportional to the sum of gas and radiation pressure. These two assumptions give extremely different predictions.

The second family of models is not strongly influenced by radiation pressure although the radiation pressure dominates in the innermost parts (e.g. Clarke 1988). Models are thermally stable in the inner parts (see Piran 1978) and are not expected to develop any extended optically thin regions which would be a natural site for x-ray emission.

The models from the first family, however, are most probably strongly stratified as the unstratified models are known to be thermally unstable (see Sec. 5.2 and 5.3). Some parts of the disks are optically thin (in a sense of effective optical depth), and therefore their emission is predominantly in the x-ray band.

5.1. OPTICALLY THICK EMISSION FROM UNSTRATIFIED DISKS

The spectra of unstratified α disks were calculated already by Shakura and Sunyaev (1973) using analytic approximations of the effects of electron scattering and Comptonization. More advanced models were subsequently published (e.g. Czerny & Elvis 1987, Laor & Netzer 1989, Sun and Malkan 1989). The best models so far (Ross, Fabian and Mineshige 1993) are based on properly solved radiative transfer, i.e. free-free, bound-free and bound-bound opacities as well as Comptonization calculated in the diffusion approximation (Kompaneetz equation) but for a constant density and constant viscous energy generation in the vertical direction. Fully self-consistent solutions, with viscous energy deposit given locally by α scaling and the hydrostatic equilibrium taken into account are not available yet for regions dominated by radiation pressure (see Riffer et al, this volume).

These more advanced spectra models are more favorable from the observational point of view. Since the spectra extend to higher frequencies lower accretion rates are required to fit high redshift quasars or those few Seyferts where the optically thick emission extends into soft x-ray band. The flattening of the spectra in the uv range also partially closes the gap with the observed slopes (see Sec. 3). Some external irradiation may then reproduce the observed spectral shape if the accretion rate is close to the critical one; such a model was used both to fit the spectrum and to account for the microlensing limit for the size of emitting region in the "Einstein cross', or Q2237+0305 (Czerny, Czerny and Jaroszyński 1993).

However, these spectra are not fully reliable as long as the possibility of strong stratification within the disk is not studied and eventually incorporated into the model.

5.2. Radial stratification

The vertically averaged α disks equations are known to have two solutions for a given accretion rate. The first is the Skakura-Sunyaev optically thick solution. The second one is the optically thin branch - a hot ion torus. It was discovered by Shapiro, Lightman and Eardley (1976) within a specific context of non-local model (cooling by external photons) but also local considerations give similar branch (e.g. Bjornsson and Svensson 1992, Kusunose and Zdziarski 1993). The two solutions merge for very high accretion rate (Liang and Wandel 1991) and there are no solutions for higher values of accretion rate at a given radius. The details depend on the assumption about the cooling mechanism, i.e. whether external soft photons or synchrotron photons are present in addition to bremstrahlung emission as the number of soft photons cooling the medium through Compton scattering is essential both for the structure of the disk and the emitted spectrum.

Radially stratified solution consists of two parts: optically thick outer part and optically thin inner part. This radius of change R_{div} from one branch of solutions to another is in principle uniquely determined by the model from the condition of marginal thermal stability of the outer branch: $P_{gas}/P_{tot} = 0.4$ (Shakura and Sunyaev 1976).

This criterion predicts rather extended ion torus and the x-ray luminosity strongly dominating the big bump luminosity. This is not expected for Seyfert galaxies and quasars, and also it might be difficult to reconcile with the mean energy index between optical and x-ray band which tends to be steeper for bright quasars.

However, this criterion changes if timescales for the development of instability are considered (Shakura & Sunyaev 1976) as well as the irradiation of cool part by the hot ion torus is taken into account as the illumination tends to suppress the instability (e.g. Czerny, Czerny and Grindlay 1986). More detailed consideration of stability are needed to improve it so arbitrarily adopted value of r_{div} is a reasonable possibility for the time being.

The existence of two temperature ion torus has been questioned as the collective plasma phenomena may result in a strong coupling between the electrons and ions. However, simple numerical experiment of changing the value of the coupling constant indicate that the torus structure adjust to such a change without decreasing the difference of temperature between ions and electrons (Kusunose & Zdziarski 1993).

5.3. Vertical stratification

If the α prescription for viscosity is used locally without averaging the disk structure in the vertical direction also solutions with strong vertical stratification can be found which is perhaps closely related to the difficulties in solving both the equations for the disk structure and the radiative transfer self-consistently without approximations (see Sec. 5.1).

We study the possibility of such stratification using two slab approximation (Czerny, Collin-Souffrin & Życki 1994). We assume that an optically thin hot corona is situated on the top of optically thick disk layer. The viscous heat generation both in the cold layer and in the hot corona are described using the same value of α. The interaction of the two slabs is described by (i) assuming the hydrostatic equilibrium at the base of the corona (ii) assuming the flux exchange between the two slabs, as in Haart and Maraschi (1992). The two-temperature corona is cooled by Comptonizing the outgoing soft photons. The pressure in the colder slab is described as a sum of gas and radiation pressure.

Such a set of equations parameterized by α, M and $\dot{M}$ give two solutions at a given radius. The fraction of energy generated in a corona is significant in one solution and much smaller in the other one. The first solution has been found by Nakamura & Osaki (1994) for the gas pressure dominated disk. However, the two solutions merge at a certain radius and no vertical stratification is predicted at larger radii.

The coronal solution region increases with increasing accretion rate; it is smaller than the division point for radial stratification although it increases faster with $\dot{m}$. However, if the criterion for radial stratification is modified (see Sec. 5.2) then for large accretion rate the coronal solution may develop in the outermost parts and continue inwards. For low accretion rate the ion torus will form unavoidably, according to the model.

5.4. OBSERVATIONAL CONSEQUENCES OF ALPHA MODELS

Since these models allow for strong stratification effects the detailed observational predictions should include the interaction of separate parts of the disk. Such a full program have not been completed yet but partial results are available.

Hot part of radially stratified disks well reproduce the slope of x-ray spectrum if synchrotron emission of soft photons is allowed (e.g. Kusunose and Zdziarski 1993). External soft photons would also give the required spectrum but it is difficult for them to penetrate the hot ion torus as its optical depth for scattering is significant (Wandel & Liang 1991).

The effect of irradiation of the outer cold part of the disk by the inner torus have been calculated by Rokaki, Collin-Souffrin & Magnan (1993) . Assuming constant value of the division point but variable x-ray luminosity the authors could explain the observed optical/uv variability of NGC 5548; however, additional power law ir/uv component was still required to fit the data.

The emission line profiles in 10% of radio galaxies are successfully fitted if the extension of the cool disk is given by radiation pressure criterion (Eracleous & Halpern 1993).

The detailed predictions for vertically stratified models are not available yet but they are expected to be qualitatively similar to coronal models (see Sec 6.2).

6. Modified models

Since the strict justification for the α scaling of viscous torque does not exist and the limits of applicability are unknown a number of different, still more ad hoc, approaches were suggested.

6.1. Irradiated disks with unspecified x-ray source

Unstratified α disk models have to be arbitrarily supplemented with a source of x-ray emission. Detailed calculations of continuum have been made for a compact source of power law radiation above an accretion disk (Ross & Fabian 1993, Matt, Ross & Fabian 1993). Iron K_α line predictions were made for the same geometry (Matt, Fabian & Ross 1993, Życki & Czerny 1993).

6.2. Coronal solutions

Hot corona with gravitational energy release (as opposite to corona caused by external irradiation) not constrained to α viscosity was considered e.g. by Liang & Thompson (1979). Attractive and simple approach was suggested by Haardt & Maraschi 1991, later followed by Kusunose & Mineshige (1994) and Haardt & Matt (1993). Model is parameterized by a fraction f of energy dissipated in a hot corona. The model gives x-ray spectra close to the observed ones in the case of f close to 1 (Haardt & Maraschi 1991). The disadvantage of the model is that the entire disk is covered by a corona and all uv radiation is strongly comptonized as well. However, the problems disappears if the corona is clumpy.

6.3. Optically thin and optically thick blobs

The presence of blobs instead of accretion disk close to the center was suggested by Rees (1987). Although a spherically symmetric cloud distribution does not seem to be consistent with observations a flattened cloud distribution is actually difficult to distinguish from accretion disk if clouds are optically thick (Malkan 1992). Only very detailed modelling (e.g. Sivron & Tsuruta 1993) may reveal some subtle differences. Optically thin clouds (Ferland, Korrista & Peterson 1990, Barvainis 1993) seem to be less convincing. Optically thin irradiated plasma can hardly be kept at the required temperature (Malkan 1992) and the Lyman edge is too large (Dumont, Collin-Souffrin, Czerny & Życki, in preparation).

7. Future prospects

Further progress is mostly expected from new observational constraints of parameters crucial for theoretical models. For example, mean density of the disk can, in principle, be determined both from the shape of the reflected x-ray component and from the polarization studies in uv (e.g. Webb et al. 1993). The overall geometry is perhaps most strongly constrained by detailed modelling of the monitored variable sources. Explicit nonstationary models may be needed and such models are available now only for vary special cases. A lot of work, both observational and at a modelling level, has to be done yet.

Acknowledgements

I would like to thank Suzy Collin-Souffrin, Jean-Pierre Lasota and Piotr Życki for fruitful discussions. This work was supported in part by grant No. 2 P30401004 financed in 1993-1995 by the Polish State Committee for Scientific Research.

8. References

Abramowicz, M., A., Czerny, B., Lasota, J.P., & Szuszkiewicz, E., 1988, Ap. J., 332, 646
Antonucci, R., 1992, in Testing the AGN Paradigm, ed. S. Holt, S. Neff & C.M. Urry, New York, AIP, p. 486
Barthel, P.D., 1992, in Physics of Active Galactic Nuclei, eds. W.J. Duschl & S.J. Wagner, Springer-Verlag, p. 637
Barvainis, R., 1992, in Testing the AGN Paradigm, ed. S. Holt, S. Neff & C.M. Urry, New York, AIP, p. 129
Bechtold, J., Green, R. F., Weymann, R. J., Schmidt, H., Easterbrook, F. B., Sherman, R. D., Wahlquist, H. D. & Heckman, T. M. 1984, ApJ, 281, 76
Björnsson, G. & Svensson, R., 1992, Ap. J., 394, 500
Clarke, C.J., 1988, MNRAS, 235, 881
Collin-Souffrin, S., 1992, in Physics of Active Galactic Nuclei, eds. W.J. Duschl & S.J. Wagner, Springer-Verlag, p. 133
Czerny, B., Czerny, M. & Grindlay, J.E., 1986
Czerny, B., Czerny, M. & Jaroszyński, M., 1993, submitted to MNRAS
Czerny, B. & Elvis, M., 1987, Ap. J., 321, 305
Edelson, R. & Malkan, M.A., 1986, Ap. J., 308, 59
Eracleous, M. & Halpern, J.P., 1993, Ap. J. Suppl. (in press)
Fabian A.C. George I.M., 1990, in Treves A., Perola G.C., Stella L., eds, Iron Line Diagnostics in X-ray Sources, Springer–Verlag, Berlin, p. 169
Frank, J., King, A.R. & Raine, D.J., 1985, Accretion power in astrophysics. Cambridge University Press, Cambridge
Haardt, F., & Maraschi, L., 1991, Ap. J. 380, L51
Haardt, F. & Matt, G., 1993, MNRAS, 261, 346
Heckman, T., in Testing the AGN Paradigm, ed. S. Holt, S. Neff & C.M. Urry, New York, AIP, p. 595
Henri, G. & Pelletier, G., 1991, Ap. J., 383, L7
Kolykhalov, P.L. & Sunyaev, R.A., 1984, Adv. Space Rev., 3, 249
Kusunose, M. & Mineshige, S., 1994, Ap. J. (in press)

Kusunose, M. & Zdziarski, A.A., 1993, submitted to Ap. J.
Laor, A. & Netzer, H., 1989, MNRAS, 238, 897
Liang, E.P. & Thompson, 1979, MNRAS, 189, 421
Liang, E.P. & Wandel, A., 1991, Ap. J., 376, 746
Lynden-Bell, D., 1969, Nature, 223, 690
Malkan, M.A., 1988, Adv. Space Res., 8, 249
Malkan, M., 1992, in Physics of Active Galactic Nuclei, eds.W.J. Duschl & S.J. Wagner, Springer-Verlag, p. 109
Matt,G., Fabian, A.C. & Ross, R.R., 1993, MNRAS, 262, 179
Mushotzky, R.F, Done, C. & Pounds, K.A., 1993, Ann. Rev. Astron. Ap. ???
Neugebauer, G., Green, R.F., Matthews, K., Schmidt, M., Soifer, B.T., & Bennet, J., 1987, Ap. J. Suppl., 63, 515
Novikov, I. & Thorne, K.S., 1973, in Black Holes, eds. C DeWitt & B. deWitt, New York, Gordon & Breach
O'Brien, P.T., Gondhalekar, P.M. & Wilson, R., 1988, MNRAS, 233, 801
Nakamura, K. & Osaki, Y., 1994, PASJ (in press)
Piran, T., 1978, Ap. J., 221, 652
Pringle, J.E., 1981, Ann. Rev. Astron. Ap., 19, 137
Rees, M.J., 1984, Ann. Rev. Astron. Ap., 22, 471
Rees, M.J., 1987, MNRAS, 228, 47P
Reimers, D. et al. 1992, Nature, 360, 561
Rokaki, E., Collin-Souffrin, S. & Magnan, C., 1993, A&A, 272, 8
Ross, R.R. & Fabian, A.C., 1993, MNRAS, 261, 74
Ross, R.R., Fabian, A.C. & Mineshige, S., 1993, MNRAS, 258, 189
Salpeter, E.E., 1964, Ap. J., 140, 796
Sanders, D.B., Phinney, E.S., Neugebauer, G., Soifer, B.T., Mathews, K., Green, R.F., 1989, Ap. J., 347, 29
Schlosman, I., Begelman, M.C. & Frank, J., 1990, Nature, 345, 679
Shakura, N.I. & Sunyaev, R.A., 1973, A&A, 24, 337
Shakura, N.I. & Sunyaev, R.A., 1976, MNRAS, 175, 613
Shapiro, S.L., Lightman, A.P. & Eardley, D.M., 1976, Ap. J., 204, 187
Shields, G.A., 1978, Nature, 272, 706
Sivron, R. & Tsuruta, S., 1993, Ap. J., 402, 420
Sun, W.-H. & Malkan, M.A., 1989, Ap. J., 346, 68
Terlevich, R., Tenorio-Tagle, G., Franco, J. & Melnick, J., 1992, MNRAS, 255, 713
Walter, R. & Fink, H.H., 1993, A&A, 274, 105
Wandel, A. & Liang, E.P., 1991, Ap. J., 380, 84
Wilkes, B.J. & Elvis, M., 1987, Ap. J., 323, 243
Zeldovich,Ya.B. & Novikov, I.D., 1964, Dokl. Acad. Nauk SSSR, 158, 811
Życki, P.T. & Czerny, B., 1993, MNRAS (in press)

SPECTRA OF QUASARS WITH EXTREME CONTINUUM PROPERTIES

SMITA MATHUR

Harvard-Smithsonian Center for Astrophysics, Cambridge, MA 02138, USA

Abstract. The quasar population as a whole covers a wide range of continuum properties; sufficient to cause a marked difference in the physical conditions of the gas in the nuclear region of a quasar. In this paper we will discuss our results on 3C351, an X-ray "quiet" quasar; and 3C212, a "red" quasar. We identify the "warm absorber" observed in the ROSAT observations of 3C351 to be the UV absorber observed by HST.

1. Introduction

The quasar population covers a wide range of continuum shapes ('SEDs', see Elvis, these proceedings). One consequence is that photoionization models must reflect the difference in the physical conditions of the line emitting gas; so using an "average quasar" obscures the physics. Multiwavelength (IR–X-ray) observations can now be gathered for a wide range of quasars; so photoionization theory can be applied to individual objects self-consistently. Here we will discuss two extreme cases—one with 'no' X-ray and one with 'no' UV, ionizing photons to look for the governing parameters in BELR physics

2. 3C351: An X-ray Quiet Quasar

3C351 is $\sim$ 10 times less X-ray bright than a typical radio loud ($\alpha_{ox} = 1.4$) quasar, so we call it 'X-ray quiet' (Figure 1a). Quasi-simultaneous optical (MMT) and X-ray (ROSAT) observations rule out variability as a cause of the low X-ray flux. The ROSAT X-ray spectrum of the object (Fiore et al, 1993) was not well fit with a simple power-law, with the residuals strongly suggesting an Oxygen edge at 0.76± 0.08 keV. A 'warm' (ionized) absorber along the line of sight instead gives a good fit with a column density, $N_H = 1.4 \pm 0.3 \times 10^{22} cm^{-2}$; and an ionization parameter, $U = 6.7^{+2.3}_{-1.2}$. The large U of the absorber implies a location close to the central ionizing source. There are two obvious sites for the X-ray absorber: the clouds in the Broad Line Region (BLR); and the clouds responsible for the high ionization UV absorption lines.

2.1. BELR CLOUDS AS X-RAY ABSORBER?

Standard photoionization models for the BLR imply U (0.01) more than two orders of magnitude smaller than the best fit value for X-ray Absorber. However, reverberation studies (Peterson, 1988) have shown than the BLR is small and stratified so that parts of the BLR have higher U than previously thought. In addition, 3C351 is X-ray quiet, requiring higher U to produce the observed emission lines.

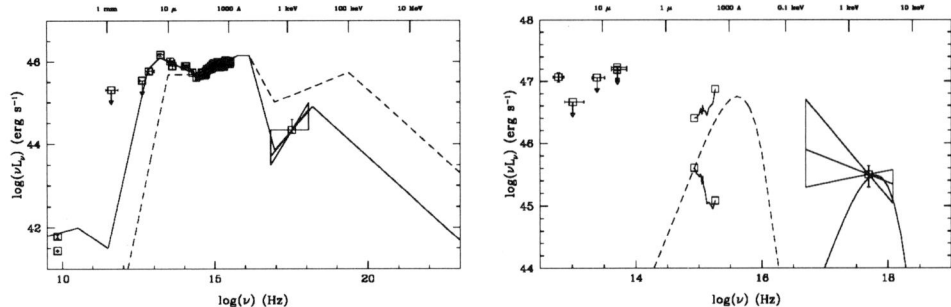

Fig. 1. SEDs for (a)3C351: —— = the best fit SED, - - - - = 'average' quasar continuum for radio-loud quasars. (b)3C212: see the text. - - - -=typical quasar UV bump.

The observed SED has a large effect on the strengths of high ionization emission lines (using CLOUDY, Ferland, 1991): OVI 1034 is weakened by almost an order of magnitude (Figure 2a). The model is compared with the observed (HST spectrum) line ratios. Even though the inferred U for OVI line is as high as 0.3, it's still much lower than that required for the X-ray absorber. So the BELR is not a plausible site for the X-ray ionized absorber. The X-ray quiet SED also affects traditional density diagnostics; *CIII] λ1909 ceases to be a density indicator* for this SED (Figure 2b) (Mathur et al, 1994) due to increased free-free heating which produces heating without ionization.

2.2. X-RAY ABSORBER AS UV ABSORBER?

The ultraviolet HST spectrum of 3C351 (Bahcall et al 1993) shows an unusually strong associated metal line absorption system. In particular, strong OVI absorption doublets are observed (c.f. OVII absorption edge in soft X-rays).

Table 1.

| Inferred Parameter | Model Parameter | |
from HST Spectrum	$log\rho_H = 3$	$log\rho_H = 5$
$f_{OVI} = -1.52$	-1.532	-1.526
$f_{NV} = -2.86$	-2.43	-2.42
$f_{CIV} = -2.81$	-2.79	-2.8

X-ray observations give strong constraints on the total $N_H = 1.4 \times 10^{22} cm^{-2}$ which allows us to solve for a consistent model of UV lines and X-ray absorber. The UV absorption lines give the following constraints: $N_H > 2\times10^{18}/f_{(OVI)} cm^{-2}$ (where $f_{(OVI)}$ is the ionization fraction of oxygen in OVI state), 'b' parameter $< 600 - 1200\ km\ s^{-1}$ (= FWHM of the lines). A small value ($\sim 4\ km\ s^{-1}$) of 'b' implies unacceptably large column density. We find that 'b'$\sim 110\ km\ s^{-1}$ satisfies

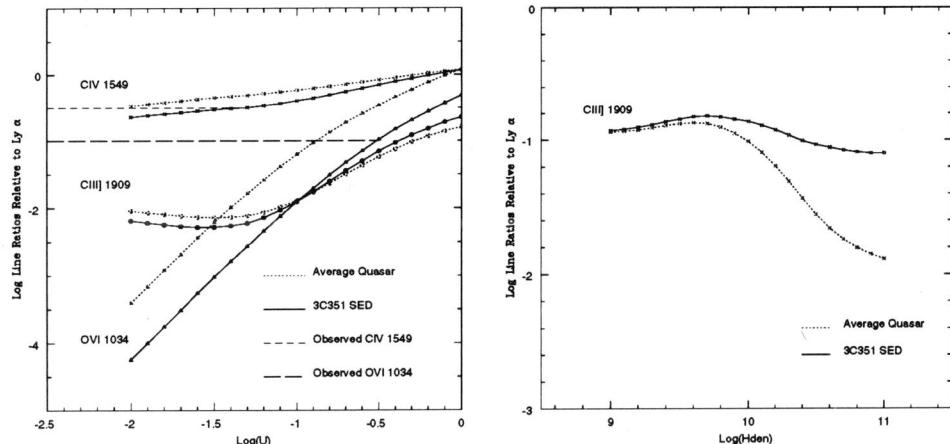

Fig. 2. Emission line strengths relative to $Ly\alpha$ as a function of (a) ionization parameter (b) density. Errors on observed line strengths are $\sim 10\%$

the constraints on the ionization fractions of all the high ionization absorption lines; OVI, CIV and NV and allows a consistent model (Table 1) for U= 6.7. We conclude that the identity of X-ray absorber as the UV absorber is likely, although we cannot rule out two separate absorbers (Mathur et al 1994). The implied physical properties of the X-ray and UV absorber are: high N_H ($1.4 \times 10^{22} cm^{-2}$), high U (6.7), an outflow velocity of ~ 2000 km s^{-1} (from UV blueshift), low density ($10^3 - 10^5 cm^{-3}$), large size ($\sim 10 - 0.1 pc$) and distance from the central source ($\sim 100 - 10 pc$), low covering factor (< 0.04), and mass loss rate $2 < \dot{M} < 20$ $M_\odot yr^{-1}$. This implies kinetic luminosity of $\sim 10^{42} - 10^{43} ergs^{-1}$, which is only $\sim 10^{-3} - 10^{-4}$ of the radiative luminosity of the quasar. These properties describe a component of nuclear material not previously recognized.

3. 3C212: A Red Quasar

The extreme case of a UV weak ionizing continuum may be seen in the case of the 'Red Quasars' (Smith & Spinrad 1980). Figure 1b. shows the SED for 3C212 which could either have a red optical-UV continuum and a X-ray black body or a dereddened optical-UV and X-ray power law (Elvis et al 1994). Clearly, the two models have extreme differences in the implied shape of their UV to soft X-ray continuum, with a factor of $\sim 10^3$ difference in the number of ionizing photons. They might be expected to predict quite different emission line ratios.

Surprisingly, in spite of these huge differences between the possible ionizing continua, the photoionization predictions provide no convincing discriminants between the models given the few observed lines. The similarity of the predictions is partly because the increased number of ionizing photons in the de-reddened power law is almost exactly matched by the increased number of emitted line photons, since both lines and continuum are dereddened by the same amount. Another limitation is that there are few observed lines to provide constraints. Only Mg II $\lambda 2798$Å and C III]$\lambda 1909$ are clearly detected (Smith & Spinrad 1980). All the Hydrogen lines lie outside the observed range (1500Å−3560Å, rest). Better diagnostics could be searched for with improved optical and infrared spectra . For example, the greatest difference between the two continua is in the EUV, so that the HeII lines, which require $\lambda < 304$Å photons for their production, should provide a clean discriminant between the two models.

4. Conclusions

An X-ray quiet SED, as in 3C351, strongly affects the governing parameters in the BELR physics. We identify the X-ray absorber in 3C351 as the UV absorber. This implies a new component of the nuclear material. The 3C212 continuum could either be red optical-UV and a X-ray black body; or a dereddened optical-UV and X-ray power law. Observations of HeII lines could discriminate between the two models. The strategy of seeking out extreme quasars does produce new insights. The study of "Quasars across electromagnetic spectrum" is indeed the key towards understanding the underlying physical processes.

Acknowledgements

Fabrizio Fiore, Martin Elvis and Belinda Wilkes have contributed to the present investigation. This work was supported by NASA grant NAGW-2201.

References

Bahcall, J. N. *et al* 1993, Ap.J.Supp., **87**, 1.
Elvis, M., Fiore, F., Mathur, S., Wilkes, B., 1994, Ap.J., submitted.
Ferland, G. F. 1991 OSU Astronomy Department Internal Report.
Fiore, F., Elvis, M., Mathur, S., Wilkes, B. J., McDowell, J. C. 1993, Ap.J., **415**, 129.
Mathur, S. *et al* 1994, in preparation.
Peterson, B. M. 1988, P.A.S.P., **100**, 18.
Smith, H. E. & Spinrad, H. 1980, Ap.J., **236**, 419.

THE PALOMAR OBSERVATORY DWARF SEYFERT SURVEY

LUIS C. HO[1], ALEXEI V. FILIPPENKO[1], and WALLACE L. W. SARGENT[2]
[1]*Department of Astronomy, University of California, Berkeley, CA 94720, U.S.A.*
[2]*Palomar Observatory, 105-24 Caltech, Pasadena, CA 91125, U.S.A.*

ABSTRACT. We describe an optical, spectroscopic survey of the nuclei of the 500 brightest galaxies in the northern sky. The primary goal is to search for low-luminosity active galactic nuclei (LLAGNs) in the centers of nearby galaxies. The results of this survey will have many astrophysical applications, including quantifying the faint end of the local AGN luminosity function and estimating the contribution of LLAGNs to the X-ray background. We summarize the statistical properties of the survey, describe our methods of analysis, and present some preliminary results based on $\sim 60\%$ of the sample.

1. Introduction

Although it has been recognized for some time that low-level nuclear activity is present in a substantial fraction of nearby galaxies (see, e.g., review by Keel 1985), both the quantitative assessment of the prevalence of nuclear activity and a coherent understanding of the physical nature of such activity are complicated in part by the lack of a complete set of reliable data. Most of the observational evidence for the apparent ubiquity of low-luminosity active galactic nuclei (LLAGNs) has come from optical spectroscopic surveys of nearby galaxies. (As an operational definition, we define LLAGNs as galactic nuclei having $M_B < -18$ mag and emission-line ratios unlike those of H II regions.) The accuracy of the detection technique, which involves measurement and modeling of weak emission lines, critically depends on several factors, not all of which could always be met by the surveys performed in the past. These include having data of adequate signal-to-noise ratio (S/N) and spectral resolution, and the proper removal of the underlying starlight which often dominates the observed spectra.

A reliable census of the space density of LLAGNs will have many important applications. For example, the local ($z \approx 0$) luminosity function of LLAGNs is very poorly constrained at the present time (e.g., Weedman 1986). The faint end of the local AGN luminosity function is of critical importance to several astrophysical issues, including the evolution of the overall AGN luminosity function and the contribution of AGNs to the cosmic X-ray background. In addition, the intrinsic low luminosity of LLAGNs, coupled with their proximity, in principle allows us to study the properties of the host galaxies in a level of detail not feasible for more luminous and/or distant AGNs. In this way, we may hope to be able to better understand the fundamental issue of what physical processes are responsible for the formation of AGNs in galaxies.

2. The Spectroscopic Database

In an effort to address some of the issues raised in §1, an optical spectroscopic survey of the nuclei of nearby galaxies was carried out at Palomar Observatory between 1982 and 1990 (see Filippenko and Sargent 1985, 1986 for details). Briefly, the nuclei of all galaxies with $B_T \leq 12.5$ mag and $\delta > 0$ were observed with the Double Spectrograph (Oke and Gunn 1982) mounted on the Hale 5 m telescope at Palomar Observatory. This statistically complete sample consists of ~ 500 objects. The resulting database contains spectra of excellent quality (S/N ≈ 100) and moderate spectral resolution (2.5-4 Å) over the wavelength regions $\sim$ 4200-5200 Å and $\sim$ 6200-6900 Å. Basic calibration steps such as bias subtraction and flatfielding, extraction of 1-D spectra ($2'' \times 4''$ effective aperture) from 2-D images, and wavelength and flux calibration follow standard procedures. As mentioned in §1, a crucial step in the analysis of the emission-line spectra is the removal of the starlight contamination. The method adopted closely follows that described in Ho, Filippenko, and Sargent (1993), and is illustrated in Figure 1 for NGC 4261 (also see Ford et al. in these proceedings). Note that most of the emission lines of this relatively weak-lined object can only be reliably measured after the subtraction of an appropriately scaled "template" galaxy (in this case NGC 2300) whose stellar population, stellar velocity dispersion, and metalicity closely match those of the object spectrum.

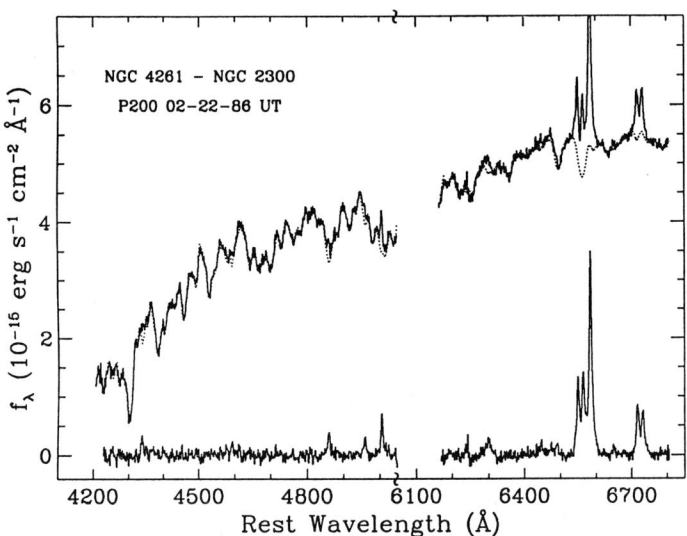

Figure 1. Illustration of starlight subtraction for NGC 4261.

3. Preliminary Results

From our uniformly observed, calibrated, and reduced data set, we aim to systematically quantify emission-line parameters such as fluxes and line widths, stellar velocity dispersions, and rotation curves. We will model the emission-line spectra with photoionization and shock calculations (e.g., Ho et al. 1993) in order to determine the excitation mechanism(s) operative in the nuclei, derive a luminosity function for the LLAGNs in the survey, and attempt to find correlations between the types of excitation mechanism and various global properties of the host galaxies. The database and subsequent applications thereof will be published in future papers.

In this contribution, we present a preliminary analysis of about 60% of the survey (324 galaxies), focusing on the occurrence of various classes of emission-line galaxies as a function of Hubble type. As illustrated in Figure 2, the redshift and morphological type distributions of this subsample are very similar to, and thus representative of, those of the entire survey. Conclusions we draw from the present subsample will likely be valid for the survey as a whole once the entire sample has been analyzed.

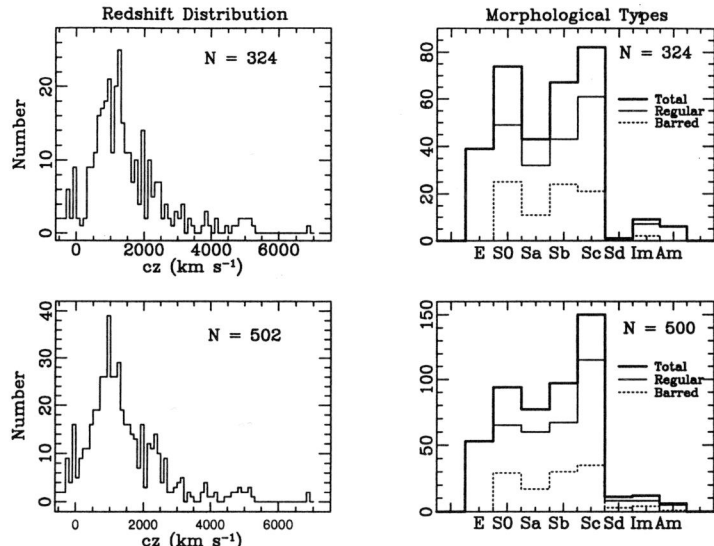

Figure 2. Redshift and morphological type distributions for objects in the subsample discussed in the text (*top*) and in the whole survey (*bottom*).

We classified the calibrated spectra containing emission lines into three "excitation classes": (1) LLAGNs, (2) H II galaxies, and (3) Seyfert 1 nuclei. The LLAGNs group includes both low-ionization nuclear emission-line regions (LINERs; Heckman 1980) and Seyfert 2 galaxies. Since we have not yet performed a careful starlight subtraction for all of the objects, we did not attempt to discriminate between the two; however, judging from the space density of Seyfert 1 galaxies and the relative space densities of Seyfert 1s and 2s (e.g., Osterbrock and Martel 1993), the vast majority of the LLAGNs should be LINERs. H II galaxies are objects whose spectra resemble those of H II regions and are thus galaxies with nuclei undergoing recent or current star formation. The distribution of excitation class as a function of Hubble type (Fig. 3; see also Table 1) reveals that (1) LLAGNs are nearly as numerous as H II galaxies, (2) the host galaxies of LLAGNs and Seyfert 1s tend to be early-type spirals (S0-Sb), and (3) nuclear star formation (i.e., H II galaxies) occurs most frequently in later type spirals (mostly Sb-Sc).

Quantitatively, Table 1 shows that as many as 40% of *all* nearby galaxies may harbor an LLAGN, comparable to the fraction of galaxies whose nuclei are experiencing recent or ongoing star formation (38%). LLAGNs make up about 50% of S0 and Sb galaxies, ∼ 70% of Sa galaxies, and ∼ 60% of all spirals. Although these results are qualitatively similar to those discovered in past surveys (see Keel 1985), they are quantitatively different (and presumably more accurate). In particular, the fraction of galaxies containing mildly active nuclei appears to be even higher than previously thought.

TABLE 1. DISTRIBUTION OF EXCITATION CLASS

Class[1]	No.	%	E[2]	S0	Sa	Sb	Sc	Sd	Im	Am
L	128	40	33	49	70	49	17	0	11	17
H	122	38	3	5	16	45	80	100	89	83
S	8	3	0	4	5	4	0	0	0	0
T	63	19	64	42	9	2	3	0	0	0

[1]L = LLAGNs, H = H II galaxies, S = Seyfert 1 galaxies, and T = "template" galaxies.
[2]Numbers listed under each Hubble type are percentages.

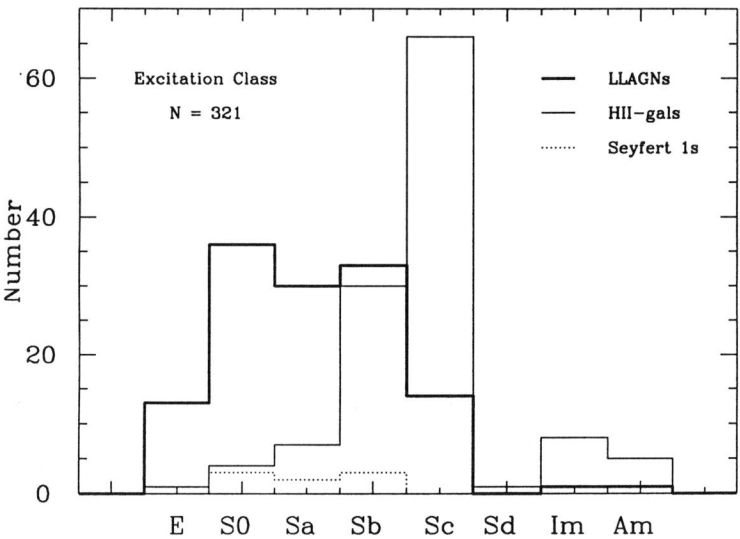

Figure 3. Distribution of excitation class as a function of Hubble type.

4. References

Filippenko, A. V., and Sargent, W. L. W. 1985, *Ap. J. Suppl.*, **57**, 503
Filippenko, A. V., and Sargent, W. L. W. 1986, in *Structure and Evolution of Active Galactic Nuclei*, ed. G. Giuricin *et al.* (Dordrecht: Reidel), p. 21
Heckman, T. M. 1980, *Astr. Ap.*, **87**, 152
Ho, L. C., Filippenko, A. V., and Sargent, W. L. W. 1993, *Ap. J.*, in press
Keel, W. C. 1985, in *Astrophysics of Active Galaxies and Quasi-Stellar Objects*, ed. J. S. Miller (Mill Valley, CA: Univ. Science Books), p. 1
Oke, J. B., and Gunn, J. E. 1982, *Pub. A.S.P.*, **94**, 586
Osterbrock, D. E., and Martel, A. 1993, *Ap. J.*, **414**, 552
Weedman, D. W. 1986, in *Structure and Evolution of Active Galactic Nuclei*, eds. G. Giuricin *et al.* (Dordrecht: Reidel), p. 215

QUASAR FORMATION IN HIERARCHICAL STRUCTURE FORMATION MODELS

M.G. HAEHNELT
Institute of Astronomy, Madingley Road, Cambridge CB3 0HA

Abstract. Hierarchical cosmogonies can consistently explain the evolution of the quasar population if quasars are short-lived and supermassive black holes form fast in the newly-formed nuclei of dark-matter haloes. Here we investigate the relevant physical processes and show that such a fast formation is plausible. The angular-momentum and the gas-supply problem for the formation/feeding of a supermassive black hole are strongly alleviated compared to a scenario in which gas is transported to the centre by tidal interaction of ready-assembled galaxies. The baryonic component of the newly-formed nucleus will cool catastrophically and settle into a self-gravitating angular momentum-supported disc of radius $\sim 100\,\mathrm{pc}$. Gravitational instabilities and/or supernovae-induced turbulence will transport the gas further to the centre within less than 10^8 yr. In nuclei of very massive dark matter haloes with sufficiently deep potential well to retain the gas against feedback processes from massive stars and supernovae, concentration of a major fraction of the gas component of the nucleus within the central 1 pc and subsequent formation of a black hole seem unavoidable. A coeval short phase of efficient star formation could explain the observed high metallicities of quasars.

1. Introduction

While the existence of supermassive black holes is still not proven beyond doubt, accretion onto black holes seems still to be the only viable explanation for the most fundamental properties of the observed activity in AGN [1]. The question of the formation of quasars is therefore basically the same as the question how supermassive black holes form in the nuclei of galaxies and is closely related to the problem of the fueling of AGN [2]. Recent investigations of the quasar luminosity function and its evolution with redshift [3, 4] indicate that quasars are rather short-lived with a life-time $t_Q \sim 10^8$ yr close to the Salpeter time scale $t_{\mathrm{Salp}} = 4 \times \epsilon_{\mathrm{rad},0.1} 10^7$ yr (the e-folding time for the mass of a black hole accreting at the Eddington limit and radiating with efficiency ϵ_{rad}). Haehnelt & Rees [4] demonstrated that the strong evolution of the quasar population can be consistently explained in hierarchical structure formation models if quasars are short-lived and supermassive black holes form fast in the cores of newly-formed dark-matter haloes.

2. The formation of a supermassive black hole in the newly-formed nucleus of a dark matter halo

The formation of the nucleus of a dark-matter halo is a complicated multi-phase process. If a halo forms from a linearly growing density fluctuation in the 'fluid' of collisionless dark matter dominating the dynamics of the universe, the inner most dense parts will decouple first from the Hubble flow and collapse until virial equilibrium is reached at a radius

$$R_{\rm vir} \sim 550 \left(\frac{M_{\rm nuc}}{10^9 \, M_\odot}\right)^{1/3} \left(\frac{1+z_{\rm form}}{1+4}\right)^{-1} \left(\frac{f_{\rm bar}}{0.1}\right)^{-1/3} \text{pc}. \qquad (1)$$

$z_{\rm form}$ is the formation redshift, $f_{\rm bar}$ is the fraction of baryons in the universe and $M_{\rm nuc}$ is the baryonic mass of the nucleus (inner 1%) of a dark matter halo with isothermal density profile.

The collisional baryonic component is heated by shocks and feedback processes of massive stars and supernovae. However, this heating is not sufficient for pressure support; the gas will shrink by a factor $\sim f_{\rm bar}/(2\lambda)^2$ in radius on the dynamical time scale of the nucleus ($t_{\rm nuc} \sim 10^8$ yr) and settle into a self-gravitating angular-momentum supported structure of radius

$$R_{\rm disc} \sim 55 \left(\frac{M_{\rm nuc}}{10^9 \, M_\odot}\right)^{1/3} \left(\frac{1+z_{\rm form}}{1+4}\right)^{-1} \left(\frac{f_{\rm bar}}{0.1}\right)^{-4/3} \left(\frac{\lambda}{0.05}\right)^2 \text{pc}. \qquad (2)$$

$\lambda = |E|^{1/2} J / G M^{5/2}$ is the usual angular momentum parameter, which describes the spin-up of the growing density inhomogeneity by tidal torques. During the collapse of the baryonic component the first stars form and some weak magnetic fields might be built up. Gravitational instabilities will keep the self-gravitating 'disc' thick ($h/r \gtrsim 0.1$) and the disc will be highly viscous due to gravitational instabilities and/or supernova induced turbulence. Magnetic fields could also contribute to a high viscosity, if the disc became strongly magnetized. The matter is transported to the centre in a few hundred dynamical times of the disc and the gas in the disc will form stars very efficiently. Together with an IMF biased to massive stars this could explain the high metallicities observed in quasars [5].

Once the gas is concentrated by a factor $\sim 10-100$ in radius star formation will become more and more difficult. At a radius

$$R_{\rm coll} = 0.8 \left(\frac{M_{\rm nuc}}{10^9 \, M_\odot}\right)^{3/7} \left(\frac{1+z_{\rm form}}{1+4}\right)^{3/7} \left(\frac{l_*}{L_\odot}\right)^{-4/7} \left(\frac{m_*}{M_\odot}\right)^{6/7} \left(\frac{h/R}{0.1}\right)^{-2/7} \text{pc} \quad (3)$$

the collisional time scale becomes smaller than the Kelvin-Helmholtz time scale $t_{\rm KH}$ of a contracting protostar of mass m_*, radius r_* and luminosity l_* and protostars will be destroyed. This assumes that a fraction $t_{\rm KH}(r_*)/t_{\rm nuc}$ of the gas is in the form of protostars of radius r_* and that the halo has an isothermal density profile.

Star formation will certainly cease if the energy liberated in collisions of stars exceeds the Eddington luminosity of the gas cloud. This occurs at

$$R_{\rm edd} = 0.07 \left(\frac{\xi}{0.05}\right)^{2/9} \left(\frac{M_{\rm nuc}}{10^9 \, M_\odot}\right)^{5/9} \left(\frac{r_*}{R_\odot}\right)^{4/9} \left(\frac{m_*}{M_\odot}\right)^{-2/9} \left(\frac{h/R}{0.1}\right)^{-2/9} \text{pc} \qquad (4)$$

[6], where ξ is the fraction of the mass typically liberated in a stellar collision.

Once the gas has been concentrated so far the formation of a black hole is unavoidable. The exact path of the formation is unclear and several intermediate

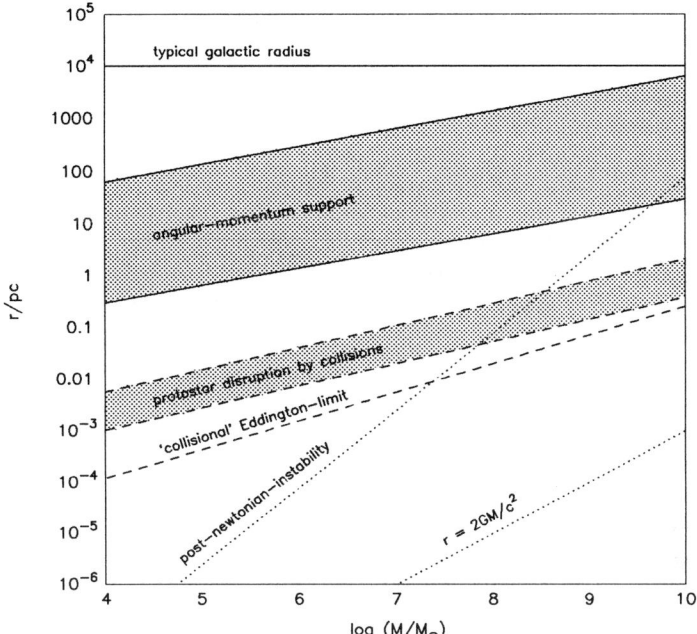

Fig. 1. Characteristic radii for the formation of a supermassive black hole in the newly-formed nucleus of a dark-matter halo as a function of baryonic mass of the nucleus. The upper shaded area shows the range of radii where angular-momentum support of the gas sets in and the lower shaded area shows the range of radii where disruption of protostars by collisions would inhibit star formation. The highest solid line and the middle dashed line are for a halo of constant density, a small formation redshift $z = 1$ (and the canonical value of the angular momentum parameter $\lambda = 0.05$ for the upper shaded area). The opposite limits of the shaded regions are for a halo with isothermal density profile, a high redshift $z = 4$ (and a small $\lambda = 0.025$). The lowest dashed line shows the radius where the luminosity produced by collisional disruption of stars equals the Eddington luminosity of the gas cloud. The dotted lines indicate the Schwarzschild radius and the radius where the post-newtonian instability of a supermassive star sets in.

stages might be involved. However, all of them would be extremely short-lived compared to the dynamical time of the nucleus in which the object is formed [7]. One possible path would be the formation of a supermassive star, which would be unstable for a radius smaller than

$$R_{\rm pni} = 2.4 \left(\frac{M_{\rm sms}}{10^9 \, M_\odot} \right)^{3/2} \, {\rm pc}, \tag{5}$$

due to a post-newtonian instability.

3. Discussion

Figure 1 shows the different characteristic radii introduced in the last section. It is easily seen how much the angular-momentum problem is alleviated compared to the situation where gas in a ready-assembled galaxy has to be transported inward from radii of order 10 kpc. In summary, the transport of the gas to the centre is so much easier in a newly-forming nucleus for the following reasons:

- A self-gravitating disc undergoing violent star formation is certainly highly viscous. Furthermore angular-momentum is not necessarily conserved in the time-varying, non-axisymmetric potential of the newly-forming nucleus.
- The ratio of the radius where angular-momentum support sets in and the radius where the gas can neither form stars nor settle into a stable gaseous configuration can be as small as $\sim 10 - 100$. A comparatively moderate loss of angular momentum is therefore required.
- The dynamical timescale of the angular-momentum supported disc is short and many dynamical times are available for the inward transport of the matter.

A supermassive black hole can therefore form fast in the newly-formed nucleus (within $t_{\rm nuc}$). The only way to avoid the formation of a black hole would be to turn all the gas of the self-gravitating disc into stars before a significant fraction has reached the centre. For early-forming, very massive, low-λ haloes this would require an abnormally high star formation efficiency. In general the black hole formation efficiency should strongly depend on the redshift of formation as the gap between the radius where angular-momentum support sets in and the radius where star formation becomes impossible decreases with increasing redshift for a halo of given mass ($R_{\rm disc}/R_{\rm coll} \propto (1+z)^{-10/7}$). A second important parameter is the depth of the potential well which determines the ability to retain the gas against feedback processes like winds of massive stars and supernovae.

Acknowledgements

I would like to thank Martin Rees for helpful discussions and acknowledge support by the Gottlieb Daimler- and Karl Benz-Foundation.

References

1. Blandford R.D., Rees M.J., 1992, in Holt S.W., ed., Testing the AGN paradigm. American Institute of Physics, New York, p1
2. Shlosman I., Begelman M.C., Frank J., 1990, Nat, 345, 679
3. Haehnelt M.G., 1993, in Akerlof C.W., Srednicki M.A., ed., Annals of the New York Academy of Sciences Vol. 688, New York, p. 526
4. Haehnelt M.G., Rees M.J., 1993, MNRAS, 263, 168
5. Hamann F., Ferland G., 1992, ApJ, 391, L53
6. Begelman M.C., Rees M.J., 1978, MNRAS, 185, 847
7. Rees M.J., 1984, ARA&A, 22, 471

Unified Schemes and Relations with other Types of Objects

ON THE DIFFERENCE BETWEEN RADIO LOUD AND RADIO QUIET AGN

KARL MANNHEIM
Universitäts-Sternwarte, Geismarlandstr. 11, D – 37083 Göttingen, Germany
E-mail: kmannhe@medusa.uni-sw.gwdg.de

Abstract. Nuclear jets containing relativistic "hot" particles close to the central engine cool dramatically by producing high energy radiation. The radiative dissipation is similar to the famous Compton drag acting upon "cold" thermal particles in a relativistic bulk flow. Highly relativistic protons induce anisotropic showers raining electromagnetic power *down* onto the putative accretion disk. Thus, the radiative signature of hot hadronic jets is x-ray irradiation of cold thermal matter. The synchrotron radio emission of the accelerated electrons is self-absorbed due to the strong magnetic fields close to the magnetic nozzle.

1. Jets and accretion disks: Castor and Pollux?

A puzzling mystery for AGN theorists is the relation between the big blue bump emission component, which is believed to originate as thermal emission from matter surrounding a supermassive black hole and the emission related to the morphological appearance of jets. Recent γ-ray observations have shown that the blazar spectrum from the subparsec jet is quite different from a thermal one and extends over almost twenty orders of magnitude in frequency with an almost constant level νS_ν. Amazingly, the properties of the big blue bump alone never show any indication of whether or not the AGN also has a powerful jet. On the other hand, BL Lacs show no sign of a big blue bump at all. On the basis of these facts one is tempted to assume that there are two hearts in AGN: one beating for the thermal processes (high entropy) and one for the nonthermal processes (low entropy). But then we remember our aim as physicists is to simplify and not to secularize the physical world. Could there be a relation between the two phenomena, in the sense that one is, perhaps, more fundamental than the other? This question has been around for some time during which the paradigm changed from the nonthermal origin of activity to the thermal one, the accretion paradigm. If the latter is correct, then viscosity must be totally robust with respect to the generation of a powerful jet.

In this context I find two new results most challenging to our understanding of AGN. Firstly, γ-ray measurements teach us that the prime radiation mechanism in jets seems to favour extremely high energies: Mkn421 was observed at an energy of 1 TeV with a $\nu S_\nu \propto$ const. spectrum. Secondly, a fairly robust relation $Q_j = 100 L_{nlr} \approx L_{bb}$ between the kinetic power of jets Q_j and the narrow line luminosity, resp. the photoionizing luminosity of the big blue bump, in radio galaxies and quasars has been found by Rawlings & Saunders (1991) and Celotti & Fabian (1993). They tell us that whatever links nonthermal and thermal components does this independent of the total luminosity and the presence of broad lines or a big blue bump. The existence of pure objects of either class then appears inconsistent.

In fact, radio quiets do have weak radio jets and a nonthermal optical/x-ray continuum. So, one should perhaps take a different point of view: assuming *all* AGN have powerful jets with $Q_j \approx L_{bb}$, could their radiative appearence be different enough to account for all AGN classes (Camenzind & Courvoisier, 1983)? In particular, this requires that the jets in radio quiets must dissipate most of their kinetic power witin the central parsec, because further out they would unescapably show strong radio emission. The momentum, however, is still there and, indeed, radio quiets do show high speed nuclear outflows, most spectacularly in broad absorption line quasars. From this heuristic starting point the following questions arise: (i) if jets in radio loud sources generally start out with relativistic speeds (which can plausibly be assumed with $\gamma_j \leq 10$), is this a prerequisite for the jets to emerge out of the central parsec without suffering much from radiative losses? If so, (ii) what is it that lets some jets start out slower, (iii) why do slower jets dissipate so much of their power in the central parsec and (iv) why is this connected to the properties of the host galaxy?

In this contribution I will attack question (iii), while leaving the answers to the other questions open. My personal guess would be that question (i) must be answered affirmatively, probably because relativistic bulk motion protects the jet from shock acceleration too close to the disk. Concerning (ii) it is quite natural for any jet forming mechanism to generate something like an inverse power law distribution for the speeds of jets, so that one easily gets 90% mildly relativistic jets and 10% relativistic jets. And finally, (iv) may have to be reversed in order: different jet properties make different host galaxies because jets can trigger star formation via cosmic rays and shocks. Remnant winds in quiets would sweep out interstellar matter above the disk leaving behind a torus.

2. Beamed γ-rays: a fingerprint from accelerated protons

In an effort to empirically answer the question whether jets contain an ordinary mixture of protons and electrons, Mannheim *et al.* (1991) investigated the radiative signature of highly relativistic protons accelerated at shock fronts. Photoproduction of pairs and pions injects electromagnetic power into the acceleration zone which is further reprocessed by an (unsaturated) synchrotron cascade. This proton initiated cascade (PIC) should operate at shocks of all sizes: starting from kpc Hot Spots (cf. Harris *et al.*, this volume) down to the shocks at the subparsec scale thought to be responsible for "proton blazar" emission (Mannheim, 1993). An essential parameter is the distance of the proton blazar from the source of the big blue bump photons, because local photons compete with thermal photons from outside the jet as a target. As the proton acceleration zone moves closer in, the γ-ray spectrum steepens as shown in Fig.(1) until cooling is entirely dominated by the anisotropic thermal target photons.

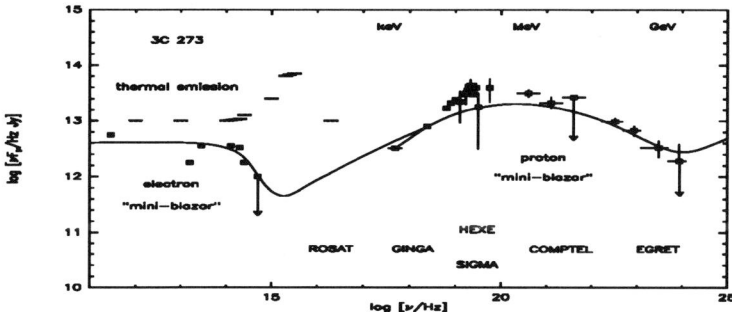

Fig. 1. The proton blazar model for 3C273 with a proton/electron ratio of $\eta = 15$, see Mannheim, *Phys. Rev. D*, Vol.48, No.4 (1993). Note the steepening of the γ-ray spectrum due to the additional blue bump target photons.

3. Dying jets: hadronic shower precipitation and nuclear winds

Protons exposed to an anisotropic target field cool via photoproduction mostly in the direction of the source of the target photons. They prefer head-on collisions with $\mu = -1$ because of the threshold condition $\gamma_p x (1 - \beta_p \mu) \geq x'_{k,th}$ where $\mu = \cos\theta$ is the cosine of the angle between proton and target photon momentum, $x = h\nu/m_e c^2$ the photon energy and $x'_{k,th}$ the photon threshold energy in the proton rest frame to create particle k$= e^{\pm}, \pi$ on the mass shell. Thus, head-on collisions require the lowest γ_p and thus produce the strongest flux for a proton distribution $n_p \propto \gamma_p^{-2}$ cooling on an almost monoenergetic photon target $x \approx 2 \times 10^{-4}$. The required proton Lorentz factor is $7 \cdot 10^5$ for pion production. Figs.(2) shows the emergence of this *natural anisotropy*. Maximum efficiency of the irradiation is obtained at the distance to the disk of $z_o = 200\sqrt{r_o/100}$ (in units of the Schwarzschild radius) where the photoproduction optical depth becomes unity for a jet radius of r_o, but $\tau_{pp} \ll 1$. At this location the magnetic field has the strength $B_o = 1.6 \cdot 10^3 M_A^{-1} m_8^{-1/2} r_o^{-1} \beta_{0.3}^{-1/2}$ and the target radiation compactness is $l_o = 3 m_8^{-1} r_o^{-1/2}$. Here it was assumed that $L_{edd} = L_{bb} + Q_j + L_B + L_{rel} \simeq L_{bb} + L_B (M_A^2 + 2)$ with equipartition $L_B = (B^2/8\pi) A_j c \beta_j = L_{rel}$ and the kinetic power $Q_j = M_A^2 L_B$. The irradiation spectrum is a powerlaw with $\alpha \simeq 1$ from eV up to GeV (because of a lack of x-ray photons to reprocess γ-rays from MeV to GeV) and $L_{pic} = L_{bb}(1 + M_A^{-2})/(0.5 + M_A^2)$. A mildly relativistic jet suffering from such severe radiative losses rapidly expands. The surviving momentum is shared to the surrounding thermal matter driving a high speed nuclear wind. Radio emission from the accelerated electrons is synchrotron-self-absorbed up to the frequency $\nu_s = 4 \cdot 10^{13} m_8^{-1/3} r_o^{-1} M_A^{-4/3} \beta_{0.3}^{-2/3}$ which makes such dying jets radio quiet. There is some radio emission associated with the remnant wind and the low emissivity reflects its large opening angle.

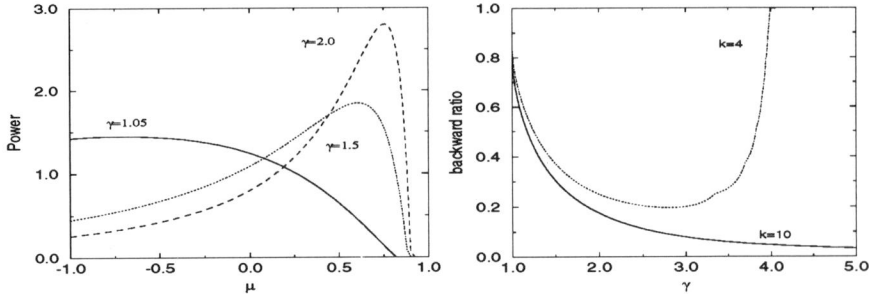

Fig. 2. **Left panel:** The angular distribution of cascade power. Note the transition from emission into the forward Doppler cone $\mu = \beta$ to emission into the backwards hemisphere when the bulk Lorentz factor γ decreases. At very small angles no emission is produced because there are no photons satisfying the threshold condition. **Right panel:** The backward/total luminosity ratio as a function of jet Lorentz factor. A maximum value of $\approx 90\%$ irradiation is possible. The label $k = 4$ denotes the curve for a proton maximum energy four times the threshold energy for head-on collisions. Thus, for $\gamma > 4$ no photoproduction ocurrs. Close to the limit the total luminosity is small, but emitted solely towards the backward hemishere.

4. Critique and conclusions

In this contribution I have proposed that powerful jets are ubiquitous in AGN. Radio quiets are equipped with mildly relativistic jets which cannot emerge from the central parsec because of severe radiative losses and rapid expansion. Relativistic protons in these jets induce electromagnetic showers which irradiate the accretion disk with hard radiation and which have very weak emission towards the observer. An *experimentum crucis* is the detection of the neutrino signature of the hadronic processes which is feasible with experiments currently under construction (Stenger *et al.*, 1992). The model needs a high efficiency of converting kinetic energy into radiation which is difficult to reconcile with statistical acceleration mechanisms (Mastichiadis & Kirk, this volume). Magnetic dissipation could resolve the problem, but could it yield proton Lorentz factors of 10^6?

References

Mannheim, K., Krülls, Biermann, P.L.: 1991, *Astron. Astrophys.* **251**, 723
Mannheim, K.: 1993, *Astron. Astrophys.* **269**, 67
Celotti, A., Fabian, A.C.: 1993, *Mon.Not.R.Astron.Soc.* **264**, 228
Rawlings, S., Saunders, R.: 1991, *Nature* **349**, 138
Camenzind, M., Courvoisier, T.J.-L.: 1983, *Ap.J.* **266**, L83
Stenger, V.J., Learned, J.G., Pakvasa, S., Tata, X. (eds.): 1992, Proc. of the Workshop on High Energy Neutrino Astrophysics, held in Honolulu, Hawaii, 23-26 March 1992, World Scientific, Singapore

HUBBLE SPACE TELESCOPE OBSERVATIONS OF NGC 4151: IMPLICATIONS FOR THE UNIFIED MODEL OF AGN

Z. TSVETANOV
Center for Astrophysical Sciences, Johns Hopkins University

I.N. EVANS
Space Telescope Science Institute

and

G.A. KRISS AND H.C. FORD
Center for Astrophysical Sciences, Johns Hopkins University

Abstract. *HST* Planetary Camera narrow band emission-line and continuum images are used to study the nuclear region of NGC 4151 at the highest possible spatial resolution. The [O III] $\lambda 5007$ image reveals a striking biconical structure with a projected opening angle of $75° \pm 10°$, and whose apex coincides with the bright, unresolved central source. The projected axis is oriented along PA $60°/240° \pm 5°$, and is aligned with the extension of the nuclear VLBI radio source. Analysis of the geometry of the narrow-line region places our line of sight well outside the ionization cones, and yet we see nearly unobscured optical and near-UV continuum and broad lines. In addition, material with significantly different column densities is required to explain the numerous optical and UV absorption lines and the soft X-ray absorption. We conclude from these data that the simplest version of the obscuring torus unification model is inconsistent with the observations and some modifications are required. We discuss some alternative collimation mechanisms that are compatible with our observations.

1. HST Imaging Results

The nearby Seyfert 1.5 galaxy NGC 4151 has been a subject of numerous studies over the whole wavelength range. We have imaged the nucleus of NGC 4151 with the Planetary Camera onboard *HST* to study the morphology of the emission-line gas at the highest avilable today spatial resolution. Full description of our observations is given in Evans et al. (1993), and here we present the most important results and implications for the Unified Models of AGN.

On the *HST* imaging scale ($0''\!.1$ resolution) the nucleus of NGC 4151 is unresolved in both the continuum and emission-line images. A point-like nuclear source remains even after the continuum subtraction (see Fig. 1). The [O III] $\lambda 5007$ emitting gas is distributed in a bi-conical structure with apices coincident with the nuclear point-like source with projected opening angle $\theta_{cone} = 75° \pm 10°$, and cone axis position angle PA$_{cone} = 60°/240° \pm 5°$. The [O III] $\lambda 5007$ emission is concentrated in filaments of discrete clouds with sizes of up to $\sim 0''\!.4$ ($\sim 20h$ pc; $H_0 = 100h^{-1}$ km s^{-1} Mpc^{-1}), implying that the clouds themselves are resolved.

On a larger (*arcsecond*) scale, the string-like Extended Narrow-Line Region (ENLR) inferred from the ground-based observations (e.g., Perez et al. 1989, Pérez-Fournon & Wilson 1990) is completely contained in the sector of the sky generated by extending the *HST* bi-conical structure to the radius of the ENLR. This suggests

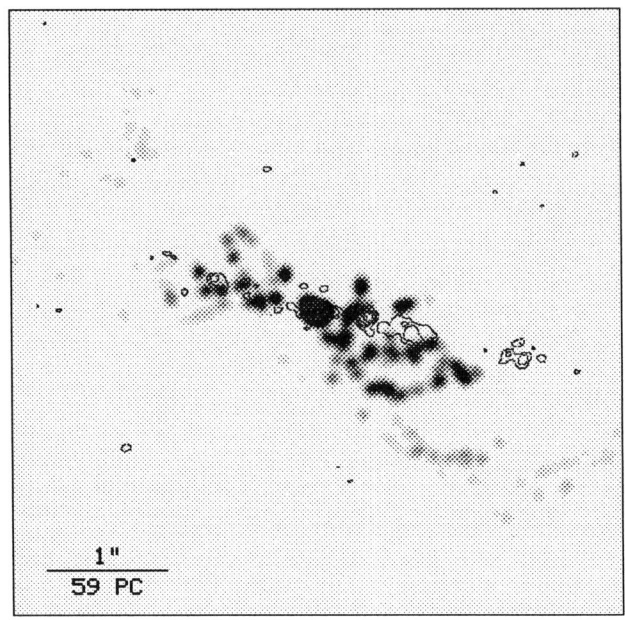

Fig. 1. MEM reconstruction of the continuum subtracted [O III] $\lambda 5007$ image sampled at 4 times the original pixel sampling. The image is scaled as square root of the intensity between 0% and 15% of the peak value. Contours represent the 5 GHz MERLIN radio map from Pedlar et al. (1993) at 2.2,4.5,9,18,36,65 and 95% of the peak value. The suspected nuclear component of the radio image has been aligned with the position of the HST unresolved nuclear source.

very strongly that the extended emission-line gas is encapsulated *physically* within the volume of space subtended by the nuclear cones.

No line-emission is visible in our images in either [O III] $\lambda 5007$ or Hα+[N II] in the direction perpendicular to the cones. The Hα+[N II] emission is due almost entirely to the unresolved nuclear point source, and there is additional weak Hα+[N II] emission associated with the extended bi-conical structure. This strongly implies that any theory aiming to explain the observed properties of NGC 4151 must incorporate a Broad-Line Region (BLR) not larger than few hundreds of an arcsecond on the sky.

Contrary to some other well studied cases (e.g., NGC 1068, Evans et al. 1991) there is little correspondence between the optical and radio emission on subarcsecond and arcsecond scales (see Fig. 1). The radio structure is misaligned by 12°–24° from our measured cone axis. The elongation of the nuclear VLBI radio source, however, is coincident with the symmetry axis of the [O III] $\lambda 5007$ cones. This suggests strongly that a common mechanism may be responsible for the orientation of both the radio plasma and the ionizing radiation field.

2. Orientation and Geometry: Contradiction with the simplest Unified Model of AGN

From the ground, the string-like ENLR can be followed to more than $\sim 30''$ from the nucleus and is oriented along PA 228°(e.g. Schulz 1990). Based on the comparison of the H I velocities with the velocities of the ionized gas it has been shown that the string lies in the disk of the host galaxy and it is almost in the plane of the sky. The suggested inclination to our line of sight is $\sim 98°$.

Previous ionization studies show that the gas in the ENLR is ionized by the nuclear radiation field as long as the string falls somewhere within the volume of the ionization cone. Kinematical studies of the gas in the ENLR imply that the SW cone is directed toward us and the NE cone is directed away from us. This orientation is also suggested by the fact that the filaments in the SW cone are generally brighter and more extended then their counterparts in the NE cone.

The geometry that minimizes the deviation between our line of sight and the nearest edge of the cone places the string along the far edge of the SW cone. With the above assumptions, and from the measured opening angle and orientation of the cones we derive a value of the true inclination of the cone axis to our l.o.s. $\phi \sim 65°$ and true cone opening angle of $\theta \sim 70°$. The inclination ϕ can be larger if the string falls totally within the cone instead of along the far edge. This implies that the *minimum* angle between our line of sight and the cone edge is $\sim 30°$. We note here that the large inclination angle is in excellent agreement with the results of the optical and X-ray observations of NGC 4151.

The inferred geometry contradicts with the simplest version of the Unified Model in which an infinitely optically thick torus obscures the nucleus and collimates the ionizing radiation field. Since our line of sight is at least 30° off the edge of the cone, it must pass through the torus. It is also not possible to "see" the BLR over the edge of the torus because of the estimated sizes of both the BLR (~ 0.003 pc) and the continuum source (~ 0.0003 pc), and the expected distance of the inner wall of the torus to the continuum source (~ 0.2 pc).

Our line of sight to the central source and the BLR is not totally unobscured. The low energy turnover of the X-ray spectrum of NGC 4151 requires an absorbing column of $N_H(\text{X-ray}) \approx 10^{23}$ cm^{-2}, and the best model suggests partial covering of $\sim 90\%$ of the source. Numerous absorption lines cover both the continuum and the broad emission lines (e.g. Kriss et al. 1992) at optical and UV wavelengths. These require a total equivalent hydrogen column density of $N_H(\text{UV}) \sim 10^{21}$ cm^{-2} and a neutral hydrogen column of $N_{HI} \sim 10^{18} - 10^{21}$ cm^{-2}.

It is clear from the geometry that an infinitely optically thick torus cannot collimate the ionizing radiation field, but the presence of a molecular torus cannot be ruled out. The larger cone opening angle we measure compared with the previous assumptions decreases the area of the sky subtended by the absorbing material by only $\sim 15\%$. This is not enough to significantly affect the overpredicted far IR luminosity. This controversy, however, is resolved in the recent theoretical studies which predict that the IR radiation from the torus is substantially anisotropic.

3. Collimating the Ionizing Radiation

In summary, our *HST* observations of NGC 4151 suggest that Unified Model must be made more complex to explain not only the collimation of the radiation into a bi-conical structure, but also to allow for a partially obscured view of the BLR and continuum source, and must explain the differing UV and X-ray column densities along our line of sight.

Consider a slight modification of the shadowing/obscuration scenario in which a physically thin, dense molecular torus is surrounded by a lower density atmosphere of neutral or semi-neutral gas. Hard X-ray heating or magnetically driven winds may "puff-up" the torus and lead naturally to such an atmosphere. The edges of the *ionization* cones will then be determined by the angle at which the path through the atmosphere gives a column density sufficient to block the ionizing ultraviolet. Columns of $N_H = 10^{20}$ cm^{-2} will be optically thick at all wavelengths past the Lyman edge up to the soft X-ray. Above the Lyman limit, and for normal gas-to-dust ratios, extinction will not have much impact up to column densities higher than several times 10^{21} cm^{-2}. Consequently, lines of sight outside the ionization cone but also outside the shadow of the opaque torus (the "twilight zone") will have a relatively clear view of the central regions at wavelengths longward of the Lyman edge. The atmosphere around the torus can explain the UV absorption lines, and higher density, fully ionized material inside the inner edge of the torus, e.g. in the BLR, may account for the higher X-ray column density.

The collimation of the ionizing radiation may not necessarily be related to the torus. A flattened configuration of optically thick broad-line clouds may produce the observed geometry. Such a distribution has been proposed to account for the X-ray observations, and by the reverberation mapping. And finally, the ionizing radiation may be intrinsically anisotropic, e.g. from a "naked" accretion disk around a supermassive black hole.

Acknowledgements

We are indebted to Alan Pedlar for the digital version of his MERLIN radio map. This work is suppoted by NASA grant NAG 5-1630 to the FOS team.

References

Evans, I.N. et al.: 1991, ApJ, 369, L27
Evans, I.N. et al.: 1993, ApJ, 417, 82
Kriss, G.A. et al.: 1992, ApJ, 393, 485
Pedlar, A. et al.: 1993, MNRAS, 263, 471
Pérez, E. et al.: 1989, MNRAS, 241, 31P
Pérez-Fournon, I., & Wilson, A.S.: 1990, ApJ, 356, 456
Schulz, H.: 1990, AJ, 99, 1442

QUASI-STEADY STATE COSMOLOGY

G. BURBIDGE
University of California, San Diego, 9500 Gilman Drive, La Jolla, California 92093-0111 USA

F. HOYLE
102 Admirals Walk, Bournemouth BH2 5HF, UK

and

J.V. NARLIKAR
Inter-University Center for Astronomy & Astrophysics Post Bag 4, Ganeshkhind Pune 411007, India

1. Introduction

The standard big bang cosmology has the universe created out of a primeval explosion that not only created matter and radiation but also spacetime itself. The big bang event itself cannot be discussed within the framework of a physical theory but the events following it are in principle considered within the scope of science. The recent developments on the frontier between particle physics and cosmology highlight the attempts to chart the history of the very early universe.

Exciting though these studies are, they have failed to resolve some of the basic issues of cosmology. These issues can be stated briefly as follows:

1. The microwave background radiation (MBR) is considered a fundamental proof of the big bang cosmology. Yet, cosmological considerations so far have failed to deduce its present temperature of $2.73K$.

2. While the Planckian spectrum of the MBR is deduced from its relic interpretation, its small scale anisotropy has not been successfully related to the observed large scale inhomogeneity of matter distribution in the universe, the recent findings of the satellite *COBE* notwithstanding.

3. There is no consistent theory of structure formation that takes in a reasonable hypothesis of dark matter and can reproduce the observed large scale structure and motions from primordial seed fluctuations.

4. The claims of explaining the large scale features of the universe in terms of discrete source populations inevitably require epicyclic hypotheses of evolution of physical properties of these sources that are post-facto rather than having any predictive power.

5. The age distribution of galaxies poses many problems for the canonical big bang model. How can we accommodate globular clusters of ages $15 - 18 Gyrs$ in a big bang universe with $k = 0$, that is required by inflation in the very early phase? Equally, it is difficult to understand the existence of very young galaxies at the present epoch, for, galaxy formation is supposed to have taken place in the early universe.

6. The phenomena in high energy astrophysics like the *QSOs,* AGN, radio sources, etc. show big outpourings of matter and energy from compact regions. However these events have no relation to the primordial big bang which is totally isolated from this relatively recent activity.
7. Finally on a theoretical issue, the big bang singularity is deduced from the equations of general relativity which are derived from an action principle. Yet, the action principle breaks down at the singularity, thus making the cycle of reasoning self-contradictory.

We believe that despite the popularity enjoyed by the big bang cosmology today, this list is sufficient to motivate an alternative approach to cosmology. (cf Arp et al. 1990, Hoyle et al. 1993 a,b).

Any alternative to big bang cosmology, should fulfill a few minimum conditions. First it must do at least as well as the big bang cosmology in explaining the MBR and light isotope abundances and in describing the observed features of discrete source surveys. Next it must try to do better than the big bang model on some of the above mentioned fronts. Finally, as a scientific theory it must make a few disaprovable predictions that distinguish it from the standard model.

In what follows we summarize a model that claims to do just that. This is the quasi steady state cosmology *(QSSC)* proposed by Hoyle et al. (1993 a,b).

2. Creation of Matter

In 1948 Bondi and Gold (1948) and Hoyle (1948) had independently proposed the steady state theory as an alternative to the big bang cosmology. Bondi and Gold had adopted the Perfect Cosmological Principle as the starting point of their approach while Hoyle had taken a field theoretic description of matter creation as the main motivation. Here we will follow the second approach but with some significant modification.

The field equations are derived from an action principle. Although Hoyle et al. (op.cit.) consider a direct particle interaction approach motivated by Mach's Principle, the following simplified derivation essentially reproduces their equations in the more familiar field theory format. Thus the classical Hilbert action leading to the Einstein equations is modified by the inclusion of a scalar field C whose derivatives with respect to the spacetime coordinates x^i are denoted by C_i. For the notation followed and further details see (Narlikar 1993). The action is given by

$$\mathcal{A} = \sum_a \int_{\Gamma_a} m_a c ds_a + \int_v \frac{c^3}{16\pi G} R\sqrt{-g} d^4x - \frac{1}{2c} f \int_v C_i C^i \sqrt{-g} d^4x + \sum_a \int_{\Gamma_a} C_i da^i \tag{1}$$

where C is a scalar field and $C_i = \partial C/\partial x^i$. f is a coupling constant. The last term of (1) is manifestly path-independent and so, at first sight it appears to contribute no new physics. The first impression, however, turns out to be false if we admit the existence of broken worldlines.

Thus, if the worldline of particle a begins at point A, then the variation of $\mathcal{A}$ with respect to that worldline gives

$$m_a \frac{da^i}{ds_a} = g^{ik} C_k \tag{2}$$

at A. In other words, the C-field balances the energy-momentum of the created particle.

The field equations likewise get modified to

$$R_{ik} - \frac{1}{2} g_{ik} R = -\frac{8\pi G}{c^4} \left[\overset{T^{ik}}{m} + \overset{T^{ik}}{c} \right] \tag{3}$$

where $\overset{T^{ik}}{c} = -f \left\{ C_i C_k - \frac{1}{2} g_{ik} C^l C_l \right\}$. \hfill (4)

Thus the energy conservation law is

$$\overset{T^{ik}}{m}{}_{ik} = - \overset{T^{ik}}{c}{}_{ik} = fC^i C^k{}_{ik}. \tag{5}$$

That is, matter creation via a nonzero left hand side of (5) is possible while conserving the overall energy and momentum.

The C-field tensor has negative stresses which lead to the expansion of spacetime, as in the case of inflation. The formalism described here is essentially that used by Hoyle and Narlikar (1962, 1966 a,b) in the 1960s to produce inflation type solution (which, of course, predated Guth's inflationary cosmology by 15 years!).

From (2) we therefore get a necessary condition for creation as

$$C_i C^i = m_a^2 c^4; \tag{6}$$

This is the 'creation threshold' which must be crossed for particle creation. How this can happen near a massive object can be seen from the following simple example.

The Schwarzschild solution for a massive object M of radius $R > 2GM/c^2$ is

$$ds^2 = c^2 dt^2 \left(1 - \frac{2GM}{c^2 r}\right) - \frac{dr^2}{1 - \frac{2GM}{c^2}r} - r^2 \left(d\theta^2 + \sin^2\theta d\phi^2\right), \tag{7}$$

for $r \geq R$. Now if the C-field does not seriously change the geometry, we would have at $r \gg R, \dot{C} \approx \alpha$ and $C' \approx 0$, so that

$$C^i C_i \equiv \left(1 - \frac{2GM}{c^2 r}\right)^{-1} \frac{\alpha^2}{c^2}. \tag{8}$$

In other words $C_i C^i$ increases towards the object and can become arbitrarily large if $R \approx 2GM/c^2$. So it is possible for the creation threshold to be reached *near* a massive collapsed object even if $C_i C^i$ is *below* the threshold far away from the object. In this way massive collapsed objects can provide new sites for matter creation. Thus, instead of a single big bang event of creation, we have mini-creation events (MCEs) near collapsed massive objects.

3. A Cosmological Solution

Since the C-field is a global cosmological field, we expect the creation phenomenon to be globally cophased. Thus, there will be phases when the creation activity is large, leading to the generation of the C-field strength in large quantities. However, the C-field growth because of its large negative stresses leads to a rapid expansion of the universe and a consequent drop in its background strength. When that happens creation is reduced and takes place only near the most collapsed massive objects thus leading to a drop in the intensity of the C-field. The reduction in C-field slows down the expansion, even leading to local contraction and so to build-up of the C-field strength. And so on!

We can describe this up and down type of activity as an oscillatory solution superposed on a steadily expanding de Sitter type solution of the field equations by a scale factor that varies with cosmic time t as follows:

$$S(t) = exp\left(\frac{t}{P}\right)\left\{1 + \alpha \cos \frac{2\pi t}{Q}\right\}. \tag{9}$$

Note that the universe has a long term secular expanding trend, but because $|\alpha| < 1$, it also executes non-singular oscillations around it. We can determine α and our present epoch $t = t_0$ by the observations of the present state of the universe. Thus an acceptable set of parameters is $\alpha = 0.75, t_0 = 0.85Q, Q = 4 \times 10^{10} yr., P = 20Q$. Although the set is not unique and there will be a *range* of acceptable values, we will work with this set to illustrate the performance of the model.

4. The Origin of Nuclei and the Microwave Background

We have as yet not said what particle is being created by the C-field. The answer is, the Planck particle whose mass is

$$m_p = \sqrt{\frac{3\hbar}{4\pi G}} \sim 10^{-5} g \tag{10}$$

This particle, however, has a very short lifetime $\sim 10^{-44}s$. It decays ultimately into the baryon octet and radiation. Most members of the octet except n and p are also short-lived and decay into protons. Only the neutron and the proton combine into stable helium nuclei. Thus approximately 25% by mass (2 out of 8 baryons) combine to form helium.

A more careful calculation gives the helium mass fraction to be around 23%, with a tiny fraction of 1-2% in the form of metals. This type of nucleosynthesis also generates ^2H, ^3H, ^3He, ^6Li, ^7Li, ^9Be, ^{10}Be etc. in small amounts that are in agreement with the observations, see Appendix I in Hoyle et al. (1993 a). In fact, the abundances of the light isotopes in this cosmology lead to a better agreement with observations than in the big bang model.

There is one further important consequence. In the big bang model the required production of deuterium imposes a stringent upper limit on the present day baryon density. This limit forces the conventional big-bang cosmologists to asume that the dark matter component of the universe must be largely nonbaryonic. In the QSSC, there is no such density limit from deuterium abundance and thus the dark matter component *can be baryonic*.

What about the microwave background? The *QSSC* obtains it in the following way. First, each Planck particle decay is like a fireball: it produces lots of energy, including baryons ($\sim 10^{19}$ per Planck particle) and radiation. More than the hot big bang, the Planck fireball can provide several interesting and realistic studies in astroparticle physics. Further, since the Planck fireballs are repeated phenomena, rather than the 'once only' type situation of the hot big bang cosmology, they are amenable to more exhaustive scientific study.

The bulk of the fireball energy goes into expansion. However, some radiation remains as a relic of the fireball. Together with the starlight generated in the preceding oscillatory cycles this energy is to be thermalized to provide the microwave background. Does it provide enough radiant energy to give a 2.7K background? Is the background thoroughly thermalized to produce a black body spectrum? Also, is it homogeneous to the extend given by *COBE* (Smoot, et al 1992) and other measurements? Quantitative studies (see Hoyle, et al. 1993 a,b) answer all these questions in the affirmative.

The starlight from several past generations of stars is sufficient to maintain a steady background of radiation whose present temperature is calculated to be $\sim 2.7K$, provided, some agency is available to thermalize it. The agency proposed is dust in the form of metallic needles, mostly of iron, which absorb the ambient radiation and reradiate it in the microwave region. Provided this has gone on long enough, the radiation spectrum will hae an accurate black body form. Calculation shows that indeed the thermalization has occurred through as many as 10^3 absorptions and re-emissions by iron whiskers – sufficient to ensure an extremely close approximation to the black body curve. The iron whiskers are typically $\sim 1mm$

in length and $10^{-6}cm$ in radius of cross section. The iron itself is produced partly from stellar nucleosynthesis in supernovae and partly fromt he decay of the Planck particle. The required density in the form of such whiskers is only $\sim 10^{-35} g\ cm^{-3}$: well within the observed cosmic abundances of iron.

The background produced will be very smooth with a patchiness of density of the order of 10^{-5}. Fluctuations of density and temperature of this or larger order get smoothed out by redistribution of iron grains by the radiation pressure. On smaller scales the dynamical smoothness-restoring forces are too small to make the radiation smooth. Thus, the *COBE* finding of $\triangle T/T \sim 10^{-5}$ is consistent with the above picture. Moreover, the characteristic scale of 10^{26} cm at the oscillatory minimum will expand to $\sim 5 \times 10^{26} cm$ at present, giving a characteristic angular scale for the above patichiness to be $\sim 10°$, in conformity with the *COBE* scale of angular inhomogeneity.

In our papers (Hoyle et al. 1993 a,b) we have discussed the redshift-apparent magnitude relation, the counts of radio sources and the angular diameter redshift relation all with in the framework of the QSSC. Because of the shortage of space, and the subject matter discussed at this meeting, we now turn directly to the cosmogony associated with the QSSC.

5. The Cosmogony Associated with the QSSC

(i) *Violent Events in Galaxies* What we see all about us is a violent universe in which from a variety of centers large amounts of energy in the form of photons and particles are being ejected. In the radio galaxies and in the QSOs and in the nuclei of many galaxies we know that we are seeing the results of non-thermal energy production often arising through the incoherent synchrotron mechanism. Also in Seyfert nuclei and the like we see the ejection of large amounts of kinetic energy in the form of hot gas moving at velocities $\sim 10^9 cm\ sec^{-1}$. There is also extensive circumstantial evidence based on geometrical configurations and optical morphology suggesting that coherent objects are ejected from the nuclei of galaxies. Finally the γ-ray burst sources, if they lie at cosmological distances may very well be direct evidence for creation.

This whole class of phenomena is in our view best interpreted as direct evidence for matter creation on a small scale - minicreation events (MCE) as was originally proposed by Ambartsumian (1958, 1965) and Jeans (1929).

Other phenomena which we believe can best be explained by the QSSC include

(ii) *The Age of the Universe* According to QSSC the universe is infinitely old but the average age of astronomical objects is $1/3P \sim 3 \times 10^{11} yrs$. This makes many clusters much older than hitherto assumed. Even our Galaxy might have age of this order with several generations of stars formed, evolved and burnt out. The dark matter component in the Galaxy may be largely made of burnt out stars.

(iii) *Hierarchy of Structures* The largest structure to form in the MCEs is the so called supercluster of mass $\sim 10^{15} - -10^{16} M_\odot$. There are, however, MCEs on smaller scales going right down to galactic nuclei with masses $\sim 10^6 - 10^7 M_\odot$. It is, however, the former that keep the universe going steady state at all times.

Finally there is the possibility that gravity wave sources may be due to creation events. Narlikar and Das Gupta (1993) have shown that such events in the mass range $100 - 1000 M_\odot$ can be detected by the laser interferometric detectors being planned. Further the gravity wave background created by such MCEs may also affect the timing mechanism of millisecond pulsars by an amount that is detectable.

References

Ambartsumian, V.A., (1958, 1965) *Proc. of the Solvay Conf. on the Structure of the Universe* 241, and, *Proc. of the Solvay Conf. on The Structure and Evolution of the Galaxies* 1.
Arp, H.C., Burbidge, G., Hoyle, F., Narlikar, J.V. and Wickramasinghe, N.C. (1990) *Nature*, **346**, 807.
Bondi, H. and Gold, T. (1948) *M.N.R.A.S.*, **108**, 252.
Hoyle, F. (1948) *M.N.R.A.S.*, **108**, 372.
Hoyle, F. and Narlikar, J.V. (1962) *Proc. Roy. Soc.* **A270**, 334.
Hoyle, F. and Narlikar, J.V. (1966) *Proc. Roy. Soc.* **A290**, 143.
Hoyle, F. and Narlikar, J.V. (1966) *Proc. Roy. Soc.* **A290**, 162.
Hoyle, F., Burbidge, G. and Narlikar, J.V. (1993 a) *Ap.J.*, **410**, 437.
Hoyle, F., Burbidge, G. and Narlikar, J.V. (1993 b), preprint.
Jeans, J. (1929) *Astronomy & Cosmogony* (Cambridge Univ. Press) 352.
Narlikar, J.V. (1993) *Introduction to Cosmology*, 2nd Edition, (Cambridge Univiversity Press).
Narlikar, J.V. and Das Gupta, P. (1993) *M.N.R.A.S.*, to be published.
Smoot, G.F., et al., (1992) *Ap.J.*, **396**, L1.

UNIFIED MODELS: RELIGION AND SCIENCE

ROBERT ANTONUCCI
Physics Department, University of California, Santa Barbara, CA 93106

September 1993

1. My Personal Religious Conversion to Unification

The Unified Model states that the classification of individual AGN is a function of orientation, and that orientation effects are key to understanding the different classes. In its most extreme form, it states that every AGN has a featureless continuum (FC) source and a broad line region (BLR), both enclosed in an opaque torus. The torus is perpendicular to the associated radio structure axis. For the powerful radio sources (in Elliptical galaxy hosts), the jets undergo bulk relativistic motion, giving rise to phenomena such as superluminal motion associated with the blazar class. All strong radio sources have diffuse double radio lobes, although in the blazars one is sometimes seen projected onto the other. To take this to the extreme, we can suppose that all opaque tori are made of dust[1] and have the same opening angle and that the radio jets are all narrow and have the same bulk-motion Γ factor.

Many issues related to Unified Models have been reviewed in Antonucci 1993. Here I will just summarize very briefly the arguments which I followed most closely and found compelling. (I take such a narrow track not just for personal aggrandizement but also to get on to some relatively new information.) Although the model seems to be qualitatively correct for most sources, well-documented deviations are known, and they have interesting consequences and applications. Three of these deviations are discussed in Section 2.

1.1. Seyfert 1's and 2's

Many Seyfert 2's show Seyfert 1 spectra when the polarized flux alone is plotted, and the optical polarization position angles lie perpendicular to the associated radio axes. This has been interpreted to mean that these Seyfert 2's have featureless continuum ("FC") sources and broad line regions (BLR's) like Seyfert 1's, but that they are enclosed in opaque tori oriented perpendicular to the radio sources, and are seen only by reflection off material in the polar directions. It follows from the

---

[1] We've recently learned that NGC 4151 has a hard-edged ionization cone, strongly suggestive of shadowing by an opaque torus (Evans et al 1993). We know both from the appearance of the cone, and the fact that the HI column density in the line of sight is $>> 10^{17}$ (Kriss et al 1992), that we are outside the cone. Thus I'm confident NGC 4151 would be a Seyfert 2 if observed below 912Å. The fact that the FC and BLR are seen directly in the optical implies that gas opacity is operating in this case, at least for our sight line (as Evans et al conclude).

Copernican Principle that at least some Seyfert 1's are equivalent objects, viewed from a polar direction.

In the case of NGC 1068, starlight subtraction indicated that the FC polarization was wavelength independent at 16% indicative of electron scattering (Miller & Antonucci 1983, McLean et al 1983). This was verified by Wuppe data (Code et al 1992) and new UV spectropolarimetry by Hubble Space Telescope (Antonucci, Hurt & Miller 1993). Specifically, the FC should dominate throughout most of the range of the HST observations, with just a little starlight contribution near 3000Å. Therefore, we should be able to overplot the polarized flux (divided by 16%) on the total flux and get a perfect match, independent of wavelength except for a modest divergence at the red end. Figure 1 shows that this expectation has been borne out.

1.2. Blazars and Normal Double Radio Sources

The same basic phenomenology was seen in the Narrow Line Radio Galaxy 3C234, before the NGC 1068 data were understood, and the torus (or occultation/reflection model) was first based on 3C234 back in 1982. There was one subtle difference which will become important for Section 2: There was some evidence that the BLR polarization was significantly higher than the FC polarization (Section VI b of Antonucci 1984).

At the same time another unification scheme was being debated for the radio loud sources. The flat-spectrum core dominant sources generally show superluminal motion. Also, in most cases, the core synchrotron spectra dominate the spectral energy distribution all the way into the optical region, leading to a blazar classification.

The beam model was proposed to account for superluminal sources: the moving components were said to be ejected relativistically from the cores, almost in the direction of Earth, producing an illusion of faster than light motions. It was further supposed that these superluminal jets were identifiable with the (apparently) weaker jets linking cores to lobes in normal double radio sources. The blazars were simply the ones seen from nearly the jet direction (Blandford & Rees 1978).

The blazar/normal double unification obviously predicted that the blazars should show diffuse radio halos or foreshortened doubles in high dynamic range maps, because they must also have the double lobes in this model. Several groups started detecting such diffuse radio emission (including Ian Browne, John Wardle, Jim Ulvestad and others). Ulvestad and I did a major search, gathering data on all the objects known at the time of the Angel and Stockman (1980) review paper on blazars. We had enough data that we could argue that this unification *must* be qualitatively correct.

Our argument went as follows. Suppose the superluminal, roughly linear radio cores of blazars are highly anisotropic and beamed towards Earth (arguably just the "conventional physics" assumption). Suppose also that the diffuse and often two-sided halos were emitting roughly isotropically. The data show that several blazars have sufficient flux *in their diffuse halos alone* to merit a place in the 3C category.

By the first assumption above, these are aimed closely in Earth's direction; there must be many more equivalent objects which are not. By the second assumption, the misdirected equivalents must be in the 3C catalog. The only things in the 3C catalog with big powerful diffuse radio sources are the normal doubles. In fact, there are about the right number so that most or all are misdirected blazars, if the blazar emission is as isotropic as in "narrow cone" models. (Cohen 1990 later tightened the noose on "wide cone" models with better VLBI maps.)

1.3. Paradoxes ... and Resolution

In the mid 1980's most radio astronomers considered the quasars and Narrow Line Radio Galaxies to be distinct classes, and when testing the beam model statistically, they considered only blazars and quasars. They found various statistical anomalies which can all be expressed as a dearth of quasars in the sky plane. For example, the asymmetry of the arcsec scale jets in normal double radio quasars could in principle be explained by beaming, but it was argued that quantitatively the jet/counterjet ratios were observationally constrained to be too large, too often. A robust and picturesque second example comes from the depolarization/jet side anticorrelation. This seemed supportive of the beam model, but, as Laing (1988) pointed out, "The sources observed here must be oriented within about 45° of the line of sight...to generate sufficient asymmetry in path length between the two lobes, and this is consistent with the observed asymmetry in the jets."

Barthel (1989) emphasized that all of these problems are greatly alleviated if we admit the possibility that luminous narrow line radio galaxies are quasars in the sky plane. Some early radio polarization mapping was consistent with this (e.g. the inner parts of both lobes, rather than all of one lobe, seem to be depolarized in 3C234: Strom et al 1985). And, of course, the spectropolarimetry arguments show that this identification must be true at some level. Recent optical polarization imaging of distant luminous radio galaxies argues very strongly that this is true for many of them (e.g. Tadhunter et al 1992, Jannuzi & Elston 1991, and di Serego Alighieri et al 1993).

My conclusion is that the Unified Model is qualitatively correct and it is now a religion.[2] However, we can rule out the simplest version on many grounds, including those discussed below.

---

[2] Our religion is called the Unification Church by Archbishop Barthel. Also on this topic, Cardinal Laing would like to receive suggestions regarding appropriate vestments.

2. Science Finds Deviations from Unification Dogma

2.1. LOW REDSHIFT FRII STATISTICS

In the simplest version, the unification of double-lobed radio quasars (and Broad Line Radio Galaxies) with Narrow Line Radio Galaxies is true for all objects of both classes, and they all have the same opening angle for the obscuring torus. This has been ruled out as follows.

The 3CRR catalog is essentially complete and completely optically identified. It is also selected mostly by lobe emission, because of the low frequency of the survey (178MHz). These properties make it ideal for comparing a sample of objects which are to be identified in unification schemes.

Barthel (1989) has compared the 3CRR Narrow Line Radio Galaxies (all FRII) and quasars in the $0.5 < Z < 1.0$ range all of which are FRII. The former have twice the space density on average, and half the projected linear size. These and other properties are consistent with the assertion that all of these objects have FC sources and BLRs enclosed in opaque tori with opening angles of 45°. Such a unification can resolve the statistical anomalies associated with the beam model as applied to quasars alone, as described above. It is now known from the polarization imaging of distant luminous narrow line radio galaxies that high perpendicular polarization is common. This is probably due to reflected light from objects which must (by the Copernican Principle) be quasars. Thus Barthel is qualitatively correct, and perhaps quantitatively correct in the parameter space he studied.

I'll cite three subsequent papers which test the Barthel pattern at lower redshift. Kapahi (1990; see also Lawrence 1993, Fig. 1) shows that for $0.25 < Z < 0.5$, the Narrow Line Radio Galaxies become more predominant numerically over the broad line objects, and the difference in average projected linear size vanishes. By contrast, Gopal Krishna and Kulkarni (1992) find that the size difference remains, and holds from $0.1 < Z < 2.0$, if we look at the upper envelope of the size distributions at each redshift. They justified the focus on the upper envelopes partially because it eliminates the complicated, contentions and pernicious Steep-Spectrum Compact (SSC) sources. They also made adjustments to the observed sizes for the power of the sources, though I think this was a smaller effect.

Singal (1993) reconsidered the issue, focusing on average size, but modifying the Kapahi analysis to exclude SSCs and core-dominant sources. His Figure 2 shows all of the issues very clearly: the "good" behavior of the size distributions at $1 < Z$, $0.5 < Z < 1$, and in the upper envelope of the $Z \lesssim 0.5$ data. But it also shows that for the relatively low luminosity, nearby objects, there are many "extra" small Narrow Line Radio Galaxies, which apparently do not participate in the unification with broad line objects in any significant way! Barthel (1993, p.c.) speculates that they may be the "optically dull" ones, without luminous high-ionization narrow lines. That's well worth checking.

A worthwhile study would be to consider an isotropically-selected sample in which the blazar/quasar unification alone leads to statistical anomalies. Would the quasar/NLRG unification hold sufficiently for that sample to solve these problems? That is the key question, and I'm optimistic about the answer.

2.2. THE STRANGE CASE OF CYGNUS A

This is the extremely luminous prototypical Narrow Line Radio Galaxy at a redshift of only 0.056. Stockton (1993) points out that you have to go out to $Z = 1.00$ before you find a more luminous radio galaxy, despite the volume factor $\cong Z^3$ and the strong cosmological evolution working for you. It may be like the extremely luminous distant radio galaxies. We are lucky to have it.

Baade and Minkowski discovered the double nucleus of Cygnus A in 1951. Later it was shown that both nuclei are spatially resolved; that the NW nucleus is almost entirely high-ionization line emission while the SE nucleus is almost entirely Featureless Continuum; and that the true center of the system as marked by the flat-spectrum radio core is between the two optical nuclei. A slightly smoothed but not deconvolved HST image of the double nucleus in the 3360Å region is shown in Figure 2.

Pierce and Stockton 1986 showed very convincingly that the dilution of Population II stellar absorption lines, and hence the FC, is spatially resolved in the SE nucleus. There is no evidence for hot stars, so they proposed that the FC is light reflected from a hidden quasar. This makes a lot of sense in the light of the unified model. Broad permitted lines do not show in polarized flux however (Goodrich and Miller 1989, Jackson and Tadhunter 1993). One could imagine that the reflection region consists of very hot electrons which broaden the reflected lines beyond recognition. However, I think the off-nuclear polarization ($\sim 1\%$) is just too low for spatially-resolved scattering. We have HST UV polarization imaging and spectra on the shelf which should help to test this further.

We recently found another puzzle, almost a paradox, regarding Cygnus A. If we see the flat-spectrum radio core through a molecular torus, we should see CO in absorption very easily. In fact, independent of the unified model, we supposedly know there are at least 50 magnitudes of extinction in front of the nucleus ((Djorgovski et al 1991, Ward et al 1991), and the X-ray absorption column is also large (Koyama 1992; D. Harris 1993, p.c.). Again, we expect to see CO absorption, but we do not (Barvainis & Antonucci 1993).

The observations could be accommodated by a torus composed of clouds which have very narrow internal velocity dispersions, which each cover a small fraction of the radio core, and which have a covering factor near one and not much larger. The space between the clouds must be very free of molecules. This stretches our imaginations just about to the breaking point. Rees (1993, p.c.) has suggested that the rotational excitation temperature could be very high despite a low kinetic temperature, owing to the strong radio source. In that case, stimulated emission largely cancels absorption, as is common with HI. We are considering this intriguing suggestion quantitatively.

2.3. WHY THE FEATURELESS CONTINUUM IS EXTENDED IN GENERAL!

A very interesting and unexpected subtlety in the data on almost all of the "reflected light" objects is the following. The polarized flux shows the broad permitted lines at their normal equivalent widths. We would therefore expect in the simplest case that this would also be true in the total-flux spectra corrected for contaminating starlight. In other words, if the FC and the BLR are seen only in reflection, then their ratio in total nonstellar flux should be normal. But that is not the case! The BLR is generally polarized *much more* than the FC, so that it hardly shows up in total nonstellar flux, while it has normal strength in polarized flux.

The size of the effect in the objects of Tran, Miller and Kay 1992 is very large. Mrk 477 is typical. The polarized flux spectrum looks just like a Seyfert 1. But, while $P(BLR) \gtrsim 20\%$, $P(FC)$ is only $\sim 1\%$! The most likely explanation is that part of the FC source is exposed and so part of the FC flux is reaching us directly, greatly diluting the fractional polarization of the FC. The exposed component is *energetically insignificant* and is visible only because the torus is blocking a large majority of the direct FC flux. However, it provides a crucial clue that the fundamental emission process is diffuse and not entirely ultracompact as in, say, accretion disk models with sizes of order the gravitational radii. Tran and Miller refer to the exposed part of the Featureless Continuum as "FC2".[3] (Sources here are Tran, Miller and Kay 1992 and Miller 1993.)

Can the exposed part of the FC come from a source unrelated to the central energetically-dominant part, for example, hot stars? Unlikely because the total-flux FC, dominated by the exposed part, has the same spectral energy distribution as the polarized flux, from the hidden part. Both have the shape of Seyfert 1 spectra rather than that of hot stars.

The shocking implication of this is that at least a minor part of the FC source is larger that the BLR! A large FC source would be a good explanation of why the observed continuum *never* shows a Ly edge due to the BLR clouds (Antonucci, Kinney, & Ford 1989; Tytler 1993 (p.c.)). The FC may actually come from clouds extending throughout the emission line regions. If their emission is thermal, as suggested, for example, by the very steep soft X-ray cutoffs,[4] the clouds are then probably optically thin emitters. Such a scenario has been proposed on other grounds by, e.g., Antonucci and Barvainis 1988; Ferland, Korista and Peterson 1990; Barvainis 1993; and Binette, Fosbury, and Parker 1993.

If the FC source is indeed optically thin and extended, with at least a small fraction arising above the occulting torus, then it should be sufficiently strong and extended in the radio to be directly mappable. Barvainis and I have proposed this observation to the VLA.

---

[3] Similarly, the constant soft X-ray excess seen in Seyfert 2's (e.g. Mulchaey et al 1993) may have low polarization, and be dominated by this exposed component in most cases.

[4] Two recent arguments that the optical/UV light is the same component as the soft X-ray excess are given by Walter and Fink 1993, and A. Orr et al, this meeting. A counterargument has been given by Lee et al 1993.

There are two other important implications of Tran et al 1992 for the torus model. First, a low total FC polarization is not inconsistent with seeing a normal Seyfert 1 spectrum in polarized flux. This encourages us to believe that the Seyfert 1/2 unification may be universal. Second, the percent polarization of the *reflected* light, as shown directly by the BLR, is generally very high, as expected for a thick obscuring torus rather than, say, a warped disk.

3. Conclusions

The Unified Model is qualitatively correct, correct to "zeroth order," but far from the whole story of active galactic nuclei.

References

Angel, J.R.P., and Stockman, H.S. 1980, ARA&A, 18, 231
Antonucci, R. 1984, ApJ, 278, 499
Antonucci, R. 1993, ARA&A, 31, 473
Antonucci, R., and Barvainis, R. 1988, ApJ Lett, 332, L13
Antonucci, R., Hurt, T., and Miller, J. 1993, ApJ, submitted
Antonucci, R., Kinney, A., and Ford, H. 1989, ApJ, 342, 64
Barthel, P. 1989, ApJ, 336, 606
Barvainis, R. 1993, ApJ, 412, 513
Barvainis, R., and Antonucci, R. 1993, AJ, submitted
Binette, L., Fosbury, R., and Parker, D. 1993, Pub Astron Soc Pac, submitted
Blandford, R., and Rees, M. 1978, in Pittsburgh Conference on BL Lac Objects, A.M. Wolfe, ed, Pittsburgh: University of Pittsburgh Press
Code, A. et al 1993, ApJ Lett, 403, L63
Cohen, M. 1990, in Parsec-Scale Radio Jets, J. Zensus and T. Pearson, eds, Cambridge: Cambridge University Press
di Serego Alighieri, S., Cimatti, A., and Fosbury, R. 1993, ApJ, 404, 584
Djorgovski, S., Weir, N., Matthews, K., & Graham, J.R. 1991, ApJ, 372, L67
Evans, I.N., Tsvetanov, Z., Kriss, G.A., Ford, H.C., Caganoff, S., and Koratkar, A.P. 1993, ApJ, in press
Ferland, G.J., Korista, K.T., and Peterson, B.M. 1990, ApJ Lett, 363, L21
Goodrich, R., and Miller, J. 1989, ApJ Lett, 346, L2
Gopal Krishna, R., and Kulkarni, V. 1992, Astron Astrophys, 257, 11
Jackson, N., and Tadhunter, C.N. 1993, Astron Astrophys, 272, 105
Januzzi, B., and Elston, R. 1991, ApJ Lett, 366, L69
Kapahi, V. 1990, in Superluminal Radio Sources, J. Zensus and T. Pearson, eds., Cambridge: Cambridge University Press, p. 304
Koyama, K. 1992, in X-Ray Emission for AGN and the Cosmic X-Ray Background, W. Brinkmann and J. Trumper, eds, Garching: MPE Press
Kriss, G.A. et al 1992, ApJ, 392, 485
Laing, R.A. 1988, Nature, 331, 149
Lawrence, A. 1991, MNRAS, 252, 586
Lee, G-H., Kriss, G.A., Zheng, W., and Davidsen, A.F. 1993, BAAS, 25, 792
McLean, I., Aspin, C., Heathcote, S., and McCaughrean, M. 1983, Nature, 304, 609
Miller, J.S. 1993, talk presented at the First Stromlo Symp.: The Physics of Active Galaxies, held in Canberra, Australia, June 28-July 2, 1993
Miller, J.S., and Antonucci, R. 1983, ApJ Lett, 271, L7
Mulchaey, J.S., Colbert, A.S., Mushotzky, R.F., and Weaver, K.A. 1993, ApJ, 414, 144
Pierce, M.J., and Stockton, A. 1986, ApJ, 305, 204

Singal, A. 1993, MNRAS, 262, L27
Stockton, A. 1993, in First Light in the Universe: Stars or QSOs, Rocca-Volmerauge et al, eds.
Strom, R.G., and Conway, R.G. 1985, Astron Astrophys Supp, 61, 547
Tadhunter, C., Scarrott, S., Draper, P., and Ralph, C. 1992, MNRAS, 256, 53
Tran, H., Miller, J.S., and Kay, L. 1992, ApJ, 397, 452
Walter, R., and Fink, H.H. 1993, Astron Astrophys, 274, 105
Ward, M.J., Blanco, P.R., Wilson, A.S., and Nishida, M. 1991, ApJ, 382, 115

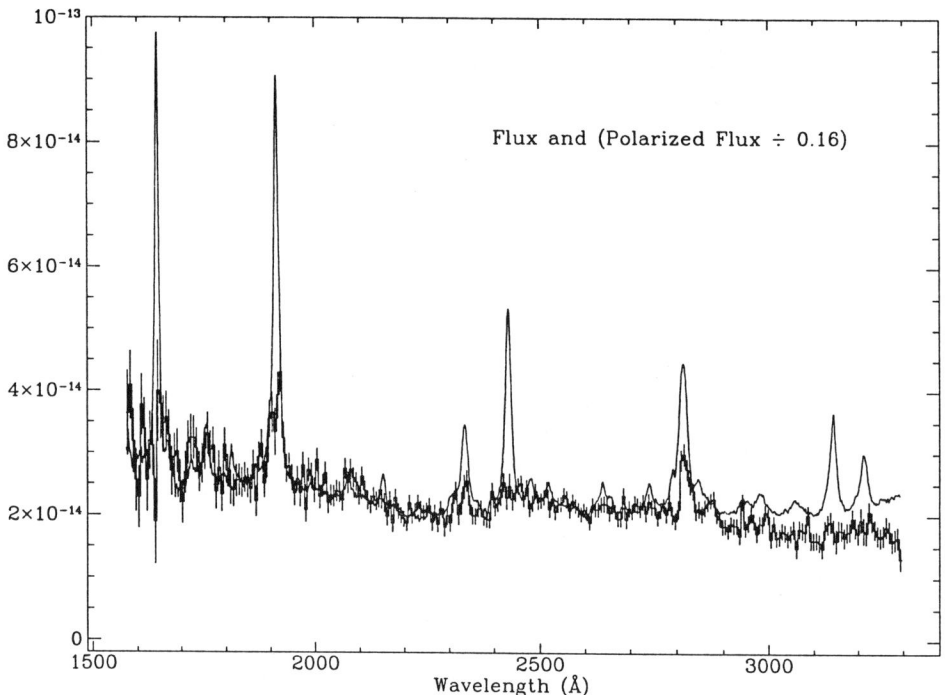

Figure 1 - Total Flux and polarized flux of NGC 1068 in the ultraviolet

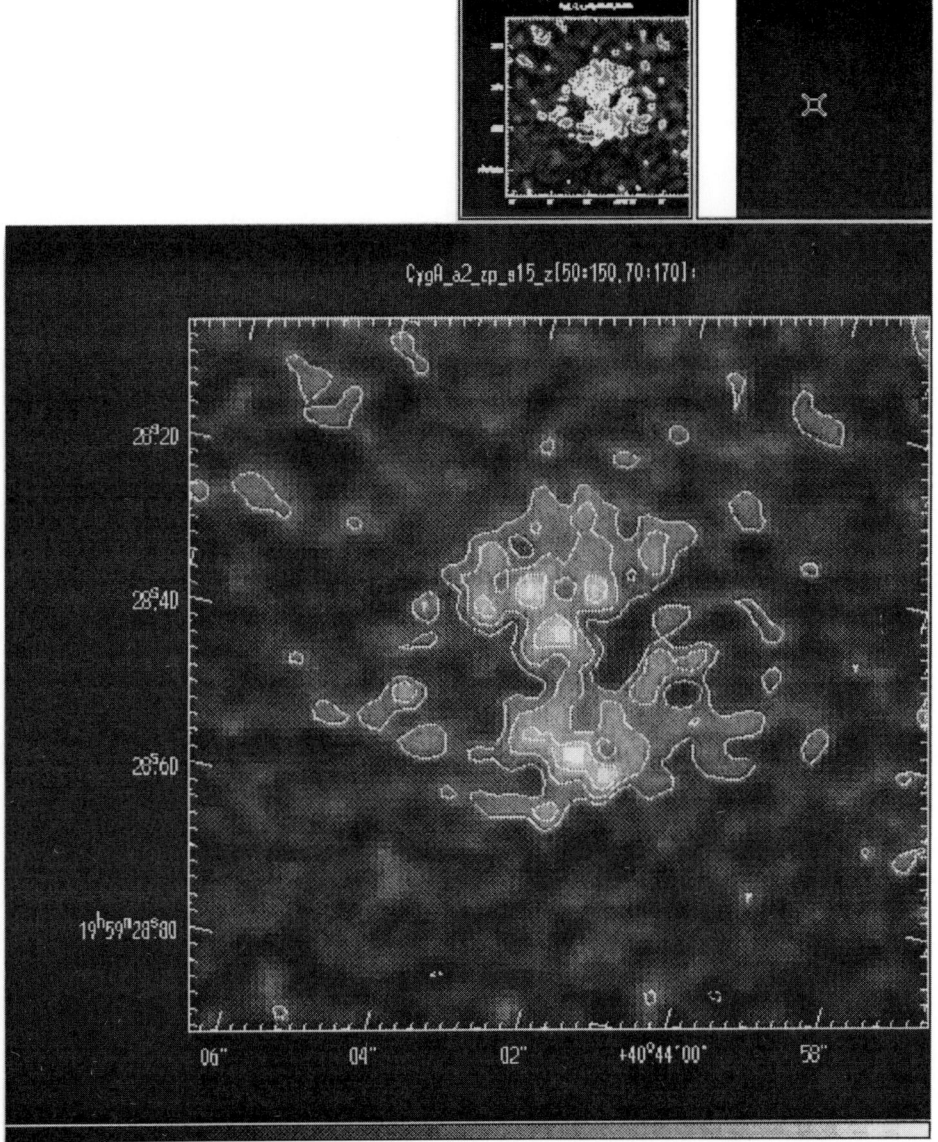

Figure 2 - Preliminary HST image of Cygnus A at 3360Å; no deconvolution, $\sigma = 0\overset{''}{.}12$ Gaussian smoothing.

X-RAY LUMINOUS NON-SEYFERT GALAXIES

M.J. WARD and D.H.HUGHES
Astrophysics, University of Oxford, Keble Road, Oxford OX1 3RH, U.K.

J.S. DUNLOP
Dept. of Chemical and Physical Sciences, Liverpool JM University, U.K.

and

P.N. APPLETON
Dept. of Physics and Astronomy, Iowa State University, U.S.A.

Abstract. There are now well established classes of X-ray emitting galaxies, which exhibit characteristic ranges of luminosity, for example Green *et al.* (1992). The *ROSAT* all sky survey, has detected a relatively small number of galaxies whose overall properties do not appear consistent with any of these classes. We discuss whether these X-ray luminous galaxies (XLGs) are the extreme end of the luminosity range of starburst galaxies, or whether they represent a new X-ray class of AGN.

Recently the *ROSAT* all sky survey has been cross-correlated with about 15,000 galaxies from the *IRAS* Point Source Catalog, resulting in detections of 244 galaxies (Boller *et al.*, 1992). In addition to the recognized classes of AGN, starbursts and normal galaxies, this survey revealed a small number of X-ray luminous non-Seyferts exhibiting soft X-ray luminosities between $10^{41.5-43}$ ergs s^{-1}. The top end of this range is comparable to Seyfert 1s, and the lower end is similar to Seyfert 2s, and the most luminous starburst galaxies detected before *ROSAT* (Green *et al.*, 1992).

First we must offer a word of caution. Published classifications of AGN types based on optical spectroscopy, can be in error for a variety of reasons. From our high quality optical spectroscopy we note three cases of misclassification in the original list of XLGs, Boller *et al.* (1992). IRAS0628+63, IRAS1537-18 and IRAS1322-38, are all in fact Seyfert 1s. In view of this, their X-ray luminosities are not exceptional. However there still remain a number of XLGs whose optical spectra do not resemble Seyferts, see Fig. 1. In Table I, we list the properties of a subset of these non-Seyferts, together with IRAS1537-18 for comparison. Their X-ray luminosities are more than an order of magnitude higher than found for normal galaxies, Fabbiano *et al.* (1992).

Fruscione and Griffiths (1991) identified a number of *Einstein* selected galaxies, as X-ray luminous starbursts. Their conclusion, based on very few examples, was that starburst galaxies could account for at least 15% of the extragalactic background at 2 keV. Some of the XLGs discovered by *ROSAT* are even more luminous than those identified by Fruscione and Griffiths. In order to quantify their activity we plot a set of emission line ratio diagrams in Fig. 2.

The X-ray emission from normal (non-AGN) galaxies originates predominantly from the bulge component in early type spirals, and from the disk component in

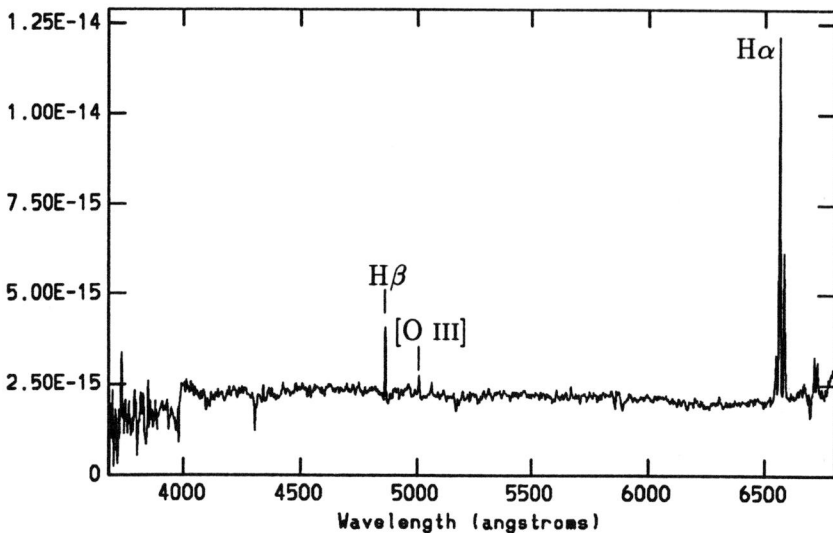

Fig. 1. Optical spectrum of IRAS 16155+6831, showing it is a typical starburst/HII region galaxy. The flux scale is ergs cm^{-2} s^{-1}. Compare with NGC 5248 in Kennicutt, (1992).

late-type spiral galaxies. However, a scaling of these stellar components shows that they fail to account for the X-ray emission from XLGs. Therefore in an attempt to explain their high X-ray luminosities we examine two propositions: either they are some variation on the theme of obscured AGN, following the Seyfert 1/Seyfert 2 unification scheme, or they are examples of powerful starburst activity at an extreme level previously unrecognized.

TABLE I

source	log L(X-ray) (0.1 − 2.4keV) ergs s^{-1}	log L(FIR) (40 − 120μm) ergs s^{-1}	log L(Hα) λ6563Å ergs s^{-1}	log L([O III]) λ5007Å ergs s^{-1}
IRAS11395+1033	42.28	44.54	40.9	40.5
IRAS12393+3520	42.80	44.22	39.8	40.1
IRAS15374−1817	42.40	44.00	40.5	40.0
IRAS16155+6831	42.39	44.36	41.3	40.1

Considering the first possibility, it is unlikely that the low energy X-ray (*Einstein* and *ROSAT*) emission is due to *transmitted* radiation from an obscured AGN. This is because a foreground extinction of $A_V \gtrsim 5$ mag would be required to obscure the broad wings on the Balmer lines (e.g. Blanco, Ward & Wright 1990). The asso-

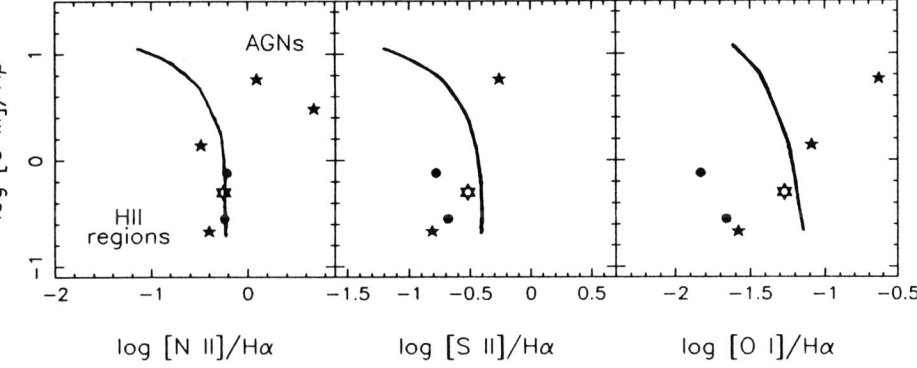

Fig. 2. Diagnostic emission line ratio diagrams. The solid line divides typical AGN from typical starburst/HII regions. Symbols are; filled stars for galaxies in table 1, filled circles for NGC 1614 and NGC 3256, previously known well-studied starburst galaxies with luminosities between $10^{41.5-42}$ ergs s^{-1}. The open star is for 1E2251-17, one of the most X-ray luminous starburst galaxies detected by *Einstein*, with 10^{43} ergs s^{-1}. Note: not all galaxies have measured ratios in all three plots.

ciated gas column of $\simeq 10^{22}$ cm^{-2} would easily extinguish the soft X-ray emission. A hidden AGN could produce detectable X-ray emission through electron-scattering as in the generally accepted model for Seyfert 2s. However, because of their high X-ray luminosities (10–100 times that of a typical Seyfert 2), the scattering *efficiency* would have to be correspondingly high, but still not high enough for the scattered Broad Line Region to be detected in the optical spectrum. This might be possible with a scattering efficiency of around 10%. Polarization measurements would settle this question.

Turning now to the super-starburst hypothesis, Fabbiano, Trinchieri & MacDonald (1984) showed that in irregular blue galaxies undergoing bursts of star formation the ratio of f_X/f_B was greater than in normal galaxies, with some notably extreme cases (Fabian & Ward, 1993).

Hence energetic, widespread starburst activity is one way in which the X-ray emission can be significantly enhanced relative to the optical. Exactly how this is achieved is still a matter of debate. Evolved low-mass X-ray binaries would contribute starlight and X-rays proportionally, so appear to be ruled out by the high L_X/L_B ratios. This is demonstrated graphically in Fig. 3, which shows the similarity of a XLG to a typical HII region galaxy, in all but its X-ray properties. Ohashi *et al.* (1990) favor thermal emission from gas at kT$\simeq$1–3 keV to fit their GINGA spectra of NGC253 and M83. The diffuse component in those galaxies could be fueled by supernovae and their remnants, but for the XLGs the numbers would need to be exceptional.

A plausible alternative would be a strong contribution from luminous, short-lived High Mass X-ray Binaries which display similar spectral characteristics. Such

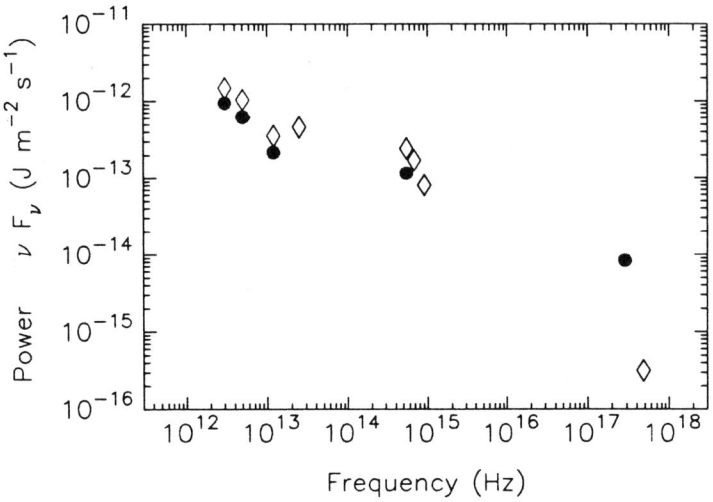

Fig. 3. The energy distribution, from far IR to X-rays, of IRAS 16155+6831 (filled circles), compared with a typical starburst galaxy NGC 5248 (open diamonds). The fluxes for IRAS 16155+6831 have been scaled up by a factor 10, to facilitate comparison.

a short-lived population would enhance the X-ray/photospheric luminosity ratios of XLGs relative to quiescent galaxies (David et al., 1992). A difficulty is the requirement that a high percentage of all O-stars form in these systems.

In summary, ROSAT has detected a small but significant number of non-Seyferts, with high X-ray luminosities. A few are misclassified AGN, others may be LINERs, but some do appear to be starburst/HII region galaxies. Unless they evolve strongly, they probably do not contribute significantly to the X-ray background. Nevertheless, their X-ray luminosities do present problems for starburst models. X-ray spectra to define absorption cut-offs and temperatures, radio observations to quantify the SNR component, and measurement of the spatial extent of the starburst, are the next logical steps in this study.

References

Blanco, P., Ward, M.J.,& Wright, G.S., 1990, *Mon. Not. R. astr. Soc.*, **242**, 4p
Boller, Th., Meurs, E.J.A., Brinkmann, W. et al., 1992, *Astron. Astrophys. Suppl.*, **261**, 57
David, L.P., Jones, C. & Forman, M.J., 1992, *Astrophys. J.*, **388**, 82
Fabian, A.C. & Ward, M.J., 1993, *Mon. Not. R. astr. Soc.*, **263**, L51
Fabbiano, G., Trinchieri, G. & MacDonald, A., 1984, *Astrophys. J.*, **284**, 65
Fabbiano, G., Kim, D.-W. & Trinchieri, G., 1990, *Astrophys. J. Suppl.*, **80**, 531
Fruscione, A. & Griffiths, R., 1991, *Astrophys. J.*, **380**, L13
Green, P.J., Anderson, S.F. & Ward, M.J., 1992, *Mon. Not. R. astr. Soc.*, **254**, 30
Kennicutt, R.C., 1992, *Astrophys. J. Suppl.*, **79**, 255
Ohashi, T. et al., 1990, *Astrophys. J.*, **365**, 180.

Poster Contributions:
Continuum studies

MODELLING THE SOFT X-RAY EXCESS IN E1615+061

M. BAŁUCIŃSKA-CHURCH
School of Physics and Space Research, University of Birmingham, U.K.

L. PIRO
Istituto di Astrofisica Spaziale, CNR, Frascati, Italy

H. FINK
Max-Planck-Institut für extraterrestrische Physik, Garching, F.R.G.

F. FIORE
Harvard-Smithsonian Center for Astrophysics, Cambridge MA, U.S.A.

M. MATSUOKA
Institute of Physical and Chemical Research (RIKEN), Tokyo, Japan

G. C. PEROLA
Istituto Astronomico dell'Universita' di Roma "La Sapienza", Italy

and

P. SOFFITTA
Columbia University, New York, U.S.A.

Summary. We report results of an international UV - X-ray campaign in 1990 - 1992 involving the IUE, Rosat and Ginga satellites to observe E1615+061, a Seyfert 1 galaxy with peculiar spectral and intensity behaviour over the last 20 years. The source has been found to be stable in its medium state during the observations. The Ginga (1 - 20 keV) spectrum of E1615+061 is adequately represented by a simple power law with a photon index $\alpha = 1.8 \pm 0.1$. However, $\alpha \sim 2$, as expected for the intrinsic power law component in a reflection model, cannot be ruled out statistically. The Rosat PSPC (0.1 - 2 keV) spectra collected during the All Sky Survey and the AO-1 phase can be well-described by a simple power law ($\alpha = 2.2 \pm 0.1$) with cold absorber ($N_H = 3.5 \pm 0.3 \cdot 10^{20}$ H/cm^2). Both the photon index being significantly different than that obtained from the Ginga spectrum and the column density being smaller than the galactic column ($N_H \sim 4.2 \cdot 10^{20}$ H/cm^2) give an indication of a soft excess over and above the hard component seen in the Ginga spectrum. E1615+061 has been observed with IUE in 1990 and in 1992. The source was stable and the colour excess E(B-V) derived from the data = 0.1 is in good agreement with that expected from the galactic absorption.

To parameterise the soft excess we fitted the Rosat data with a two-component model consisting of a power law, and a blackbody or thermal bremsstrahlung, with a single galactic absorption term. The column density and the slope of the power law were kept constant. The blackbody temperature was 80 ± 6 eV and 63 ± 12 eV for photon index equal to 1.8 and 2.0, respectively, whereas the bremsstrahlung temperature was 220 ± 40 eV and 115 ± 30 eV for the two cases.

An attempt to model the soft excess seen in the Rosat PSPC spectrum has been made assuming that the soft excess is the high energy tail of a disc spectrum which peaks in the UV part of the spectrum. Additionally it was assumed that there is a hard component contributing to the spectrum from UV to X-rays with parameters as described by the Ginga spectrum. The best fit parameters: the mass of the central source and the mass accretion rate were around $5 \pm 1 \cdot 10^6$ M$_\odot$ and 0.2 ± 0.04 M$_\odot$/yr, respectively.

Our modelling shows that the soft X-ray excess can be described ($\chi^2_{red} < 1.2$) as the high energy tail of an accretion disk spectrum if the intrinsic power law is quite steep ($\alpha = 2$). The main contribution to the residuals in the Rosat PSPC range comes from 0.3 - 0.6 keV, with a tendency for these residuals to increase when the slope gets flatter. The accretion luminosity is $\sim 6.5 \cdot 10^{44}$ erg/s for the best fit parameters, i.e. about the Eddington luminosity.

HST, IUE AND ROSAT OBSERVATIONS OF HIGH Z QSOS

D. L. BAND, R. D. COHEN, P. R. BLANCO, V. T. JUNKKARINEN,
E. M. BURBIDGE and R. E. ROTHSCHILD
CASS, UC San Diego, La Jolla, CA 92093

and

G. A. REICHERT
USRA, NASA-Goddard SFC, Code 668, Greenbelt, MD 20771

Abstract. The optical through X-ray spectra of 3 QSOs between $z=1.32$ and 3.53 show these objects to be very similar to low z AGN.

Luminous AGN were more prevalent in the past. Their redshift moves the "big blue bump," unobservable in nearby AGN, into the observable UV. Therefore we have constructed spectra of three high z QSOs, with contemporaneous ROSAT, IUE, HST and Lick 3m telescope observations. PHL 1377 ($z=1.436$) and OQ 172 ($z=3.530$) are radio loud QSOs (RLQs) while PG1522+101 ($z=1.321$) is a radio quiet QSO (RQQ). These three AGN were selected because of their unusually low column depth of intervening gas: PHL 1377 and PG1522 are observable to 520Å (rest) while OQ 172 is observable down to 320Å.

The emission line spectra are typical, with fairly weak Fe II UV multiplets for two objects. The far UV spectra clearly bend towards the X-ray emission, constraining the extent of the "big blue bump," and ruling out large quasi-thermal components; our objects do not show an extra ionizing component unobservable in low z AGN.

The ROSAT spectra do not require absorption beyond the Galactic N_H, contrary to the increased absorption reported in other high redshift QSOs; this may be a consequence of the low column of intervening gas. All three spectra are fit satisfactorily by a power law with Galactic absorption. However the source with the greatest number of counts, PHL 1377, is better fit by a broken power law with a softer low energy component, as has been seen in many other AGN.

Our objects have X-ray properties similar to those of lower z RLQs and RQQs. The two RLQs have harder power law spectra. The higher z RLQ has the hardest spectrum while the lower z RLQ has the soft low-energy component, consistent with X-ray spectra hardening with increasing energy. On the other hand, the X-ray spectrum of the RLQ PG 1522 is surprisingly steep. The RLQs have $\alpha_{ro} \sim 0.5-0.7$ and $\alpha_{ox} \sim 1.3$ while the RQQ has $\alpha_{ro} < -0.15$ and $\alpha_{ox} \sim 1.8$, consistent with the trends seen in large X-ray surveys of QSOs.

With total 1μ-10 keV luminosities of $\sim 2 \times 10^{47}$ ergs-s^{-1}, these QSOs are among the most luminous AGN, but they have very different energy distributions. However the sources are by no means the brightest sources discovered.

A future publication will report these results more extensively. This research was supported by NASA grant NAG5-2059 and a SERC fellowship (PRB).

GROUND-BASED OBSERVATIONS OF PKS2155-304 IN NOVEMBER 1991

A. BLECHA[1], T. J.-L. COURVOISIER[1], H. D. ALLER[2], M. F. ALLER[2],
P. BOUCHET[3], P. BRATSCHI[1], M. T. CARINI[4], M. DONAHUE[5],
E. D. FEIGELSON[6], A. V. FILIPPENKO[7], I. S. GLASS[8], J. HEIDT[9],
P. A. HUGHES[2], R. I. KOLLGAARD[6], T. MATHESON[7], H. R. MILLER[10],
J. C. NOBLE[10], P. S. SMITH[11] and S. WAGNER[9]

[1] *Geneva Observatory, CH-1290 Sauverny, Switzerland*
[2] *Astronomy Department, University of Michigan, Ann Arbor, Michigan 48109-1090, US*
[3] *European Southern Observatory, Casilla 19001, Santiago 19, Chil*
[4] *Computer Sciences Corporation, NASA/GSFC, Greenbelt, MD 20771, US*
[5] *Carnegie Observatories, 813 Santa Barbara St., Pasadena, CA 91101, US*
[6] *Department of Astronomy and Astrophysics, 525 Davey Lab, Pennsylvania State University, University Park, PA 16802, USA*
[7] *Department of Astronomy, University of California, Berkeley, CA 94720, USA*
[8] *South Africa Astronomical Observatory, PO Box 9, Observatory 7935, South Africa*
[9] *Landessternwarte Heidelberg-Königstuhl, Königstuhl, D-69117 Heidelberg, Germany*
[10] *Department of Physics, Georgia State University, Atlanta, GA 30303*
[11] *Steward Observatory, University of Arizona, Tucson, AZ 85721, USA*

Abstract. We present ground-based data of the BL Lac object PKS 2155-305 obtained during a large international campaign spanning the electro-magnetic spectrum from the radio waves to X-rays in November 1991. For the complete description of the observations and data analysis we refer to the paper by Courvoisier et al. 1993, and references therein. The ground-based data include radio, infrared JHKL and UBVRI fluxes as well as optical and near IR polarimetry.

The broad-band optical and near IR data from U to I exhibit the same behaviour in all bands: the flux nearly doubled over the well-covered period of 23 days. The cross-correlation function does not reveal any significant changes in the light-curves. Though significant variations in 24 hours have been recorded, the cumulated Fourier power spectrum drops to the noise level for periods shorter than 2.5 days. The spectral index remained constant.

The polarised flux varied by a larger factor than the total flux and did not follow the same pattern. The degree of polarisation and polarisation angle are nearly independent of the wavelength and are strongly correlated in all filters.

In the radio domain the spectral index increased from -0.1 on November 5 to +0.02 on 25-th.

The absence of the lag between the optical and infrared bands and the polarisation variations are consistent with a model in which the variability is caused by micro-lensing of the source (Stickel, Fried and Kühr 1988). One would, however, expect in this model that the variation in the polarisation and the total flux are tightly correlated contrary to what is observed.

The constant shape of the continuum spectral energy suggests that only the number of electrons whose emission is beamed towards the observer changes, rather than the arrival of fresh electrons that are being accelerated.

The variability of the polarisation may be explained by changes in the geometry of the magnetic field (dominant direction). This is consistent with the observed variations of the polarisation angle.

References

Courvoisier T. J.-L. et al. 1993 Ap. J. submitted
Stickel M., Fried J.W. and Kühr H., 1988, A&A 191, L16

ELECTRON ENERGY DISTRIBUTIONS OF AGNS IN THE THIN SYNCHROTRON LIMIT. II. PEAKED ELECTRON ENERGY SPECTRA OF BLAZARS AND OVV QUASARS

IRENE CRUZ-GONZÁLEZ, LUIS SALAS and LUIS CARRASCO
Instituto de Astronomía UNAM, Apdo. 70-264, Mexico DF 04510, Mexico

Abstract.
We analyse multifrequency quasi-simultaneous observations of AGNs with the Inverse Synchrotron Transform (IST) method described in Salas et al. (1993) and in Paper I (this conference).

The observational spectral energy distribution of blazars and OVV quasars used as applications of the IST method were obtained from multifrequency quasi-simultaneous observations of Landau et al. (1986) and Brown et al. (1989). The application of the method requires good spectral coverage, since the presence of gaps larger than $\approx 0.75 dex$ can introduce ficticious peaks. We present the analysis for: 1749+096, 0735+178, 1308+326, 3C273, 3C345, 3C279, OJ287, OJ049, 3C216, 4C39.25.

The observed spectra are transformed to the source rest frame, the inverse transform is calculated over the whole spectrum to derive the electron energy spectra, which gives the amount of energy emitted in each characteristic frequency. Our results show that the electron energy spectra, $\Psi(\nu_c)$, of blazars and OVV quasars consists mainly of 4 or 5 well defined peaks superimposed on a down curved continuum. This indicates that the source radiates mainly around certain values of the electron's critical frequency (or energy, since $\nu_c \propto B\,E^2$). These peaks are consistent with injection of fresh relativistic particles, occuring in pulses of quasi-monoenergetic groups.

The separation between successive peaks (Δp) is ≈ 1.2 dex in critical frequency, and is characteristic for each object. The peaks in the radio region contain approximately the same number of electrons, consistent with evolution given by the continuity equation for constant B. While in the OIR region, the energy radiated increases towards lower frequencies.

A comparison of Δp with L_{bol} and L_{radio}, and jet parameters such as, jet velocity and B_{1pc} (inferred from the relativistic jet models of Blandford and Königl 1979) given by Cruz-González and Carrillo (1991), shows that Δp increases with jet velocity, stronger B fields imply smaller Δp, and that it is not correlated with L_{bol}, but is apparently correlated with L_{radio}.

The Inverse Synchrotron Transform method applied in this paper is a powerful tool in the analysis of spectral energy distributions of AGNs.

Key words: active galactic nuclei, multiwavelength emission, emission processes

References

Biretta,J.A., Moore,R.L., Cohen,M.H., 1986, Astrophysical Journal **308**, 93.
Brown, L.M.J. et al.: 1989, Astrophysical Journal **340**, 129.
Cruz-González,I., Carrillo,R., 1991, Rev.Mex.Astron.Astrofis **22**, 217.
Landau et al.: 1986, Astrophysical Journal **308**, 78.
Salas, L., Cruz-González, I., Carrasco, L. 1993, in preparation.

SIMULTANEOUS OPTICAL TO NEAR-IR OBSERVATIONS OF BLAZARS.

R. FALOMO and M. BERSANELLI
Osservatorio Astronomico di Padova, v. Osservatorio 5, 35122 Padova, Italy
Istituto di Fisica Cosmica, CNR, v. Bassini 15, 20133 Milano, Italy

We obtained simultaneous optical spectrophotometry (4000-8000 Å) and near-IR photometry (JHKL) for 34 blazars, with repeated observations for 21 of them. All the data were taken at ESO and cover a period of 3 years with a typical time interval of 6 months. The observations were corrected for interstellar extinction and analyzed using a uniform procedure. This allows to form a large and homogeneous data set of optical-near-IR measurements of blazars emission.

For each observation we constructed a composite spectral flux distribution (SFD) covering the range 8×10^{13}–7×10^{14} Hz. We find that, with very few exceptions, the SFD of each object is well described either by a single power law ($f_\nu \propto \nu^{-\alpha}$) or by a power law plus the contribution of a giant elliptical host galaxy (see Fig. 1). Our results show that, contrary to other findings, *negative* ($d\mathrm{Log}(f_\nu)/d\mathrm{Log}(\nu) < 0$) spectral curvatures in the optical to near-IR, when observed, can be easily ascribed to the contribution from a host galaxy and the effect of galactic extinction.

For our sample we find that the average spectral index of the non-thermal emission is $\langle \alpha \rangle = 1.08 \pm 0.06$. Moreover the X-ray selected objects exhibit a significantly (99% c.l. for a KS test) flatter spectra than radio selected ones.

For ~30% of the sources we have repeated observations. The optical and near-IR emission are strongly correlated and usually maintain the spectral index ($\langle \Delta \alpha_{max} \rangle = 0.17 \pm 0.04$) even under substantial flux variations. No evidence of correlation between α and flux level is observed. A full report of these observations is given elsewhere [1].

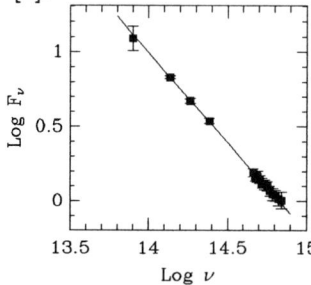 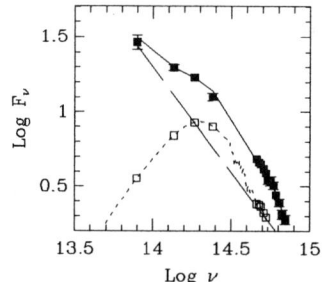

Fig. 1. Representative SFD for two objects: 0048–097 (left), and 0521–365 (right).

References
[1] Falomo R., Bersanelli M., Bouchet P., and Tanzi E.G. 1993, AJ, 106, 11

OBSERVATIONS OF PKS 2155−304 WITH THE EXTREME ULTRAVIOLET EXPLORER

A. FRUSCIONE, C.S. BOWYER and T.E. CARONE
Center for EUV Astrophysics, University of California, Berkeley.

S.M. KAHN
Dept. of Physics and Astronomy, and Space Science Lab., University of California, Berkeley.

A. KÖNIGL
Department of Astronomy and Astrophysics, University of Chicago.

and

H.L. MARSHALL
Massachusetts Institute of Technology, Cambridge.

We have studied the first extreme ultraviolet spectroscopic data and a high accuracy light curve for the BL Lac object PKS 2155−304 observed with the Extreme Ultraviolet Explorer (*EUVE*) on July 21-22, 1992. This target was observed with the Deep Survey Spectrometer telescope for approximateley 30,000 sec during the in-orbit-calibration phase of the mission, allowing to obtain simultaneous image and spectrum.

The average observed count rate in the Lexan/B band (65–190Å) is 0.582±0.0047 count/sec. The EUV light curve shows variations of about 10% over a time scale of 0.3 days during the 1.3 days observing period. Variations at the same level have been reported for the ultraviolet band. The EUV flux measured by *EUVE* from PKS2155−304 is 1.7 times fainter than the flux measured by the *ROSAT* Wide Field Camera during the EUV survey in 1990. Simultaneous optical data indicate however that the overall shape of the spectrum has remained unchanged between the two observations.

PKS 2155−304 was detected in the short-wavelength spectrometer (70–190 Å) with detectable flux out to 105 Å. We modeled the spectrum with a power law plus absorption by interstellar HI, HeI (which dominates the absorption at the EUV wavelengths) and HeII. We applied various models with varying slope and HeI/HI ratio. The best fit is given by a model with an EUV energy index 1.6 and HeI/HI=0.1. From the model fit we can establish tight constraints on the amount of neutral He in the line of sight of the BL Lac. For an energy index of 1.6 and $N(HI)=1.36\times 10^{20}$ cm^{-2} (the Galactic value) the 1σ interval for HeI/HI is 9.8-10.4%.

No intrinsic absorption in the BL Lac is needed to explain the spectral data and both the spectrum and the broad-band data are consitent with a simple spectral model in the extrapolation of the X-ray band.

THE MULTIWAVELENGTH SPECTRA OF CEN A, NGC 4151, 3C273, 3C279

Sten Odenwald (Applied Research Corporation),
Neil Gehrels (NASA/Goddard Space Flight Center),
Sethanne Howard (BDM/NASA)

We present SEDs for Cen A, NGC 4151, 3C273 and 3C279 with new results from the CGRO. The spectra cover 16 orders of magnitude in ν, radio to γ-ray, over 20 years of observations. These 4 galaxies represent a Seyfert, a FSRG and 2 blazars. The γ-ray data indicate two classes of AGN. At γ-ray energies Type 1 Seyferts and radio galaxies have quantitatively different properties from blazars.

The solid line indicates CGRO data. Data are not simultaneous. The spectral differences are clear, not only at low energies but also at high energies. NGC 4151 cuts off at high energies. 3C273 and 3C379 remain hard. Cen A is in-between the Seyfert 1 type cutoff and the blazar flattening.

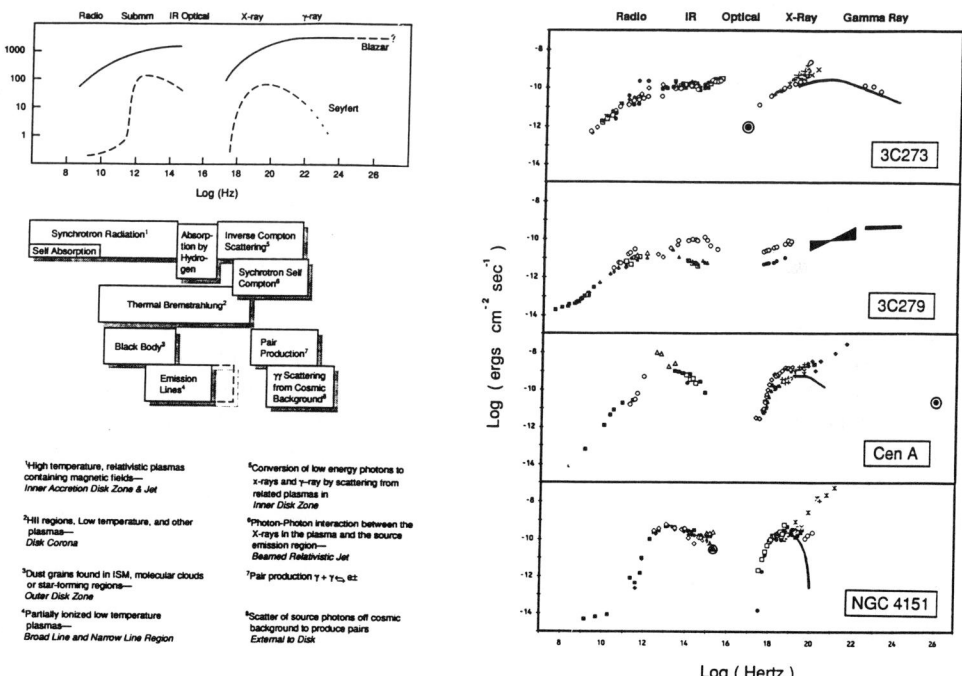

SUBMILLIMETRE SPECTRAL INDICES OF RADIO-QUIET QUASARS

D. H. HUGHES
Astrophysics, Dept. of Physics., Oxford University, U.K.

J. S. DUNLOP
Dept. of Chemical & Physical Sciences, Liverpool John Moores University, U.K.

E. I. ROBSON
Joint Astronomy Centre, Hilo, Hawaii, U.S.A.

and

W. K. GEAR
Royal Observatory, Edinburgh, U.K.

Hughes *etal.* (1993) have made submillimetre continuum observations of 10 IRAS selected radio-quiet quasars (RQQs). Three RQQs, I Zw 1, Mrk1014 and Mrk376, have been detected at 800 and 450μm using the ^{3}He bolometer UKT14 on t he 15-m James Clerk Maxwell telescope. These submillimetre data, together with existing 1.3 mm observations (Chini *etal.* 1989) demonstrate that the measured submillimetre spectral indices, $\langle \alpha_{sm} \rangle = 3.8 \pm 0.5$, significantly exceed the critical theoretical limit of $\alpha_{sm} = 2.5$ predicted for the self-absorption of synchrotron emission. This result is independent of any contributions to the 100μm IRAS fluxes from *cirrus* emission in the host galaxies, extended circumnucl ear starformation and FIR emission from companion or confusing sources. All current non-thermal models (de Kool & Begelman 1989, Schlickeiser *etal.* 1991) are rejected in favour of the alternative explanation that the FIR luminosity is dominated by thermal emission from warm (45–60 K) dust grains. The submillimetre optical depth and source-size for the thermal emission cannot yet be constrained by these data and, as a result, no discrimination can be made between dust heated by an extended (> 1 kpc) starburst region or a central compact luminosity source. However ground-based imaging observations at mid-IR wavelengths and FIR photometry (60 – 200μm) with the KAO are currently in progress specifically to address this problem. The high gas masses (> $10^{10} M_\odot$) in RQQs inferred from the submillimetre continuum observations are in agreement with the H_2 masses determined from CO measurements. Alternatively the results show that the M_{H_2}/M_{dust} ratio measured in RQQs ($\sim 370 \pm 150$) is consistent with that measured in spiral galaxies and ultra-luminous IRAS galaxies.

References

Chini,R., Kreysa,E. & Biermann,P.L., 1989, A&A, 219, 87
de Kool,M. & Begelman,M.C. 1989. *Nature*, 338, 484
Hughes,D.H., Robson,E.I., J.S. Dunlop & Gear,W.K., 1993. *MNRAS*, 263, 607
Schlickeiser,R., Biermann,P.L. & Crusius-Wätzel,A., 1991. A&A, 247, 283

EUVE OBSERVATIONS OF NGC 5548

J.S. KAASTRA
SRON, P.O. Box 9504, 2300 RA Leiden, The Netherlands

R. MEWE, J. HEISE, F.J.M. ALKEMADE
SRON, Sorbonnelaan 2, 3584 CA Utrecht, The Netherlands

C.J. SCHRIJVER
Sterrekundig Instituut, Princetonplein 5, Utrecht, The Netherlands

and

T. CARONE
SSL, University of California at Berkeley, USA

Abstract.
The first extreme ultraviolet spectrum of NGC 5548 obtained by EUVE is presented.

Key words: NGC 5548, EUV-emission

1. Introduction

NGC 5548 is one of the brightest Seyfert galaxies in the soft X-ray band. The spectrum and 0.5 day variations of the soft X-rays can be interpreted by correlated temperature - luminosity changes of an accretion disk (Kaastra & Barr 1989) or alternatively by ionisation of a warm absorber (Nandra et al. 1993).

2. Observations

Here we study the EUV spectrum of NGC 5548 using the Extreme UltraViolet Explorer (EUVE) launched in 1992. EUVE is sensitive to photons from 16–180 eV with a high spectral resolution ($E/\Delta E \sim 200\text{--}400$). Due to interstellar absorption only the hardest part of the spectrum of NGC 5548 is observable (above 100 eV).

NGC 5548 was observed during 3 periods: March 10–24, April 26 – May 4 and May 12–14. It was detected only during the first epoch with a count rate of $(1.6\pm0.4)\ 10^{-3}$ counts/s. The 126–182 eV flux, corrected for absorption ($N_H = 1.5\ 10^{20}\ \text{cm}^{-2}$) is $(6.7\pm1.9)\ 10^{-15}\ \text{Wm}^{-2}$ and 3 times larger if $N_H = 2\ 10^{20}\ \text{cm}^{-2}$. For the second and third observation the 3σ upper limits are $2.6\ 10^{-15}\ \text{Wm}^{-2}$ and $2.1\ 10^{-15}\ \text{Wm}^{-2}$, respectively. The spectrum shows no significant line features, except maybe a weak feature near 90.2 Å (significance 3.8σ), which however is difficult to interpret and might be an instrumental artefact.

References

Kaastra, J.S. & Barr, P. 1989, A&A 226, 59
Nandra, P. et al. 1993, MNRAS 260, 504

INSTANTANEOUS FLUX SPECTRA MEASUREMENTS OF THE ABOUT 100 AGNS AT 6 FREQUENCIES FROM 1 GHZ TO 22 GHZ

N.A.NIZELSKI, YU.A.KOVALEV and A.B.BERLIN
Astro Space Center, Lebedev Physical Institute, Profsoyuznaya street, 84/32 Moscow 117810
Russia

Abstract. Preliminary spectral results are presented for the 100 objects, selected from a VLBI-servey at 6-7 wave lengths of 31 cm, 13 cm, 8.2 cm, 7.6 cm, 3.9 cm, 2.7 cm, and, for the declinations greater than +13 degrees, at 1.38 cm. The most of presented spectra may be resulted by the emission of a halo-jet structure of AGNs.

Key words: AGNs, Radio Flux Spectra

Measurements was made at the radio teleskop RATAN-600 in 1989, from 26 November to 4 December for a list of 116 sources as a part of spectra monitoring programme (*Kovalev, 1991*). The list is a complete sample of sources with declinations from -29 to +42 degrees and correlated flux densities greater than 0.8 Jy at the frequency 2.3 GHz, selected from the VLBI survey (*Morabito et al., 1986*).

Because of horizontal lokalization of horns we have multi frequency response for "instantaneous" spectrum diring 1.5-2.5 minutes, when a source moves on the sky by the Earth rotation. Calibrate sources are 0237-23, 0624-05, 1328+30 and 2105+42 as the main, and 0518+16, 0134+32 and 2037+42 as the additional. Observations of all calibrate sources did not differ from the other sources of the list.

As can be seen, most of instantaneous spectra can be divided by two main components. First of these has a maximum at a lower frequency (less than 1 GHz) and is quickly decreased to the higher frequencies. Second has a maximum at the about 20 GHz or at a higher frequency and is a quasi flat with slow decreasing to the lower frequencies. We conclude that these may be resulted by the emission of a halo-jet structure of AGNs.

References

Kovalev, Yu.A. (1991). Soobshenia SAO, **64**, 60 (in russian).
Morabito, D.D. *et al.* (1986). Astron. J. **91**, 1038.

SIMULTANEOUS OBSERVATIONS OF THE CONTINUUM EMISSION OF THE QUASAR 3C 273 FROM RADIO TO γ-RAY ENERGIES

G. G. LICHTI, T. BALONEK, T. J.-L. COURVOISIER, N. JOHNSON,
M. McCONNELL, C. von MONTIGNY, W. PACIESAS, E. I. ROBSON,
A. SADUN, C. SCHALINSKI, A. G. SMITH, R. STAUBERT, H. STEPPE [+],
B. N. SWANENBURG, M. J. L. TURNER, M.-H. ULRICH, O. R. WILLIAMS

ABSTRACT. From June 15 to 28, 1991 the Compton Gamma-Ray Observatory (CGRO) observed the radio-loud quasar 3C 273. All four CGRO instruments detected radiation from this quasar in their relevant energy range (from 20 keV to 5 GeV). Simultaneous and quasi-simultaneous observations (spanning the time period May 27 - July 25, 1991) by instruments sensitive at other wavelengths have also been obtained. The data from all these observations spanning the frequency range from $\sim 10^9$ Hz to $\sim 10^{26}$ Hz were collected and analysed. The resulting energy-density spectrum is shown in the figure below. It shows two maxima, one in the UV, another one at low-energy γ-rays which have nearly the same strength (the corresponding luminosities per decade of frequency for $H_0 = 60 (km/s)/Mpc$ are $3.2 \cdot 10^{46}$ erg/s and $2.7 \cdot 10^{46}$ erg/s, respectively). A break of the spectrum at low-energy γ-rays is evident. From a detailed analysis a break energy of (2 ± 1.5) MeV could be derived corresponding to a frequency of $(4.8\pm 3.6)\cdot 10^{20}$ Hz. The observed spectral break between X- and γ-rays is ~ 0.8, much higher than the value of 0.5 predicted by some models. A more detailed paper on this topic is in preparation (Lichti et al.).

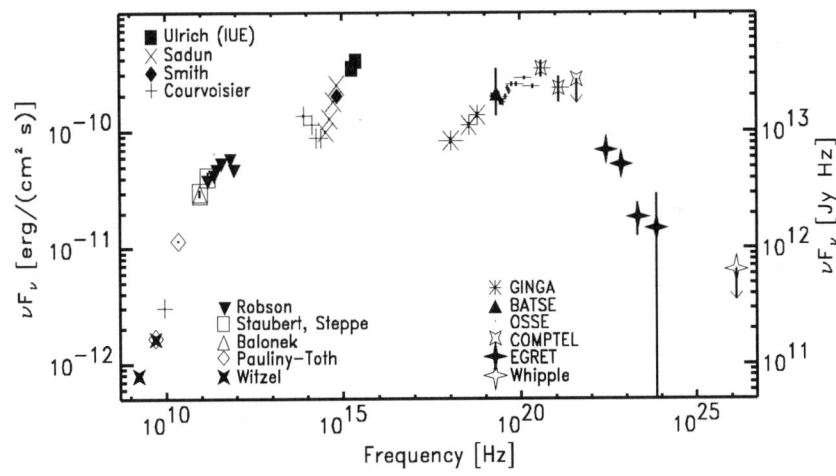

Energy-density spectrum of 3C 273 from quasi-simultaneous observations.

Acknowledgements

We thank I. Pauliny-Toth for supplying his radio data on 3C 273. We are grateful to T. Weekes for giving us the upper limit at ultra-high energies prior to publication.

[+] Hans Steppe perished on June 30, 1993 during a hiking tour in the Alps.

SIMULTANEOUS LONG TERM MILLIMETRE AND SUB-MILLIMETRE MONITORING OF BLAZARS*

S.J. LITCHFIELD, J.A. STEVENS and E.I. ROBSON **
Centre for Astrophysics, University of Central Lancashire, Preston PR1 2HE, UK

We have monitored a sample of 17 blazars at 375, 270, 230 and 150 GHz with the James Clerk Maxwell Telescope (JCMT). We also have 230, 150, 90 GHz data from IRAM and SEST, and 37, 22 GHz from Metsähovi Radio Research Station, Finland. The data shows a range of flaring time-scales, from months for e.g. 0851+202 to years in the case of e.g. 1641+399. Superimposed shorter time-scale variability ('flickering' behaviour) is seen in many sources from our sample. We make no conclusive statements on whether BL Lacs as a class are preferentially variable on shorter time-scales. All sources show variability on many time-scales regardless of classification.

For sources showing well defined flares of longer duration, there is a clear (Spearman rank-order) correlation between increasing 270 GHz flux and a flatter 375–150 GHz spectral slope (calculated by a weighted least squares fit). This can be interpreted as injection of a flatter electron distribution (which subsequently steepens), or may be related to the passage of the synchrotron self-absorption turnover past the viewing window. There is a clear separation of populations between the BL Lacs and the OVVs: the BL Lacs have a flatter mean spectral index (a 1-d K-S test gives a 98.8% confidence result). This agrees with the results of Gear et al. 1993; see also Stevens et al. in these proceedings.

The shock model of Marscher & Gear (1985) predicts three phases of flare evolution, with direct observational consequences in terms of flare amplitudes and delays in attainment of maximum flux. Specifically, in the initial growth phase the flare amplitude should increase as frequency decreases, and there should be no lag between frequencies. In subsequent evolutionary phases the amplitude of the flares should remain roughly constant or fall with decreasing frequency and there should be lags between frequencies. We compare this with our sources calculating lags using the DCF of Edelson & Krolik (1988), and find reasonable agreement between the model and observations.

References

Edelson, R. A., & Krolik, J. H., 1988, Astrophysical Journal, 333, 646
Gear, W. K. et al. 1993, Monthly Notices of the RAS, in press
Marscher, A. P, & Gear, W.K., 1985, Astrophysical Journal, 298, 114

---

* Stevens et al. 1993, in preparation
** JAC, 660 N. A'ohōkū Place, University Park, Hilo, Hawaii 96720, USA

NEAR IR - OBSERVATIONS OF THE JET IN M87

M. NEUMANN, K. MEISENHEIMER, H.-J. RÖSER and M. STICKEL
Max Planck - Institut für Astronomie,
Königstuhl 17, D - 69117 Heidelberg

The central region of M87 in the K' (2.1 μm) - band was imaged using the 2.2m telescope on Calar Alto, Spain. In addition to the jet our K' exposure (Fig. 1) shows an arclike structure ~ 24 arcsec southeast of the nucleus. This is the IR counterpart to the feature detected by Stiavelli et al. (1992) in the optical and to the brightness maximum of the southeastern radiolobe designated as "theta" (Hines et al. 1989). Its radio to IR spectral index is $\alpha = -0.88 \pm 0.01$ ($S_\nu \propto \nu^\alpha$), consistent with the radio spectral index of $\alpha = -0.84 \pm 0.07$ (Salter et al. 1989) of the whole radio lobe measured between 5 GHz and 230 GHz.

A second object can possibly be assigned to a bright radio feature at the southern limb of the eastern radio lobe called "eta" by Hines et al. (1989). This apparent association will be investigated with further IR observations of higher spatial resolution and higher SNR.

A detailed photometric analysis of the jet will be presented elsewhere.

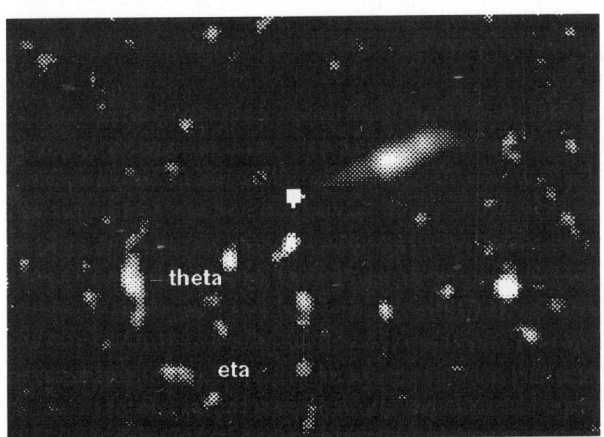

Fig. 1. 540 sec K' - exposure of M87 after subtraction of a smooth galaxy model. North is up and east to the left. The jet has a length of ~ 20 arcsec. Intensity range is $2\sigma \ldots 507\sigma$ for the jet and $2\sigma \ldots 39\sigma$ for all other features including theta.

References

Hines, D.C., Owen, F.N., Eilek, J.A., 1989, ApJ 347, 713
Salter, C.J., Chini, R., Haslam, C.G.T., Junor, W., Kreysa, E., Mezger, P.G., Wink, J.E., Zylka, R., 1989, A&A 220, 42
Stiavelli, M., Biretta, J., Møller, P., Zeilinger, W.W., 1992, Nature 355, 802

SIMULTANEOUS OBSERVATIONS OF SEYFERT 1 GALAXIES WITH IUE, ROSAT AND GINGA.

A. Orr[1], R. Walter[2], T. J.-L Courvoisier[1], H.H. Fink[2], F. Makino[3], C. Otani[3] and W. Wamsteker[4]

[1] *Observatoire de Genève, Chemin des Maillettes 51, CH-1290 Sauverny, Switzerland*
[2] *MPE, Giessenbachstraße, D-85748 Garching b. München, Germany*
[3] *ISAS, 3-1-1 Yoshinodai, Sagamihara, Kanagawa 299, Japan*
[4] *IUE Observatory, European Space Agency, P.O. Box 54065, 28080 Madrid, Spain*

Abstract. Simultaneous observations of 8 Seyfert 1 type AGN (Fairall-9, Mrk 590, NGC 4051, 3C 273, NGC 5548, Mrk 841, Q 1821+643 and 3C 390.3) obtained with *ROSAT* and *IUE* (RIASS program), and for 5 sources (Fairall-9, NGC 4051, 3C 273, Mrk 841 and Q 1821+643) with *Ginga*, have been analysed with the aim of describing the UV to soft X-ray spectral component in these sources.

1 Soft X-ray Excesses

The *ROSAT* spectra were fitted with models constructed by adding a hard X-ray power law (when possible, the *Ginga* observations) and a soft X-ray excess component (power law, modified black body thin disk, thermal bremsstrahlung or black body). We find that: 1) for a given source, the resulting integrated soft X-ray excess fluxes obtained with different models are very similar, with deviations smaller than 20%; 2) a soft X-ray excess is detected in all sources except 3C 390.3; 3) the presence and the strength of a soft X-ray excess does not depend on the luminosity of the object.

2 Ultraviolet to Soft X-ray Bump

We compared the simultaneous soft X-ray and UV fluxes and found that *the fluxes of the soft X-ray excesses integrated in the rest frame of the source are correlated to the UV fluxes*. We described the spectral shape of the bump by simultaneously fitting the UV and soft X-ray spectra. Our parametrizations are able to reproduce the observed fluxes. The maximum of the UV to soft X-ray bump component depends on the description, but not on the object. When described by a thermal cut-off the maximum temperature is $T_{max} \sim 5\ 10^5$ K. This "universal" maximum temperature can be compared to different emission processes, but its constancy among objects varying in luminosity by a factor of 10^4 cannot be understood easily by current accretion disk, comptonization or reprocessing models.

These results are published in: R. Walter et al., submitted to A & A; R. Walter, H. H. Fink, A & A, 274:105 (93).

UV VARIABILITY OF A LARGE SAMPLE OF AGN

S.PALTANI
Institut d'Astronomie de l'Université de Lausanne, CH-1290 Chavannes-des-Bois

and

T. J.-L. COURVOISIER
Observatoire de Genève, CH-1290 Sauverny

The IUE-ULDA database (Version 3.0, complete until the end of 1988) contains about 3500 spectra of more than 500 AGN of different classes (see Courvoisier and Paltani, IUE-ULDA Access Guide No. 4 A & B, ESA SP 1153 A & B 1992). We selected 67 objects for which the variability properties could be investigated. For each object we estimated the amplitude of the variability by means of the standard deviation $\sigma(\lambda)$ of the flux at different frequencies divided by the mean flux.

We characterise the variability as a function of wavelength using a linear representation:

$$\sigma(\lambda) \approx \nabla_\lambda(\sigma_f) \cdot \frac{\lambda - 2000 \text{Å}}{1000} + \sigma_{f,2000}$$

$\sigma_{f,2000}$ is the variability at 2000 Å expressed as a fraction of the mean flux and $\nabla_\lambda(\sigma_f)$ is the wavelength gradient of the variability in fraction/(1000 Å).

The main conclusions of this study are:

- $\sigma_{f,2000}$ (the variability) is found to be larger than 20 % in about 80 % of the objects. Variability appears to be the rule in the UV domain.
- Most of the objects have a negative value of $\nabla_\lambda(\sigma_f)$, indicating that the variability is larger at small wavelength than at long wavelength.
- We tried to compare the distributions of $\sigma_{f,2000}$ and $\nabla_\lambda(\sigma_f)$ among different subclasses of AGN: BL Lac objects, Seyfert 1 galaxies, radio-quiet and radio-loud quasars. The distribution of $\sigma_{f,2000}$ is probably different for BL Lacs objects from the one of the other subclasses, but mean variabilities are within 1 σ from each other. Mean values of $\nabla_\lambda(\sigma_f)$ are also within 1 σ from each other.
- We examined the relation between $\sigma_{f,2000}$ and the luminosity. We found that there is a weak trend for the variability to decrease with the increase of the luminosity. The correlation coefficient indicates that the trend has a 90 % probability of being real. In this case, the mean variability changes from 42 % to 31 % from low to high luminosity AGN over 5 decades. Even if this correlation is confirmed, the small value of the slope makes the models in which the luminosity is due to a number of independent events (supernovae, outbursts) inconsistent with the observations. In these models the variability should decrease with the square root of the luminosity.

A more complete study, based on ULDA 4.0, complete until the end of 1991, is in progress and will appear in a paper.

MULTIWAVELENGTH ENERGY DISTRIBUTIONS OF ULTRALUMINOUS IRAS GALAXIES

D.RIGOPOULOU and A.LAWRENCE
Physics Dept., QMW, Mile End Rd., London E1 4NS, U.K.

Abstract. Ultraluminous IRAS Galaxies (ULG's) have luminosities comparable to quasars while their space density is much higher than that of active galaxies. Much debate has centered around the origin of the energy source for these objects, whether this is a burst of star formation or a hidden quasar. The sample studied here is the Sanders et al.(1988) sample, 10 objects with $L_{FIR} \geq 10^{12} L_\odot$. We discuss our new observations at X-ray and submm wavelengths together with other published data for some of the objects [1]. Some useful ideas can be gained from comparisons of the shape of the spectral energy distributions (SED's) of the ultraluminous objects with other "archetype" objects such as typical starbursts i.e. M82 or type 2 AGN i.e. NGC1068.

05189-2524:
ROSAT data for this galaxy revealed a strong source with a count rate of 0.039 cnts/sec. We fitted the spectrum with a simple model, the absorbed power law, with an energy index $\alpha=1.28\pm0.37$. The SED of 05189-2524 is compared with that of NGC 1068. Emission from the starburst disk dominates at longer wavelengths while at 20 microns and shortward nuclear emission starts to dominate spectrum. Judging from the shape of the SED of 05189-2524 we may argue that a similar situation holds for this galaxy although the relative strength of the two components is not the same as in NGC 1068. This might be a more evolved system where the starburst episode has started to fade away and the Seyfert nucleus starts to emerge through the dust. We must also stress the fact that the emission lines of 05189 look like those of Seyfert 2 [2] and we find a relatively strong X-ray emission.

14348-1447:
Our ROSAT data give evidence for a weak source. We achieved a 3 σ detection at the radio position. The multiwavelength energy distribution of 14348-1447 is compared with that of ARP220. The shapes of the two SED's are different especially in the optical-UV area. The rise in the SED of ARP 220 at the optical-UV region which is due to the hot newly formed stars is not observed in the SED of 14348. Different degrees of obscuration or a different energy source might be the cause of this discrepancy.

15250+3609:
From our ROSAT data there is evidence for a 2.5 σ source at the radio position with a low count rate of 2.7×10^{-3} counts/sec. The object is compared with the starburst galaxy M82. The overall shape is pretty similar but the ratio $S_{60\mu m}/S_{1KeV}$ is different. The above ratio is equal to 3.12×10^8 for M82 (consistent with the values found by [3]) compared to 2.48×10^9 for 15250+3609. If a starburst episode is responsible for the amounts of energy released at IR wavelengths then these bursts must occur at the inner parts of the galaxy and are hidden from our sight of view by massive molecular clouds and gas, the latter absorbing all the soft x-rays.

References

1. Rigopoulou, D., and Lawrence, A. 1993, in preparation.
2. Sanders, D.B., Soifer, B.T., Elias, J.H., Madore, B.F., Matthews, K., Neugebauer, G., and Scoville N.Z. 1988, Ap.J., 325,74
3. Weedman, D.W. 1987 in Star Formation in Galaxies, ed. C.J.Persson (Washington, DC:US Government Printing Office)

MULTIFREQUENCY OBSERVATIONS OF THE JET OF 3C 273

H.-J. RÖSER, K. MEISENHEIMER and M. NEUMANN
Max-Planck-Institut für Astronomie, Königstuhl 17, D69 117 Heidelberg, Germany

and

R.G. CONWAY
The Nuffield Astronomy Laboratories, Jodrell Bank, Macclesfield, Cheshire SK11 9DL, U.K.

Previously we have studied the jet of the quasar 3C 273 at optical and radio frequencies ([3], [1]). In our first set of X-ray data with 17.2 ksec integration time obtained with the ROSAT HRI, the jet is easily visible extending out from the bright quasar core. The total number of counts in the jet lies in the range 200 to 300, depending on the details of the background model. This corresponds to an X-ray flux $f_\nu(2.9 \times 10^{17}\,\text{Hz}) = 65\ldots140\,\text{nJy}$ (lower limit, synchrotron radiation $\alpha = -0.8\ldots$ upper limit, bremsstrahlung $\alpha = 0$, NHI $= 1.8 \times 10^{20}\,\text{cm}^{-2}$), in good agreement with the value derived from the EINSTEIN observations [2].

Are the X-rays of synchrotron origin? From an extrapolation of the radio/optical spectra we expect synchrotron X-ray flux only from the inner region around knot A (13″ from the core). All other parts show a synchrotron cutoff spectrum. The maximum level consistent with data from other wavebands would be $\approx 30\,\text{nJy}$ in the ROSAT band, well below what is observed. Thus, to explain the observations as synchrotron emission would require a completely different component, with vanishing intensity at frequencies below the X-ray regime.

The X-ray flux also exceeds the expected inverse Compton-scattered radiation both from the microwave background photons and from the quasar's own (isotropic) radiation field. Only a more contrived model (e.g. beaming) may be compatible with the observations. For a thermal bremsstrahlung scenario, densities of the order $1\,\text{cm}^{-3}$ in a 5×10^7 K plasma are required, resulting in a mass of $\approx 10^9\,M_\odot$ in a volume corresponding to a diameter of about 1″. These values are difficult to reconcile with the very low rotation measure at radio wavelengths [1].

Thus, none of the straightforward emission models seems to reproduce the X-ray properties and we have to think of a more refined picture to model this source.

References

1. R.G. Conway, S.T. Garrington, R.A. Perley, and J.A. Biretta. Synchrotron radiation from the jet of 3C 273. II. – The radio structure and polarization. *Astron. Astrophys.*, 267:347–362, 1993.
2. D.E. Harris and C.P. Stern. X-ray emission associated with the jet in 3C 273. *Astrophys. J.*, 313:136–140, 1987.
3. H.-J. Röser and K. Meisenheimer. The synchrotron light from the jet of 3C 273. *Astron. Astrophys.*, 252:458–474, 1991.

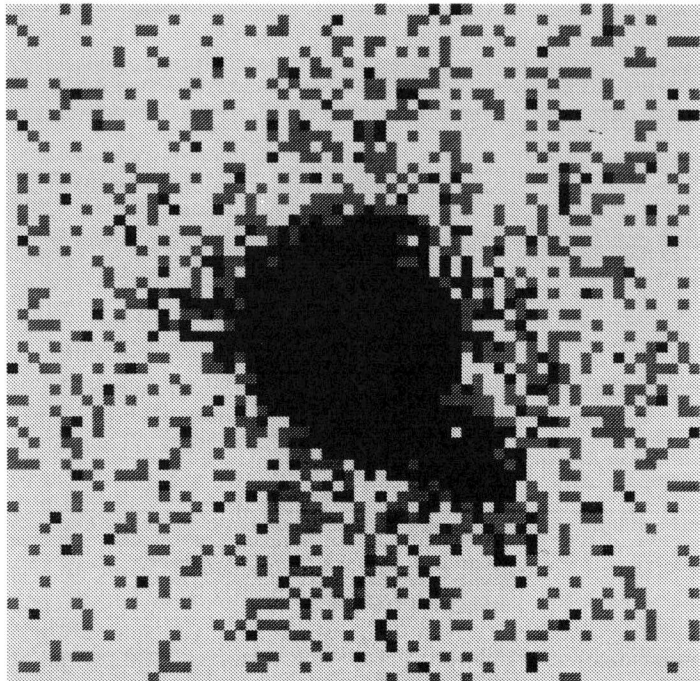

Fig. 1. ROSAT HRI image of the quasar 3C 273. In comparison with the EINSTEIN image the jet is clearly visible at PA 222, extending out from the bright quasar core. The field shown here is $63'' \times 63''$ in size and the HRI resolution is $2''.5$.

THE NATURE OF THE NIR EMISSION FROM SEYFERT 1 GALAXIES: NGC 4051

M. SALVATI and G. CALAMAI
Osservatorio Astrofisico di Arcetri, L. E. Fermi 5, I-50125 Firenze, Italy

L.K. HUNT
CAISMI, L. E. Fermi 5, I-50125 Firenze, Italy

G. DEL ZANNA, E. GIANNUZZO, F. MANNUCCI and R.M. STANGA
Dipartimento di Astronomia e Scienza dello Spazio, L. E. Fermi 5, I-50125 Firenze, Italy

M. KIDGER
IAC, Via Lactea, E-38200 La Laguna, Tenerife, Spain

and

W. WAMSTEKER
VILSPA, Ap.do 50727, E-28080 Madrid, Spain

The near infrared emission (NIR) of radio quiet active galactic nuclei is commonly attributed to dust, absorbing and re-emitting some of the nuclear UV radiation. The dust can survive only at distances larger than a few light months, and any rapid variation of the nuclear radiation, on timescales of a day or less, will be completely smeared out.

In order to test this paradigm, we have undertaken a high time resolution monitoring program of nearby Seyfert 1 galaxies in the K band (2.2 μ). The target galaxies are all known from the EXOSAT long looks to be highly variable in the X rays on timescales from minutes to days: if a fraction of the NIR emission is direct nuclear radiation, NIR variations on similar timescales are expected. The observing technique, described in detail in Hunt et al. (1992, A&A 257, 434), allows a relative photometric accuracy of 1 to 2% according to the quality of the night, and a sampling time of 10 to 20 minutes. In about 14 hours of total exposure, no *rapid* variation larger than 4% has been detected.

We have instead detected a *slow* outburst in NGC 4051, which flared and dimmed by more than a factor of 2 in 6 months: such a timescale is compatible with realistic dust distributions. However the UV flux, although sparsely sampled, is less variable than the near infrared. The available data suggest that dust reprocessing of variable UV radiation from the nucleus is an acceptable explanation only if the light travel time to the dust is shorter than the UV variability timescale, and the NIR to UV ratio remains nearly constant. Alternatively, the NIR could be direct non–thermal emission from the nucleus. In either case NGC 4051 does not fit into the same scheme as Fairall 9, NGC 3783, and NGC 1566, where the UV variations are the dominant ones, and are reprocessed into NIR variations after a non–negligible delay. A supernova in a heavily obscured starburst is excluded.

THE ORIGIN OF AGN-IR EMISSION

L. G. STENHOLM
Astronomical Observatory
Uppsala University
Box 515
S 751 20 Uppsala
Sweden

and

Swedish Defence Research Establishment
S 172 90 Sundbyberg
Sweden

Abstract. Advanced multi dimension radiative transfer calculations for an AGN source with a dust disk shows that the AGN-IR emission can be due to reradiation from heated dust in a thick disk. The models produces a weak silicate feature for a wide range of physical conditions, in agreement with the observations.

Key words: Radiative Transfer, Dust, IR, AGN

1. Summary

The AGN spectrum from IR to X-rays is included in the model. The model is divided in two parts a central engine producing the optical to X-ray emission and a disk of small single sized silicate grains. The radiative transfer is solved 3 dimensionally. The temperature structure is derived in the grey approximation and the spectrum is covered by 80 frequencies ranging from millimeter to X-rays. One model needs 20 to 30 CPU hours on a SPARC 2.

The models show that geometrically tick disks with inner boundary densities varying with at least two orders of magnitude and with varying radial dependencies, produces spectra which are in reasonable agreement with observed AGN. The silicate feature is weak due to a substantial optical thickness in the near-IR. The silicate feature varies little with the axilal angle of the line of sight, as long as disk volumes directly heated by the central source can be seen directly by the observer. This also coincidences with the angular range in wich the central engine can be seen without substantial obscuration.

2. Acknowledgements

This work was supported by the Swedish Natural Science Research Council.

ON THE CORRELATION OF THE FIR AND RADIO RADIATION OF SPIRAL GALAXIES

H.M.Tovmassian

INAOE, AP 51 y 216, 72000, Puebla, Pue, Mexico

The comparison of the Byurakan classification of central parts of galaxies with the results of their radio observations showed that there is a definite correlation: the radio emission is more often observed in those galaxies which were suggested to have active nuclei.

In this talk the data on the IRAS bright galaxy sample (Soifer et al. 1987) are compared with the Byurakan classes of spiral galaxies. Total number of spiral galaxies in the Byurakan list brighter than $13.^m5$ is 525 and that of in the IRAS list common to the Byurakan catalogue is 128. The ratios of relative numbers of galaxies of the IRAS galaxies of different Byurakan types to that of galaxies from Byurakan list, normalized to unity for galaxies of type 3, are shown below:

Byurakan type	3	1	4	2	5	2s
IRAS/Byurakan	1.00	1.38	2.89	3.33	3.80	6.67

Thus galaxies of types 4, 2, 5 and 2s, which have optical manifestations of nuclear activity and more often have radio emission, appreciably more often have also FIR radiation. The FIR luminosity itself is depending on the degree of activity expressed by their Byurakan types. The mean values of the $m_V - m_{60}$ colors of galaxies are changing from -1.32 for type 3 galaxies to -0.42 for 2s galaxies when moving from type 1 towards types 3, 4, 5, 2 and 2s. The same trend is seen in in the case of $m_V - m_{100}$ colors: from -0.33 to 0.35 (The FIR stellar magnitudes are in arbitrary units).

Thus the FIR emission of spiral galaxies is definitely correlated with their Byurakan types. And since the radio emission of spiral galaxies is also correlated with their Byurakan types (Tovmassian 1982) *it is natural to expect that the FIR and radio radiations should also be correlated.* It shows that the Byurakan classification based on the optical appearances of the central parts of galaxies revealed the galaxies with an active nuclei and star burst processes.

The author is indebted to CONACYT for support of this work.

REFERENCES

Sofier B.T., Sanders D.B., Madore B.F., Neugebauer G., Danielson G.E., Elias J.H., Lonsdale C. & Rice W.L. 1987. ApJ, **320**, 238.

Tovmassian H.M. 1982. Astrofisika, **18**, 25.

THE SOFT X-RAY EXCESSES OBSERVED WITH ROSAT

R. WALTER, H.H. FINK
Max Planck Inst. für extraterrestr. Physik, Giessenbachstraße, D-85748 Garching b. München

Abstract. The properties of the soft X-ray excesses of bright Seyfert 1 galaxies and Quasars are described using the observations obtained with the PSPC (0.1-2.4 keV) detector of the XRT telescope aboard ROSAT during the ROSAT all sky survey (RASS). The sample consists of 58 Seyfert 1 type AGN detected with more than 300 counts during the RASS and observed at least once with *IUE*.

The soft X-ray photon indices of our sample members range from 1.6 to 3.4 in a wide distribution ($<\Gamma> = 2.50$, $\sigma = 0.48$). The width of the distribution is considerably larger than the mean statistical uncertainty on the individual spectral slopes ($\sigma = 0.33$). Excepting for IC 4329A and Mrk 766, the mean contribution of absorbing cold matter intrinsic to the Seyfert galaxies of our sample to the absorbing column density is less than $10^{20} cm^{-2}$. In IC 4329A and Mrk 766 intrinsic absorbtion is observed at soft X-ray. Both sources are also strongly reddened by dust.

An excess of soft X-ray flux is detected in 90% of the sources above the exptrapolation of the hard X-ray power law. It can be shown that the PSPC spectral slope is a measure of the strength of the soft X-ray excess. If the reddened sources are excluded, a correlation appears between the strength of the ultraviolet blue bump and the soft X-ray photon index (figure 1). The ratio of the ultraviolet to infrared fluxes and the ultraviolet spectral slope are also related to the strength of the blue bump. The observations are compatible with a model where most of the spectral variations arising among the sources studied are driven by the strength of the bump component, which varies by a factor of 100 from object to object. A bump model consisting of a power law with a high energy cutoff at 80 eV can fit most of the sources. In any case, the spectral energy distribution of the ultraviolet to soft X-ray bump is characterised by $\nu F_{\nu(1375\text{Å})} = (1-5) \int_{\epsilon>150eV} F_\epsilon d\epsilon$.

Further reading : R. Walter, H. H. Fink, 1993, A&A 274, 105.

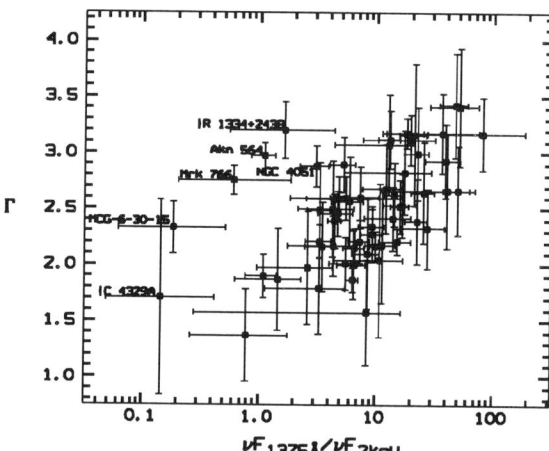

Fig. 1. Relation between the spectral slope Γ measured in the PSPC band and the logarithm of the ratio of the ultraviolet (1375Å) and the 2 keV flux. Named sources are reddened.

Poster Contributions:
Emission Processes

THE STRUCTURE OF RELATIVISTIC MHD JETS

STEFAN APPL and MAX CAMENZIND
Landessternwarte Königstuhl, 69117 Heidelberg, Germany

An analytic model for a stationary force-free relativistic magnetized jet is presented [2]. In those models a poloidal current provides the tension that balances the internal pressure. Our ignorance of the detailed conditions in the collimation zone is accounted for by specifying the current distribution as an (essentially free) function of the flux surfaces. The asymptotically cylindrical Grad-Schlüter-Shafranov (GSS) equation is solved for a nonlinear current distribution, $I(\Psi) = c(1 - \exp(-b\Psi))$ which covers both diffuse ($b \ll 1$) and sharp ($b \gg 1$) pinches, and weak (small c) and strong (large c) currents.

It is the first equilibrium that properly accounts for the relativistic effects: the electric field reduces the collimating tension of the toroidal magnetic field. Consequently, relativistic jets are more difficult to collimate only by means of their self-generated fields [1]. It is therefore proposed that in the innermost part jets are collimated with the help of external pressure. In the course of propagation the ambient pressure falls off, and the jet expands until it is held again by the tension of its toroidal magnetic field. The light cylinder (LC), which in the force-free aproximation coincides with the Alfvén surface is a critical point in the GSS equation. Regularity already completely determines the structure of the jets. Prescribing boundary conditions outside the LC overdetermines the problem. This probably also holds for the 2D case and would explain why attempts to numerically solve the pulsar equation produce kinks at the LC (e.g. Michel 1982, Rev. Mod. Phys. 54, 1).

The plasma flows within the nested magnetic surfaces. Beyond the light cylinder it decouples from the magnetic field, and rotates essentially with constant angular momentum. Sufficiently far from the light cylinder plasma rotation becomes unimportant and the material streams with a constant poloidal velocity u_p.

The jet structure is characterized by a current-carrying core and a current-free envelope. The core radius R_c is the relevant scale for any physical quantity in the jet. It is related to the poloidal jet velocity u_p and the light cylinder R_L by $R_c = u_p R_L \simeq 10^{16} \gamma_{jet} M_9$ cm. It is shown how the form and the strength of the current determine the shape and the profile of the jet. A sharp pinch with a strong current generally produces narrower jets. A minimum current is required for an equilibrium to exist. Less current is needed for sharp pinches and relativistic velocities.

The jet luminosity is in electromagnetic form, and it is not homogeneously distributed across the jet. The Poynting flux is instead concentrated at the core radius. This provides an off-axis energy reservoir, which, if it can be tapped, could be responsible for the origin of the VLBI knots. The rest frame magnetic field in the core is essentially parallel to the axis, and toroidal in the envelope. Its pitch angle is given by $\tan\theta' = R/R_c$.

The propagation of a jet is treated as a sequence of quasi-equilibria along the jet's path. Self-confinement and confinement by external pressure is considered. We conclude that not much current is leaking during the propagation, but it tends to concentrate in the innermost flux surfaces. This situation applies most likely to the powerful jet sources.

References

1. Appl, S., Camenzind, M. 1993a, A&A **270**, 71
2. Appl, S., Camenzind, M. 1993b, A&A **274**, 699

ANISOTROPIC EFFECTS IN THOMSON CASCADE MODELS OF ACTIVE GALACTIC NUCLEI

MATTHEW G. BARING*
NASA Goddard Space Flight Center,
Lab. for High Energy Astrophysics, Code 665,
Greenbelt, MD 20771, U.S.A.

Abstract. CGRO observations of high energy γ-rays from active galactic nuclei (AGN) suggest that beaming may be prevalent in these sources. Anisotropic effects in AGN continuum spectra produced by the inverse Compton scattering mechanism are outlined, in particular the resulting spectral breaks and modification of spectral indices that depend strongly on the electron anisotropy and the observational viewing perspective.

1. Thomson Cascade Spectra

The recent EGRET observations of energetic γ-rays coming from blazars, together with the Whipple detection of TeV γ-rays from Mkn421 has bolstered support for the hypothesis that many active galactic nuclei are strongly beamed sources. These high energy data contrast the turnovers seen (or inferred) by other instruments aboard the Compton Gamma-Ray Observatory (CGRO) in Seyferts, indicating either that they are possibly a distinct class of AGNs, or perhaps that observational orientation may provide the difference between hard and softer emitters. Prior to the launch of CGRO, pair cascade models of AGNs largely focussed in Seyfert-type spectra, generating the continuum through inverse Compton scattering of UV photons by isotropic relativistic electrons. The new observations indicate that this electron injection is probably beamed in many sources; some of the consequences of such anisotropic injections are outlined here.

Moderate degrees of beaming suppress two-photon pair production to the point that the cascades can be mediated purely by Compton scattering. In such situations, collisions between relativistic electron beams and soft photons can generate flatter spectra at X-ray energies near the direction of beaming but with much lower intensity than for isotropic electrons. Spectral steepening by an index of order unity occurs at soft gamma-ray energies, for beaming angles of a few degrees, above which the intensity approximates the isotropic case. Steeper, softer spectra are seen at large viewing angles relative to the beam axis. Solutions obtained to the photon kinetic equation, in cases where multiple scatterings of the photons occur, reveal that electron beaming can generate spectra steeper than the familiar isotropic case, since energy gains in successive upscatterings are reduced once the photons become beamed quasi-parallel to the electrons. These electron anisotropy effects may well help to determine AGN beaming and viewing angles from refined spectral models of hard X-ray and gamma-ray emission.

* NAS-NRC Research Associate.

FREE-FREE EMISSION AND THE BIG BLUE BUMP

RICHARD BARVAINIS
MIT/Haystack Observatory, Westford MA 01886 USA

Abstract.
Optically thin thermal emission may be a viable alternative to optically thick accretion disks for the origin of the optical/UV Big Blue Bump. In this contribution I summarize a number of points favoring the optically thin, or free-free, hypothesis. The free-free model is consistent with the most stringent quasar size and luminosity constraints provided sources are composed of a large number of very small and dense cloudlets or filaments.

The most luminous continuum feature in quasars is the optical/UV Big Blue Bump (BBB). The origin of the BBB has been ascribed to thermal emission from geometrically thin, optically thick accretion disks, or to synchrotron from the vicinity of a supermassive black hole. It has also been suggested more recently that the BBB might be optically thin thermal emission, i.e. predominantly free-free.

The poster outlined here summarized observational evidence favoring the free-free mechanism, and briefly discussed some theoretical constraints on the nature of the sources. Detailed arguments and references can be found in Barvainis (1993, ApJ 412, 513). Some points in favor of the free-free hypothesis are:

1) Optical/UV slope matches mean of quasars ($< \alpha_{\rm opt} > \approx -0.3$)
2) Compatible with thermal dust emission in IR.
3) Predicts low polarization.
4) T_B can be high ($> 10^5$ K).
5) Optically thin continuum in Mrk 509 inferred from ultrabroad wings.
6) Optical/UV variations predicted to be broadband, simultaneous.
7) Reprocessed X-rays can be consistent with optically-thin opt/UV.
8) Lyman edge very weak for hot sources ($T \sim 10^6$ K).

Some theoretical constraints: Source temperatures must be in the range 10^5 – few $\times 10^6$ K in order to satisfy both the optical/UV spectral shape and the soft X-ray excess.

In order to satisfy quasar size/luminosity constraints (in particular a μlensing constraint on Q2237+0305), sources must be composed of a large number of very small and dense clouds ($r \sim 10^7$ cm, $n \sim 10^{15}$ cm^{-3}). Past critics of the free-free model have considered single uniform sources, which certainly do not work.

For free-free sources with $T \sim 10^6$ K, the Lyman edge is intrinsically weak. There is evidence from UV slopes that high luminosity sources have high temperatures. The Lyman edge has been searched for only in moderate to high L sources, where T may be high and the Lyman edge intrinsically weak.

RADIATIVE SIGNATURES OF NEUTRON BEAMS IN AGN

IFEANYI E. EKEJIUBA
*Department of Physics & Astronomy, Georgia State University, Atlanta, Georgia 30303, USA
and Department of Physics, Federal University of Technology, Yola, Nigeria*

Abstract. The escape of relativistic neutrons from their production region can have various consequences for the morphology of active galaxies. The phenomena of luminosity gaps and radio jet lighting in extragalactic radio sources (EGRSs) fit into the model that employs relativistic neutrons as the vector for particle and energy transport out of the central engines of AGNs. The central radio gaps reveal themselves as regions of relativistic neutron beam transport. The relativistic neutrons, which decay in flight after traveling for $\sim 10^3 \gamma_n$ s, produce secondaries which are responsible for the radio jet lighting and the associated phenomena in EGRSs.

1. The Radiative Signatures of Relativistic Neutrons

The accretion disc region favors the production of relativistic neutrons with the following reactions as the dominant production mechanisms (Ekejiuba & Okeke 1993a): $^4He + \gamma \rightarrow 2p + 2n; p + p \rightarrow p + n + \pi^+ \rightarrow p + n + e^+ + \nu; p + \gamma \rightarrow n + \pi^+ \rightarrow n + e^+ + \nu$ where p, n, π, e and ν refer to proton, neutron, pion, electron and neutrino, respectively. The relativistic neutrons produced escape beyond the accretion zone and travel ballistically after an initial collimation of the precursor proton beam and decay after traveling for $\sim 10^{13} \gamma_n$ s to produce protons, electrons and neutrinos: $n \rightarrow p + e^- + \bar{\nu}$.

The interactions of the secondary particles with entrained ambient matter and magnetic field produce the observed radiations in AGNs. Three channels of interaction dominate: (a) synchrotron emission of the decay-produced electrons, with $\gamma_e \sim 2k\gamma_n$, where k is the neutron- to electron-energy conversion efficiency (e.g., Ekejiuba & Okeke 1993b); (b) interaction of the decay-produced relativistic protons with entrained ambient (cold!) protons to liberate relativistic electrons, which spiral in the background magnetic field producing synchrotron emissions, with $\gamma_e \sim 60\gamma_p$ where γ_p is the energy of the leading proton; and (c) elastic Coulomb interactions between relativistic protons and non-relativistic or mildly relativistic electrons during which the electrons are boosted to the relativistic range adequate for synchrotron radiation, with $\gamma_e \sim 10\gamma_p^{1/3}$ (e.g., Ekejiuba & Okeke 1993b). Due to their neutrality, the relativistic neutrons create a luminosity gap (the so-called *central radio gap*) in the region where the collimated neutron beam is traveling before the neutrons decay.

2. References

Ekejiuba, I.E. & Okeke, P.N. 1993a, ApJ, 413, 110.
Ekejiuba, I.E. & Okeke, P.N. 1993b, ApJ, in press

INTERACTION OF JETS WITH CLOUDS IN EXTRAGALACTIC RADIO SOURCES

V. FEDORENKO[1], A. ZENTSOVA[1], T. J.-L. COURVOISIER[2] and S. PALTANI[3]
[1] Ioffe Phys. Tech. Institute, St.-Petersburg, Russia
[2] Observatoire de Genève, CH-1290 Sauverny
[3] Institut d'Astronomie de l'Université de Lausanne, CH-1290 Chavannes-des-Bois

Several points indicate that extragalactic jets can interact with dense gaseous obstacles which occur on their ways. Examples of these interactions are the knotty structure of the radio and optical jet in M 87 [1] and in other objects [2]. These observations have been interpreted by Blandford & Königl [3] in terms of collision of a jet with supernova remnants. We have reanalysed this idea taking into account new observations and improvements in the theory of diffusive shock acceleration. We find that the model [3] requires a very high supernova birthrate (~ 1 SN/year), which is not observed. It is more probable that the "obstacles" are formed by the stellar winds from the red giants. We estimate that the value of the magnetic field is $\sim 10^{-5}$ G in the interaction region (r=1kpc) (paper in preparation).

Seyfert galaxies are an other class of objects with jet-cloud interactions. Their NLR and their VLA images nearly coincide [4]. We have developped a model in which the non-relativistic jets (or anisotropic winds) collide with multiple NLR clouds. The resulting shocks produce particle acceleration and subsequent synchrotron emission. We have shown that such a model naturally explains the observational correlation between radio and optical properties of those objects [5].

We made a theoretical investigation of the hydrodynamics of 2-D jet-cloud interactions using a generalisation of [3]. We consider the penetration of the BLR clouds inside the jet. Depending on the physical conditions, the clouds may move through the jet without any deviation, be reflected, or be captured and entrained by the jet. These conditions are examined analytically and numerically (paper in preparation).

Finally, we consider a model for compact (VLBI) radio sources. The main idea is that the jets interact with BLR clouds, so that multiple (about 10^6) shock waves permanently exist. Particles are accelerated in these shocks and form "tracks" (like cometary tails). We calculate the kinetic properties of the particles in the "tracks", and their synchrotron emission. Along these lines, we interpret the superluminal compact radio sources (paper in preparation).

References
1. J. A. Biretta, C. P. Stern, D. F. Harris., 1991, Astrophys. J., 101, 1632
2. W. H. Keel, 1988, Astrophys. J., 329, 532
3. R. D. Blandford, A. Königl, 1979, Astrophys. Lett., 20, 15
4. A. S. Wilson, A. C. Willis, 1980, Astrophys. J., 240, 429
5. V. N. Fedorenko, A. S. Zentsova, 1991, Sov. Astron., 35(1), 7

A STUDY OF GAMMA SPECTRAL BREAK IN AGN

A. MARCOWITH, G. HENRI and G. PELLETIER
Laboratoire d'Astrophysique
Observatoire de Grenoble BP53X
F38 041 GRENOBLE Cedex FRANCE

Since its launch, CGRO has detected more than 20 γ-ray emitting AGN, most of them associated with powerful, radio-loud, flat-spectrum objects, exhibiting VLBI superluminal motions. In the case of 3C279, the huge value of the apparent luminosity ($\sim 10^{48} erg.s^{-1}$) and the variability time-scale of a few days (Hartmann et al., 1992) gives a very large compacity $\ell_{app} \simeq 200$, that is, the medium should be completely thick to γ-rays. This contradiction can be explained if the γ-rays originate from a relativistic jet pointing at a small angle with respect to the line of sight(Maraschi et al., 1992). However, the still large value of compacity suggests the existence of an inner, more compact region where pair production can take place efficiently (Henri et al., 1993). This supports the so-called "two-flow" model, where the superluminal motion is attributed to the expansion of a relativistic pair plasma heated by a MHD jet from an accretion disk (Sol et al., 1989). Hence we propose to interpret the spectral break observed in many objects around a few MeV (Lichti et al., 1993) by an opacity effect due to *photon- photon absorption by pair production*.

Assuming that photons of energy $\varepsilon_1 m_e c^2$ interact only with X photons of energy $m_e c^2/\varepsilon_1$, and for a cylindrical jet of transverse radius $r(z)$, the opacity to pair production is $\tau_{\gamma\gamma}(z) \simeq \sigma_T \dot{n}(1/\varepsilon_1) r^2(z)/c\varepsilon_1$ where $\dot{n}(1/\varepsilon_1)$ is the density of photons produced per second by Inverse Compton process at energy $1/\varepsilon_1$. The total luminosity can then be evaluated as $L_\gamma \simeq m_e c^2 \int_{z_\gamma}^{\infty} \varepsilon_1 \dot{n}(\varepsilon_1, z) S(z) dz$, where z_γ is defined by $\tau_{\gamma\gamma}(z_\gamma) = 1$. The jet can become optically thin to γ-rays only by a decrease of soft photons and/or pair flux along z. Then one obtains $L_\gamma = L_{thin} \varepsilon_1^{-\Delta\alpha}$, where $\Delta\alpha = \frac{s-1}{2}$ for a power-law distribution of electron with an index s and an exponential decrease of soft photons. The break energy is equal to 511 keV blueshifted by the Doppler factor. Pair annihilation photons can produce a bump at this break energy. A crucial prediction of the model is that variability at high energy should lag the low energy one, in contrast with optically thin models.

References

Hartmann, R. C., et al.: 1992,*ApJ Lett.*, **385**, L1
Lichti, G. G., et al.: 1993,*these proceedings*
Henri, G. , Pelletier, G., and Roland, J.: 1993,*ApJ Lett.*, **404**, L41
Maraschi, L., Ghisellini, G., and Celotti, A. : 1992,*ApJ*, **397**, L5
Sol, H., Pelletier, G., and Asseo, E.: 1989,*MNRAS*, **237**, 411

THE HARD X-RAY REFLECTION ON COLD MATTER

E. JOURDAIN AND J.P. ROQUES
CESR, BP4946, 31029 Toulouse, Cedex, France

1. INTRODUCTION

We have simulated the reflection on cold matter (1,2,3&5) in a variety of situations to determine which informations can actually be inferred from observations. We modelled a semi-infinite plane parallel medium of solar abundance matter (4), semi-isotropically illuminated by a X/γ ray source. The spectra are calculated from a Monte-Carlo method without any approximation in the cross-sections. Θ is the angle over which the reflecting matter is seen (90°=face-on), Θ=all means a spatially integrated spectrum. Fref is the ratio of the reflected over direct component.

2. DISCUSSION

The presence of a reflected component introduces a hump between 10 and 500 keV (1,2,3). More precisely, our study has shown that:
* small values of Θ, Fref or Nh (10^{-24} cm^{-2}) can explain the no detection of this phenomenon even when cold matter is present in the vicinity of the source.
* we can obtain rather identical spectra for several couples (Fref, Θ) (fig.1).
* the 2 breaks occur at 10 keV and 30 keV almost independently of the parameter values. However, the 2nd slope change is smooth and can mimic a thermal law. Moreover, a comptonized primary spectrum may hidde this feature.
* between 30 and 60 keV, the composite (direct+reflected) spectrum seems to have a slope close to that of the primary one. A deficit of photons occurs below and above this energy range.
* above 30 keV, the reflection effect is hard to interpret due to the curved shape of the spectrum. A broken power law fit will give results depending on the energy band chosen.

3. CONCLUSION

It is very likely that cold matter and reflection exist in AGNs central regions. It is thus crucial to get data from 2 to ~ 60 keV with the same instrument and a good sensitivity to determine how the spectrum is affected and deduce informations on the source environment. However, as Fref and Θ act similarly, their values will be generally model dependent. The primary spectrum must also be investigated.

References:

1) George I.&Fabian A.,1991,MNRAS,249,352
2) Guibert P.&Rees M.,1988,MNRAS, 233, 475
3) Lightman A.& White T., 1988, APJ, 335, 57
4) Morrisson & McCammon,1983,APJ,270, 119
5) Pounds K. A.et al., 1990, NATURE, 44, 132

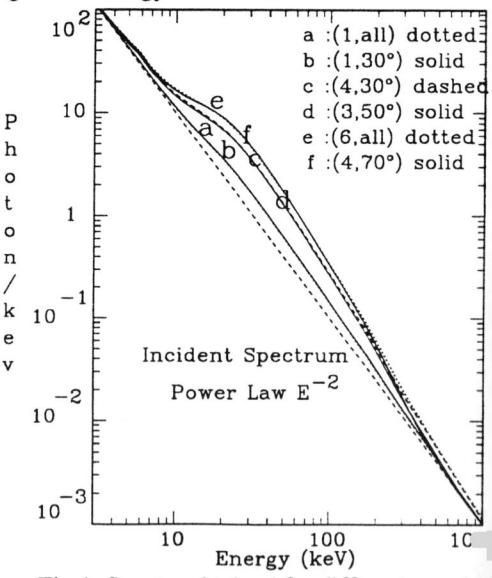

Fig.1: Spectra obtained for different couples (Fref, Θ) as indicated for each label

a :(1,all) dotted
b :(1,30°) solid
c :(4,30°) dashed
d :(3,50°) solid
e :(6,all) dotted
f :(4,70°) solid

Incident Spectrum Power Law E^{-2}

QUASI-SPHERICAL ACCRETION ONTO THE BLACK HOLE : THE VIRIAL REGIME

V S BEREZINSKY
Laboratori Nazionali del Gran Sasso, INFN, 67010 Assergi (AQ), Italy

and

I LAPIDUS*
Institute of Astronomy, University of Cambridge, Madingley Road, Cambridge CB3 0HA, UK

1. The Model

We are studying the slow ($\dot{m} < 1$) spherical accretion of a gas onto a supermassive ($M \approx 10^8 \, M_\odot$) black hole in the presence of a strong tangled magnetic field. In the core with radius $2.5 \, r_g < r < 10 \, r_g$ protons are isotropized due to scattering in magnetic field, but are not thermalized since the characteristic time of pp - Coulomb scattering is less than the infall time. A proton moves in the electron gas with a friction due to pe - scattering, gradually transferring energy to electrons. The standard equations for the proton gas allow the virial regime of accretion when the kinetic energy of the proton is a function of a distance only $E_k(r) = (2/5) \, m_p c^2 \, (r_g/r)$. The model is relevant to the slow subsonic settling of matter onto the black hole, as, for example, in the upstream region after the shock standing at a distance $r \sim 20 \, r_g$ (Mészáros and Ostriker 1983). Electrons are thermalized and are cooling predominantly by bremsstrahlung radiation. For $\dot{m} \lesssim 0.1$ the core is transparent for bremsstrahlung photons. In agreement with Park (1990) the $e^+ e^-$ - pair production is found to be insignificant. The equilibrium between the energy release in pe - scattering and the bremsstrahlung radiation results in the almost isothermal core with the temperature $T_e \approx 4 \, m_e$, which slightly increases towards the inner edge of the core. The only role of magnetic field is the isotropization of the proton gas, as the synchrotron radiation is strongly self-absorbed. Therefore the model is insensitive to the precise value of H.

The spectrum is characterized by a strong emission with $E_\gamma \sim 2 \, MeV$, and by a broad line at $E_\gamma \lesssim 70 \, MeV$ which is a sign of pion production in pp - collisions. The predicted fluxes are detectable by COMPTEL and EGRET detectors of the Compton GRO.

2. References

Mészáros, P., and Ostriker, J. P. 1983, ApJ, 273, L59
Park, M.-G. 1990, ApJ, 354, 83

* The Royal Astronomical Society Sir Norman Lockyer Fellow

THE EMISSION SPECTRA OF RADIOWEAK QUASARS.
I. THE FIR EMISSION

MARTINA NIEMEYER and PETER L. BIERMANN
Max-Planck-Institut für Radioastronomie
Auf dem Hügel 69
D-53121 Bonn, Germany

The far-infrared (FIR) spectra of radioweak quasars show a steep rise from the mm to the FIR wavelengths with an spectral index often $\alpha \geq 2.5$, and a slow decrease beyond the FIR with $\alpha \simeq -1$. A comparison of the FIR luminosity with both radio and X-ray emission demonstrates, that only the active nuclei can provide the energy to heat the dust. We propose that the heating is due to the central engine via relativistic particles. These particles originate from a putative source on the symmetry axis close to the central engine and diffuse through a tenuous galactic halo to heat dusty molecular clouds in a disk configuration. We demonstrate that this mechanism can produce the FIR spectra with reasonable assumptions, and compare them with observations (Chini et al. 1989). We examine the diffusion through the halo and the energy deposition in the disk for two types of source (pointsource, line source, intensity decrease with $z^{-\beta}$) and a diffusion coefficient, dependent on the spherical radius r ($D \propto r^{\gamma}$). We obtain the total energy deposition in the disk as a function of the disk radius. For the calculation of the FIR spectra, one must determine the dust temperature. We construct a heat balance that takes into account the absorbed and emitted emission of energy. The calculated dust temperature distribution depends on two parameter: the radial dependence of the diffusion coefficient and the z-dependence of the line source. Then we calculate the FIR spectra of a disk in the wavelength range 10 to 1300 μm. We determine the luminosity of a finite disk, where the limits are determined by the maximal and minimal temperature ($T_{D,\max} = 1500$K given by the dust destruction, $T_{D,\min} = 20$K given by a transition to dust heating from young stars in the extended disk).

We find that for a diffusion coefficient in the region above the disk, which scales linearly with the spherical radius r, a line source, which intensity decreases by z^{-1} and a total source luminosity of 3 to 10 times the observed infrared luminosity, we can reproduce and interpret the spectra of radioweak quasars in the mm to the near infrared range.

References

Chini R., Kreysa E., Biermann P.L.: 1989, *Astronomy and Astrophysics* **219**, pp. 87–97
Niemeyer M., Biermann P.L.: 1993, *Astronomy and Astrophysics* in press

THE NUCLEUS OF OUR HOME GALAXY: A REMNANT OF AN ACTIVE OR A STARBURST GALAXY?

Leonid M. OZERNOY
Institute for Computational Sciences and Informatics
George Mason U., Fairfax, VA 22030-4444, USA
and
Lab. for Astronomy and Solar Physics, NASA/GSFC
Greenbelt, MD 20771, USA
e-mail: ozernoy@hubble.gmu.edu

Abstract. To resolve the above dilemma, two essentially different approaches are undertaken: First, a new, detailed analysis of the entire radiation spectrum of Sgr A*, from radio band up to gamma-rays, is reviewed, which enables us to put substantial constraints on the mass of a putative black hole. The derived upper limit turns out to be too small to allow the black hole to serve as an 'engine' for a Seyfert galaxy. Second, analyses of recent data on the 10 KeV gas in the central 200 pc and on star formation history at the Galactic center both make a star burst the likely episode in a recent past. Taken together, the two approaches seem to indicate that the history of the central part of our Galaxy can be better described as that of a starburst, rather than a Seyfert, galaxy.

Keywords: The Milky Way galaxy; Sgr A*; Sgr A West; IRS 16; starburst galaxies

1. The Problem

Since none of the galactic nuclei is as close to us as the nucleus of the Milky Way galaxy, it is tempting to find out what key element(s) is lacking to result in its pronounced difference from both active galactic nuclei and those of starburst galaxies: unsufficient gas supply, absence of a supermassive black hole, or both?

2. Constraints on the Central Black Hole Mass

Several different methods have been recently implemented to constrain or evaluate the mass of a putative black hole at the Galactic center:

2.1. Dynamics of the Gaseous Rotating Disk in the Innermost, Central Parsec.
Provided that the so called 'mini-spiral' (Sgr A West) is a density wave generated by a hydrodynamical instability, the spiral morphology can put interesting constraints to the shape of the gravitational potential assumed to be due to a central point mass plus an extended stellar nucleus [1]. Analytical techniques and numerical simulations both show that the presence of a jump in surface density or/and a kink in rotational velocity of the circum-nuclear disk describe successfully the morphology of the mini-spiral. This enables us to constrain significantly the parameters of the gravitational potential. The observed pattern might be a superposition of the first three spiral modes, and its comparatively large pitch ($|dr/d\varphi| \gtrsim 0.1$ pc/rad in the region $r \sim 1$ pc) indicates that the total enclosed mass within $r \lesssim 0.2$ pc does not exceed $(1.7 - 4.7) \times 10^5 M_\odot$. After subtracting the inferred core mass of the central

stellar cluster found to be 5×10^5 $(10^{\pm 0.3})M_\odot$ (ref. [2]) one concludes that there is no point mass concentration exceeding $10^5 M_\odot$.

2.2. Stellar Collisions in the Core Dominated by a Central Black Hole. Presence of a $\sim 10^6 M_\odot$ BH in the stellar core would enhance considerably star-star collisions, especially in the density spike around the BH. Lower velocity collisions result in building-up massive stars whereas high-velocity collisions are disruptive. Gas released from the latter is largely bound to the BH and is gradually accreted onto the hole. It turns out that the accretion luminosity would exceed the upper limit of the central point source's bolometric luminosity of $\sim 10^{39}$ erg/s unless its mass $M_{bh} \lesssim 10^5 M_\odot$ [3].

2.3. Production of Cosmic Rays and Gamma-rays by a Wind-Accreting Black Hole. An unavoidable source of accretion by the galactic-center BH is the wind from IRS 16, a nearby group of hot, massive stars. Since the density and velocity of the accreting matter are known from observations, the accretion rate is basically a function of the putative BH mass only. Provided the available estimates of the high rate of accretion are correct, one can expect a shock to form around the BH, and efficient particle acceleration will occur. We calculate the expected cosmic-ray flux produced by the BH as a function of its mass, M_{bh}. The inferred contribution to the cosmic ray pool contradicts the existence of a supermassive, $10^6 M_\odot$ black hole, and may rule out even much less massive ones [4]. The accompanying γ-ray flux produces a similarly stringent limit when compared to the recent *EGRET* measurement of the Galactic center.

2.4. Production of X-rays by a Wind-Accreting Black Hole. The accelerated protons, by colliding with the ambient gas, produce charged pions which, in turn, are responsible for the production of relativistic electrons and positrons. Synchrotron radiation of the latter, in the magnetic field of the accreting wind, results in X-ray emission which expected flux is measurable by available X-ray facilities. This flux turns out to be a function of M_{bh} as well. Comparing these results with recent X-ray observations by *SIGMA/ART-P*, which show Sgr A* to have a relatively low activity level, one can conclude [5] that the putative BH at the Galactic center cannot have a mass greater than about $10^4 M_\odot$.

2.5. Radiation Spectrum of the Accreting Wind. Approximate analytical expressions are obtained [6] to estimate the BH mass from its spectral luminosity, assumed to result from the Bondi accretion of the IRS 16 wind. To fit the radio spectrum of Sgr A*, $M_{bh} \simeq 5 \times 10^2 M_\odot$ is required. The results of the above calculations are in contradiction with ref. [7] where a similar although numerical aproach have been used resulting in a different value of $M_{bh} \approx 10^6 M_\odot$. Such a high value of M would make the temperature of the accretion flow only marginally relativistic for electrons instead of providing them the Lorentz-factor $\gamma \sim 10^2 - 10^3$ apparently required by the radio data. The bremsstrahlung emission, that would be the dominant in X-ray band, turns out to be at odds with the recent X-ray data which seem to indicate that the X-ray counterpart to Sgr A* has a nonthermal, power law spectrum.

A comparatively low value for the mass of a putative black hole at the Galactic center, constrained/derived above by four different methods, is consistent with the upper limits to Sgr A* mass, which the author [8] found earlier by four other methods – (i) tidal disruption of stars by a BH; (ii) displacement between Sgr A* and IRS 16; (iii) electron-positron pair production by a BH via electromagnetic cascade; and

(iv) wind diagnostics of Sgr A*.

3. The Galactic Center as a Scaled-down Version of Starburst Galaxies

3.1. Ultra Hot Gas at the Galactic Center. Several years ago both continuum [9] and Fe-line emission observations [10] indicated the presence of a high-temperature ($T \sim 10$ KeV) plasma within the central 100 pc of the Galaxy, but there have only been a few attempts so far to duscuss the origin and implications of these results. Some authors [11] objected to this interpretation since that gas would be highly non-stationary; instead, they argued for underlying discrete X-ray sources. However, what at first sight might seem unusual turns out to be just one of the numerous sequences of a recent starburst in the central region of the Galaxy [12]. In particular, that hot, rarefied gas appears to be an interior of the 200-pc expanding molecular ring at the Galactic center.

3.2. Starburst Interpretation. In ref. [12], a single explosion, due to either multiple supernovae or a very massive star, was considered. A Sedov-Taylor solution able to fit the observed radius, temperature, and density of the bubble requires $\sim 10^{54}$ erg for the energy release some $(4 - 8) \times 10^3$ yrs ago in a very rarefied environment. Another possibility to form a hot superbubble would be *sequential SN explosions* [13]. In this case, a very low initial gas density required by a single-explosion model, would be naturally produced by a wind from an OB association or/and first SN explosions. In this solution, the derived SN rate is as low as 0.04 yr^{-1}, which is, by a factor of 5, less than in a single-explosion solution, though the total energy release during the characteristic time $R/c_s \sim 10^5$ yr is $\sim 10^{54}$ ergs, *i.e.* about the same. Nevertheless, the sequential SN model results, as a whole, in less restrictive requirements.

3.3. The Galactic Center Hot Bubble As a Scaled-Down Version of Those in Other Starburst Galaxies. It is instructive to compare the parameters of the Galactic center hot bubble with the M 82 and NGC 253 cases. In those objects, the observed features such as a large wind bubble, powerful far-IR, X-ray and other emissions, very young supernova remnants all seem to be formed in consequence of an ongoing starburst. In all three cases, the temperatures of the hot gas are almost the same, and this is a clear and simple signature of the starburst model. Indeed, if the wind flows into a standing shock, the temperature of the wind, during the Sedov-Taylor phase, is given by: $T_s \approx (3/16)(m_H/k)v_w \approx 10^8 (E_{SN}/10^{51} \text{ ergs})(M_{ej}/10 \text{ M}_\odot)^{-1}$ K (e.g. [14]). Evidently, this implies that T_s may vary in a rather limited range until the gas is subject to cool. Meanwhile the other parameters of starbursts differ substantially. As compared to the Galactic center, NGC 253 and M 82 have a much larger mass of gas in their hot superbubbles and, respectively, a much larger total thermal energy as well (by a factor of 10 and 10^3, correspondingly). At a larger SN production rate, both NGC 253 and M 82 have a much larger age ($\sim 10^7$ yrs) of ongoing starburst.

3.4. Isotope Ratios. Observations of various isotopic species toward the starburst nuclei of M 82 and NGC 253 have indicated that $^{12}C/^{13}C$, $^{16}O/^{18}O$, and $^{18}O/^{17}O$ ratios are $\sim$40–50, $\sim$150–200, and $\gtrsim 8$, which significantly differ from the corresponding ratios of 20, 250, and 3.5 in the Galactic center [15]. This seems to be consistent both with a much smaller massive star formation rate at the Galactic center and with a smaller age of the starburst. In the light of all this, a rather small number of young stars in the central parsec [16] should not be a surprise.

4. Conclusions: AGN/Starburst Dilemma

A detailed analysis outlined above indicates that a well-spread belief into a supermassive black hole at the center of the Milky Way galaxy has a rather shaky basis. The current constraints seem to rule out the BH mass as large as $10^6 M_\odot$. Therefore, the main reason why our Galaxy is not a Seyfert one seems to be the lack of a proper 'engine', rather than an unsufficient mass supply. Incidentally, the mass flux onto the central part of the Galaxy is estimated to be $\sim 10^{-2} M_\odot$/yr [17], which is rather large even on AGN scale. However, a lion fraction of this flux is neither going to 'feed the monster' (otherwise it would result in accretion luminosity much exceeding the available upper limit) nor provide a fresh fuel for continuous star formation (otherwise the total mass of stars within the central parcec would appreciably exceed the observed one). The current mass flux onto the galactic nucleus might be just a transient phenomenon responsible for feeding the starburst.

To sum up, the main conclusions of the present report are the following:
(i) The Galactic nucleus does not seem to be a prototype for activity in galactic nuclei; (ii) A recent starburst at the Galactic center has neither resulted from, nor was induced by, a supermassive black hole; and (iii) Rather than being a 'dormant' version of AGN, the Galactic nucleus is a scaled-down version of starburst nuclei.

Acknowledgement. This report was prepared during my visit to École Normale Supérieure, Paris. It is my pleasure to thank the staff of the Laboratoire de Radioastronomie Millimétrique, and especially M. Signore, for their hospitality.

References

[1]. Ozernoy, L., Blitz, L., Fridman, A., Khoruzhii, O., & Lyakhovich, V. 1993, Bull. AAS **25**, 891; Ap. J. (to be submitted)
[2]. Eckart, A. et al. 1993, Ap. J. (Letters) **407**, L 77
[3]. Ozernoy, L.M. 1993, In The Nuclei of Normal Galaxies: Lessons From the Galactic Center. R. Genzel & A. Harris, Eds. (in press)
[4]. Johnson, P.A., Ozernoy, L.M., & Protheroe, R.J. 1993, In Proc. 23rd Internat. Cosmic Ray Conf. (in press)
[5]. Mastichiadis, A. & Ozernoy, L. 1993, Ap. J. (submitted)
[6]. Ozernoy, L.M. 1993, In Back to the Galaxy. S.S. Holt and F. Verter, Eds. AIP Conf. Proc. **278**, 69
[7]. Melia, F. 1992, Ap. J. **387**, L25
[8]. Ozernoy, L.M. 1992, In Testing the AGN Paradigm. S.S. Holt et al., Eds. AIP Conf. Proc. **254**, 40; Idem, **254**, 44
[9]. Skinner, G.K. et al. 1987, Nature, **330**, 544
[10]. Koyama, K. et al. 1990, Nature, **343**, 148
[11]. Markevitch, M., Sunyaev, R., & Pavlinsky, M. 1993, Nature, **364**, 40
[12]. Ozernoy, L., Titarchuk, L., & Ramaty, R. 1993, In Back to the Galaxy S.S Holt and F. Verter, Eds. AIP Conf. Proc. **278**, 73
[13]. Ozernoy, L., Titarchuk, L., & Ramaty, R. 1993, Ap. J. (to be submitted)
[14]. Chevalier, R.A. & Legg, A.W. 1985, Nature, **317**, 44
[15]. Henkel, C. & Mauersberger, R. 1993, Astr. Astrophys., **274**, 730
[16]. Tamblin, P. & Rieke, G.H. 1993, Ap. J. (in press)
[17]. Blitz, L., Binney, J., Lo, K.Y., & Ho, P.T.P. 1993, Nature, **361**, 417

DUST DISTRIBUTION IN IRAS SEYFERT GALAXIES

M. G. PASTORIZA,[*] CHARLES BONATTO[*], EDUARDO BICA and
T. STORCHI-BERGMANN[*]
Instituto de Física, UFRGS, C.P. 15051, Porto Alegre, RS, Brasil

Observational evidences of dust in the nuclear region of AGNs are substantial (Rudy 1984, ApJ, 284, 33; Jones et al. 1984, PASP, 96, 692). The ionization cones observed in several Seyfert galaxies has been interpreted as shadowing effects by a dust obscuring torus which hides the broad emission line region (BLR) and the central source (Wilson 1992; Storchi-Bergmann, Mulchaey and Wilson 1992, ApJ 395, L73). A large sample of optical and far-IR data for IRAS Seyfert galaxies has been analysed together with dust emission models (Bonatto and Pastoriza 1993), where it has been concluded that the same dust emission model can be applied to both Seyfert types. In order to further study the effects of dust in the spectra of active galactic nuclei, we have obtained spectrophotometry of 21 IRAS Seyfert galaxies in the range 3500-7200 Å and analyse them in conjuction with their IRAS fluxes. The stellar population type is derived from comparisons with normal galaxy templates using dilution effects in the K CaII line as discriminator. For 55% of the sample the population is of late type. For the rest, blue continua due to recent star formation and/or power-law may amount up to 30% at 4000Å. We conclude that the bulge stellar populations of IRAS Seyfert galaxies are similar to those of normal spirals, except that they are more reddened by $E(B-V)_i \sim 0.20$. Population-subtracted emission line ratios indicate on average stronger reddening for the narrow-line region ($E(B-V)_l \sim 0.8$. From photoionization models a power-law index for the ionizing continuum $\alpha=1.5$, and a metallicity larger than solar are obtained. The most luminous IRAS galaxy of the sample (IRAS555) is discuss in detail: in order to be compatible with the observed IRAS fluxes and the optical stellar continuum, the ionizing continuum must be reddened by $A_V > 10$ magnitudes. Consequently a dust structure in this galaxy appears to be increasingly affecting stars and gas towards the galaxy center.

---

[*] Visiting Astronomer at the Cerro Tololo Interame rican Observatory of the National Optical Observatories, operated by AURA und er contract with the National Science Foundation

AUGER PROCESS FOLLOWING 1S–PHOTOIONIZATION: NE III AND NE IV LINE PRODUCTION

D. PETRINI and F.X. DE ARAÚJO
Observatoire de Nice,
B.P. 229, 06304 Nice Cedex 4
France

Removal of a 1s electron from atomic neon by highly energetic photons, gives rise mainly to the $1s2s^22p^6$ state of singly ionized neon. This state c decay mostly by ejection of an Auger electron to form various excited terms of Ne^{+2}. This process dominates largely the radiative $K\alpha$ decay ($\omega_K = 0.018$). The main route to Ne^{+2} is $1s^22s^22p^4\ ^1D$ (61%) and there is no Auger decay to the ground final term (Krause et al, 1971). Shake up processes, i.e excitations of outer shells accompagning the 1s–electron ejection, produce mainly $1s2s^22p^5$ np ^2S states (n=3,4), 5% relatively to all processes (Krause, 1971). These np states radiationless decay producing numerous Ne^{+2} states. Shake off processes, i.e. ionizations of outer shells by the photoelectron, are not negligible, nearly 14% relatively to all processes (Krause, 1971) for photons with energy greater than 1 keV. Shake off process leads by radiationless transitions to three or four times ionized neon. The $1s2s2p^5$ state is strongly favoured by shake off process and decay to the main configurations of $Ne^{+3}\ 1s^22s^m2p^n$ (Krause, 1971). In an analagous way, the case of oxygen presents the same interesting features but there is actually a lack of experimental and theoretical data (Caldwell and Krause, 1993, Petrini and Araújo, 1993).

In the study of active galactic nuclei, where high energetic photons are largely present, no interest has been devoted to these features. A thin plasma excited by a soft X–ray source (energy greater than 1 keV) will produce directly, by the double process 1s–photoionization followed by Auger decays, Ne III, Ne IV and Ne V allowed and forbidden lines. Their relative intensities will depend basically on Auger rates, transition probabilities and radiative cascades. It has to be noted that these relative intensities have a weak dependency on incident photon energies.

References

Caldwell C.D. & Krause M.O., 1993 (submitted to Phys. Rev. Letters)
Krause, M.O., 1971, J. Phys. (Paris) **32**, C4–67
Krause, M.O., Carlson T.A. & Moddeman W.E., 1971, J. Phys. (Paris) **32**, C4–139
Petrini, D. & Araùjo F.X., 1993, A&AS (to be published)

DO FLUID WAVES PROPAGATE IN MILDLY RELATIVISTIC THERMAL PAIR PLASMAS?

P. PIETRINI
Dipartimento di Astronomia e Scienza dello Spazio, Università di Firenze, I-50125 Firenze, Italy

and

J.H. KROLIK
Department of Physics and Astronomy, Johns Hopkins University, Baltimore, MD 21218, USA

Relativistic pair plasmas are implicated in the physics of the central regions of AGNs, and the observed variability of these sources can be related to the dynamics and changes in structure of these plasmas. To this respect a study of the behaviour of waves to which the pair plasma reacts as a fluid is quite relevant. We analyze the linear response to perturbations of a simple thermal mildly relativistic pair plasma system.

The equilibrium plasma is modeled as a spherical homogeneous cloud, where protons, electrons and pairs are at the same temperature, $\Theta \equiv kT/m_e c^2$, which is (trans-)relativistic for electrons and positrons; all the photons are generated within the cloud by bremsstrahlung and, possibly, Double Compton processes and subsequently they are Comptonized to form a Wien peak at the plasma temperature [3], [1], [2].

Differently from previous stability analysis [1], we allow for fluid motions in the perturbed state, to study the behavior of travelling waves as well. Therefore, we couple a momentum equation to proton continuity, pair balance and energy balance equations. Restricting the analysis to wavelengths λ, such that $\lambda \ll \lambda_{phot} \equiv 1/(n_p + 2n_+)\sigma_T$, the photon mean free path in the Thomson limit, no perturbation on the radiation field is to be considered.

Four distinct modes are sustained by the fluid-like plasma; two of them are isobaric modes, which turn out to be damped and basically non-propagating. The other two are sound waves, modified by pair and radiative effects. Their basic property is that they are generally strongly damped, on timescales shorter than the wave crossing time, and, under certain conditions, they do not propagate at all. This implies that it is difficult to propagate the information on pressure perturbations in the plasma and any pressure change which is created is smoothed out in a region whose dimension is of the order of a damping length around the origin of the pressure change itself.

1. Björnsson, G. & Svensson, R. 1991 MNRAS, 249, 177
2. Pietrini, P. & Krolik, J.H. 1994, ApJ, in press
3. Svensson, R. 1984 MNRAS, 209, 175

ELECTRON ENERGY DISTRIBUTIONS OF AGNS IN THE THIN SYNCHROTRON LIMIT. I. THE METHOD

LUIS SALAS, IRENE CRUZ-GONZÁLEZ and LUIS CARRASCO
Instituto de Astronomía UNAM, Apdo 70-264 Mexico DF 04510, Mexico

Abstract. We develop a new method called "Inverse Synchrotron Transform" (IST) to study the spectral energy distributions of AGNs. We demonstrate that it is possible to use Bayes Theorem for conditional probabilities to derive a self-consistent solution for the electron energy distribution (EEDs), starting from the observed spectral energy distributions (SEDs) and the assumption that the only physical process involved is thin synchrotron radiation. We test the IST method and find that it allows to distinguish among different EEDs that produce SEDs which nevertheless seem very similar. We apply the method to multifrequency simultaneous observations of AGNs (paper II, this conference).

Key words: active galactic nuclei, multiwavelength emission, emission processes

1. Results

Synchrotron emission has been associated with the non-thermal component in most extragalactic sources due to the observed polarization degree and power-law spectral characteristic. There is agreement that in the most violent sources the dominant component is non-thermal. The continuum distributions for the later sources were interpreted as due to synchrotron and synchrotron self-Compton components [3, 4]. In all these models it is assumed that the EED is a power-law. A power-law EED ($N(E) \propto E^{-p}$), is a convenient form since it represents a solution of the continuity equation under steady state conditions. However, there is now increasing evidence that there is no regularity in the high variability of some AGNs, suggesting one to explore non-stationary solutions for the continuity equation.

We propose to abandon the a priori assumption that N(E) is necessarily a power-law, relaxing completely the assumption of any particular form for $N(E)$. Instead, we calculate the most probable $N(E)$ which is consistent with the observed spectra. The spectrum of the emitted radiation by a thin synchrotron source $W(\nu)$, can be written as an integral transform from a space that is a combination of electron's energy and magnetic field (electron's critical frequency $\nu_c \propto B\gamma^2$), to radiation's frequency space, $W(\nu) = \int_0^\infty P_s(\nu,\nu_c)\Psi(\nu_c)\,d\nu_c$. The Kernel of this integral transform is normalized in such way that it can be interpreted as a conditional probability. This allows the use of Bayessian inference techniques to invert this equation through the Richardson-Lucy algorithm [5, 2] in a way analogous to the implementation given by Salas [6] for thermal radiation. Tests of the method with sets of artificial data, allow to distinguish between different EEDs, such as power-laws and superposition of delta functions, that produce similar spectral energy distributions, even when the input is moderately noisy.

References

1. Blumenthal,G.R., Gould,R.J., 1970, *Rev. Mod. Phys.* **42**, 237.
2. Lucy,L.B., 1974, Astronomical Journal **79**, 745.
3. Jones,T.W., O'Dell,S.L., Stein,W.A., 1974, Astrophysical Journal **188**, 353.
4. Jones,T.W., O'Dell,S.L., Stein,W.A., 1974, Astrophysical Journal **192**, 261.
5. Richardson,W.H., 1972, *J. Opt. Soc. Am.* **62**,55.
6. Salas,L., 1992, Astrophysical Journal, **385**, 288

A SIMPLE MHD MODEL FOR ONE-SIDED JETS

G. BODO, E. TRUSSONI
Osservatorio Astronomico di Torino, Italy

G. CHAGELISHVILI
Abastumani Astrophysical Obervatory, Georgian Republic

The one-sideness observed in several extragalactic jets is usually explained by the Doppler boosting effect in a pair of relativistic beams, moving almost parallel to the line of sight. However jets could be intrinsically asymmetric if magnetically driven from the accretion disk surrounding the central black hole. Wang et al. (1992) have shown that a combination of odd and even magnetic structures could lead to different Poynting fluxes in the two hemispheres of the inner nucleus. However, without any peculiar assumption, it can be shown that intrinsic one-sideness can be simply related to the symmetry properties of the MHD equations.

The magnetic configuration in the acceleration region can derive its poloidal component B_p from a primordial large scale magnetic field, convected into the central region. This poloidal component is antisymmetric with respect to the equatorial plane of the disk. The disk differential (along the vertical direction) rotation leads to the formation of a toroidal magnetic field (B_t), which changes sign across the equator (antisymmetric). On the other hand, the development of a turbulent dynamo process (Chagelishvili et al. 1990) can originate a further toroidal magnetic component ($B_{t,d}$) in the disk, much larger than B_t and symmetric with respect to the equatorial plane.

Outflow solutions in standard MHD models (Camenzind 1989) usually make use of an antisymmetric toroidal component of the magnetic field. As the MHD equations are invariant by changing the sign of both velocity and magnetic field, in this case ($B_{t,d} = 0$) we expect symmetric acceleration (i.e. double jets) from both sides of the equator. If conversely the dynamo works ($B_{t,d} \neq 0$), in one hemisphere the same previous outflow solution is found. On the other side of the disk outflowing solutions are also found but, as can be seen from the Bernoulli equation, either the flow in not magnetically driven, or the Poynting flux is opposite to the outflow velocity. In the former case a new acceleration process must be proposed, while in the second case we can expect the solution to be unstable.

References

Camenzind M. 1989, in Accretion Disks and Magnetic Fields in Astrophysics, ed. G. Belvedere (Dordrecht:Kluwer), 129
Chagelishvili G. et al. 1990, in Plasma Astrophysics, ESA SP-311, 273
Wang J.C.L., Sulkanen M.E. & Lovelace R.V.E. 1992, ApJ 390, 46

SPECTRA OF DISTANT QUASARS AND VERIFICATION OF POSSIBLE VARIATION OF FUNDAMENTAL CONSTANTS OVER COSMOLOGICAL TIME-SCALES

D.A. VARSHALOVICH and A.Y. POTEKHIN
*Department of Theoretical Astrophysics, A.F. Ioffe Institute of Physics and Technology,
Politekhnicheskaya 26, St.Petersburg, 194021, Russia*

Abstract. Constraints on possible variation rate of the fine-structure constant, $|\dot\alpha/\alpha| < 4 \times 10^{-14}$ yr^{-1}, and the electron-proton mass ratio $\mu = m_e/m_p$, $|\dot\mu/\mu| < 3 \times 10^{-13}$ yr^{-1}, over cosmological time scales are obtained from analyses of quasar spectroscopic data.

The problem of possible time variation of the fundamental physical constants was discussed by many authors (see below-cited papers for the references). An analysis of high-redshift quasar spectra makes it possible to check if the constants changed during $\sim 10^{10}$ yrs. Compared to previous works, we have performed more accurate analyses based on a more complete set of spectroscopic data, which enabled us to derive the most reliable upper limits on the possible time variation of the fine-structure constant $\alpha = e^2/\hbar c$ and the electron-proton mass ratio $\mu = m_e/m_p$.

The rate of the possible variation of α is estimated from a statistical analysis of the relative fine splitting $\delta\lambda/\lambda$ of 1414 pairs of doublet absorption wavelengths of alkalilike ions in quasar spectra at redshifts $z = 0.2 - 3.7$, compiled from data published in 1980–1992. If α were z-dependent, then the ratio $\frac{(\delta\lambda/\lambda)_z}{(\delta\lambda/\lambda)_0} = (\alpha_z/\alpha_0)^2$ would vary with z. However our analysis (Potekhin and Varshalovich 1993) revealed no statistically significant variation. The estimate of the variation rate reads

$$\alpha^{-1} d\alpha/dz = (-0.6 \pm 2.8) \times 10^{-4}. \tag{1}$$

At 95% significance level, an upper bound on this rate $|\alpha^{-1} d\alpha/dz| < 5.6 \times 10^{-4}$ is imposed. In the standard cosmological model with parameters $H_0 = 75$ km s^{-1} Mpc^{-1}, $q_0 = \frac{1}{2}$ ($\Omega = 1$) and $\Lambda_0 = 0$ this corresponds to the restriction $|\dot\alpha/\alpha| < 4 \times 10^{-14}$ yr^{-1}.

The rate of the possible variation of μ is estimated from a comparison of wavelengths λ for different electron-vibro-rotational lines of molecular hydrogen H$_2$ at $z = 2.811$ in the spectrum of quasar PKS 0528 − 250. If μ were z-dependent, then the ratio $\frac{(\lambda_i/\lambda_k)_z}{(\lambda_i/\lambda_k)_0} \approx 1 + K_{ik}(\Delta\mu/\mu)$ would deviate from unity. However our analysis (Varshalovich and Levshakov 1993) revealed no statistically significant deviation. The estimate of the variation is

$$(\Delta\mu/\mu)_{z=2.811} = (1 \pm 2) \times 10^{-3}. \tag{2}$$

The 95%-significance upper bound on the variation rate is $|\mu^{-1}د\mu/dz| < 1.8\times10^{-3}$. In the standard cosmological model with the above-mentioned parameters this corresponds to the restriction $|\dot\mu/\mu| < 3 \times 10^{-13}$ yr^{-1}.

References

Potekhin A.Y., Varshalovich D.A. 1993, Astron. Astrophys. (in press)
Varshalovich D.A., Levshakov S.A. 1993, Pisma v Zh. Eksp. Teor. Fiz. (Sov. Phys.–JETP Lett.) **58**, no. 4

Poster Contributions:
X-Rays and Higher Energies

X-RAY LOUD AGN WITH OPTICAL STARBURST OR SEYFERT 2 PROPERTIES

N. BADE
Hamburger Sternwarte, Gojenbergsweg 112, D-21029 Hamburg, Germany

and

S. SCHAEIDT
MPI für Extraterrestrische Physik, D-85740 Garching, Germany

Abstract. The Hamburg Sternwarte is conducting a large area identification program of ROSAT All Sky Survey (RASS) sources on objective prism plates taken with the Schmidt telescopes on Calar Alto and La Silla (Bade et al., 1992, MPE-Report 235, 377). With follow-up observations redshifts and a more detailed classification of the emission line spectrum of 284 AGN were derived until August 1992.

14 objects show not the typical spectra of QSO's and Seyfert 1 galaxies with a BLR. With our spectral resolution of 9 – 12 Å no difference in line width between the Balmer and the forbidden lines could be detected. From the emission line ratios these galaxies have to be classified as starburst or Seyfert 2 galaxies, but in some cases various line ratios give ambiguous results. Two objects (RX J05047−2541 and RX J10119−1635) have strong Balmer absorption lines and Balmer continuum emission down to the Balmer discontinuity indicating the classification as a post starburst galaxy.

In contrast to already known starburst or Seyfert 2 galaxies all objects (except RX J11178+2918) are X-ray loud with $L_X > 10^{43}\,\mathrm{erg\,s^{-1}}$ ($H_0 = 50\,\mathrm{km\,s^{-1}\,Mpc^{-1}}$, $q_0 = 0$), 5 objects showing even soft X-ray emission in excess of $\log(L_X) = 44$. The analysis of the RASS data of 5 X-ray brighter objects yielded a good description of the photon spectra with a steep power law (Γ around 2.5) modified only by low energy absorption from our own galaxy.

These results resemble the soft X-ray properties of Seyfert 1 galaxies proposing an AGN in these galaxies. This assumption is supported by the short variability time scale of one object (RX J17260+7431, Schaeidt, 1993, Ph. D. thesis) that can only be understood if the X-ray emission comes from a compact object.

In combination we have a contradictory classification for the two wavelength regions. Similar emission line galaxies have also been found by identifying serendipitous EINSTEIN sources (Stocke et al., 1991, ApJS 76, 813) but there the authors suspected a broad line base which could only be tested with higher quality spectra. A broad line base cannot be ruled out for all of our emission line spectra. Therefore we plan spectroscopy with higher resolution for these objects. Perhaps we are dealing with optically hidden AGN. Support for this assumption is given by the fact that two of the most X-ray luminous sources of this sample are also IRAS sources suggesting large amounts of absorbing dust around the possible AGN. Boller et al., 1993, have found 10 of these objects within the IRAS point source catalogue.

ROSAT X-ray Observations of Pair Mrk474/NGC 5682

Hongguang Bi
MPI für Extrater. Physik, 85748 Garching, FRG & Beijing Obs., Beijing 100080, CHINA

Abstract. Like radio and optical observations of AGN/Galaxy pairs (e.g. Carilli, C & van Gorkom, J. 1992, ApJ, 399, 373), X-ray observations of the pairs can reveal absorption at high energy band, and also, possible galactic gaseous X-ray emissions that are very useful in constructing halo models. ROSAT/PSPC X-ray spectra of 4 AGNs in the well-known pairs 3C232/NGC 3067, 3C275.1/NGC 4651, 3C309.1/NGC 5832 and Mrk474/NGC 5682 are reported here. Especially, we have detected an extragalactic HI of NGC 5862 in the Mrk474 spectrum.

Key words: AGN/Galaxy pairs, AGN X-ray spectrum, extragalactic HI

1 Observations

X-ray spectra of the AGNs are fitted by power law of $f(E)dE = A(\frac{E}{1\text{keV}})^{-\Gamma}dE$ plus a neutral hydrogen absorption using χ^2 method. The resulting column densities N_{HI} are consistent with absorptions in our galaxy in all AGNs except for Mrk474, partly because the 3 3CR QSOs do not have enough counts (≤ 300) for good spectral calculation. Mrk474 that has ~ 10830 total counts can be fitted by $N_{HI} = 3.61 \pm 0.23 \times 10^{20}$ cm^{-2} (comparing to the Galactic $N_{HI} = 2.01 \times 10^{20}$ cm^{-2}) and photon index $\Gamma = -2.31 \pm 0.06$. Note that the power law fitting has systematic positive residuals below 0.5 keV, another blackbody component is then added that results in even larger an HI of $5.5 \pm 1.0 \times 10^{20}$ cm^{-2}. Therefore, an extragalactic absorption of $N_H = 1.3 \sim 4.5 \times 10^{20}$ cm^{-2} is evidenced.

An optical image of the pair is in Fig.1a, the bar gives the angular scale of 1 arc minute (Arp, H., Baldwin, J.A. & Wampler, E.J. 1975, ApJ, 198, L3). Fig. 1b shows a deconvolved ROSAT X-ray image. The X-ray center is coincident with M474. The nearby galaxy NGC5682 has an average $N_H = 9.2 \times 10^{20}$ cm^{-2} within the linear diameter 13.8 kpc (Bottinelli, L., et al. 1984, A&AS, 56, 381). Assuming the HI radial density drops as $\sim r^{-2}$, we estimate that at the M474 impact distance 13 kpc it is about 3 times smaller that the average : 3.1×10^{20} cm^{-2}, which is very close to that derived from the ROSAT X-ray spectrum.

However, the X-ray configuration of the pair shows interesting structures to be studied carefully. In Fig.1b, the emission consists of at least four component : the nucleus (M474), two sources at (-48″,-40″) and (82″,-61″) relative to M474 (the first one is identified as an optical source, marked by the arrow), and a **diffuse emission** at 50″ NE from M474 that may in turn be a composition of 3 or more point sources. No optical counterpart can be identified to such a diffuse feature.

Acknowledgements. I am grateful to Drs. J. Trümper and H. Zimmermann for stimulating discussions and hospitality when I visited MPE. Dr. H. Arp is the PI and Dr. J. Sulentic is another Co-I of the 4 AGN/NGC observations. I thank them for allowing me use the preliminary results here.

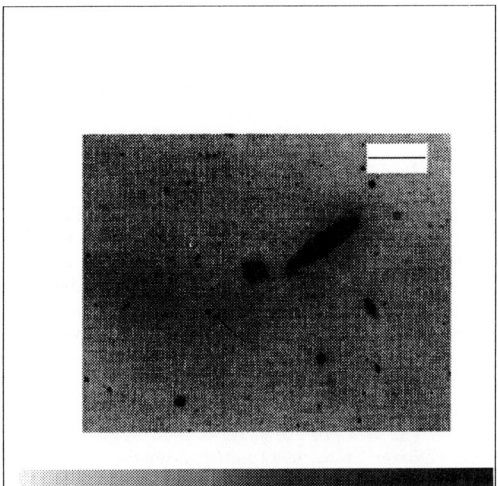

Fig_1a - Mrk474/NGC5682:optical

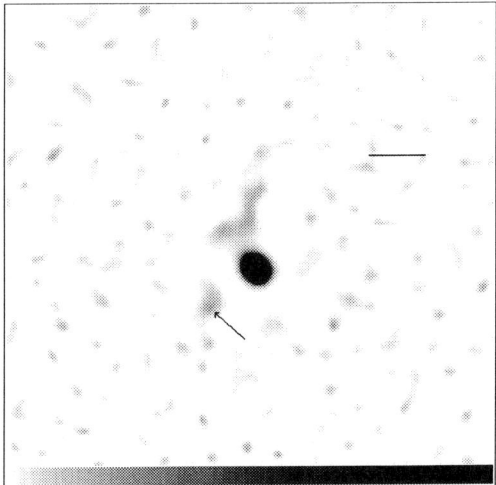

Fig_1b - Mrk474/NGC5682:deconvolution

ROSAT OBSERVATIONS OF AN OPTICAL QUASAR SURVEY FIELD

H. BRUNNER, T. DÖRRER, P. FRIEDRICH, G. LAMER, R. STAUBERT
Astronomy Institute, University of Tübingen, Germany

Abstract. Deep (T~35 ksec) pointed ROSAT observations of a $2.2° \times 2.2°$ optical quasar survey field (149 quasars; $m_{lim} = 20.5$; Crampton et al., 1989) have yielded a detection rate (3 σ) of $\sim 60\%$ (86 quasars; limiting sensitivity $\sim 5 \cdot 10^{-15}$ erg cm^{-2} s^{-1} keV^{-1} at 1 keV). See Fig. 1 for the distribution of the ROSAT PSPC source count rates and Fig. 2a, b for the fraction of quasars detected in X-rays as a function of redshift and optical magnitude. 46 quasars were bright enough to perform spectral power law fits. The mean energy power law index drops from ~ 1.4 at $z = 0$ to ~ 0.9 at $z > 2$ (Fig. 4; only the 20 brightest sources are plotted). This is interpreted as being due to a break in the spectrum between a soft, thermal accretion disk and a hard power law component, occuring at a source frame energy around 1 keV (Fig. 5). Mean accretion disk model parameters are derived (M = $5 \cdot 10^8$ M$_\odot$, $\dot{M} = 0.65$ M$_{Edd}$, $\alpha_{visc.}=0.5$), using an optically thin α-accretion disk model (Dörrer et al., 1992 and references therein). Model predictions for the decline of the X-ray spectral index with redshift are plotted in Fig. 4. The α_{ox} distribution (Fig. 3; dotted line: X-ray upper limits) and the optical number-redshift relation (Fig. 6; dotted line: X-ray number-redshift relation) is modeled using the accretion disk parameters as determined from the X-ray spectral data and assuming a constant comoving volume density ($H_0 = 100$ km/s Mpc, $q_0 = 0.5$) and statistical orientation of the inclination angles of the model source population.

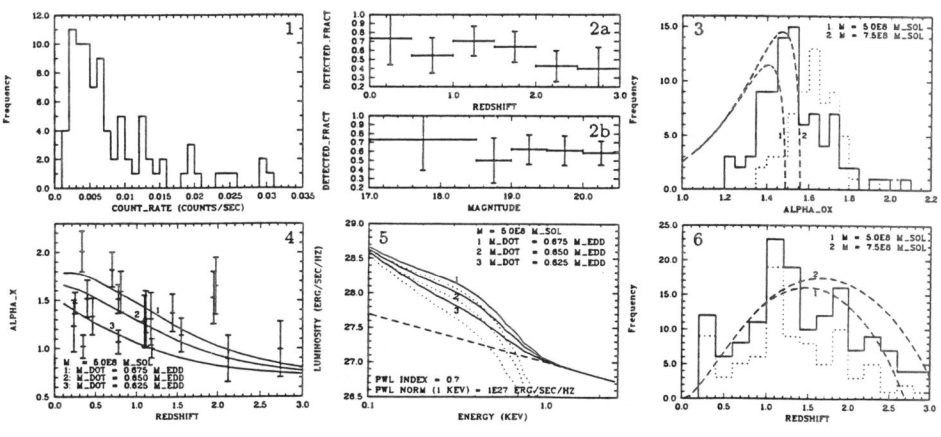

References

Crampton, D., Cowley, A. P. and Hartwick, F. D. A.: 1989, *Ap.J.* **345**, 59
Dörrer, T., Friedrich, P., Brunner, H., Staubert, R. et al.: 1992, *X-ray emission from active galactic nuclei and the cosmic X-ray background*, eds. Brinkmann and Trümper, *MPE report* **235**, 130

ROSAT SPECTRA OF QUASARS

P. BÜHLER [1,2], T.J.-L. COURVOISIER [1], R. STAUBERT [3], H. BRUNNER [3] and G. LAMER [3]
[1] Observatoire de Genève, 1290 Sauverny, Switzerland
[2] Paul Scherrer Institut, 5232 Villigen-PSI, Switzerland
[3] Astronomisches Institut der Universität Tübingen, 72076 Tübingen, Germany

X-ray observations of AGN with Einstein, EXOSAT and Ginga have shown, that the spectra of quasars in the energy range 2 to 10 keV can be approximately described by a single power law model with a photon index of 1.7 to 2.0. They also suggested that a soft X-ray excess component (below $\approx$ 1 keV) is a common feature in many quasars. In order to investigate whether a soft excess is characteristic for a certain class of objects we analysed the data of the pointed ROSAT PSPC observations of the six radio-loud quasars PG0007+106, PKS0135-247, QSO0537-286, QSO0923+392, PG1225+317, 3C273 and the radio-quiet quasar PG0804+761. In a first step the observed spectra were fitted with an absorbed single power law model. The hydrogen column density was fixed to its galactic value and the normalisation at 1 keV and the spectral index α were the free fit parameters. In order to decide whether a soft component is present in a source, the resulting power law index was compared with the hard X-ray power law index (2-10 keV) determined in the past with other instruments. A steep ROSAT PSPC spectrum indicates the presence of an additional soft X-ray component. In four cases (PKS0135-247, PG0804+761, QSO0923+392, 3C273) we find that the spectra in the PSPC band are considerably steeper than the spectra above 2 keV and therefore suggest the presence of a soft excess. In order to quantify the contribution of the soft excess these spectra were successively fitted with a model containing a hard power law component and an additional soft component described either by a power law, thermal bremsstrahlung or black body model. For the other three members of our sample (0007+106, 0537-286, 1225+317) the fitted power law index is not enhanced. This means that no soft component has been detected, but not necessarily that it does not exist. There are two effects which render more difficult the detection of a soft component in ROSAT spectra, the absorption of photons by interstellar material and the shift of the spectra towards lower energies due to the redshift. Both processes have first an effect on the soft part of the observed spectrum and it is therefore evident, that this leads to a decrease of the sensitivity for soft X-rays of the emitted spectrum. For the three quasars in our sample, where no soft excess has been detected, either the column density (0007+106) or the redshift (0537-286, 1225+317) is especially large and therefore an eventually present soft component could have remained undetected. In these cases we calculated upper limits for the strength of such a soft component (P. Bühler et al., to be published in A&A.)

ROSAT OBSERVATIONS OF EINSTEIN EMSS AGNS

M. CAPPI and G.G.C. PALUMBO
Dipartimento di Astronomia, Università di Bologna

R. DELLA CECA
Johns Hopkins University, Baltimore, U.S.A.

and

T. MACCACARO
Osservatorio astronomico di Brera, Milano, Italy

Active Galactic Nuclei (AGN) can be studied in the ROSAT energy band also when serendipitously observed during long pointed observation of unrelated targets. From the available ROSAT database three Extended Medium Sensitivity Survey (EMSS) AGNs, detected with more than 500 counts, have been studied. One AGN (MS 1803.6+6738 = Kaz 102) was observed twice and in one field the target is the BL Lac 0716+714.
-Spectral studies reveal that all three AGNs are well described by a single power law. Values of relevance are summarized in the table where they are compared to previous values from Einstein and the ROSAT survey (Walter R. & Fink H.H., 1993, *A&A*, **274**, 105). All four sources have a N_H value well below the estimated value given by Stark et al. (1992, *ApJS*, **79**, 77) and Elvis et al. (1989, *ApJ*, **97**, 777). This probably is an indication that a soft excess is present.
-Variability studies show flux variations for the BL Lac on time scales of days, while the photon index appears rather constant. The high count rate for this object allows to attribute a probability of variability greater than 99% (running a K-S test against constancy). In the case of Kaz 102, the comparison of previous measurements with the present one does not show evidence of long term (years) variability.

Name	z	N_H^{Gal} [a]	N_H^{Fit} [a]	Γ_{Fit} [b]	$Flux$ [c]	χ_ν^2	Γ_{MSS} [b]	Γ_{RASS} [b]
MS0719.9+7100	0.125	3.4	3.2	2.59±0.36	0.54	0.99	$4.44^{+9.99}_{-1.49}$	—
MS1617.9+1731	0.116	4.3	3.11	1.77±0.31	1.62	1.25	$2.23^{+0.44}_{-0.47}$	—
MS1803.6+6738 I	0.136	5.0	4.85	2.55±0.29	2.31	0.79	$1.72^{+0.26}_{-0.28}$	2.21±0.11
MS1803.6+6738 II	0.136	5.0	3.76	2.10±0.29	3.67	1.00	$1.72^{+0.26}_{-0.28}$	2.21±0.11
BL Lac 0716+714	≥ 0.2	3.4	2.46	2.54±0.08	7.11	1.16	—	—

[a] In units of $10^{20} cm^{-2}$.
[b] Photon indexes.
[c] Fluxes in the (0.1-2.4 KeV) band in units of $10^{-12} erg cm^{-2} s^{-1}$. Statistical errors are about 20%.

SUPERNOVA FRAGMENTS AND THE ORIGIN OF THE RAPID X-RAY VARIABILITY

R. CID FERNANDES
Institute of Astronomy, Cambridge, U.K.

R. TERLEVICH
Royal Greenwich Observatory, Cambridge, U.K.

G. TENORIO-TAGLE
Instituto de Astrofísica de Canarias, Tenerife, Spain

J. FRANCO
Instituto de Astronomía UNAM, México D. F., México

and

M. ROZYCZKA
Warsaw University Observatory, Warszaw, Poland

Abstract. The Starburst model for Radio Quiet Active Galactic Nuclei proved able to explain the origin of the broad line region, the variability characteristics of line and continuum in Seyfert galaxies, X-ray spectra, the luminosity function of QSOs and etc. But can we understand the rapid X-ray variability observed in several AGN with supernovae?

Key words: Active Galaxies, Starburst, Supernova Remnants, X-ray

In the Starburst model of AGN, strongly radiative compact supernova remnants (cSNRs) are the source of the broad emission lines, the high energy continuum and optical-UV variability. There is growing evidence that both the SN ejecta and its circumstellar medium are very clumpy and inhomogeneous. We have examined the role played by inhomogeneities in the SN ejecta on the evolution of otherwise normal cSNRs. The hydrodynamical models show the evolution of fragments into dense, fast moving "tortillas" (or "pancakes"), as they are processed by the reverse shock. The high densities and small cooling times result in isothermal shocks leading to large compression factors along the direction of motion. Dense fragments are not significantly decelerated by the reverse shock. They fly through the hot cavity between the reverse and outer shocks to ram against the thin shell at the outer edge of the remnant, producing a short, luminous burst of X-ray emission.

Our preliminary results indicate that a collection of pancake-shell collisions is capable of producing light curves and power spectra similar to those observed in AGN. Details of our computations will be presented elsewhere (Rozyczka et al. and Terlevich et al. 1993—submitted).

As a test of this model we encourage observers to look for short term X-ray variability in *bona fine* cSNRs. Also, high spatial resolution radio monitoring of the nuclei of nearby AGN should be able to detect individual supernovae as they go off in the nuclear cluster. This is a decisive observational test on the nature of the central engine.

ROSAT SELECTED INTERACTING GALAXIES WITH NARROW EMISSION LINES

D. ENGELS, N. BADE and J. STUDT
Hamburger Sternwarte, Gojenbergsweg 112, D-21029 Hamburg, Germany

and

H. FINK
MPI für Extraterrestrische Physik, D-85740 Garching, Germany

In the context of an identification program of sources from the ROSAT All-Sky Survey (RASS) on Schmidt objective prism plates (Bade et al. 1992a,b) we discovered two galaxy pairs, which contain a narrow-line Seyfert 1 component with an X-ray luminosity of $L_x \sim 10^{44}$ erg s^{-1} and an HII–region galaxy. Apparently they are interacting. Their redshifts are $0.1 < z < 0.3$ and their brightnesses $17.5 < B < 19.5$. A third one was found among EINSTEIN sources. Typical separations between the components are 10". Near the pairs other galaxies were found, and although their physical association is not confirmed spectroscopically it is quite probable that they form a small cluster of galaxies. ROSAT HRI observations indicate that the X-ray emission is not extended and originate from the AGN alone. It is remarkable that the AGN in all physical pairs identified so far have rather narrow permitted emission lines with linewidths ≤ 1500 km s^{-1}.

Optically selected galaxies in interacting systems show enhanced nuclear activity compared to isolated galaxies. Interaction between galaxies is thought therefore to be an important process leading to the formation of an active galactic nucleus (Fricke & Kollatschny 1989). It was proposed that ultraluminous IRAS galaxies with bolometric luminosities similar to QSO's may form the beginning of the evolutionary sequence leading to ordinary QSO's. They are to a high degree interacting systems (Sanders et al. 1988).

The credibility of this evolutionary scenario would be strengthened considerably if transition objects could be found. We speculate that our X-ray selected interacting galaxies may be in such an intermediate state: An accretion disc responsible for the X-ray emission has already formed, while the obscuring dust has dissipated so far that the X-ray emission can leave the center of the AGN unhindered. The narrowness of the permitted emission lines leads to the assumption that their width is evolving with age. The broad-line region responsible for the optical emission of the presumed transition objects has then not reached the full velocity dispersion observed in ordinary AGN.

Bade N., Dahlem M., Engels D., et al., 1992a, MPE-Report 235, 377
Bade N., Engels D., Fink H., et al., 1992b, A&A 254, L21
Fricke K.J., Kollatschny W., 1989, IAU Symposium 134, p. 425
Sanders D.B., et al., 1988, ApJ 325, 74

SOFT X-RAY EMISSION OF QUASARS

N. SCHARTEL, R. WALTER, H.H. FINK
Max Planck Inst. für extraterrestr. Physik, Giessenbachstraße, D-85748 Garching b. München

Abstract. From a list of known quasars compiled from various catalogues we selected all sources detected by the PSPC (0.1 − 2.4 keV) aboard ROSAT with more than 80 counts during the all sky survey. A sample of 102 sources resulted. At higher redshifts most of the selected sources are radio-loud. At a redshift smaller than 0.50 we found 54 radio-quiet quasars and 30 radio-loud sources. For this reduced sample the mean spectral index of the radio-quiet sources ($< \Gamma > = 2.53$) and that of the radio-loud ones ($< \Gamma > = 2.26$) are clearly different with a significance of 3.3 σ.

About 2/3 of the bright quasars observed with *Einstein* also belong to our sample. The spectra observed with *ROSAT* are sytematically steeper than the ones observed with *Einstein* yielding a $< \Gamma_{ROSAT} - \Gamma_{Einstein} >$ of 0.66 ± 0.18 for radio quiet and of 0.68 ± 0.19 for radio-loud sources, respectively.

For radio loud quasars, the mean spectral slope decreases from 2.3 to 1.5 when the redshift increases beyond 0.5 (figure 1). The fact that high redshift sources show a photon index of about 1.5, which is similar to the mean index observed with *Einstein* for radio-loud sources, suggests that this decrease towards higher redshifts can be interpreted by the shift of the soft X-ray excess outside of the ROSAT spectral band when the redshift increases. The solid lines in figure 1 represent theoretical pathes of the photon index as a function of the redshift as derived from simulations assuming a power law plus black body model spectrum for the quasars X-ray emission. In curve No 1 the powerlaw index is fixed to 1.4. To be compatible with the observation the temperature of the blackbody component must range between 50 and 70 eV. Curve No 2 asssumes the same model with a powerlaw index fixed to 1.8 to account for radio quiet sources.

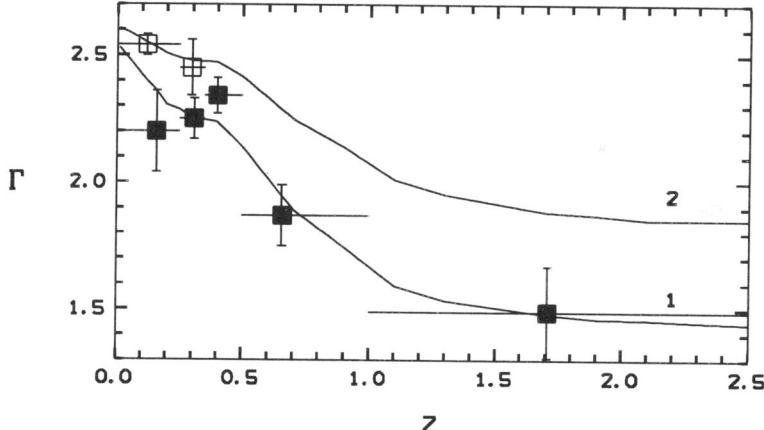

Fig. 1. Mean photon index in redshift bins plotted as a function of the redshift. Black and white symbols are for radio-loud and radio quiet-sources, respectively. The lines show the results of the simulations of the redshift effect on power law plus black body model spectra. The two models differ only in the photon index of the power law component.

ROSAT All Sky Survey AGN spectra: Constraints on accretion disks

P. Friedrich, T. Dörrer, H. Brunner, R. Staubert
Astronomy Institute, University of Tübingen, Germany

Abstract. We found that ROSAT spectra of a sample of 89 AGN are generally steeper than 0.7. The excess above a hard X-ray power law spectrum in this energy range which has been found already with *Einstein* and EXOSAT for some AGN is now seen very clearly in most sources. Our α-disk models (Dörrer et al., 1992 and references therein) which include Comptonization and relativistic corrections are in agreement with the measured soft excesses when the ($\dot{M}_{Edd.}, \alpha$) parameter space is restricted to $\alpha > 0.4$ and $\dot{M}_{Edd.} \in [0.4, 0.8]$ ($\dot{M}_{Edd.}$: Eddington accretion rate).

We attempt to model this soft X-ray excess emission in terms of thermal emission from a thin α-accretion disk which contributes both to the UV and soft X-ray emission (*big blue bump*). We have selected a sample of 89 QSO and Seyfert I galaxies for which soft X-ray (ROSAT All Sky Survey) and UV measurements (IUE) are available. Hard X-ray power law slopes as measured by EXOSAT, *Einstein* and GINGA were taken from the literature.

Fig. 1 shows the 68% and 90% confidence contours for the distributions of the spectral index α_{soft} in the ROSAT range and the index α_{hard} of hard X-rays (> 2keV), both for QSOs and Seyfert I galaxies. We have calculated the excess count rates in the spectral bands from 0.07 to 0.4 keV and from 0.4 to 1.0 keV over the hard power law (see above). A re-normalization of the hard power law to the ROSAT 1.0 – 2.4 keV count rate was applied to allow for variability. From the excess count rates *excess hardness ratios* were calculated. These were compared to predictions from our accretion disk model. The derived disk parameters are given in Fig. 2. We have also performed individual spectral fits for a number of bright AGN and determined their disk model parameters. Results from this agree well with the hardness ratio method.

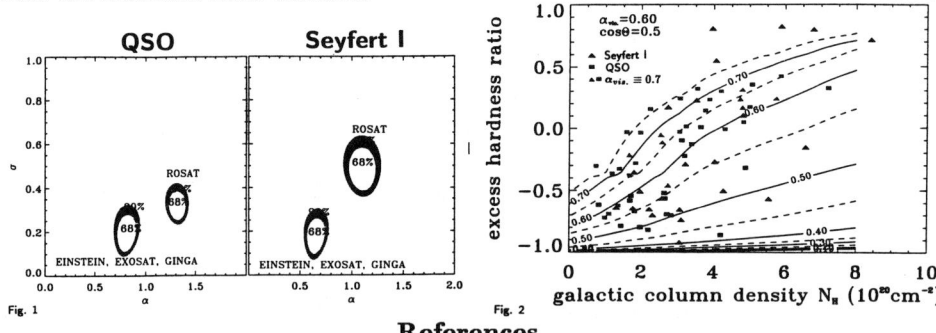

Fig. 1   Fig. 2

References

Dörrer, T., Friedrich, P., Brunner, H., Staubert, R., Pussel, V., Riffert, H., Ruder, H.: 1992, *X-ray emission from active galactic nuclei and the cosmic X-ray background*, eds. Brinkmann and Trümper, *MPE report* **235**, 130

X-RAY DETECTION OF THE NUCLEAR SOURCE IN THE CYGNUS A GALAXY

D.E. HARRIS
Center for Astrophysics, Cambridge, MA, USA

R.A. PERLEY
National Radio Astronomy Observatory, Socorro, NM, USA

and

C.L. CARILLI
Sterrewacht te Leiden, Leiden, the Netherlands

Abstract. From a ROSAT HRI observation of Cygnus A (42 ksec), we detect x-ray emission from the galaxy identified with the radio source. This was accomplished by subtracting a modified King model: (Surface brightness $\propto$ to $[1+(r/a)^2]^{(0.5-3\beta)}$) in order to study residual features once the main body of emission from diffuse gas had been deleted. The central source was present for all acceptable values of the core radius, a, and exponent, β. Details of the image processing, an evaluation of emission from the radio hotspots, and a study of the effect of the radio lobes on the gas distribution will be presented elsewhere.

Key words: X-ray Emission

Although an unresolved x-ray source coincident with the optical galaxy has not heretofore been detected directly, Arnaud et al. (1987) made a strong case for a power-law component in their spectral analysis of data from several missions. To explain the failure to detect the core source with the Einstein HRI, they hypothesized a column density intrinsic to the source of $N_H = 8.2 \times 10^{22}$ cm^{-2}, a number which corresponds to 50 magnitudes of optical extinction, A_v, (using $N_H = 4.8 \times 10^{21} \times$ E(B-V) and $A_v = 3.1$ E(B-V)). Obviously the only way to accommodate this amount of absorption would be to site the emission at the center of the galaxy.

Only a rough estimate of the intensity can be obtained because it is difficult to make a correction for the background and for obvious reasons, the residual intensity is dependent on the amplitude of the subtracted King model. For King models normalized from 87% to 100% of the counts from the cluster gas within r=125", the intensity of the core ranges from 272 to 184 counts. After an 18% correction for scattering of photons beyond the measuring area, we thus assign a value of 270 +/-60 counts to the central source. Most of these can be represented by an unresolved component, but there appears to be some extended emission present in the measuring circle of radius 10".

Transforming the countrate to flux and luminosity presents another large uncertainty because of the unknown amount of absorption. For no intrinsic absorption (i.e. $N_H = 3.3 \times 10^{21}$ cm^{-2}, the galactic value) our countrate would be consistent with a 2-10keV luminosity of 3.8×10^{42} erg/s. For an additional 8.2×10^{22} cm^{-2} at the source, a 2-10 keV luminosity of 1.6×10^{45} erg/s would be required. Adjusting the intrinsic column density to obtain the Arnaud et al. (1987) luminosity (converted to H$_o$=75 km/s/Mpc) of 2.3×10^{44} erg/s requires N$_H$(at the source) = $(4.6 \pm 0.4) \times 10^{22}$ cm^{-2}, which corresponds to A$_v$ = 29.7 magnitudes. Although our data provide no constraints on the amount of absorption if the luminosity is not known, it is obvious that for agreement with Arnaud's spectrum, either the absorption is significantly less than 8.2×10^{22} or the source is 7 times more luminous than it was ten years ago. Since both possibilities appear plausible, we searched for corroborating evidence.

If the core source is heavily absorbed, only the highest energy photons within the ROSAT HRI band will survive the natural highpass filter provided by the ISM of Cyg A. For N$_H$ = 5×10^{22} cm^{-2}, the transmission at 2 keV is 12%. We examined the Pulse Height distributions of counts for the core source (circles with r=5" and r=10") and compared them to the distribution in an annulus from 15" to 25". The expected signature for significant absorption is present in the data at very low significance (< 1σ). The median channel number (i.e. half the counts in higher channels, half in lower) is marginally higher for the source than for the "background" annulus.

We searched for variability by examining the data from 3 epochs (Nov91, Dec91, and May92). The count rate for each data segment was measured for circles centered on the emission peak with radii of 2.5, 5, 7.5, and 10". As the area of the measuring circle increases, there is more dilution from the cluster gas but for small circles, aspect jitter means that many counts from the core source will be outside the circle. There is no convincing evidence (i.e.> 2σ) for variation greater than 20% on a 1 to 6 month timescale.

Both of these tests are inconclusive, partly because of the difficulty of isolating the counts from the core source from those of the cluster gas in the same direction. While our data have adequate spatial resolution to demonstrate the reality of the core source, a substantial improvement in spectral resolution and a longer time base will be required to answer these questions.

References

Arnaud, K.A., Johnston, R.M., Fabian, A.C., Crawford, C.S., Nulsen, P.E., Shafer, R.A., and Mushotzky, R.F. 1987, M.N.R.A.S., 227, 241

ROSAT OBSERVATIONS OF BRIGHT BL LACERTAE OBJECTS

G. LAMER, H. BRUNNER, R. STAUBERT
Astronomisches Institut der Universität Tübingen, Germany

Abstract. We have compiled a sample of 23 X-ray and radio selected BL Lacertae objects which have been observed with the Position Sensitive Proportional Counter (PSPC) on board of the ROSAT Satellite. The sample consists of three parts:
In Table I results from 4 objects observed for their known rapid X-ray variability are presented. 5 objects are the BL Lac subset of a complete sample of flat spectrum radio sources with 5 GHz flux densities > 1 Jy . Detailed results from this sample will be published in Brunner et al. 1993. The data of the 14 remainig objects were collected from the ROSAT data archive to supplement the sample. The whole sample contains 7 X-ray selected objects (XBLs, $\alpha_{OX} < 1.2$) and 16 radio selected objects (RBLs).
The X-ray spectra of the sources are well described by single power laws with galactic absorption. The X-ray energy indices α_X are widely dispersed around a mean of 1.34. Significant X-ray flux variability and correlated spectral variability was detected on timescales down to hours. The object H 1218+304 was found to be rapidly variable within each of three observations. Its spectral hardness is correlated with the flux level (see Table I).
We calculated the intrinsic distributions of the spectral indices α_X for the XBL and RBL samples and of the differences between ROSAT and EXOSAT ME spectral indices $\alpha_{PSPC} - \alpha_{ME}$ (only XBL sample) using a maximum likelihood fit. There is no significant difference in the mean spectral indices between the X-ray and radio selected subsamples. The mean values $<\alpha_X>$ are 1.34 for XBLs and 1.33 for RBLs. The spectra of the X-ray selected objects slightly steepen at higher X-ray energies ($<\alpha_{PSPC} - \alpha_{ME}> = -0.11$). This supports the view that the X-ray emission of XBLs is supplied by synchrotron radiation. The steepening of the X-ray spectrum is then due to a cutoff in the energy distribution of the electrons.

TABLE I

Object	date	N_H gal.[1]	N_H [2]	F_{1keV} [μJy]	α 0.1-2.4 keV	variab. ampl.(Δ t)
0548-322	92/3/6	2.94	2.76 ± .14	9.59 ± .12	0.95 ± .05	-
1218+304	91/6/15-16		2.61 ± .09	12.72 ± .12	1.19 ± .03	15%(1 d)
1218+304	92/6/18-22	1.78	2.51 ± .10	10.05 ± .11	1.22 ± .03	50%(6 mon)
1218+304	92/12/8-10		2.85 ± .25	5.11 ± .14	1.47 ± .07	-
3C 371	92/4/9	4.61	3.43 ± .81	0.32 ± .02	0.87 ± .25	-
2005-489	92/4/27-29	5.44	4.09 ± .17	5.19 ± .07	1.92 ± .04	50%(6 mon)
2005-489	92/10/28-11/1	5.44	3.58 ± .21	2.81 ± .06	1.91 ± .05	-

[1] [$10^{20} cm^{-2}$], Elvis et al. 1989, Stark et al. 1992. [2] [$10^{20} cm^{-2}$]

References

Brunner, H., Lamer, G., Worrall, D. M., et al., 1993, in preparation
Elvis, M., Lockman, F. J., and Wilkes, B. J., 1989, *AJ*, **97**, 777
Stark, A. A., Gammie, C. F., Wilson, R. W., et al., 1992, *ApJS*, **79**, 77

CAN AGN (ACTIVE GALACTIC NUCLEI) ALONE MAKE THE COSMIC X-RAY BACKGROUND?

DARRYL LEITER* and ELIHU BOLDT
*Laboratory for High Energy Astrophysics,
NASA/Goddard Space Flight Center, Greenbelt, MD 20771*
* NRC Senior Resident Research Associate

Recent ROSAT X-ray observations of AGN have yielded important new information[1] about the analytic structure of the AGN X-ray luminosity function and its evolution out to z = 3. Using the luminosity evolution obtained within the cosmological context of $\Omega=0$, we find[that AGN could readily make up the CXB (cosmic X-ray background)[2,3,4]. However, in this case we find that accounting for the CXB with accretion-powered AGN emission is incompatible with the observed mass function for present-epoch black hole galactic nuclei (both active and dormant)[5]. On the other hand, we find that the luminosity evolution obtained with ROSAT for such AGN within the cosmological context of $\Omega=1$ is indeed compatible with the present-epoch black hole galactic nuclei mass function. This apparently acceptable solution, though, definitely falls short of accounting for all the CXB, even when considering unified models for AGN. This difficulty can be resolved by noting that the underlying supermassive black holes which already exist at the onset of the canonical AGN phenomenon of supply-limited accretion must have undergone a previous growth phase where the accretion would be expected to be Eddington-limited. In this likely scenario (i.e., for $\Omega=1$) the residual CXB, that over and above the foreground of canonical AGN, can be naturally explained by the characteristic X-ray emission from highly compact PAG (precursor active galaxy) sources associated with these numerous black holes, at redshifts just beyond the earliest AGN[6].

1. Boyle et al. 1993, MNRAS, 260,49.
2. Madau, P., Ghisellini, G., & Fabian, A., 1993, ApJ., 410, L7.
3. Zdziarski, A., Zycki, P., & Krolik, J., 1993, ApJ., 414,L81.
4. Comastri, A. Setti, G., Zamoriani, G., & Hasinger, G., 1993, First Stromlo Symposium, Canberra, Australia.
5. Boldt, E., & Leiter, D., 1993, First Stromlo Symposium, Canberra, Australia.
6. Leiter, D., & Boldt, E., 1993, submitted to ApJ. Letters.

ACCELERATION EFFICIENCY IN NONTHERMAL SOURCES AND THE SOFT GAMMA-RAYS FROM NGC 4151 OBSERVED BY OSSE AND SIGMA

ANDRZEJ MACIOŁEK-NIEDŹWIECKI and ANDRZEJ A. ZDZIARSKI
Copernicus Astronomical Center, Bartycka 18, 00-716 Warsaw, POLAND

and

ALAN P. LIGHTMAN
Dept. of Physics, MIT, Cambridge, MA 02139

1. ABSTRACT

We show that the recent observations of the Seyfert galaxy NGC 4151 in hard X-rays and soft γ-rays by the OSSE and SIGMA detectors onboard *CGRO* and *GRANAT*, respectively, are well explained by a nonthermal model with acceleration of relativistic electrons at an efficiency of $\lesssim 50\%$ and with the remaining power dissipated thermally in the source (the standard nonthermal $e^{\pm}$ pair model assumed 100% efficiency). Such an acceleration efficiency is generally expected on physical grounds. The resulting model unifies previously proposed purely thermal and purely nonthermal models. The pure nonthermal model for NGC 4151 appears to be ruled out. The pure thermal model gives a worse fit to the data than our hybrid nonthermal/thermal model. Our results are presented in Zdziarski, Lightman, & Maciołek-Niedźwiecki (1993).

References

Zdziarski, A. A., Lightman, A. P., & Maciołek-Niedźwiecki, A. 1993, *ApJLetters*, 414, L93

X-RAYS FROM PHOTOIONIZED ACCRETION DISCS

G. MATT and A.C. FABIAN
Institute of Astronomy, Madingley Road, Cambridge, CB3 0HA, U.K.

and

R.R. ROSS
Physics Department, College of the Holy Cross, Worcester, MA 01610, USA

The presence of iron lines and high energy excesses in the X-ray spectra of Seyfert galaxies has been firmly established by *Ginga* (e.g. Nandra & Pounds 1993 and references therein). These features are generally interpreted as signatures of the reprocessing of the primary X-rays by matter in the neighbourhood of the central black hole, probably distributed in an accretion disc (Lightman & White 1988, George & Fabian 1991, Matt, Perola & Piro 1991).

The illumination of the disc by the X-rays can significantly alter the thermal and ionization structure of the surface layers, modifying the lines and continuum emission properties. We have studied in detail these effects; the main results can be summarized as follows:

i) While for low–ionization matter the only important line in the X-ray band is the iron fluorescence one, many other lines are expected from highly ionized matter (Ross & Fabian 1993).

ii) The properties of the iron line itself are modified if the matter is ionized. The fluorescence yield increases with the ionization state, and the importance of re-absorption diminishes with the ionization of the matter. On the other hand, the FeXVII-FeXXIII line photons are destroyed by resonant trapping. Therefore, the line intensity can be either much smaller or much greater than for the neutral case (Ross & Fabian 1993, Matt, Fabian & Ross 1993a,b).

iii) If the matter is highly ionized the disc could be significantly reflective even in the soft X-rays. The resulting emerging continuum spectra are quite complex, and have been calculated in great detail for a single slab by Ross & Fabian (1993). Matt, Fabian & Ross (1993c) extended the calculation to the whole disc, and estimated the angular dependence on the flux. They also calculated the polarization; both the degree and the angle resulted to be strongly energy–dependent.

References

George I.M., Fabian A.C., 1991, MNRAS, 249, 352
Lightman A.P., White T.R., 1988, ApJ, 335, 57
Matt G., Fabian A.C., Ross R.R., 1993a, MNRAS, 262, 179
Matt G., Fabian A.C., Ross R.R., 1993b, MNRAS, submitted
Matt G., Fabian A.C., Ross R.R., 1993c, MNRAS, in press
Matt G., Perola G.C., Piro L., 1991, A&A, 247, 25
Nandra K., Pounds K.A., 1993, MNRAS, submitted
Ross R.R., Fabian A.C., 1993, MNRAS, 261, 74

GAMMA RAYS AND JETS IN ACTIVE GALACTIC NUCLEI

DONALD MEYER

Physics Department, University of Michigan
Ann Arbor, MI 48109-1120 USA

ABSTRACT: It will be shown that the jets observed in AGN's are the result of the high energy gamma rays recently observed by the EGRET and Whipple collaborations interacting with a medium of density typical of that expected in galaxies. All of the observed properties of the jets follow from this interaction.

In all of the blazars from which gamma rays have been observed by EGRET the high energy gamma rays contribute an energy comparable to or greater than that of the lower energy radiation. Gamma rays in the energy range of interest (100 MeV to 5 TeV) interact with atoms almost entirely by producing electron pairs with a small component of Compton electrons at the low end of the energy range. This production occurs with close to 100% efficiency. Synchrotron radiation is emitted close to the forward direction by these pairs in the magnetic field produced by the Compton electrons. A substantial fraction of the synchrotron radiation is emitted by the pairs in distances of the order of parsecs. Most of the gammas will interact in one radiation length of the medium (13 atoms/cm^3 for 10^5 pc.). It is now simple to see how all of the jet properties arise. The jets remain collimated because the gamma rays travel in straight lines. The low energy synchrotron radiation is produced in approximately the right quantities and in the correct frequency range by the magnetic field produced by the Compton electrons (the pairs are statistically charge symmetric). All of the synchrotron energy needed is available from the gammas so there is no need for a high energy charged particle beam, an engine to accelerate particles or a beam confining mechanism. Electron pairs are produced the entire length of the jet as the gammas interact. The kinks and bends occur because the direction of the gamma beam wiggles slightly with time. The irregular structure observed along the jets is caused by the chaotic interaction between the pairs, the Compton electrons and the magnetic field as well as by intensity variations. The charged particles and magnetic field act rather like a two dimensional undamped plasma perpendicular to the beam direction. Given the observed presence of the gamma rays and their well known interactions, these phenomena must occur with an efficiency close to 100% which precludes the need for any other source for the phenomena observed in jets.

Spectral Variability of 3C 273 at soft X-rays

R.Staubert, T.J.-L.Courvoisier, H.Fink, M.-H.Ulrich, H.Brunner, S.Friedrich, K.Otterbein

ABSTRACT.
We report on the results of three observations of 3C 273 by ROSAT (Staubert et al. 1993).

OBSERVATIONS AND RESULTS

	Pointed observation [#]	Survey [&]	Pointed observation
date	18 June 1990	18-21 Dec. 1990	14-15 Dec. 1991
integr. time / source cts	916 s / 7377	497 s / 2801	6243 s / ~50000
spectral fits (0.08-2.4 keV) with single power law:			
a) N_H variable:			
N_H [10^{20} cm^{-2}]	(1.67 +- 0.26) [*]	(1.27 +- 0.36) [*]	(1.56 +- 0.11) [*]
photon index	1.78 +- 0.10 [*]	2.05 +- 0.17 [*]	1.89 +- 0.05 [*]
χ^2_{red} (dof)	1.27 (29)	1.0 (42)	2.18 (26)
b) N_H fixed to galactic value of $1.8 \cdot 10^{20}$ cm^{-2}:			
photon index	1.84 +- 0.05 [*]	2.29 +- 0.08 [*]	1.98 +- 0.014 [*]
χ^2_{red} (dof)	1.27 (30)	1.26 (43)	2.89 (27)

[#] contemporaneous with observations by IUE, [&] contemporaneous with observations by IUE and Ginga, [*] all uncertainties are 90% joint confidence limits

CONCLUSIONS. Soft X-ray spectra (0.1-2.4 keV) of 3C 273 have been measured by ROSAT in June '90, in Dec. '90 (contemporaneous with Ginga and IUE) during the All Sky Survey and in Dec. '91 (Staubert et al. 1991, Staubert 1992). Marginally acceptable spectral fits are found for a single power law model with the column density fixed to the galactic value $N_H = 1.8 \cdot 10^{20}$ cm^{-2} for the June '90 and the Dec. '90 data. A single power law fit is not acceptable for the Dec. '91 data. If N_H is a free parameter the best fit values are lower than the galactic value, particularly so for the observation of Dec. '91. This is indicative of a soft excess. Under the assumption of a single power law model, there are significant variations in power law index from one observation to the next, demonstrating spectral variability on time scales 5 to 12 month. The power law spectra found by ROSAT are significantly steeper than those observed in the 2-10 keV range (with an average photon index of about 1.47 and a spread from 1.41 to 1.54; Turner et al. 1990). This again indicates the existence of a 'soft excess' component. The simultaneous ROSAT and Ginga observations of Dec. '90 clearly show two spectral components. A double power law fit reproduces the standard hard component and establishes an additional soft component with a photon index of 3.5 (Staubert 1992). For the June '90 and Dec. '91 observations no simultaneous hard X-ray measurements are available. However, a double power law model yields an acceptable fit to the Dec. '91 data with photon indices of 1.54 for the hard component and 2.27 for the soft component. Since the ROSAT energy range (0.1-2.4 keV) is rather small for double power law fits, the parameter space for such fits has been explored for all three ROSAT observations, stepping through a grid of parameters (I_0 / α) which define the underlying hard component: a) the spectral variability of the soft component is evident (under all reasonable assumptions for the hard component); b) for the Dec. '90 data the combined ROSAT/Ginga spectrum provides a calibration: fitting the ROSAT data alone (with a double power law) underestimates the steepness of the soft component.

References
Staubert et al. 1991, Proc. *"Testing the AGN Paradigm"*, Maryland, Oct. 1991, p. 366
Staubert 1992, Proc. *"X-Ray Emission from AGN and the Cosmic X-ray Background"*, Garching, Nov. 1991, MPE Rep. 235, p. 42
Turner et al. 1990, MNRAS 244, 310
Staubert et al. 1993, in preparation, to be submitted to A&A

X-RAY OBSERVATIONS OF BLAZARS WITH GINGA AND ASCA

M. TASHIRO, K. MAKISHIMA and Y. KOHMURA
The University of Tokyo, 7-3-1 Hongo, Bunkyo-ku, Tokyo, Japan

T. OHASHI
Tokyo Mitropolitan University, 1-1 Minami-Osawa, Hachioji, Tokyo, Japan

C. OTANI, T. KII, R. FUJIMOTO and F. MAKINO
Institute of Space and Astronautical Science, 3-1-1 Yoshinodai, Sagamihara, Kanagawa, Japan

and

GINGA TEAM and ASCA TEAM

1. *Ginga* Observations of Highly Variable BL Lacs

Among 13 BL Lacs observed with *Ginga*, 1H 0323+022, Mkn 421 and PKS 2155-304 exhibited significant variablity during each (typically one day) observation [1]. On the flux-hardness plane, the data points obtained from each source draw a sort of clockwise hysteresis motion. It means that the spectrum hardens before the source gets brighter, while the spectrum softens before the source becomes fainter. Such a soft-lag behavior, first pointed out for PKS2155-304 by Sembay et al. [2]. These properties were also confirmed with the discrete cross correlation function technique.

The soft-lag property is predicted by the synchrotron emission model. Suppose that synchrotron emission arises from relativistic electrons, continuously injected with a fixed spectral shape into a region of homogeneous magnetic field. In this case, flux of the more energetic electrons are expected to respond more quickly to the changes in the electron injection rate because they have shorter life times, thus producing the larger time lags for the softer X-rays.

2. *ASCA* Observation of Mkn 421

We observed Mkn 421 on May 10-11, 1993 as one PV observations of *ASCA*. The source was in a low state on 10 and flared up on 11. Using PHA ratio technique, we confirmed that convex or flat spectrum is exhibited in the low state, while flat or concave spectrum in the flaring state. Although the results are still preliminary, it can support not only the similar results obtained by Tashiro with *Ginga* [1], but also the synchrotron emission model mentioned above.

References

1. Tashiro, M.: 1993, Ph. D. Thesis, *the University of Tokyo*
2. Sembay, S. et al. : 1993, *ApJ* **404**, pp. 112-123

ROSAT SPECTRA AND LIGHTCURVES OF BRIGHT BL LACERTAE OBJECTS AT LOW INTERSTELLAR ABSORPTION

H.-C. THOMAS and H. H. FINK
Max-Planck-Institut für Astrophysik/Extraterrestrische Physik, Garching, Germany

Abstract.
A programme to obtain soft X-ray spectra of bright BL Lac's from pointed observations with ROSAT has been started. So far 13 objects have been observed and another 6 have been accepted for observation. Available data for the following sources were reduced: OQ 530, OJ 287, B2 0912+29, GB 1011+496, ON 231, B2 1215+30, B2 1308+32, and Mrk 421.

In this sample the most outstanding observation clearly is that of Mrk 421. This source was observed with the ROSAT PSPC for a total of 34 ksec between May 5 and May 7, 1992, at a mean countrate of 159 cts/s (see Fink *et al.*, 1991, *A&A* **246**, L6 for ROSAT survey results). At the time of the observations Mrk 421 was also very luminous in the optical bands (see poster by S. Kikuchi, this conference). Within the ROSAT band the spectrum is flatter at the soft end (photon index for broken power law model between -2.2 and -2.0) than at the hard end (photon index between -2.6 and -2.4). The countrate increases during the first 5 orbits with a maximum e-folding time of 1.5 days, reaching a maximum luminosity of $4 \times 10^{45} erg/s$ in the energy range of 0.1 to 2.4 keV (for $H_0 = 75 km/s/Mpc$). It is followed by a slow decline until a second rise starts which is much stronger in harder X-rays than at low energies. Plotting the hardness ratio (defined as the difference in the countrates above and below 0.5 keV divided by the total countrate) versus the total countrate the second rise displays a much steeper gradient than the first one (see figure 1a).

For most of the other BL Lac's the countrates are too low to detect variability on these timescales of hours to days, except for B2 1215+30 and OQ 530, where the chance probability for the observed variations is 4×10^{-4} and 10^{-10}, respectively. Also no clear correlation between hardness ratio and countrate could be found. Spectral fits of single power law models with absorption result in photon indizes ranging from -3.1 to -1.9 with no detectable absorption above the galactic value for 5 of these sources, and some additional absorption for GB 1011+496 and B2 1215+30 (see figure 1b).

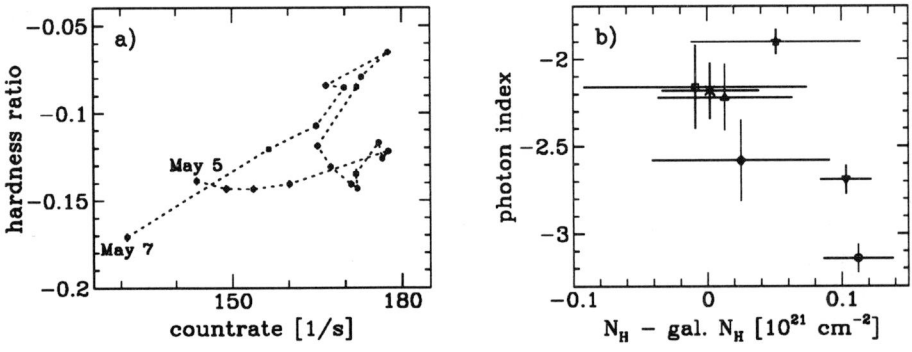

Fig. 1. a) hardness ratio vs. countrate for Mrk 421, b) photon index vs. intrinsic absorption for the other 7 BL Lac's

NEW X-RAY RESULTS ON RADIO GALAXIES

D.M. WORRALL and M. BIRKINSHAW
Harvard-Smithsonian Center for Astrophysics, Cambridge, MA 02138, U.S.A.

Abstract.
Prior to ROSAT, separation of X-ray components in radio galaxies has been limited to a few well-known sources, e.g., M 87 and Cen A. Now, from ROSAT PSPC measurements of the first six objects in our study of low-power radio galaxies, we find that both resolved (thermal) and unresolved X-ray emission in a single source is typical (Worrall & Birkinshaw 1993; ApJ, submitted). The angular size of, and fraction of luminosity in, the resolved X-ray emission varies between objects. There is evidence to relate the unresolved X-ray emission with the inner radio jet.

Our joint X-ray spatial fits to unresolved and resolved emission (characterized by a thermal β model) in the six radio galaxies are better than those to either component alone. Spectral fits to two components (a two-temperature gas, or a one-temperature gas plus a power law) are better than those to one component. One source, NGC 326, is anomalous: its X-ray-emitting gas is very extended (of cluster dimension) and asymmetric; more complex models are required here. Two X-ray components characterize each of the other sources adequately, as shown by the self-consistency of our spatial and spectral fits (Table 1).

The X-ray data alone do not determine whether the unresolved component is thermal or non-thermal, although for NGC 6251, where the unresolved component is dominant, we have used gas-confinement properties to argue for a non-thermal origin (Birkinshaw & Worrall 1993; ApJ, 412, 568). A proportionality between the X-ray power-law (from our two-component fit) and radio-core luminosity densities (Fig. 1) further supports an origin for most of the unresolved X-ray emission as non-thermal radiation from the inner regions of a parsec-scale radio jet.

Table 1
% counts in unresolved/total (spatial fit) & power law/total (spectral fit)

Galaxy	Spatial Fit[a]	Spectral Fit[a]
NGC 4261	51 ± 4	52 ± 6
NGC 315[b]	61 ± 7	57 ± 9
...	41 ± 5	54 ± 7
4C 35.03	29 ± 6	52 ± 15
NGC 6251	93 ± 5	78 ± 6
NGC 2484	65 ± 9	76 ± 15
NGC 326	10 ± 2	25 ± 2

a. 1σ statistical errors for best-fit model; systematic errors in model parameters not included.
b. Two PSPC exposures listed separately.

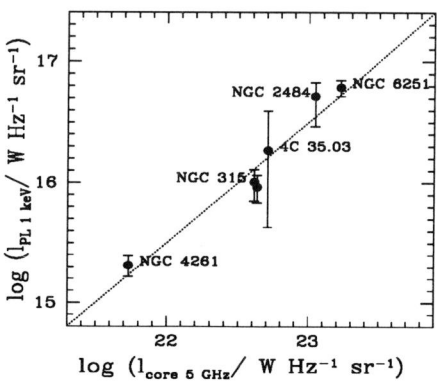

Fig. 1. X-ray power-law and radio core luminosity densities; correlation supports association of the unresolved X-ray component with the inner radio jet. (NGC 326 excluded; see text.)

Poster Contributions:
Variability

CENTIARCSECOND STRUCTURE AND VARIABILITY OF AGN

D.R. ALTSCHULER and L.I. GURVITS*
NAIC**, Arecibo Observatory, P.O. Box 995, Arecibo, Puerto Rico 00613

W. ALEF and D. GRAHAM
MPIfR, Auf dem Hügel 69, D-53121 Bonn, Germany

B. DENNISON
Dept. of Physics, Virginia Polytech. Institute and State University, Blacksburg, VA 24061-0431, USA

J.E. CARSON
MIT, Cambridge MA 02139, USA

and

A.S. TROTTER
Harvard University, Cambridge, MA 02138, USA

One possible approach to distinguish between "intrinsic" and "external" mechanisms for variability implies a study of possible correlations between flux variability and the fine structure of sources. We present results of snapshot global VLBI observations at 327 MHz of a sample of 16 extragalactic radio sources (1 radio galaxy, 9 quasars, 4 BL Lacs, 2 unidentified) selected on the basis of the bi-monthly flux monitoring at five frequencies (1400, 880, 606, 430, and 318 MHz). The observed source sample is presented in the Table I.

TABLE I
The sample of low frequency variable extragalactic radio sources

0116+319	0735+178	1422+202	2050+363
0235+164	0851+202	1611+343	2145+067
0333+321	1055+018	1633+382	2230+114
0723−008	1117+146	1901+319	2251+158

The VLBI data show clear evidences of significant scattering at baselines 2 − 6 Mλ and longer. All 16 sources are mapped. The main qualitative result confirms the existence of an anticorrelation between source typical size and their indices of variability. The observational data and the discussion are presented in full in Altschuler et al., 1994, AJ, in preparation.

---

* On leave from Astro Space Center of P.N. Lebedev Physical Institute, Moscow, Russia
** NAIC is operated by Cornell University under the Cooperative Agreement with the National Science Foundation

3C 345: THE PERIODS IN THE OPTICAL VARIABILITY AND FURTHER CONFIRMATION OF A CONNECTION BETWEEN OPTICAL OUTBURSTS AND SUPERLUMINAL COMPONENTS OF RADIOJET

M.K.BABADZHANYANTS and E.T.BELOKON'
Astronomical Institute of St.Petersburg University Peterhof, St.Petersburg, 198904, Russia

We present new results of our monitoring program for the superluminal quasar 3C 345 having been in continuous operation since 1968. The photographic B-band observations newly reported were made in 1984-91 during 218 nights. The optical light curve (1965-91) containing all available B-band observations was obtained.

It was supposed earlier (Babadzhanyants and Belokon, 1985) that in 60-70s the appearance of the superluminal jet components of 3C 345 are connected with the large amplitude optical outbursts. Now we analyze the structure of the light variations in 80s for comparison with the new VLBI observations.

The Fourier analysis was performed using only 1979-90 data set. It supposes 695-day period with confidence. The mean light curve obtained by means of the phase diagram for the 695-day period using all 1979-90 observations (N=506) represents the outburst having two-peaked main maximum and secondary one. This mean curve turned out quite identical both in form and duration to the 1967-68 outburst having been observed by Kinman et al. (1968) in detail.

Moreover, the power spectrum obtained for united (1965-69 + 1979-90) time intervals containing nearly equal number of data points supposes practically the same period (702-days), mean light curve remaining the same as earlier. We mark that determined periodic component corresponds completely to 800-days harmonic revealed by Barbieri et al. (1977) who had analyzed 3C 345 light curve up to 1976 only. Besides, a new large optical outburst beginning in 1991 shows much more reality of nearly 5-years period suspected by number of authors repeatedly.

The "birth-times" of the new superluminal components C4, C5, C6 (T.Krichbaum private communication) and C6' by Tang et al. (1989) observed in 80s coincide with the beginnings of revealed periodical outbursts quite similarly as for the events in 60-70s reported earlier.

We are grateful for support from Russian FFR grant 93-02-17237

References

Babadzhanyants, M.K. and Belokon, E.T.: 1985, *Astrofizika*Vol. 23, pp. 459-471. English translation: *Astrophysics*Vol. 23, pp. 639-649
Barbieri, C., Romano, G., di Serego, S., Zambon, M.: 1977, *A&A*Vol. 59, pp. 419-426
Kinman, T.D., Lamla, E., Ciurla, T., Harlan, E., Wirtanen, C.A.: 1968, *Ap.J.*Vol. 152, pp. 357-374
Tang, G., Ronnang, B. and Baath, L.: 1989, *A&A*Vol. 216, pp. 31-38

THE RECENT LIGHTCURVE OF 3C 345

K.J. SCHRAMM[1], U. BORGEEST[2] and J. V.LINDE[2]
[1] Université de Liège, Inst. d'Astrophysique, Belgium
[2] Hamburger Sternwarte, Germany

and

S.J. WAGNER and J. HEIDT
Landessternwarte Heidelberg, Germany

Abstract. We present the lightcurve of 3C 345 (1641+399, $z = 0.595$) in Johnson R. The data until summer 1992 are analysed and discussed in detail in Schramm et al. (A&A, Nov. 1993). The more recent lightcurve is almost flat ($R \simeq 16.9$), giving new constraints on variability models, see Camenzind, *this proceedings*.

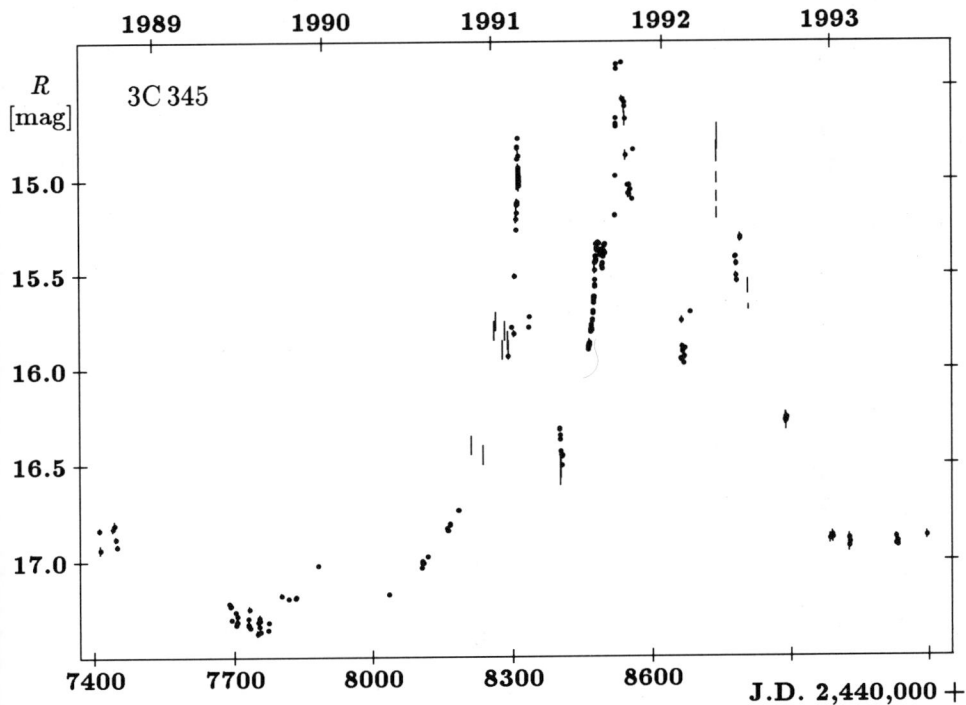

Fig. 1. Dots: data collected at Calar Alto, Spain, or LSW Heidelberg; data from other observers (for refs. see Schramm *et al.*) are indicated by the error bars only.

References

Schramm K.-J., Borgeest U., Camenzind M., Wagner S.J. et al., 1993, *A&A, to appear in Nov. 93*

VARIABLE IR AND OPTICAL SOURCES IN BLAZARS

V.A.Hagen-Thorn
St.-Petersburg University, Bibliotechnaya Pl. 2, Petrodvoretz, 198904, St.-Petersburg, Russia
E-mail: vayak@astro. lgu.spb.su

As was shown by Choloniewski (1981), in the frame of two-component model (constant component + variable source) the confrontation of the fluxes observed in various spectral bands permits to find relative spectral energy distribution (if it is unchanged) of the variable source. In this case the observed points are settled on straight lines the slopes of which give the flux ratios in various bands for the variable source.

The published multicolour data for 3C 345 and OJ 287 in the outbursts of 1983-84 are analyzed with this technique. The spectra of variable sources for the whole spectral range investigated are given in Fig.I (3C 345) and Fig.2 (OJ 287) as dots. The solid lines give the calculated spectra of homogeneous synchrotron sources (in Fig.I for three values of critical frequency ν_c (ν_c = const $H_\perp E^2_{max}$) differing by a factor 2). The agreement between observed and calculated spectra is quite well. Taking into account the high optical and IR polarization observed the synchrotron nature of variable sources is beyond doubt. As one can see from Fig.I the value $H_\perp E^2_{max}$ is determined very exactly.

For details see the papers submitted to MN RAS (3C 345) and A&A.

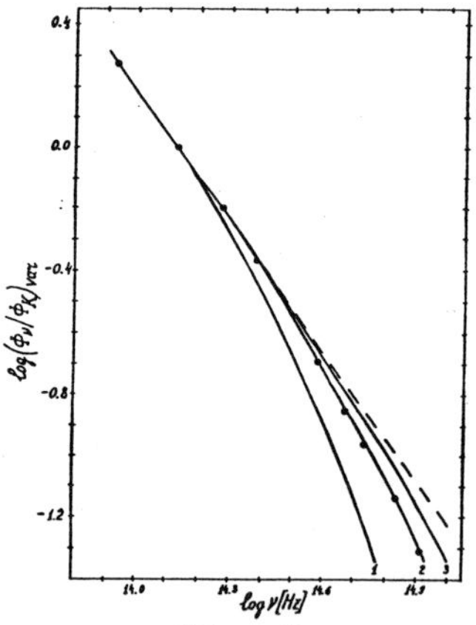

Fig.I

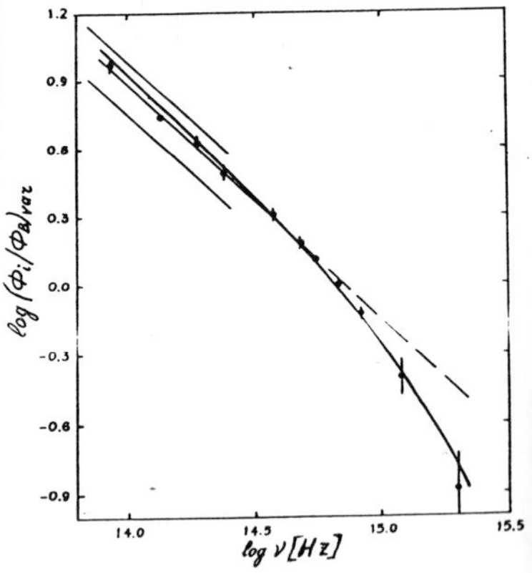

Fig.2

References.

Choloniewski J. 1981, Acta Astronomica, v.31, p.293.

T. J.-L. Courvoisier and A. Blecha: Multi-Wavelength Continuum Emission of AGN, 392.

OPTICAL MICROVARIABILITY IN COMPLETE SAMPLES OF BL LAC OBJECTS

JOCHEN HEIDT
Landessternwarte Königstuhl, 69117 Heidelberg, Germany

A radio-selected (1 Jy, Stickel et al., 1991) and a x-ray selected (EXOSAT, Giommi et al., 1991) sample of BL Lac objects was monitored by performing relative CCD photometry in order to examine the duty cycle, the dominant time-scales and the typical amplitudes of the variability. The samples consist of 34 and 11 objects, respectively.

28 (82%) of the radio-selected BL Lac objects (RBLs) were variable at any given epoch, whereas only 6 (55%) x-ray selected BL Lac objects (XBLs) displayed variability behaviour. One RBL and two XBLs were variable on time-scales longer than the length of the specific campaign (typically 7 nights). The amplitudes are 10-20% on average in RBLs, in most extreme cases 70% (on time-scales as short as 1-2 days). The mean amplitude in the XBLs was 5%, but never higher than 10%. Several RBLs were observed during two or more campaigns. None of these objects changed its variability behaviour. This implies that the duty cycle in these objects remain constant and may be a characteristic property of these objects. The non-variable objects were the optically faintest during the observations among both samples. It seems that there is a correlation between duty cycle and apparent brightness (in only two objects of each sample the errors are too large to detect microvariability). This, however, has to be confirmed by repeated measurements of the duty cycle in these objects.

The fraction of variable RBLs is higher than that of variable XBLs and they show larger amplitudes of the variability. This implies that on average the contribution of the nonthermal flux from the jet to the total flux is in RBLs higher than in XBLs and supports the unified scheme picture, in which the XBLs and RBLs are FR I radio galaxies seen at different angles between jet axis and the line-of-sight.

Acknowledgements

This work was supported by the DFG (Sonderforschungsbereich 328).

References

! Giommi, P., et al.: 1991, *Ap.J.*, **378**, 77
! Stickel, M., et al.: 1991, *Ap.J.*, **374**, 431

LONG-TERM OPTICAL BEHAVIOR OF BL LAC OBJECTS

S.KIKUCHI

National Astronomical Observatory, Mitaka, Tokyo 181, Japan

Abstract. On the basis of results of optical photometry and polarimetry of OJ287 and Mrk421 since 1981, we find two states of activities for both objects. In OJ287, a color-magnitude relation which is consistent with the standard jet-shock model is seen after the outburst in 1983, while before 1982 the optical colors were almost constant. In Mrk421, the colors became redder in 1985-87, and the change of the preferred direction of polarization was associated on smaller time scale in 1987.

We have been making optical polarimetry and photometry of bright BL Lac object OJ287 and Mrk421 with the multichannel polarimeter on the 91cm telescope since late in 1980, and the results were already given (Kikuchi 1992). In OJ287, we find a relation that the spectrum steepens as the brightness decreases after the outburst in 1983. This is explained by a standard jet-shock model (Marscher and Gear 1985). On much smaller time scale, this trend is more distinct, and we consider that the relation predicted by the jet-shock model has been observed in a course of evolution of a single flare.

However, photometric results in 1970s (Kikuchi et al. 1976, Hagen-Thorn 1980) and also in 1980-82 show that the colors in the optical region are almost constant and independent from the brightness. These two branches on the color-magnitude diagram are crossing around B-V=0.45. This leads that a basic, in other words, a less variable component, which is related probably with the emission from a jet, has become less dominant since 1983. It should be noted that in the outburst in 1972, much larger energy than in 1983 has emitted. If the constant color branch suggests that acceleration processes are well developed in a jet, it is likely that a large scaled outburst will occur only when the colors becomes blue, i.e. B-V is around 0.4.

On the other hand, Mrk421 became redder rather drastically in 1985-87 followed by a change of the preferred direction of polarization by about 40 deg in 1987. It will be useful in future to watch carefully the variations of polarized light, since rather sudden increases of polarized flux followed by slow decreases were observed at the starts of flares in 1982, 87 and 92.

References

Kikuchi, S.: 1992, in *Variability of Blazars*, ed. E.Valtaoja and M.Valtonen , p.289
Kikuchi, S., Mikami, Y., Konno, M., and Inoue, M.: 1976, *Publications of the ASJ* **28**, 117
Hagen-Thorn, V. A.: 1980, *Astrophysics and Space Science* **73**, 267
Marsher, A. P., and Gear, W. K.: 1985, *Astrophysical Journal* **298**, 114

LINE PROFILE VARIATIONS IN AGN

W. KOLLATSCHNY and M. DIETRICH
Universitätssternwarte, Geismarlandstr. 11, D-37083 Göttingen, F.R.G.

An international collaboration is monitoring the variable Seyfert galaxy NGC 5548 in the optical spectral range since 1988. In Fig. 1 (left) the Hα light curves of the blue wing (-6000 until -1000 km s^{-1}) and the red wing ($+1000$ until $+6000$ km s^{-1}) are shown for the first year of the monitoring campaign from Dec.1988 until Oct.1989. It can be seen that these line components of the Hα profile have different amplitudes in the light curves. The mean Hα and Hβ difference spectra with respect to the minimum state are plotted in Fig. 1 (right) for the same period. The relative strength of the blue component at $v_{rel} = -2000$ km s^{-1} is different with respect to the core of the line profiles. Therefore, these components originate under different physical conditions or in regions with different dust content.
This work has been supported by DFG grant Ko 857/13-1

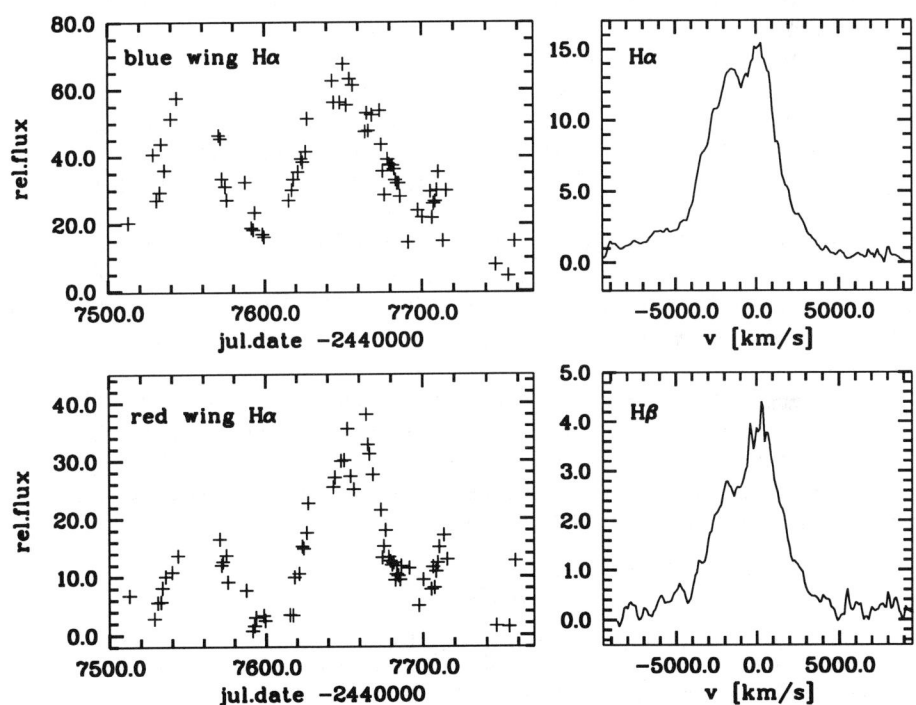

Fig. 1. Light curve of the blue and red wing of the Hα emission line (left). Mean difference spectra of the Hα and Hβ emission lines for the first year of the monitoring campaign of NGC 5548 (right).

STRUCTURE FUNCTION ANALYSIS APPLIED TO THE METSÄHOVI SAMPLE

MARKKU LAINELA
Tuorla Observatory, University of Turku
SF-21500 Piikkiö, Finland

and

ESKO VALTAOJA
Metsähovi Radio Research Station
Metsähovintie
SF-02540 Kylmälä, Finland

Our monitoring program in Metsähovi, Finland, consists of a complete sample of bright, flat-spectrum extragalactic compact radio sources. For the structure function analysis we have selected 42 best observed objects from this sample. The structure function analysis gives us two parameters, the slope value α and the maximum variability timescale T_{max}.

In our sample the slope value a varies between 0.6 and 3.9, but in most cases $\alpha \leq 1.5$. The mean value for α is 1.2 at 22 GHz and 1.3 at 37 GHz. All sources with $\alpha > 1.5$ have flux curves that are dominated by one or a few prominent outbursts that steepen the structure function. It is likely that over a greater timespan even these sources would have structure functions with smaller slopes. Thus, one can say as a general conclusion that in most, if not all, cases the typical noise type at 22 and 37 GHz is shot noise (i.e., $\alpha \approx 1$). This is consistent with there being one basic physical mechanism of millimeter variability for all compact radio sources and together with the successfully modelling at centimeter wavelengths our results support the idea that shock-in-jet models can explain most of the time variability in all compact radio sources.

The time scales of variability support an orientation-dependent model, where HPQs have smaller average viewing angles than LPQs. In many scenarios one assumes that in BL Lacs the jet is pointing almost directly towards us and so T_{max} for BL Lacs should be the smallest among AGN. However, we do not find this kind effect in our data, on the contrary there seems to be a wide spread in T_{max}. One interesting possibility is that BL Lacs really consist of two different classes of objects (Valtaoja et al. 1992).

References

Lainela, M., & Valtaoja, E. 1993, ApJ (in press)
Valtaoja, E., Teräsranta, H., Lainela, M., & Teerikorpi, P. 1992, in Variability of Blazars, ed. E. Valtaoja & M. Valtonen (Cambridge: Cambridge Univ. Press), 70

A JET MODEL INTERPRETATION OF MULTI FREQUENCY FLUX OBSERVATIONS OF RADIO OUTBURSTS IN THE AGN 0235+16

Y.Y. KOVALEV, G.M. LARIONOV
Astro Space Center, Lebedev Physical Institute
Profsojuznaja street, 84/32
117810, Moscow
Russia

There are analyzed 1.5 - years observations of a series of bursts in the quasar 0235+16 at 8 frequencies between 0.3 GHz and 15 GHz from July 1981 to December 1982, obtained by Altschuler et al. (1984) and Aller et al. (1985). These observations are compared with a Hedgehog model (see Kovalev and Mikhailutsa, 1980, with full references).

Results for 3 of 7 fitted spectra are presented on the Fig.1. Points reflect the real simultaneous observations, but crosses - interpolated data. Solid lines are model results. All 7 observational spectra are in agreement with in this model calculated spectra by the χ^2 - criteria on the validity level of 0.05 (the level of faithfulness is 0.95). Observations can be explained by the synchrotron emission of a narrow jet of relativistic particles in a strong radial magnetic field of the active nucleus.

The variability of the spectra is explained by a time variability of this particles flow. The jet starts at a distance of ~1.E19 cm from the center of the AGN. On the preliminary estimations, the observed brightness temperature and a magnetic field near the start of the jet can be equal or less than 4.E11 K and 0.5 Gauss, respectively.

References

Aller,H.D., Aller,M.F., Latimer,E., Hodge,P.E. (1985). Astrophys. J. Suppl. V.59. P.513.

Altschuler,D.R., Broderick,J.J., Condon,J.J., Dennison,B., Mitchell,K.J., O'Dell,S.L., Payne,H.E. (1984). Astron. J. V.89. P.1784.

Kovalev, Yu.A., Mikhailutsa V.P. (1980). Sov.Astron. V.24. P.400.

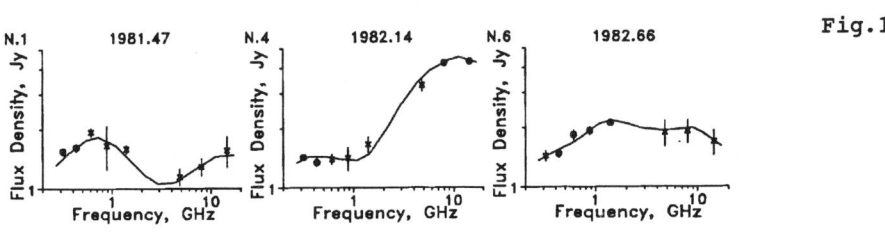

Fig.1.

TIME SCALE OF RADIO SOURCES ACTIVITY FROM THE STATISTICAL MODEL OF VARIABILITY

P. MAGDZIARZ and J. MACHALSKI
Astronomical Observatory, Jagiellonian University
ul. Orla 171,
PL-30244 Cracow, Poland

The lack of a plateau in the <u>average</u> structure function (hereafter SF) for the <u>complete</u> flux-limited sample of Seielstad *et al.* (1983) [Fig.1] suggests that the <u>mean</u> time scale of radio sources variability at 10.8 GHz should be longer than the 3.8-yr time base of those observations. A comparison of the variability time scales derived from the SF analysis with those predicted by the statistical model of Magdziarz & Machalski (1993, AA, in print) [Fig.2] confirms consistency of that model. The mean time scale of variability of $<\tau>=6\pm2$ yr was found in both cases for $\nu=10.8$ GHz.

The mean variability time scale characterizing the <u>total</u> population of radio sources seems to be longer than that (about 2 yr) commonly accepted from observations of <u>selected</u> variables (*e.g.* Altschuler 1989, Fund.Cosm.Phys. **14**,37; Hughes *et al.* 1992, ApJ **396**,469). This is consistent with the selection effects if the shot noise character of variability (Cruise & Dodds 1985, MNRAS **215**,417; Hughes *et al.* 1992, ApJ **396**,469), giving a wide range of time scales, is taken into account, and means that many sources can likely vary at time scales exceeding 40–50 yr. The linear character of intermediate time-lags SF found by Hughes *et al.* for UMRAO sources agrees with predictions of our statistical model, and confirms the universal physical nature of the variability. This is also consistent with predictions of the standard shock model of Marsher & Gear 1985, ApJ **298**,114 (*cf.* also Valtaoja *et al.* 1992, AA **254**,71).

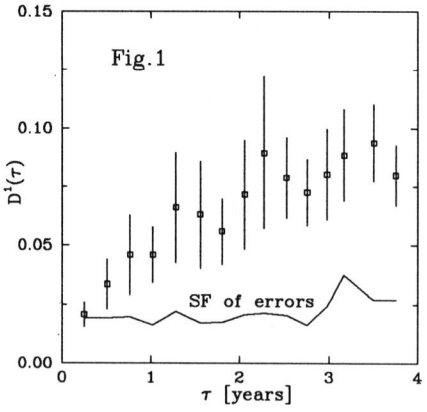
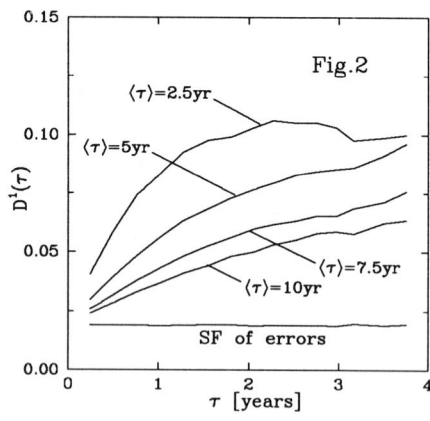

HOW LARGE ARE QUASAR BROAD-LINE-REGIONS? RESULTS FROM A PROGRAM TO MONITOR THE PG QUASARS

DAN MAOZ and BUELL T. JANNUZI
Institute for Advanced Study, Princeton, NJ, 08540, USA

PAUL S. SMITH
Steward Observatory, University of Arizona, Tucson, AZ 85721, USA

and

SHAI KASPI and HAGAI NETZER
Wise Observatory, Tel-Aviv University, Tel-Aviv 69978, Israel

We have monitored spectrophotometrically a subsample (28) of the Palomar-Green Bright Quasar Sample for two years in order to measure the sizes of the broad-line regions of high-luminosity AGNs. Half of the quasars showed optical continuum variations with amplitudes in the range 20-75%. In most objects with continuum variations, we detect correlated variations in the broad $H\alpha$ and $H\beta$ emission lines. The amplitude of the line variations is usually 2-4 times smaller than the optical continuum fluctuations. The lines respond to the continuum variations with a lag that is smaller than or comparable to our typical sampling interval (a few months). This suggests that the quasars have broad-line regions smaller than about 1 lt-year. The figures below show spectra and light curves for one of the quasars. Two of the quasars monitored show no detectable line variations despite relatively large-amplitude continuum changes. This could be a stronger manifestation of the low-amplitude line-response phenomenon we observe in the other quasars. Further details appear in Maoz et al. (1994, ApJ, Jan 20, in press).

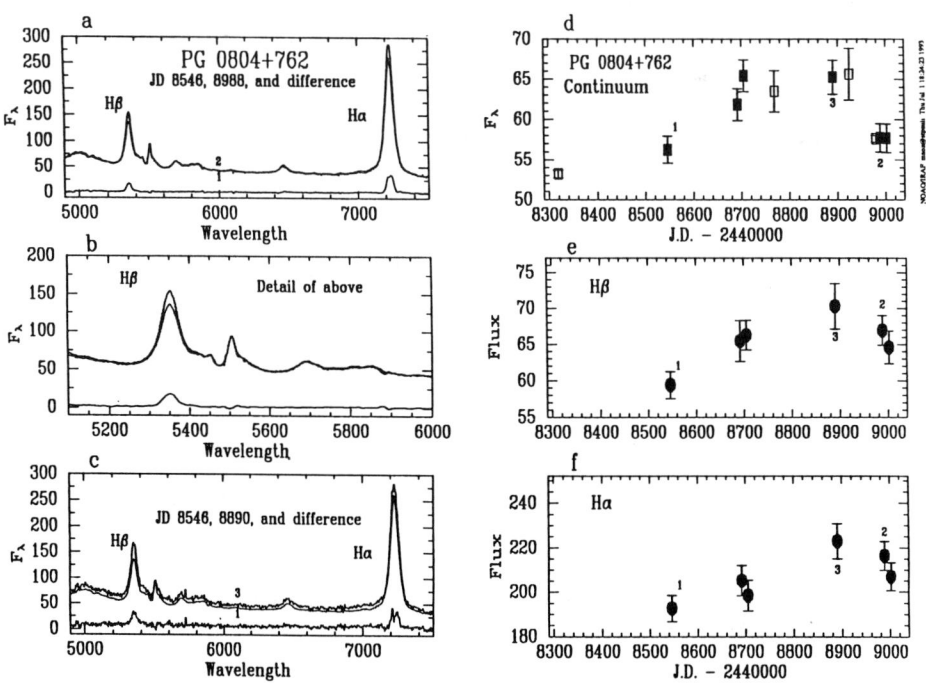

SPECTRAL VARIABILITY OF SIX BRIGHT BL LAC OBJECTS IN THE NEAR IR

E. MASSARO, R.NESCI AND G.C. PEROLA
Istituto Astronomico, Università di Roma "La Sapienza", Roma (Italy)

and

D. LORENZETTI AND L.SPINOGLIO
IFSI, CNR, Frascati (Italy)

The results of several photometric campaigns (1986-1993) in the near IR of the six bright BL Lac objects 3C 66A, PKS 0422+004, PKS 0735+178, PKS 0754+100, PKS 0829+046 and OQ 530 are presented. The observations were carried out at the 1.5 m Italian IR Telescope at Gornergrat (TIRGO - 3150 m a.s.l.) equipped with a IR photometer using a InSb detector cooled at the solid nitrogen temperature and the standard J, H, K filters. The principal aims of our program are the search of rapid variability (typical time scales of 1 day or less) and of correlations between the flux level (typically in the J band, where the largest variation amplitude is found) and the spectral slope. In the following we summarize some relevant results.

Variations ranging from 0.4 to about 1.1 mag have been observed for all the sources in the three bands. Flux changes of ~ 0.3 mag at distance of two or three days were found in several occasions but, on shorter time scales, variations greater than ~ 0.2 mag are not frequent events. Remarkable episodes were the ≤ 1 day brightening by 0.25 mag in K and 0.16 mag in H, while J did not change, of PKS 0754+100 and the change by 0.4 mag in J, 0.14 mag in H, without a detectable variation in K, which occurred in PKS 0735+178 on a time scale of ~ 1 hour.

Our data do indicate that the correlation between the J flux and the spectral slope of the NIR emission, reported in previous papers, cannot be considered a general property of the BL Lac objects. Two sources (PKS 0829+046 and OQ 530) show large changes of the slope without evidence of correlation with the J flux. A practically constant spectral index is observed in PKS 0754+100 (with the only exception of the event reported above), despite the J flux varies by a factor of 2. A possible correlation can be envisaged for PKS 0422+004 by considering the average spectral index and flux on a time scale of a few years. Finally, 3C 66A and PKS 0735+178 have a mild correlation: the weighted coefficients of linear correlation are 0.54 and 0.66 corresponding to probabilities to have a chance effect less than 0.03 and 0.01, respectively.

SPECTRAL VARIABILITY OF NGC 4151 IN 1972-1991

V.L. OKNYANSKIJ and V.M. LYUTYI
Shternberg State Astronomical Institute, Universitetskij Prospect 13, Moscow, 119899, Russia,
e-mail: OKNYAN@SAI.MSK.SU

and

K.K. CHUVAEV
Crimean Astrophysical Observatory, Nauchny,334413, Crimea, Ukraine

After a minimum in the spring of 1984, when the broad permitted lines were found to become much weaker, the spectrum of NGC 4151 became much more like a Sy 2 (Lyutyi et al.,1984; Chuvaev and Oknyanskij, 1989; Penston and Perez, 1984). Recently, this object returned to the high state after about 8 yr in the semiquiescent phase (Oknyanskij et al., 1991).

The purpose of the present work was to investigate the variability of different emission components in the spectrum of NGC 4151 over a long series of coordinated spectral (2.6-m telescope) and photometrical (60-cm telescope) observations which were made with the same equipment during several hundreds nights in last two decades. Additionally, we use in our investigations other published (or obtained from private communications) spectral, IR (see details of near IR data reduction in Oknyanskij, 1993), UV (IUE archive) and X-ray data . Using all these data we drew combined light curves and then investigated them with the Fourier and cross-correlation analysis.

The main results are the following:

1. To obtain clear power spectra, we applied the CLEAN algorithm to optical continuum light curves in two time intervals: 1968-1977 and 1978-1988. We revealed that the semiregular component $1/66^d$ was present only during the first interval.

2. Cross-correlation functions for the lines H_β, H_α vs. F(U) during "low" and "active" states are very different. For the "low" state we surely found time delays between the optical continuum and the line variations about 12±5 days for H_β and H_α. For "active" state the values of time delays are about the same but can't be estimated so exactly in view of the broader profile of peaks in the cross-correlation functions.

3. Measurable time delay between the UV, X-ray (2-10 keV) and optical continuum light curves is abdent, but near IR flux F(K) variations appear to lag those in optical F(U) by 18 ± 6 days.

4. From the modeling we concluded that the IR emission region can be better modeled as a thin disk viewed nearly edge-on than as a thick spherical shell.

References

Chuvaev K.K., Oknyanskij V.L., 1989, *Sov. Astron.* **Vol.33**, p.1
Lyutyi V.M., Oknyanskij V.L., Chuvaev K.K., 1984, *Sov. Astron. Lett.* **Vol. 10**, p.335
Lytyi V.M.: 1977, *Astron. Zhurn.* **Vol. 54**, p.1153
Oknyanskij V.L., Lyutyi V.M., Chuvaev K.K., 1991, *Sov. Astron. Lett.* **Vol. 17**, p.100
Oknyanskij, V.L., 1993, *Pis'ma v Astron. Zhurn.* **Vol. 19**, N 11, p.1021
Penston M.V., Perez E., 1984 *MNRAS* **Vol. 211**, 33P

LONG TERM X-RAY VARIABILITY OF NGC 4151

I. E. PAPADAKIS and I. M. M^{C}HARDY
Physics Department, Southampton University,
University Road, Southampton, SO2 5NH, UK

Introduction. Short time scale X-ray power spectra of AGN are in general well fitted by a power law with slopes between -1 and -2 but we expect these slopes to flatten at low frequencies (indication of such a flattening has already been seen in NGC 5506 [1]). We have searched for such a low-frequency break in the power spectrum of NGC 4151 by investigating its long term X-ray light curve (2-10 keV). To construct this light curve we used Ariel V SSI, OSO-8, HEAO-1, Ariel VI, EXOSAT ME and GINGA LAC data.

Method. We have developed a new method to estimate the power spectrum (ps) of this unevenly sampled light curve. This method is based on the "unbiased discrete correlation" technique [2] which we use to estimate the auto-covariance function of the light curve, $DACF(k)$, at lag k. The power spectrum is then estimated by computing the Fourier transform of $DACF(k)$. In this way the estimated ps is not affected by the "window function" of the uneven sampling pattern.

We have calculated the ps for 8 different bin sizes ($\Delta k = 3, 6, 10, 20, 30, 40, 60,$ and 80 days). Small bin sizes help us estimate the ps at high frequencies but we have to use larger bin sizes to estimate the ps at low frequencies. We smoothed the 8 individual ps using a simple rectangular window and finally built up the overall ps using contributions from all of them.

Results. We found that our longterm X-ray power spectrum of NGC 4151 is consistent with a power law model with slope of ~ -2.4 which flattens at frequencies below 3×10^{-7} Hz. This frequency corresponds to a time scale of ~ 38 days.

It is interesting to compare the break frequencies (bf) in Cyg X-1 and NGC 4151. We found that $\mathrm{bf}_{4151}/\mathrm{bf}_{CygX-1} \sim 7 \times 10^{-6} - 8 \times 10^{-7}$ (for Cyg X-1, bf $\sim 0.04 - 0.37$ Hz [3]). If bf in these objects scales proportionally with the central black hole mass, then the black hole mass in NGC 4151 should be between 1.4×10^6 $M_\odot$ and 1.2×10^7 $M_\odot$ (assuming a 10 $M_\odot$ black hole in Cyg X-1). If the break frequency corresponds to the viscous time scale of the disk at ~ 5 Schwarzschild radii (where most of the X-ray luminosity comes from), then for a 10^7 $M_\odot$ black hole the viscosity parameter should be larger than 0.01.

References.
[1] McHardy, I.M. 1989, in "Two Topics in X-ray Astronomy", Proc. 23rd ESLAB Symp., Italy, p. 1111
[2] Edelson, R.A. and Krolik, J.H. 1988, ApJ, 333, 646
[3] Belloni, T. and Hasinger, G. 1990, A&A, 227, L33

VARIABILITY OF THE ULTRAVIOLET CONTINUUM AND EMISSION LINES OF NGC 3783

G.A. REICHERT
Universities Space Research Association
NASA-GSFC, Code 668, Greenbelt, MD 20771

On behalf of the International AGN Watch, I report on the results of intensive ultraviolet spectral monitoring of the Seyfert 1 galaxy NGC 3783. The nucleus of NGC 3783 was observed with the *International Ultraviolet Explorer* satellite on a regular basis for a total of seven months, once every 4 days for the first 172 days and once every other day for the final 50 days. Significant variability was observed in both continuum and emission-line fluxes. The light curves for the continuum fluxes exhibited two well-defined local minima or "dips," the first lasting $\lesssim 20$ days and the second $\lesssim 4$ days, with additional episodes of relatively rapid flickering of approximately the same amplitude. As in the case of NGC 5548 (the only other Seyfert galaxy that has been the subject of such an intensive, sustained monitoring effort), the largest continuum variations were seen at the shortest wavelengths, so that the continuum became "harder" when brighter. The variations in the continuum occurred simultaneously at all wavelengths ($\Delta t < 2$ days). Generally, the amplitude of variability of the emission lines was lower than (or comparable to) that of the continuum. Apart from Mg II (which varied little) and N V (which is relatively weak and badly blended with Lyα) the light curves of the emission lines are very similar to the continuum light curves, in each case with a small systematic delay or "lag." As for NGC 5548, the highest ionization lines seem to respond with shorter lags than the lower ionization lines. The lags found for NGC 3783 are considerably shorter than those obtained for NGC 5548, with values of (formally) ~ 0 days for He II+O III], and ~ 4 days for Lyα, and C IV. The data further suggest lags of ~ 4 days for Si IV+O IV], and $8-30$ days for Si III]+C III]. Uncertainties in these quantities are likely to be of order $2-3$ days for the stronger features (Lyα, C IV), and $3-4$ days for the weaker ones (He II+O III], Si IV+O IV], Si III]+C III]). Mg II lagged the 1460 Å continuum by ~ 9 days, although this result depends on the method of measuring the line flux, and may in fact be due to variability of the underlying Fe II lines. Correlation analysis further shows that the power density spectrum contains substantial unresolved power over time scales of $\lesssim 2$ days, and that the character of the continuum variability may change with time.

This research was supported by NASA grant NAG5-1824 to Ohio State University and NASA contract NAS5-32474 to Computer Sciences Corporation.

COORDINATED MULTIFREQUENCY OBSERVATIONS OF OJ 287 AND MK 421 (MARCH 1993)

A. SILLANPÄÄ and L.O. TAKALO
Tuorla Observatory, Tuorla,
21500 Piikkiö, Finland

E. VALTAOJA and H. TERÄSRANTA
Metsähovi Radio Research Station, Finland

YU.S. EFIMOV and N. SHAKHOVSKOY
Crimean Astrophysical Observatory, Ukraine

and

J. HEIDT and H. BOCK
Landessternwarte Königstuhl, Germany

We present some very preliminary results of the multifrequency microvariability campaign of two BL Lac objects OJ 287 and Mk 421 carried out mainly during four days in March 1993. During these four days also IUE was observing these blazars. In this poster we present only a small part of the whole data: radio observations at Metsähovi, CCD observations on La Palma and Heidelberg and UBVRI photopolarimetric observations at Crimea. Because of the very bad global weather conditions and also because of the faintnes of OJ 287 our optical observations were very limited.

In the radio bands both objects were quite stable during the campaign, but in the optical region both were more variable. The brightness of Mk 421 increased about 30 % in two days (in R-band). OJ 287 was very weak and also more stable but a typical sinusoidal wave behaviour was visible in a four day timescale. The polarization properties of Mk 421 were normal: the polarization level varied slightly and was between 1.5 and 4 percent, the position angle varied between 10 and 40 degrees.

More detailed results will be presented in the forthcoming papers.

NGC 4151: AN ACCURATE UPPER LIMIT TO THE DIMENSIONS OF THE CENTRAL CONTINUUM SOURCE ?

NICOLAOS H. SOLOMOS
Physics Dept., Hellenic Naval Academy, Piraeus- 17503, Greece

Abstract. Simultaneous (5672 Å and 3400-3900 Å)) optical continuum photometric observations of the Seyfert galaxy NGC4151, were carried out within ≤4 week periods, down to a time resolution of < 1hr. A new active phase (1990) is detected with onset and slope captured in 1989, [3]. Implications of the variability record from 1983 [2], to 1990 (Fig-1), are discussed.

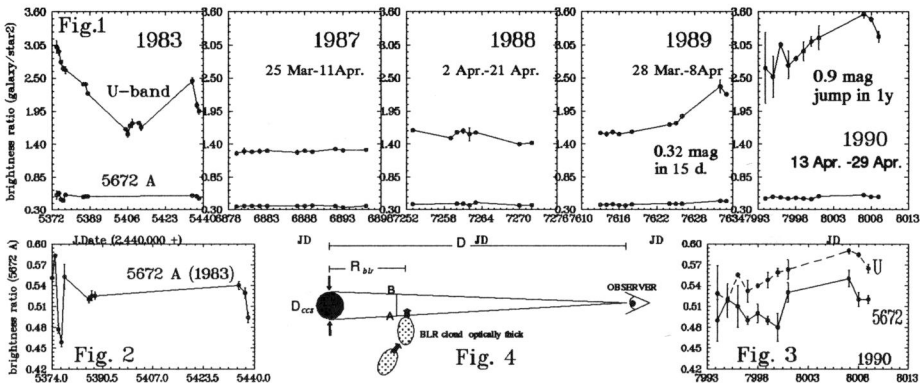

UV/Opt. variations decoupled? The close correlation between the variations of the 5672 Å and near-UV continua that appears well established during low states of activity (1987-1988), turned to at least a 5d episode of anticorrelation while on the passage to the new 1990 maximum (overlay Fig-3). Such interesting aspects of the variability are very difficult to be explained within the framework of the current accretion disk models. Instead, if the UV emission is partly originated in shocks created by the motion of supersonic clouds, then the independent rise of U could be of 'meteorological' nature, reflecting fluctuations in the cloud formation rate.

A size estimation of the central continuum source. As pointed out by Ulrich [4] rapid UV/Opt. variability cannot be attributed to corresponding changes in the infall rate of the accreted gas. Therefore, variations like the rapid (20% level) transient decrease of the 5672 Å continuum flux (∼2 days) (Fig.2), need to find alternative explanations. A passage of a rotating opaque BLR cloud moving with velocity V_c could cause the observed feature in the 5672 Å lightcurve. If the proposed picture is correct then the geometry of the very simple occultation model of Fig-4, gives the diameter of the central source in NGC 4151 as: $D_{ccs} = D.AB(D - R_{blr})^{-1}$ where D is its distance, R_{blr} the BLR radius, $AB = V_c.t$, with t the duration of the transition to the diminution (∼1 day). If R_{blr} is considered to be the luminosity weighted radius estimated by the line-continuum cross-correlation studies (9 ld for NGC4151, [1]) and for a typical V_c of 5.10^3 Km/sec, then the upper limit of the size of the central source is 2.8 AU (4.2×10^{13} cm) implying a black hole mass of $1.4 \times 10^7 M_\odot$.

References

1. Maoz, D. et al. (1990): Ap.J. 351, 75.
2. Meaburn et.al. (1985): in *'Active Galactic Nuclei'*, Ed.J.Dyson, Manchester Univ. Press.
3. Solomos et. al. (1990): A&A 229, 80.
4. Ulrich M. (1991): in *'Variability of Active Galaxies'*, Eds. W.J.Duschl et. al., LNP vol.377, Heidelberg.

VARIABILITY OF NGC 3783: FIRST RESULTS FROM AN INTENSIVE GROUND-BASED MONITORING CAMPAIGN

GIOVANNA M. STIRPE
Osservatorio Astronomico di Bologna, Via Zamboni 33, 40126 Bologna, Italy

The International AGN Watch collaboration undertook an intensive monitoring campaign of the bright Seyfert 1 galaxy NGC 3783 between December 1991 and August 1992, in order to study the variations of the continuum and broad emission lines. Spectroscopic and photometric observations took place at several ground-based observatories and formed the optical/IR counterpart to the UV observations conducted with the IUE (Reichert et al. 1993).

The data allowed to derive the light curves of the optical continuum and of the $H\beta$ flux with a median sampling of about 2 days. Both curves strongly resemble the UV continuum light curves, and are characterized by a series of rapid, low-amplitude minima, which have counterparts in the UV curves. The optical continuum varied by up to 50% and the $H\beta$ flux by up to 30%.

Cross correlation analysis shows that there is no significant delay between the optical and UV continuum curves, which confirms the trend observed between different UV continuum bands. The light curve of $H\beta$ presents a lag of 8 days with respect to the UV continuum. This lag is longer that those of the high ionization lines (Reichert et al. 1993), indicating that the broad line region (BLR) is stratified. All line vs. continuum lags are shorter than the corresponding ones in NGC 5548, a brighter Seyfert 1 monitored by the collaboration: this indicates that the BLR of NGC 3783 is smaller than that of NGC 5548.

The IR continuum light curve, while displaying some variability ($\sim$20% in the K band), does not resemble the UV and optical curves. It is possible that the lower sampling rate and the lower amplitude of the IR variations smoothed the IR light curve, thus hiding any counterparts to the features observed in the other bands.

A detailed description of the results obtained during the ground-based campaign is presented in Stirpe et al. (1993).

Acknowledgements

This campaign is the result of the combined efforts of many colleagues, who generously contributed to it with their data and time.

References

Reichert, G.A., et al. 1993, ApJ, in press
Stirpe, G.M., et al. 1993, ApJ, submitted

TWO YEAR UV-OPTICAL MONITORING OF THE SEYFERT 1 GALAXY MARKARIAN 335

WEI-HSIN SUN[1,2], CHRIS R. SHRADER[2,3], TRACEY J. TURNER[4],
MATTHEW A. MALKAN[5], BRADLEY M. PETERSON[6], PAUL M.N. HINTZEN[7],
YOJI KONDO[8], SUNG-NAN LIN[1], TING-CHANG LIN[1] and
REMINGTON P.S. STONE[9]

[1] *Inst. of Astron. and Dept. of Phys., National Central Univ., Chung-Li, Taiwan 32054, ROC*
[2] *Guest Observer with the International Ultraviolet Explorer Satellite*
[3] *GRO Science Support Center, NASA/GSFC, Code 668.1, Greenbelt, MD 20771, USA*
[4] *Laboratory for High Energy Astrophysics, NASA/GSFC, Code 665, Greenbelt, MD 20771, USA*
[5] *Dept. of Astro., UCLA, Los Angeles, CA 90024, USA*
[6] *Dept. of Astron., Ohio State Univ., Columbus, OH 43210, USA*
[7] *Dept. of Phys., Univ. of Nevada, Las Vegas, NV 89154, USA*
[8] *Lab. for Astron. and Solar Phys., NASA/GSFC, Code 684, Greenbelt, MD 20771, USA*
[9] *Lick Obs., Univ. of California, Mt. Hamilton, CA 95140, USA*

We report on the results of a UV-Optical spectral monitoring of the bright Seyfert 1 galactic nuclei Mkn 335. This campaign began in June, 1989, and ended in June, 1991. Ultraviolet spectra of fourteen epochs at nearly uniform sampling of 30-day intervals, except when the object was inaccessible from the IUE satellite, have been obtained, of which twelve were coordinated with quasi-simultaneous ground-based optical observations made at Lick Observatory.

Continuum variation at 20 − 30% have been observed in both the UV and the optical. Cross-Correlation analysis of line and continuum variability showed that H_β lags behind the UV continuum by approximately 30 days while H_α lags behind the UV continuum by 44 days. We fit the multi-epoch UV-Optical continua with (i) a series of power-laws with different slopes and strength, and (ii) the improved accretion disk models. The single power-law in general fits the continua well and we used these fits to derive the emission line properties. In accretion disk fitting, we used a fixed black hole mass, obtained from the averaged spectrum, and varying accretion rate. These single rate accretion disk models provide acceptable fits to almost all the spectra. A series of accretion rates have been inferred which indicated a factor of two variation in rates between the high and low states. The integrated power-law energy correlates well with that estimated from accretion rate. We also found marginal continuum color variation as the luminosity changes. However, the spectral index does not become harder monotonically when the object brightens, in contradiction to what has been suggested previously. We argue that this behavior may be an observational test for hypotheses of the nature of the AGN variability. This work is supported by the National Science Council in Taiwan, ROC, through the grant NSC 82-0208-M-008-011.

Subject headings: galaxies: individual (Mkn 335) — galaxies: Seyfert — quasars: emission lines

OJ 287: A BLAZAR WITH EVERYTHING

L.O. TAKALO
Tuorla Observatory, Tuorla,
21500 Piikkiö, Finland

Abstract.
We have collected all the observations available to us of blazar OJ 287. Here we will present preliminary results from our investigation of these data. The photometric light curves show large outbursts occuring (quasi)simultaneously in all frequences. The largest outbursts occured during 1972 and 1983. The B-band light curve can be extended to the year 1894. This being the longest available observational data set of all blazars. Polarization light curves in optical show random variations in all timescales. In the radio bands the polarization observations show well defined structure in the light curves. Light curves and simple correlation analysis based on the light curves on different frequences will be presented.

OJ 287 is one of the best observed blazars, and also one of the most exiting ones. It shows all the characteristic blazar behaviours, but it has some features that are observed only in it. One of these unigue features are the observed periodic variations, in time scales from tens of minutes to years.

We have collected all the available data on OJ 287 with the intention of doing a "complete" study of its behaviour and comparison at different wavelenghts.

The historical light curve clearly shows outbursts at 1913, 1937, 1947, 1959, 1972 and 1983. Based on these data several groups have calculated that the outbursts occur with a 11.6 year period. OJ 287 is the only blazar, in which this kind of periodicity is so well observed. Based on this periodicity one can predict that the next outburst should occur during 1994!

Comparing the light curves of the last two outbursts (1972 and 1983) indicate very similar behaviour, with the double maxima and slow decline with small (periodic) variations. The time separation between the two maxima is about one and a half years.

The polarization is seen to show random variations in time scales from days to months. During the 1983 outburst the polarization level was almost zero. But right after the outburst it was over 30 %. The "average" polarization seems to around 10 %. Position angle behaviour is really remarkable. Around the average position angle of 90 degrees the variations are fast and large. No preferred position angle can be seen.

The variability has smaller amplitude in the radio bands than in the optical, and the 1983 outbursts is not so noticeable. The polarization level is smaller than the optical one (average being 5 %). The average position angle is around 100 degrees, similar to the optical one.

We have presented preliminary results of a "complete" study of blazars OJ 287 in all frequences. OJ 287 show (periodic) variability in all frequences in time scales from minutes to years. Being the only blazar with this kind of behavior.

LONG TERM MONITORING OF AGN WITH THE METSÄHOVI AND SEST TELESCOPES

H. TERÄSRANTA, M. TORNIKOSKI and E. VALTAOJA
Metsähovi Radio Research Station, Helsinki University of Technology, SF-02540 Kylmälä, Finland

The monitoring of extragalactic sources with the Metsähovi radio telescope started in 1980. Since then we have made more than 21000 observations at 12, 22, 37, 77 and 87 GHz. Currently we have 84 sources in our sample which should be observed once a month at 22 and 37 GHz. The sample is limited to declinations above -10 degrees. In most cases our sampling is dense enough to cover all the outbursts at these two frequencies, because our earlier observations have shown the timescales of a typical outburst at 22 and 37 GHz to be from 0.3 to several years.

Our observations with the SEST telescope on La Silla mountain in Chile started in 1988. The observations have been made with a 90 GHz Schottky diode mixer receiver and a bolometer whose center frequency is around 230 GHz. With our SEST observations we both collect information of relatively little observed Southern sources and get high frequency radio data of equatorial sources which can be observed at other frequencies in other observatories.

We have now started a long term project at SEST in observing 51 near equator sources between the declinations -25 and +20 degrees. Of those sources 32 are common with the Metsähovi sample. Although getting a large portion of the Finnish quota spread over the "Swedish" months, i.e. the even months, it is still difficult to obtain a monthly sampling of these sources. This is mostly due to the limit of not observing sources within 50 degrees from the Sun, which will typically cause a 3-month gap in the flux curves. The bolometer is also utilized only during the colder months and thus the sampling will unfortunately be poorest at 230 GHz where the sources will show fastest variations.

Our two projects will give new information on amplitudes and time scales of outbursts from 22 to 230 GHz and are useful e.g. for developing and testing shock models. Also multifrequency studies are easily joined in at least with the Metsähovi telescope. Our data will help to plan VLBI observations at higher frequencies, like 100 GHz, as the source can be observed while in high state and thus visible.

The VLBI maps can also be interpreted better because our database reaches more than 10 years back (at 22 and 37 GHz). From this September a great new effort will be done in participating in the multifrequency campaign related to the Compton gamma ray satellite.

CONNECTION BETWEEN OPTICAL AND HIGH FREQUENCY RADIO VARIABILITY IN AGN

M. TORNIKOSKI and E. VALTAOJA
Metsähovi Radio Research Station, SF-02540 Kylmälä, Finland

and

A.G. SMITH and A.D. NAIR
Astronomy Dept., University of Florida, Gainesville, Fl-32611, USA

We have been searching for correlated optical and radio variability in large temporal data sets of 22 extragalactic radio sources. The optical data were obtained with the 76-cm reflector at the Rosemary Hill Observatory in Florida, USA. The radio data were obtained at two different sites: 22, 37 and some of the 90 GHz data at the Metsähovi Radio Research Station, Finland, and 90 and 230 GHz data at the Swedish-ESO Submillimetre Telescope (SEST) on La Silla, Chile. Because the SEST data unfortunately reaches only back to 1988, the 90 and 230 GHz data were complemented by the IRAM data from Steppe et al. (A&AS 75, 1988 and A&AS 96, 1992).

For the correlation analysis we used the discrete correlation function method described in Edelson & Krolik (ApJ 341, 1988), suitable for unevenly sampled data sets. Even though the data streams covered 10 years or more, we ran the analysis collecting only information of the correlations with time lags of less than about 1000 days. We also inspected the optical and radio flux curves visually both for discrete events and long-term behaviour.

We found radio-optical connections in a significant fraction of our sample. In several cases optical flares were found to occur simultaneously with the high frequency radio flares. The radio-optical connection was very clear in 10 cases, in 6 of which the time delays were practically zero, i.e. less than our bin size of 30-50 days. In 5 of the remaining sources at least some optical flares seemed to be connected with radio flares, often the only notable events in the radio data sets. All of the sources in which strong radio-optical correlation was found were either highly polarized quasars or BL Lac objects, and they were all among the most frequently observed sources.

It is possible that there are several types of optical events, some of which are not directly connected to radio events. We suggest however that the converse may be true, with optical flares occuring synchronously with all the high radio frequency flares and with time delays of zero to several hundred days at lower radio frequencies. Since optical flares are typically very rapid, it is easy to miss many such events due to undersampling. We believe that strong radio-optical connections would be detected in many more sources if the sampling density were high enough to ensure detecting all the notable flares at both optical and high radio frequencies.

OPTICAL VARIABILITY OF FAINT QSOS AND AGNS

D. TRÈVESE
Istituto Astronomico, Università di Roma "La Sapienza",
Via G. M. Lancisi 29, I-00161 Roma (Italy)

and

R. G. KRON
Fermi National Accelerator Laboratory,
MS 127, Box 500, Batavia, Illinois 60510, USA

We report some results of a new analysis (Trèvese et al. 1993, ApJ (submitted) (T93)) of the variability of the faint QSO sample of SA57 (Koo, Kron and Cudworth, 1986, PASP, **98**, 285), concerning the intrinsic time scales of variability and the dependence of the amplitude of variability on absolute magnitude and redshift. Prime focus plates of SA57 have been obtained with the Mayall 4-m telescope at the Kitt Peak National Observatory since 1974. The present analysis is based on 14 B_J plates spanning 15 years at 11 independent epochs. The digitization, object selection, image classification and photometry are described in Koo (1986, ApJ, **311**, 651) and Trèvese et al. (1989, AJ, **98**, 108 (T89)). The magnitude limit of the sample is $B_J = 22.5$ and stellar objects are selected with a threshold in image size. This criteria give a sample of 694 objects in the field, 34 of which are QSOs whose spectra have been measured with the Mayall 4-m telescope. The r.m.s. error in the B_J band is 0.05 mag at $B_J \approx 22$ mag. All but one of these QSOs appear to be variable according to T93. From the light curves of the individual QSOs is possible to derive an ensemble statistics defining the first order *structure function* : $S^2(\tau) \equiv \langle [m(t+\tau) - m(t)]^2 \rangle$, where $m(t)$ is the magnitude at the epoch t, τ is the time lag and the angular brackets indicate the ensemble average, over the 34 QSOs in the present case. In the rest-frame, a fit of the structure function with the function: $S^2(\tau) = A \times [1 - exp(-\tau/T)] + 2\sigma_n^2$, σ_n being the r.m.s noise, gives $A \approx 0.2$ mag^2 and $T \approx 1$ year. The *ensemble autocorrelation function* $C(\tau)$ of our sample is negative from $\tau \approx 1$ year up to $\tau \approx 6$ years. This anticorrelation means that the variability occurs on characteristic time scales up to ≈ 6 years (see T93). Splitting the sample in low/high redshift halves and low/high luminosity halves we obtain that the amplitude of variability : a) increases for increasing redshift, b) decreases for increasing luminosity. The fact that higher redshift objects appear more variable than low redshift ones could be due, at least in part, to an observational bias deriving from spectral variability. In fact, the observing wavelength corresponds to shorter rest frame wavelengths at higher redshifts and this implies a larger variability if luminosity changes are larger, on average, at shorter intrinsic wavelengths. The fact that fainter objects appear more variable than brighter ones could support the notion that the variability of brighter objects is produced by uncorrelated varying subunits.

MULTIFREQUENCY VARIABILITY OF BLAZARS 3C 279 and 3C 345

James R. Webb
Florida International University
NASA/JOVE Fellow

The multifrequency spectra of Blazars 3C 279 and 3C 345 are presented and their implications for theoretical models are discussed. The spectra resulted from two separate but complimentary observational programs; one a serendipitous discovery by CGRO that extended the spectrum up to GEV energies, while the other a program designed to investigate the spectral variability of a Blazar during an outburst.

The quasi-simultaneous (2 weeks) broadband spectrum of 3C 279 between 10^9 and 10^{24} Hz was compiled after the CGRO detector EGRET observed a large increase in gamma-ray flux from the source. Other observations made independently throughout the observing community were gathered to form the most complete quasi-simultaneous spectrum of a Blazar to date. This spectrum and a second one resulting from observations made two months later when the gamma-ray flux had dropped significantly, reveals that a jet model with varying upper energy electron cutoff can adequately explain the differences between the two spectra.

Target of Opportunity multifrequency observations of 3C 345 made during an outburst in 1991 resulted in 10 nearly simultaneous (1-5 days) spectra between 10^8 and 10^{18} Hz. The spectral changes can be modelled by a broadband SSC jet with the high energy cutoff of the electron distribution varying with time, plus a thermal accretion disk with a varying accretion rate.

Collaborators

3C 279:
   B. Hartman, J. Maddox, D. Kniffen, V. Schoenfeld, K. Bennett, W. Hermsen,
J. Ryan, T. Ohashi, M.J. L. Turner, J. Kurfess, A. Sadun, I. Robson,
T. Balonek, F. Makino, T. Kii, R. Fujimoto, M. Aller, H. Aller, P. Hughes,
E. Valtaoja, H. Terasranta, M. Tornikoski, A. Marscher, S. Bloom, H. Netzer

3C 345:
   C.R. Shrader, T. J. Balonek, D. M. Crenshaw, D. Kazanas, A. G. Smith,
A. D. Nair, R. J. Leacock, S. Clements, P.P. Gombola, A. Sadun, H. R.
Miller, I. Robson, R. Fujimoto, F. Makino, T. Kii, M. Aller, H. Aller,
P. Hughes, E. Valtaoja, H. Terasranta, E. Salonen, M. Tornikoski, W. Chism

References:

3C 279
   Hartman *et al.* 1994, in preparation.

3C 345
   Webb *et al.* 1994, Accepted to Ap. J, Feb. 20th 1994 issue.

OPTICAL MICROVARIABILITY IN RADIO QUIET QUASARS

PAUL J. WIITA
Department of Physics & Astronomy, Georgia State University, Atlanta, Georgia 30303, USA

GOPAL-KRISHNA
National Centre for Radio Astrophysics, TIFR, Post Bag 3, Pune 411007, India

and

RAM SAGAR
Indian Institute of Astrophysics, Bangalore 560034, India

Abstract. We have observed 11 radio quiet QSOs (RQQSOs) to see if they exhibit intra-night variability in the optical. The detection of such microvariability would support models in which fluctuations on accretion disks are dominant, while if it is never present, models based on relativistic jets would be favored. Although several of these RQQSOs show hints of microvariability, we cannot claim to have discovered this phenomenon in this class of objects. Several of the comparison stars have clearly shown rapid variability.

1. Observations and Results

By using CCDs as N-star photometers we have begun a search for small (1–2%) intra-night variations in QSO flux in RQQSOs. This microvariability is common in blazars (e.g., Carini et al. 1992), and most models attribute it to shocks propagating down relativistic jets. Another class of models explains these fluctuations in terms of flares on accretion disks. Since RQQSOs should lack relativistic jets, the detection of microvariability in one of these objects would support the latter hypothesis, while its consistent absence would support the former.

We have recently reported the first observations on 7 RQQSOs (Gopal-Krishna et al. 1993a,b). Here we summarize additional observations, made on the 2.3m Vainu Bappu Telescope in Kavalur, India, on 2 of those objects and on 4 additional objects. Observations were made in the V and R bands, with exposures ranging from 7 to 15 minutes. The following 7 RQQSOs showed no evidence of intra-night variations: 0530-379, 0540-389, 0838+35, 0946+301, 1248+401, 1338+416 and 1630+377. These four objects showed hints of optical microvariability, but in none was this phenomenon clearly detected: 1206+459, 1254+047, 1352+011, and 1522+102. At least 6 of the comparison stars were found to be rapidly variable (by 4–5% over an hour or two), confusing the situation.

2. References

Carini, M.T., Miller, H.R., Noble, J.C., Goodrich, B.D. 1992, AJ, 104, 15
Gopal-Krishna, Wiita, P.J., Altieri, B. A&A, 271, 89
Gopal-Krishna, Sagar, R., Wiita, P.J. MNRAS, 262, 963

AN EIGHT-MONTH MONITORING CAMPAIGN ON A SAMPLE OF AGN

CLÁUDIA WINGE[*,**] and BRADLEY M. PETERSON
Dept. of Astronomy, The Ohio State University, USA

M.G. PASTORIZA[**] and T. STORCHI-BERGMANN[**]
Depto. de Astronomia, UFRGS, Brasil

and

J. BALDWIN
CTIO, Chile

We present the preliminary results of an 8-month monitoring campaign carried out on 6 AGN during the period December 1991 – July 1992. All but one of our targets showed continuum and/or line variability. The data were obtained using the 2D-Frutti + Cassegrain spectrograph at the CTIO 1.0-m telescope, and reduced following standard procedures. The slit width was 5″ and the nuclear spectra were extracted in a 10″ aperture. The wavelength coverage is 3500 – 7200Å, with 8Å resolution. The data were flux calibrated using standard stars and then normalized using the [O III] λ5007Å line flux for each object. NGC 6814: our spectra reveal that this object is still in a low state of activity and within the S/N ratio of our data, no variability was observed during this campaign. The stellar population is dominant in the nuclear spectrum and a synthesis using the star cluster library of Bica (1988) indicates a mainly old ($\sim$ 86% of the continuum flux at 5870Å due to a population with age $\geq$ 10 Gyr), $[Z/Z_\odot] \geq 0.3$ stellar content, with an intrinsic reddening of E(B–V)=0.20. NGC 3227: using an off-nuclear spectrum corresponding to the two 5″x10″ regions 20.4″ E/W of the nucleus, we obtained also a mainly old (77% at 5870Å with age $\geq$ 10 Gyr), $[Z/Z_\odot]=0.3$) synthetic stellar population, which contributes $\sim$ 43% of the nuclear light at 5600Å. The cross correlation of the 4245Å continuum and Hβ light curves results in a 18±3 -day lag. IC 4329A: our data show evidence of variability as a slow and constant increase in both continuum and lines fluxes, but no isolated event was detected. ESO141-G55: the light curves show small variations in the continuum, but no noticeable line variability. Akn 120 and Fairall 9: the data consists of two sets of spectra, separated by $\sim$ 6 months. Within each set little or no variability was detected, but strong line and continuum variations occurred between them.

---

[*] On leave from UFRGS, Brasil
[**] Visiting Astronomer at the Cerro Tololo Interamerican Observatory of the National Optical Observatories, operated by AURA under contract with the National Science Foundation

Poster Contributions:
Radio Emission (Maps)

OPTICALLY QUIET QUASARS – RADIO AND OPTICAL INVESTIGATIONS

CHIDI E. AKUJOR[1], R.W. PORCAS, A.R. PATNAIK[2] and A. ARDEBERG[3]
[1] Onsala/Jodrell Bank
[2] MPIFR Bonn
[3] Lund Observatory

The activity in the nuclei of bright galaxies could arise from a number of sources: they may contain compact variable radio sources, they may be strong IR/optical/X-ray continuum sources, or they may have strong emission lines. Usually but not always these properties go together as in radio loud quasars. However, there appears to exist a class of objects which resemble quasars in radio structure and brightness – have strong flat spectrum cores but appear fainter than would be suggested by their radio brightness – unidentified on sky surveys (Zensus & Porcas, 1985, in J.Dyson, ed, AGN). These are called 'optically quiet quasars' (OQQs).

In a recent large sample of bright flat spectrum sources (Patnaik et al. 1992, MN, 254, 655), we found that a significant fraction (20%) are OQQs (that is, unidentified optically). We are currently investigating a sample of 153 OQQs using the VLA, VLBI to find their arcsec and milliarcsec radio structures and optical imaging with the NOT to search for faint optical counterparts. This will enable us to classify and determine the nature of OQQs and their relationship with other types of powerful radio sources.

The VLA observations at 8.4 GHz show that the core fractional polarisation of OQQs (mean 1.8%) is lower than those of 'normal' quasars (mean 2.4 %). Observations with VLBI at 5 GHz and VLBA at 8 GHz of a core sample of 30 strongest OQQs are being undertaken. We find a mixed bag of radio structures: unresolved sources, core–jets, doubles, triples. Such milliarcs radio structures are similar to those found in 'normal' quasars (see Akujor & Porcas, in J. Roland et al. eds, EGRSs...Beams to Jets, p 134). NOT images of 75% of the objects have been made. About half of the objects are identified with red stellar objects, the difference between the m_R and $m_V \sim 1 mag$ suggests that some of them are at very high redshifts. Also, a number of OQQs have been detected in IR measurements (Beicham et al. 1981, ApJ, 247, 780)

These preliminary results suggest the possible scenarios are: (i) OQQs may be a distinct class with very steep optical–IR spectra, (ii) may be high redshift quasars and hence faint, (iii) nuclei of variable flat-spectrum radio galaxies i.e. potential BL Lac objects, e.g. 2309+45 which although very bright $m \sim 18$ is not visible on POSS. (iv) active nuclei obscured by dust, so they owe their color to dust absorption, and (v) gravitationally lensed – the reddening arising from absorption in dust associated with the lensing galaxy (e.g. 0218+357 has been shown to be a lensed system, Patnaik et al. 1992 MNRAS, 261,435).

CORRELATED FLUX DENSITY OUTBURSTS AND STRUCTURAL VARIATIONS IN A SAMPLE OF LOW FREQUENCY VARIABLE RADIO SOURCES

M. BONDI *
Nuffield Radio Astronomy Laboratories, Jodrell Bank, Macclesfield, U.K.

L. PADRIELLI, R. FANTI, L. GREGORINI and F. MANTOVANI
Istituto di Radioastronomia, Bologna, Italy

J.D. ROMNEY
National Radio Astronomy Observatory, Charlottesville, USA

N. BARTEL
Harvard Smithsonian Center for Astrophysics, Cambridge, USA

K.W. WEILER
Naval Research Laboratory, Washington, USA

and

G.D. NICOLSON
National Institute for Telecommunication Research, South Africa

September 22, 1993

Snapshot VLBI observations at 18 cm have been obtained with a global array at three epochs (1980.1, 1981.8, 1987.9) in order to investigate flux density and/or structural variations for a sample of 21 low frequency variable sources (Padrielli *et al.* 1987 *Astron Astrophys. Suppl. Ser.*, **67**, 63; Bondi *et al.* 1993 *in preparation*).

We have calculated the Doppler factors implied by the structural variations and compared with the ones obtained by the flux density outbursts at low frequency (408 MHz) during the epoch of VLBI experiments. In such a way we can check, quantitatively, which low frequency bursts can be associated with the superluminal motion and expansion of synchrotron plasmoids, and hence be the relic of an higher frequency variability. Few sources (e.g. BL Lac and 0202+149) show low frequency flux density outbursts that can always be related to observed (or to an upper limit of) structural variations at 18 cm. Others (e.g. 1127-145 and 1611+343) have strong low frequency variability without any structural changes at 18 cm in a period of about 8 years. These outbursts have to be extrinsic and can be explained in the framework of refractive scintillation theory. Most of the sources (e.g. 1055+018, 3C345, 3C454.3) show both the components.

Finally, the previous comparison between the VLBI observations of the first two epochs had proposed three new superluminal candidates (0224 + 671, 0605 − 085, and 1510 − 089). The new data confirm as superluminal candidates only 0224 + 671 and 0605 − 085, and suggest that also 0607 − 157 and 1055 + 018 might be added.

* On leave from Istituto di Radioastronomia, Bologna.

RELATIVISTIC JETS WITH STELLAR WIND ENTRAINMENT.

MARK BOWMAN
NRAL, University of Manchester, Jodrell Bank, Nr Macclesfield, Cheshire SK11 9DL, UK.

Observations suggest that the flow speeds in extragalactic radio jets are relativistic on parsec scales. This evidence includes rapid variability, brightness temperatures in excess of the inverse Compton limit and even the motion of individual features with apparent superluminal velocities. In contrast, FR-1 flow speeds appear to be non-relativistic on kiloparsec scales (the case is not clear for FR-2 sources); consequently, the possibility of jet deceleration has been explored.

Simulations of steady state jets with both relativistic bulk motion and relativistic equation of state are presented. The effects of entrainment from the stellar winds within the jet are included as source terms in the equations of motion. As cool material is entrained, the flow is forced to decelerate. At the same time dissipation tends to increase the fluid temperature (however, in the jets which were initially hottest [i.e. thermal energy much greater than rest energy] the dissipation is not sufficient to overcome adiabatic cooling). The sub-adiabatic decline in jet pressure and fall in Mach number causes the jet to expand rapidly, as is often observed in real sources.

By expressing the relativistic equation for the evolution of particle number density in a dimensionless form, a parameter can be obtained that expresses the relative importance of entrainment on the jet dynamics. This is:

$$\kappa = \frac{qz_c kT}{Pm^*\Gamma_L c} = \frac{qz_c}{\rho\Gamma_L c}, \qquad (1)$$

where q is the level of entrainment, z_c is the core radius of the host galaxy, k is Boltzman's constant, T is the fluid temperature, P is the jet proper pressure, m^* is the average particle rest mass, Γ_L is the Lorentz factor of the flow, ρ is the rest mass density and c is the speed of light. Clearly, initially hot, light jets are more effectively decelerated than equivalent cooler, denser flows. If all jets are assumed to be formed in similiar environments and therefore have similiar escape velocities from the AGN, then the likely deceleration of FR-1 sources to non-relativistic speeds, is indicated to be due to a difference in initial fluid temperature, between these and FR-2 sources. Alternatively, if the two classes of object have initially similiar temperatures then FR-2 sources must have higher initial Lorentz factors. Succintly, the second form of equation (1) indicates that deceleration increases as the jet 'flux' decreases. The effects of stellar wind entrainment on jet opening angles and temperature evolution will be discussed elsewhere.

Acknowledgment

I would like to express my gratitude to Sergey Komissarov and Patrick Leahy for their useful input on this work. I acknowledge the support of the SERC and the NRAL.

HIGH RESOLUTION HI OBSERVATIONS OF NGC 1068

ELIAS BRINKS
NRAO*, Socorro, NM 87801, USA

EVAN D. SKILLMAN
Univ. of Minnesota, MN 55455, USA

and

ROBERTO J. TERLEVICH and ELENA TERLEVICH
RGO, Cambridge CB3 0EZ, UK

1. Observations and Results

Surprisingly few Seyfert galaxies have been mapped at near optical resolution in the 21-cm line of neutral atomic hydrogen, despite the fact that studies of the gaseous component hold out the possibility of identifying the cause of infall of gas to or outflow from the central region. We therefore decided to observe the Seyfert 2 galaxy NGC 1068 with the NRAO-Very Large Array. The spatial resolution of the final images is about 8 " or 700 parsec at a distance of 18 Mpc; the velocity resolution is 5.2 km s^{-1}.

Several features are apparent in our maps. The HI column density map divides into 3 distinct regions: a central minimum (r $\leq$ 30 "); a high surface brightness inner ring (30 "$\leq$ r $\leq$ 80 "); and a lower surface brightness outer extension which is detected out to $\sim$ 200 ". The inner minimum is partly due to the fact that HI is absorbed against the strong continuum of the central source (see below). The very high surface brightness bar and inner spiral structure of NGC 1068 are all within a radius of 15 " of the nucleus, and therefore coincide with the central HI minimum. The pronounced CO torus extends from 10 to 25 " (0.9 to 2.4 kpc; Planesas *et al.* 1991, ApJ 369, 364).

A tilted-ring analysis of the HI velocity field shows that the inclination of NGC 1068 remains constant throughout the disk. The position angle varies, though, from 95° in the inner part of the ring to 115° further out, indicating substantial warping of the disk. The rotation curve shows a very fast rise to a maximum of 220 km s^{-1}, and a gradual decline after which it stays constant at a velocity of roughly 150 km s^{-1}.

NGC 1068 has a strong central radio continuum source. It is extended over about 1' and has a peak value of about 4.5 Jy. We find broad ($\sim$ 300 km s^{-1} wide absorption at the systemic velocity reflecting the fast rotation within the inner 10" of this galaxy. In addition redshifted HI indicating infall is seen at a velocity of about 175 km s^{-1} with respect to systemic.

* The National Radio Astronomy Observatory is operated by Associated Universities, Inc., under contract with the NSF

ASTROPHYSICAL RESULTS FROM GEODETIC VLBI CAMPAIGNS

S. BRITZEN, A. WITZEL and A.-M. GONTIER
Max-Planck-Institut für Radioastronomie, Bonn, Germany

C.J. SCHALINSKI
Institut de Radio Astronomie Millimétrique, Grenoble, France

and

J. CAMPBELL
Geodätisches Institut der Universität Bonn, Germany

Geodetic VLBI observations of extragalactic radio sources –designed to measure earth orientation parameters– are performed regularly since more than 10 years now. In different campaigns about 20 Quasars and BL Lacs are being monitored at 2.3 and 8.4 GHz. The high duty-cycle of these observations (every month for IRIS-S) and the resulting enormous data base allow to study the milliarcsecond structures and their variations simultaneously at two frequencies on short timescales. Since the accuracy of baseline determination and measurement of earth rotation parameters has reached a limit where source structure contributions become significant, there is substantial interest to monitor and correct for the source structure "effects". Therefore we perform a regular analysis of sources selected from IRIS-S (high north-south resolution and almost circular beam for low declination sources) and EUROPE (high sensitivity due to large telescopes) campaigns covering a maximum time base of 10 years (1983-1993). The (undesired) structure of the sources leads to an occasional updating of the source catalogue in order to have "pointlike" sources which are observed 10-15 times for three to seven minutes during 24 hours. Meanwhile about 50 sources have been observed in geodetic campaigns. The comparatively poor data quality requires careful calibration with a source whose structure is well-known from high dynamic range observations (e.g. the blazar 1803+784). The dynamic range of these maps is limited to 1:50. The merit of this database lies obviously in the monthly persecution of structural changes in extragalactic radio sources over at best 10 years time and the filling of the gaps between high dynamic range observations. Sources which are presently of particular interest to us, include 1803+784, 4C39.25, 0528+134, 3C454.3 and OJ287.

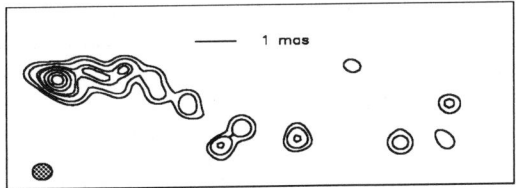

Fig. 1. 8.4 GHz CLEAN map of 1803+784 (IRIS, May 1988). FWHM of the restoring beam is 0.5 mas. Contours are 5, 10, 20, 25, 40, 60 and 80 % of the peak brightness.

ON THE AGE OF GPS RADIO SOURCES

JOEL C. CARVALHO
Departamento de Fisica - UFRN - C.P. 1641
59072-970 Natal - RN - Brazil

The GHz-Peaked-Spectrum (GPS) radio sources are galaxies and quasars with compact structures on scales of tens to hundreds of parsecs and spectrum peaking around 1 GHz. Phillips and Mutel (1982) suggested that the double GPS sources (also called compact doubles, CD) are an early stage of the evolution of the classical double sources of Fanaroff-Riley class II type. We have presented a model (Carvalho, 1985) in which we show that the properties of the CD sources are consistent with such a scenario. Their ages would be of the order of 10^4 yr.

Recent observations of diffuse extended radio emission around some (20%) GPS sources, indicate that this sub-class may not be as young as initially thought. O'Dea et al. (1991) suggested that they would be smothered radio galaxies, a consequence of a recent increasing in the gas density near the center of the host galaxy due to an interaction or merger with a companion. Indeed, numerical simulations (De Young, 1990) show that jets can be drastically decelerated by dense clouds which are later destroyed by them. This would keep the source confined to the inner region of the galaxy. Thus, if the medium surrounding the radio source is very cloudy, the propagation will be slow and consequently the age will be larger than that deduced from a uniform medium (O'Dea et al., 1991). We have constructed a simplified model and analytically calculated the average velocity of the jet in a clumpy medium to found

$$V = \left[1 + R_c \lambda \frac{R_c}{R_j}\sqrt{\frac{n_c}{n_j}}\right]^{-1} V_j.$$

Here R is the radius, n the particle number density, λ is the number of clouds per unit length and the indices c and j refer to the cloud and jet respectively. The jet velocity in a homogeneous medium is V_j.

For characteristic values of the quantities we find $V \sim 10^{-2} - 10^{-1} V_j$. We calculated the age of the GPS sources using this velocity and found it to be of the order of $10^5 - 10^6$ yr. This is near the range of ages of extended radio sources ($10^6 - 10^7$ yr).

References

Carvalho, J.C. 1985, Mont. Not. R. Astron. Soc.**215** 463.
De Young, D.S. 1990, Astrophys. J.**371** 69.
O'Dea, C.P., Baum, S.A. and Stanghellini, C. 1991, Astrophys. J.**380** 66.
Phillips, R.B. and Mutel, R.L. 1982, Astron. Astrophys.**106** 21.

VLBI MORPHOLOGY OF GHZ-PEAKED SPECTRUM RADIO GALAXIES

D. DALLACASA,[*] C. FANTI[**] and R. FANTI[**]
Istituto di Radioastronomia-CNR, Bologna, Italy

1. Discussion

GHz-Peaked Spectrum (GPS) radio sources are intrinsically small ($< 1\ kpc$) and unbeamed objects. The galaxies considered here (0316+161, 0404+768, 0428+205, 1323+321, 1358+624, 1819+39, 1829+29) have been selected from the Peacock and Wall (1981) catalogue, and belong to a complete sample of Compact Steep-Spectrum (CSS) radio sources (Fanti et al., 1990). Their radio spectra show a turnover which could be explained in terms of synchrotron self-absorption. It occurs at frequencies ranging from about 100 MHz to 5 GHz and for this reason they do not appear in the 3CR catalogue.

The 3CR CSS galaxies generally show Double morphology (lobes) on the 1 to 10 kpc scale (Fanti et al., 1990). Faint cores are found at high frequencies (e.g. 15 GHz) for most of them (van Breugel et al., 1984 and 1992), but in all cases they account for only a very tiny fraction of the total flux density ($< 0.5\%$). Jets are rare, so that the radio emission is generally lobe–dominated.

GPS galaxies instead show in most cases cores (or candidates), and bright jets even at 1.66 GHz (Dallacasa et al., submitted), while the fractional flux density in the lobes is generally less relevant than in 3CR CSS's. The low–frequency selected 3CR CSS's are probably biassed towards lobe dominated morphologies, with overall sizes and component sizes larger than in GPS radio sources and with the turnover in the spectrum occurring at lower frequencies. The evolutionary scenario in which the GPS galaxies will become CSS's when the source expands and the central engine progressively turns off, is not adequate, since extended radio emission has been found in $\sim 20\%$ of GPS's, and the age of GPS's turns to be of the same order of the extended radio sources when modelling the source expansion in a clumpy medium (see also Carvalho, these proceedings).

References

van Breugel W.J.M., Miley G.K., Heckman T.M.: 1984, *Astron. J.*, **89**, 5.
van Breugel W.J.M., Fanti C. and R., Stanghellini C., Schilizzi R.T., Spencer R.E.: 1992, *Astron. Astrophys.*, **256**, 56.
Fanti R. and C., Schilizzi R.T., Spencer R.E., Nan Rendong, Parma P., van Breugel W.J.M, Venturi T.: 1990, *Astron. Astrophys.*, **231**, 333.
Peacock J.A. and Wall J.V.: 1981, *M.N.R.A.S.*, **194**, 331.

[*] also Dipartimento di Astronomia, Bologna. Present address: NFRA, Dwingeloo, NL.
[**] also Dipartimento di Fisica, Bologna.

THE MILLIARCSECOND STRUCTURE OF FOUR SEYFERT GALAXIES AT λ 18 CM

T. GHOSH (NAIC,PUERTO RICO), R.T. SCHILIZZI (NFRA,NL),
G.K. MILEY (LEIDEN, NL), A.G. DEBRUYN (NFRA, NL), M.J. KUKULA (JB, UK),
A. PEDLAR (JB, UK), D. GRAHAM (MPIFR, GERMANY) and
D.J. SAIKIA (NCRA, INDIA)

The major contending scenarios capable of explaining various aspects of the Seyfert phenomenon in AGNs are (i) the Super-massive Black-Hole model and (ii) the Starburst model. Detailed optical emission-line and radio images of Seyfert nuclei, and their mutual correlations, provide important clues in evaluating the claims of each. Using the EVN at λ 18 cm, we have mapped four Seyfert galaxies, Mkn 1, 3, 231 and 463 at a resolution of $\sim$ 25 mas. The maps, and comparisons with images at other wavelengths, will be presented elsewhere (Ghosh et al. 1993, in preparation). Here, we present parameters derived from elliptical-Gaussian fits to all discernible components (Table 1).

Table 1

Source[*]	Dist mas	PA °	S_i mJy	Fitted size mas × mas, °	Source	Dist mas	PA °	S_i mJy	Fitted size mas × mas, °	
Mkn 1			44	10 × 7, 78	Mkn 231, 1	0		75	< 12.8	
Mkn 3, 2	366	77	36	46 × 44, 142	2	25	187	39	25 × 3, 41	
3	233	79	30	40 × 38, 147	3	65	202	23	29 × 14, 9	
4	0		11	37 × 16, 127	4	136	183	3	< 16.7	
5a	188	261	18	73 × 21, 108						
5b	300	264	8	33 × 29, 140	Mkn 463, 1a	0		60	23 × 16, 140	
5c	320	264	5		1b	51	196	16	37 × 26, 37	
5d	330	264	3		1c	89	193	3	< 18.6	[*] For
6a	569	260	17	61 × 48, 50	2a	267	172	5	14 × 12, 10	
6b	580	261	2		2b	284	177	5	24 × 7, 134	
6c	592	260	4		2c	310	174	2	< 10.3	
7	741	262	4	19 × 16, 7	2d	330	179	3	19 × 8, 3	
8a	1250	267	109	53 × 34, 26	3a	1247	178	8	46 × 42, 37	
8b	1286	266	49	47 × 30, 54	3b	1280	178	2	< 24.8	
8c	1281	270	13	< 44.9						
9	646	138	8	36 × 26, 79						

Mkn3, components are numbered as in Kukula et al. 1993

Three of the galaxies observed, Mkn 3, 231 and 463, show collimated emission indicating jet-like structures containing non-thermal knots. This implies radio-galaxy/quasar-type phenomena in the central regions. For Mkn 1, the structure does not rule out the possibility of a star-burst origin for the radio emission. However, the brightness temperature of the source appears to be the highest of any in this study at $\approx 10^9$ K!

References

Kukula, M.J., Ghosh, T., Pedlar, A., Schilizzi, R.T., Miley, G.K.
deBruyn, A.G., Saikia, D.J., 1993, MNRAS (in Press)

RADIO SOURCE STRUCTURE AND UNIFIED MODELS

J. B. HUTCHINGS
*Dominion Astrophysical Observatory, 5071 W. Saanich Rd
Victoria, B.C. V8X 4M6, Canada*

and

S. G. NEFF
*NASA Goddard Space Flight Center, Greenbelt
MD 20771, USA*

Abstract. The results are presented of a program to study the radio structure of radio galaxies, and to compare them with quasars. The samples are matched in redshift and luminosity in the redshift range 0.2 to 1.0, and mapped with the same VLA configuration. The quasar results have been published in earlier papers and the radio galaxy results are in press. We compare 100 quasars with 80 radio galaxies. The comparison shows the following radio galaxies differ from quasars in the following ways:

1) The overall sizes are larger for radio galaxies by 2 to 3 times.
2) There are few large bend angles in radio galaxies.
3) The lobe length ratio distribution differs.
4) The radio galaxies have no population of one-sided sources.
5) The lobe flux ratio has less range in the radio galaxies.
6) Radio galaxies have few core-dominated sources and many with no detectable core: few with comparable core and lobe flux.
7) The most luminous sources are lobe-dominated and more powerful than the quasars. The most luminous quasars are unresolved cores.

Points 1,2,3 can be fit with a single 'unified' model which has a distribution of sizes and bend angles up to $25°$, with small expansion velocities. The quasars are viewed within a cone angle of $50°$ and the radio galaxies outside the cone. However, the model does incorporate a source growth scenario and points 4 through 7 imply that there are instrinsic differences between the source types as well as orientation. We also discuss the limitations on beaming and bulk motions implied by the results.

Key words: radio galaxies, quasars, radio maps, unified models

DIRECTIVITY PATTERN SIMULATION OF THE JET RADIO EMISSION IN AN AGN MODEL

Y.Y. KOVALEV
Astro Space Center, Lebedev Physical Institute
Profsojuznaja street, 84/32
117810, Moscow
Russia

Recently the 1.5 years flux observations of a quasar 0235+16 at 8 frequencies have been explained by the emission of a jet in a strong radial magnetic field of an AGN (in according with the Hedgehog model, suggested by N.S.Kardashev in 1969; see poster by Kovalev Y.Y. and Larionov G.M. to this Symposium).
Results of the numerical calculations are presented on the Fig.1 as a function on the angle θ from the jet axes, $0.5^\circ < \theta < 90^\circ$, at the normalized frequencies ν/ν_{m00}=4; 1; 1/4 and 1/16 (labeled near the curves). Directivity pattern will have a minimum at θ=0° (do not showed on the Fig.1), caused by the jet emission, observed inside the angles $0^\circ < \theta < \varphi$, where φ is a jet corner angle. As a result a maximum at $\theta \sim \varphi$ or $\theta > \varphi$ will be obtained at all frequencies.

A possibility to observe such sources, orientated inside the beam to an observer in this model, is ~ 0.09 at a frequency ν_{m00}. It is at ~ 30 time greater than in the relativistic jet model (Blandford and Koenigl, 1979) at this frequency, but is less than in it at frequencies less than ν_{m00}/16. A frequency ν_{m00} is a preferable frequency for a search of objects for this model. It is estimated as 18 GHz.

References
Blandford R.D. and Koenigl A. 1979. Asrophys. J. V.232. P.34.

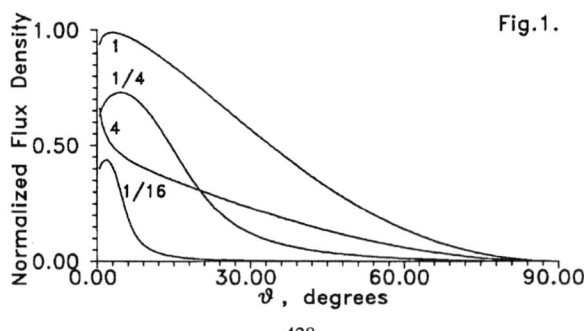

Fig.1.

NEUTRAL HYDROGEN IN THE SEYFERT GALAXY NGC 3227

CAROLE G. MUNDELL, ALAN PEDLAR and DAVE J. AXON
Nuffield Radio Astronomy Laboratories, Jodrell Bank, Macclesfield, Cheshire SK11 9DL, U.K.

We present VLA C and D array neutral hydrogen observations of the Seyfert galaxy NCG 3227. The neutral gas is highly disturbed due to the interaction with the nearby elliptical, NGC 3226 and D array observations (with resolution of $40''$ and velocity resolution of 40 kms^{-1}) revealed 2 extensive plumes stretching 88 kpc south and 44kpc north of NGC3227. Complex dynamics were observed in the disk, which subsequent C array observations (resolution of $19''$ and 20 kms^{-1}) resolved into two components: the galactic disk, which is not significantly disrupted by the interaction and is undergoing predominantly solid body rotation out to a radius of 7 kpc, and an anomalous velocity cloud (with a velocity of 170 kms^{-1} above systemic) which appears to be associated with the northern plume.

There is evidence of a bar in the integrated hydrogen emission of the disk, and a 'hole' in the HI emission can be seen at the centre of the galaxy. (We are, however, unable to determine, from C array observations, whether this is due to absorption against the 20 mJy continuum source or to the absence of neutral gas). There is no obvious evidence of high velocity gas associated with the Seyfert nucleus.

Neutral hydrogen masses, derived from flux densities measured from naturally weighted data, in the individual galaxy components are as follows:

- Northern Plume $0.082 \times 10^9 M_\odot$
- Southern Plume $0.167 \times 10^9 M_\odot$
- Cloud $0.316 \times 10^9 M_\odot$
- Disk $0.53 \times 10^9 M_\odot$

Rubin & Ford, 1968, (*Astrophys.J.*, **154** 431.), conducted an extensive optical study of the NGC 3226/7 pair. They used single slit spectroscopy to study the kinematics of the excited gas in the system, finding evidence for gas outflow from the nucleus at a velocity similar to that of the anomalous velocity HI cloud in our observations. They also detected an 'arm', stretching from NGC 3227 to its companion, with a mean velocity of 500 kms^{-1}. Single dish, neutral hydrogen observations (Heckman, Balick & Sullivan, 1978. (*Astrophys.J.*, **224**, 745) seemed to substantiate the presence of the arm when a weak emission wing, extending down to 500 kms^{-1}, was seen in the profile taken 3'N of the nucleus. Our observations, however, show no evidence for neutral gas emission at these velocities.

Extended ionised gas emission has recently been observed on the INT and its dynamics and structure are currently being compared with the HI observations.

MERLIN OBSERVATIONS OF SEYFERT NUCLEI

A. PEDLAR & M.J. KUKULA

NRAL, University of Manchester, Jodrell Bank, Nr Macclesfield, Cheshire, UK. SK11 9DL

Radio emission from Seyfert nuclei appears to be intimately related to narrow line region (NLR) of ionised gas. Both regions have an extent of a few hundred parsecs corresponding to typically a few arcsec, and are only marginally resolved by ground based observations. HST and adaptive optics are giving optical images of the ionised gas with angular resolutions of order 0.1". It is essential that high quality radio images are available with similar resolution so that models relating the two regions can be tested. The extended MERLIN is ideally suited for this task. It has angular resolutions of 0.05" at 5GHz and 0.13" at 1.5GHz and sensitivities of a few 10s of μJy. In this contribution we shall summarise the results on two objects.

The 5Ghz MERLIN image of **NGC4151** (Pedlar et al MNRAS 263 471, 1993) confirms the picture of a two sided jet along Pa$\sim 77°$. By comparison with a MERLIN 1.5GHz image it is clear that most of the source has a steep radio spectral index. However the eastern component of the central double has a relatively flat index, consistent with it being coincident with the optical continuum nucleus. The radio collimation does not appear to align with the elongation of [OIII] emission on similar scales seen with the HST.

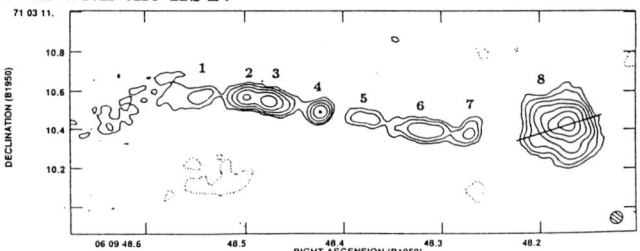

Fig. 1. MERLIN 5GHz image of Markarian 3 from Kukula et al. (MNRAS in press)

Markarian 3 shows the finest example of a radio jet in a Seyfert nucleus. We identify component 4 (See Fig. 1) as the nucleus both from its relatively flat spectral index spectral index and its compact (< 30mas) structure. This implies we have a two sided jet extending $\sim$300pc either side of the nucleus. The inner jets are quite linear, but bend significantly beyond 150pc resulting in an 'S' type symetry. The radio image is strikingly similar to HST images in [OIII] (Boksenberg and Catchpole -Private communication), although the [OIII] 'S' is measurably smaller than the radio. We are investigating a model in which a bowshock, of the type discussed by Taylor et al (MNRAS 255 351, 1991), is moving into a rotating, rather than stationary, ambient medium could produce this effect.

THE CENTRAL REGION OF NGC 1365. SEST AND VLA OBSERVATIONS OF CO AND THE RADIO CONTINUUM

AA. SANDQVIST, S. JÖRSÄTER and P. O. LINDBLAD
Stockholm Observatory, S-133 36 Saltsjöbaden, Sweden

Abstract. The barred spiral Seyfert galaxy NGC 1365 has been observed in the radio continuum at wavelengths of 2, 6 and 20 cm, using scaled arrays of the VLA, and complete maps have been made in the $J = 1 - 0$ and $J = 2 - 1$ CO emission lines using the SEST. MEM maps of the 6 and 20 cm emission, as well as a spectral index map, have been produced with a resolution of $2".3 \times 0".9$, and the 2-cm map has a resolution of $0".25 \times 0".10$. The dominant continuum features are a number of unresolved sources with relatively flat non-thermal spectral indices (−0.3 to −0.5), immersed in an incomplete circumnuclear ring, which is superimposed upon a background that extends into the bar along the prominent dust lanes. The ring has angular dimensions of $8" \times 20"$, which corresponds to a linear dimension of the order of 1 kpc. There is clear evidence of a *jet*, about 5" long, originating at the position of the Seyfert nucleus and extending in a southeastern direction, closely along the minor axis of the galaxy. The jet has a steep non-thermal spectral index (−1.0) and is aligned along the axis of a conical shell of [OIII] emission. The CO molecular gas peaks at the nucleus and is strongly concentrated to the nucleus and bar regions with a certain enhancement along the bar. The total molecular hydrogen gas mass in the observed region is $2 \times 10^{10} M_\odot$, with $6 \times 10^9 M_\odot$ lying within 2.2 kpc of the nucleus. A full presentation of the results will be published in *Astronomy and Astrophysics* in 1994.

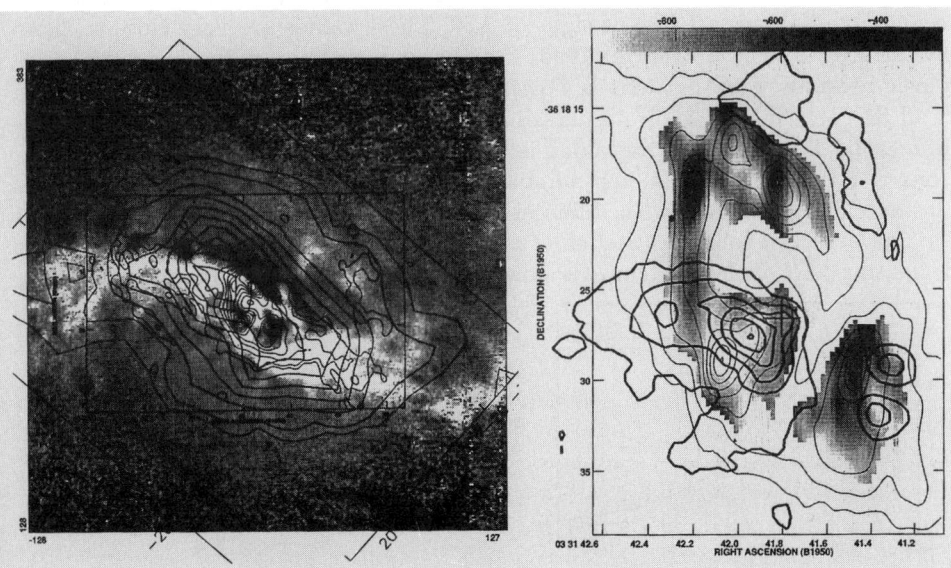

Fig. 1. *NGC 1365. Bar Region* (left): $J = 2 - 1$ CO T_A^*–thick lines; 20 cm continuum–thin lines; $B - \text{Gunn}\,z$ colour index–grays. *Central Region* (right): [OIII]–thick lines; 20 cm continuum–thin lines; spectral index–grays; nucleus–cross.

HELICAL MOTION AND NON-ADIABATIC EXPANSION IN THE JET OF 3C345

W. STEFFEN , T.P. KRICHBAUM and A. WITZEL
Max-Planck-Institut für Radioastronomie, Auf dem Hügel 69, 53121 Bonn, Germany

J.A. ZENSUS
National Radio Astronomy Observatory, P.O. Box O, Socorro, NM 87801, USA

and

S.J. QIAN
Beijing Astronomical Observatory, Chinese Academy of Sciences, Beijing 100080, China

We developed a kinematic jet model for the motion and flux density evolution of the high frequency VLBI jet components C4 and C5 in the quasar 3C345 (Zensus et al. 1994) assuming the conservation of three basic quantities: the Lorentz factor, the angular momentum, and the opening angle of the jet. This model is a simplified description of the helical motion in a conical jet expected from the magnetodynamical model of Camenzind (1986) which is based on a black hole surrounded by a magnetized accretion disc. Our best fit yields Lorentz factors of 5.7 and 5.0 for components C4 and C5, respectively, and an angle between the jet axis and the observer's line of sight of 7.5°. These values are very close to those obtained by Unwin & Wehrle (1992) from component motion further out. An intrinsic bending of the jet axis is necessary to account for the common bent path of all jet components at core separations larger than about 4 mas. We found that differential Doppler boosting alone is not able to explain the flux density variations of component C4. A non-adiabatic expansion model of an inhomogeneous plasma cloud combined with differential Doppler boosting on a helical path fits the flux density evolution (Steffen et al. 1994). We find that the expansion in the decreasing part of the lightcurve is slower than expected from adiabatic expansion.

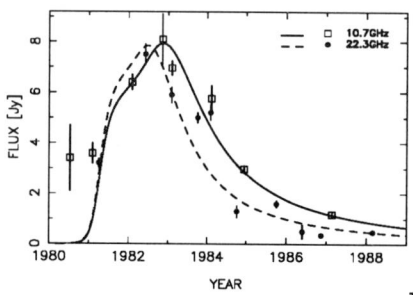

Flux density evolution of component C4 and a model using a non-adiabatic expansion of a plasma cloud in a tangled magnetic field compared with the data. Filled circles (dashed line) respresent the data (model) at 22.3 GHz and the open squares (solid line) show the data (model) at 10.7 GHz. The Doppler boosting factor has a maximum in 1981.25.

References

Camenzind M., 1986, A&A, **156**, 136.
Steffen W., Zensus A., Krichbaum T.P., Witzel A., Qian S.J., 1994, A&A, submitted.
Unwin S.C., Wehrle A.E., 1992, ApJ, **398**, 74.
Zensus J.A., Cohen M.H. , Unwin S.C., 1994, ApJ, submitted.

INTERPRETATION OF SOME PROPERTIES OF EXTRAGALACTIC JETS IN THE CONTEXT OF A TWO COMPONENT MODEL

LOURDES VICENTE
Observatoire de Paris-Meudon, DARC, France

We have studied a sample of extragalactic jets within the context of a two component model. This model supposes the existence of two flows in the extragalactic jets. An electron-positron fast beam coming from the internal regions of the accretion disk, responsible for the VLBI jet at the parsec scale and for the observed superluminal speeds. A second slow component, responsible for the jet observed at larger scales. The fast beam is destroyed when the parallel magnetic field is smaller than a critical value, $B_{II} < B_{crit} = 3.2 \times 10^{-3} n_p^{1/2}$, where n_p is the density of the background thermal plasma of the slow component.

The present sample includes 9 FRIs, 12 WATs, 1 FRII, 1 CSS and 1 Core-Halo. The critical zone was determined by careful examination of published radio maps looking for regions along the jets where there was an important morphological change, like bright knots, gaps or strong sudden polarization changes.

The identification of the critical zone fixes the internal density n_p from the magnetic field, deduced from standard equipartition argument. This allows us to estimate the mass outflows and energy flux for the slow and fast components, $\dot{M}_0 = S_0 n_p m_e m_e V_0$, $\dot{M}_f = S_f \gamma n_f m_e c$, $K_0 = S_0 V_0 (n_p m_p V_0^2/2 + B^2/6 + 5 n_p KT/2)$ and $K_f = S_f \gamma^2 (\gamma - 1) n_f m_e c^3$. Cross sectional areas S_0 and S_f are deduced from the radio maps. We assumed $\gamma = 4$ and $T = 10^8 K$. We have taken the density of the fast beam $n_f = (n_p \times n_{syn})^{1/2}$ to ensure $n_f/n_p \simeq 0.01$ and $n_{syn}/n_f = 0.01$, where n_{syn} is the proper density of the emitting particles of the fast component. Two different approaches have been made for the bulk velocity: $V_0 = V_A = 0.02c$ where V_A is the Alfven velocity in the critical region and a maximal limit, $V_0 = \frac{V_A}{tg\alpha}$ where α is the opening angle of the slow component.

The results give new light on the classification of AGNs. A smaller ratio K_f/K_0 for the WATs than for the FRIs suggests a stronger influence of the fast beam in the FRIs. The energy flux distribution versus the distance to the critical zone shows that WATs extend themselves through a large domain of critical distances (10 − 110 kpcs), always keeping small values of $K_f/K_0 = 3$ (for $V_0 = V_A$) while FRIs are grouped near the short critical distances < 10 kpcs together with the remaining sources studied; however, the domain of K_f/K_0 (5-25 for $V_0 = V_A$) is much more extended. We found a certain tendency to higher nucleus activity (Lx) for higher K_fs though we did not find data for all the sources. A larger sample should be studied to further investigate the possibility of a correlation between the large scale inflow $\dot{M}_{cool}$ and the ejection from the slow component.

Poster Contributions: Line Studies

BLOATED STARS AS BLR CLOUDS: NUMERIC RESULTS

TAL ALEXANDER and HAGAI NETZER
School of Physics and Astronomy, Tel Aviv University, Tel Aviv 69978, Israel

Abstract.
The 'Bloated Stars Scenario' proposes that AGN broad line emission originates in the winds or envelopes of bloated stars (BS) (see *e.g.* Kazanas 1989 and references therein). Its main advantage over BLR cloud models is the gravitational confinement of the gas and its major difficulty the large estimated number of BSs and the resulting high collisional and evolutionary mass loss rates (see *e.g.* Begelman & Sikura 1991). Previous work on this model did not include detailed calculations of the line spectrum, modeled solar neighborhood super giants (SG) and used very simplified stellar distribution functions for the nucleus. Here (Alexander & Netzer, 1993) we calculate the emission line ratios by applying a detailed numerical photoionization code (Rees, Netzer & Ferland, 1989) to the wind and by assuming a detailed nucleus model (Murphy, Cohn & Durisen, 1990). Allowing for the yet unknown effects of the AGN's extreme conditions on stars and stellar evolution, we study a wide range of simplified wind structures rather than confine ourselves to normal SGs. Our model consists of a spherically symmetric outflowing wind that emanates from the surface of the BS ($R_\star = 10^{13}$ cm, $M_\star = 0.8 M_\odot$, $\dot M = 10^{-6} M_\odot$/yr) whose size and edge density are determined by various processes: Comptonization by the central continuum source (calculated self consistently for our $L_{\rm ion} = 10^{46}$ erg/s model continuum by the photoionization code), tidal disruption by the black hole ($M_{bh} = 8 \times 10^7 M_\odot$) and the limit set by the assumption that the wind's mass $\leq 0.2 M_\odot$. This results in a large range of wind sizes, from 10^{13} to 10^{16} cm. We find that the line emission spectrum is mainly determined by the conditions at the edge of the wind rather than by its internal structure. Comptonization results in a very high ionization parameter at the edge which produces an excess of unobserved broad high excitation forbidden lines. The finite mass constraint limits the wind's size, increases the edge density and thus improves the results. Studying power-law wind structures ($v(R) = v_\star(R/R_\star)^{-\alpha}$ where $v_\star$ is the wind's base velocity at the BS's surface), we find that slow, decelerating, mass-constrained flows ($v_\star = 50$ m/s, $\alpha = 0.5$) with high gas densities (10^8 to 10^{12} cm^{-3}) are as successful as cloud models in reproducing the overall observed line spectrum. The Mg II $\lambda 2798$ and N V $\lambda 1240$ lines are however under-produced in our models. The denser the winds, the more efficient they are as BLR clouds. By calculating the $L\alpha$ emission from the wind we adjust the number of BSs so as to obtain the BLR's observed $EW(L\alpha)$. We find that only $\sim 5 \times 10^4$ BSs with dense winds ($v_\star = 50$ m/s, $\alpha = 0.5$) are required in the inner 1/3 pc (~ 0.005 of the total stellar population). This small fraction approaches that of SGs in the solar neighborhood. The calculated mass loss from such a small number of BSs is consistent with the observational constraints. We find that the required number of BSs, and consequently their mass loss rate, are a very sensitive functions of the wind's density structure (a $\sim 10^4$ factor between the slow $v_\star = 50$ m/s, $\alpha = 0.5$ model and the fast $v_\star = 50$ km/s, $\alpha = -2$ model). In particular, high mass loss rules out SG-like BSs ($v_\star = 10$ km/s, $\alpha = 0$). We conclude that BSs with dense winds can reproduce the BLR line spectrum and be supported by the stellar population without excessive mass loss and collisional destruction rates. The question whether such hitherto unobserved stars actually exist in the BLR remains open.

References

Alexander T., Netzer H.: 1993, Submitted to *MNRAS*
Begelman M. C., Sikura M.: 1991, in Holt S. S., Neff S. G., Urry C. M., eds, Proc. AIP Conf. 254, 'Testing the AGN Paradigm'. AIP, New York, p. 568
Kazanas D.: 1989, *ApJ*, **347**, 74
Murphy B. W., Cohn H. N., Durisen R. H., 1990: *ApJ*, **370**, 60
Rees M. J., Netzer H., Ferland G. J.: 1989, *ApJ*, **347**, 640

TYPE TRANSITIONS IN STARBURSTS POWERED AGN

ITZIAR ARETXAGA[1] and ROBERTO TERLEVICH[2]
[1] *Dpto. de Física Teórica, C-XI. Universidad Autónoma de Madrid. Cantoblanco 28049 Madrid. Spain.*
[2] *Royal Greenwich Observatory. Madingley Road, CB3 0EZ Cambridge. U.K.*

There is mounting evidence that type transitions are a common property among Active Galactic Nuclei: the broad lines in, at least, eleven Seyfert galaxies have ocassionaly appeared and subsequently disappeared leading to the reclasification of their nuclei from type 1–1.5 to type 1.8–2 or viceversa.

Type transitions are an outstanding property in the evolution of high metallicity massive starbursts due to the stochastic nature of the supernova explosions that take place in them. In the Starburst theory, the variability observed in radio quiet AGN is thought to be produced by the supernova (SN) and the compact supernova remnant (cSNR) activity of such a cluster (Terlevich et al. 1992, *MNRAS*, 255, 713). According to the theory, the luminosity of these systems is directly related to its SN rate (Aretxaga & Terlevich 1993, *Astrophys. Spa. Sci.*, 205, 69). For low and medium luminosity AGN ($\overline{M_B} \lesssim -22.5$ mag), the low SN rates derived ($\nu_{SN} \lesssim 1$ yr^{-1}) suggests a non negligible probability that the broad components of the emission lines are undetectable at some epoch. The stochastic nature of the SN explosions allows to obtain periods of time in which there is a lack of new explosions, while the existing remnants are too old to produce broad lines.

The AGN in which activity transitions have been observed are all below the predicted luminosity cutoff ($\overline{M_B} \approx -22.5$ mag), as can be seen in Figure 1.

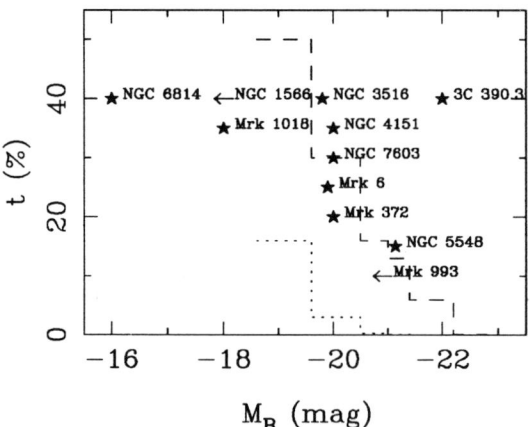

Fig. 1. Fraction of time spent as Seyfert of type 1.9 or 2 by theoretical stellar clusters of a given mean luminosity: the dashed line corresponds to type 1.9 stages and the dotted line to type 2 stages. The symbols represent the luminosities of the AGN known to have experienced type transitions: the stars indicate mean luminosities and the arrows upper limits to mean luminosities.

VARIATION OF BROAD OPTICAL EMISSION LINES IN AGN'S.

N.G. BOCHKAREV
Sternberg Astronomical Institute, Moscow 119899, Russia

A.I. SHAPOVALOVA
Special Astrophysical Observatory, N.Arkhyz, Stavropolskii Krai, 357147, Russia

and

L.S.NAZAROVA
Royal Greenwich Observatory, Cambridge, CB30EZ, UK

Abstract.
AGN optical spectra (4000-5000Å) have been obtained with the TV scanner of the 6-m telescope at Special Astrophysical Observatory (N.Arkhiz-North Caucasus) from 1986 to now. Most spectra were obtained with dispersion of 1 Å/channel and with 3-4Å spectral resolution and signal-to-noise ratio S/N=10-25 for individual spectra. NGC 4151, NGC 3516, NGC 5548 and NGC 7469 are the main sources of the programme and are also being used in international AGN watch programmes for diagnosis, estimation of the size and structure of the BLR.
NGC4151:
In 1987 the broad components of the $H\beta$ and $H\gamma$ lines decreased strongly (on about 65% as compared with 1986) and fell to a level comparable to that of 1984. In 1988 they increased up approximately to the level of 1986. During the photometrically low state of NGC 4151 satellites of the strong $H\beta, H\gamma, H\delta$ and He II4686Å lines were suspected at radial velocities of +9400 km/s and -7500 km/s. The data are could be interprite a two-jet model with jet velocity of 8400-24000 km/s. (Bochkarev et al.,1989). Analysis of the spectra shows that the broad component of HeII λ 4686Å changed its asymmetry over 6 years. Bochkarev et al.(1992), Nazarova et al.(1993) discussed a possible explanation of profile HeII λ 4686Å behavior by involving a precession of cones of ionizing radiation.
NGC5548:
Cross-correction of the optical emission line light curves with the UV continuum light curve reveals that the lines respond to continuum variations with a time delay from 7 days for HeII 4686Å to 18 days for $H\beta$. $H\gamma$ responds more rapidly than the lower order Balmer lines and there is a general trend towards higher amplitude variations in higher-order lines. The time lags for the line variations are consistent with the general pattern of radial ionization stratification of the BLR. During 1988-89 Fe II in blend 4130Å varies with an amplitude of aproximately 20% about the mean, similar the amplitude of the $H\beta$ variations and Balmer continuum.
NGC3516:
Variations of $H\beta$ during 1986-88 were small (20%), the shape of the profile (including the assymetry) did not change, but $H\gamma$ - to $H\beta$ ratios of the profile integrated intensities of broad component increased from 0.43 in 1986 to 0.8 in 1988. The 30% variations of continuum on characteristic time scale 10-15 days during the spring of 1988 were not accompanied by any noticeable changes in the line profile (Bochkarev et al.,1990).

Key words: data analysis-galaxies:NGC4151, NGC5548, NGC3515-galaxies: Seyfert-lines:profile-spectroscopy

References

N.G.Bochkarev, A.I.Shapovalova, S.A.Zhekov,1989,Preprint Special Astrophysical obs.,N35
N.G.Bochkarev, A.I.Shapovalova, S.A.Zhekov,1990,Preprint Special Astrophysical obs.,N45
N.G.Bochkarev, L.S.Nazarova, A.I.Shapovalova, 1992,Astron. and Astroiphys.Trans.,v.1, No.3.
L.S.Nazarova, N.G.Bochkarev, A.I.Shapovalova,(in press)

HST IMAGES OF THE SEYFERT GALAXY NGC 5929 AND ITS COMPANION NGC 5930[1]

G. A. BOWER, A. S. WILSON[2] and J. S. MULCHAEY[2]
Space Telescope Science Institute, 3700 San Martin Drive, Baltimore, MD 21218 USA

G. K. MILEY
Leiden Observatory, Sterrewacht, Postbus 9513, 2300 RA Leiden, The Netherlands

and

T. M. HECKMAN[3] and J. H. KROLIK
Department of Physics & Astronomy, Johns Hopkins University, Baltimore, MD 21218 USA

Images of the Seyfert 2 galaxy NGC 5929 and its interacting companion NGC 5930 have been obtained with HST's Planetary Camera. This interacting pair is also known as Arp 90. Each galaxy was imaged in the wavelength regions of [O III] $\lambda\lambda$ 4959, 5007, Hα+[N II] $\lambda\lambda$ 6548, 6583, and the green and red continua. The nuclei of both galaxies contain emission line gas, enhanced in the images by using the appropriate continuum image to remove the contribution of the continuum light in the on-band images.

Previous ground-based observations of the Seyfert galaxy NGC 5929 include [O III] and Hα+[N II] images, showing that its nucleus contains an elongated region of high-excitation emission line gas. In these HST images, this gas is clearly separated into two distinct regions separated by about $1.''1$ (138 h^{-1} pc, $h = H_0/100$ km s^{-1} Mpc^{-1}). The nucleus, as defined by the peak in the continuum, lies halfway between these two distinct emission line regions. The HST continuum images reveal a dust lane lying $0.''3$ SE of the nucleus with a length of $\sim 1''$ and $N_H \approx 4 \times 10^{21}$ cm^{-2}. We find no direct evidence of the reddening and/or obscuration effects with the characteristics of a dusty torus, which, in the context of a "unified model", is expected to obscure the AGN in type 2 Seyfert galaxies. The correspondence between the emission line gas and the radio morphology suggests that the structure of the NLR in NGC 5929 is governed by matter ejected from the AGN.

The nuclear emission line gas in NGC 5930 is concentrated into a ring around the nucleus with a diameter of $\sim 250\ h^{-1}$ pc and L(Hα+[N II]) $\approx 1.6 \times 10^{40}\ h^{-2}$ erg s^{-1}. These observations of Arp 90 present an opportunity to examine the details of the possible role of galaxy interactions in the triggering of an AGN.

---

[1] Based on observations with the NASA/ESA Hubble Space Telescope, obtained at the Space Telescope Science Institute, which is operated by AURA, Inc., under NASA contract NAS5-26555.
[2] also Astronomy Department, University of Maryland, College Park, MD 20742 USA
[3] also STScI

VELOCITIES WITHIN 1 ARCSEC OF THE NUCLEUS OF NGC 4151 AS REVEALED BY THE HST FAINT OBJECT CAMERA

R.M. CATCHPOLE and A. BOKSENBERG
Royal Greenwich Observatory, Madingley Road, Cambridge CB3 0EZ, UK.

and

THE FAINT OBJECT CAMERA TEAM
Space Telescope Science Institute, 3700 San Martin Drive, Baltimore, MD 21218

We have obtained a longslit spectrum at a position angle (PA) of 84.6° and passing within 0.38 arcsec of the nucleus of NGC 4151, using the FOC f/48 camera on the Hubble Space Telescope. The spectrum shows strong emission lines including [OII] λ 3727 and [OIII] $\lambda\lambda$ 4959, 5007. By fitting with Gaussian velocity profiles, we resolve the emission lines, within 1 arcsec of the nucleus, into a high and low velocity component. The low velocity component has a total range in radial velocity of 200 km s^{-1} and appears to be associated with material comprising the knots seen in the FOC, F501N [O III] image of NGC 4151, illustrated in Boksenberg (1993). The much weaker high velocity system has a range of 1000 km s^{-1}, is more smoothly distributed in brightness and shows a peak brightness close to the nucleus. Because the slit did not intersect the nucleus it is possible to determine the PA at which the two velocity systems cross the zero velocity axis. This is at PA -26° for the low velocity system and PA +32° for the high velocity system. These PA values may be subject to a systematic error as the zero velocity is defined by the mean position of the line, in the absence of any external calibration.

The low radial velocity system can be interpreted either as outflow within oppositely directed cones or as Kepplerian motion. For radial outflow the PA of zero velocity indicates that the axis of symmetry of outflow (PA $-116°$) lies close to the axis of the biconical structures (PA $-115°$) seen in the FOC F501N image, while for Kepplerian motion the velocity amplitude implies, for an inclination of 21° for the plane of NGC 4151, a mass within 30pc ($H_0 = 75$ km s^{-1} Mpc^{-1}) of $10^9 M_\odot$ and a line of nodes $\Omega = 64°$. This is greater than the value of $\Omega = 29°$ found by Pedlar et al. (1992) 10 arcsec from the nucleus.

The high velocity system can not be interpreted as Kepplerian motion as it would imply too large an enclosed mass; and if interpreted as radial outflow, the PA of the axis of symmetry ($-58°$) lies in the direction of the rotation axis of NGC 4151, 10 arcsec from the nucleus (PA $-61°$). These results will be discussed in more detail in a forthcoming paper.

References

Boksenberg, A., 1993. ESO Conference and Workshop Proceedings **44**, 61.
Pedlar,A., Howley,P., Axon, D.J., & Unger, S.W.,1992. MNRAS, **259**, 369.

SPECTROSCOPIC STUDIES OF EMISSION LINE GALAXIES

M. S. CHUN, E. C. SUNG and H. K. MOON
Department of Astronomy and Atmospheric Science,
Yonsei University, Seoul, Korea

and

Y. I. BYUN
Institute for Astronomy, University of Hawaii,
Honolulu, Hawaii, U.S.A.

Abstract.
Spectroscopic Observations were made to study 42 emission line objects. The analysis of these long slit spectra shows that 15 out of 42 galaxies are blue compact galaxies (BCGs). 21 of them are starforming or HII galaxies and 3 were found to be normal galaxies.

1. Observation

We observed emission line objects during July 8-14, 1991 and April 3-8, 1992. The Mount Stromlo Observatory(MSO) 74-inch telescope equipped with the spectrograph and Photon Counting Array(PCA) was used during these observational runs. The spectral range covered 3,500-5,500Å and 4,800-7,000Å.

2. Data Reduction and Results

We used NOAO. IRAF 2.0 package for data reduction in the standard way. We listed identifications of these objects in TABLE 1. Among the 42 objects, 9 galaxies show the spectra of typical starburst galaxy. The radial velocity difference between ESO 513-IG11 and ESO 513-G10 is found to be 250km/sec, and it is conceivable that they consist of an interacting system as a faint patch between the two galaxies can be seen on ESO red plate. ESO 105-IG11 is thought to be an iE type blue compact galaxy which has two highly excited HII regions in the nucleus and faint outer envelope of an elliptical shape.

TABLE I
Identifications of the Galaxies

object	identifications	object	identifications
ESO 122-IG02	starburst galaxy	ESO 270-IG22	starburst galaxy(?)
124-I12	starburst galaxy	CTS 1033	BCG
060-IG03	starburst galaxy	1034	HII galaxy
036-IG03	no emission lines	ESO 386-G09	Seyfert galaxy(?)
566-IG08	HII galaxy	513-IG11 a	BCG
435-G20	BCG	513-IG11 b	BCG
CTS 1010	HII galaxy	513-IG11 c	spiral galaxy
1011	BCG	042-IG04	BCG or HII galaxy
1012	BCG	CTS 1037	BCG
ESO 436-IG42	starburst or HII galaxy	ESO 102-G14	BCG, high excitation
264-IG13	interacting galaxy	140-G09 a	BCG or HII galaxy
CTS 1020	BCG	140-G09 b	BCG or HII galaxy
ESO 376-IG17	starburst galaxy	281-G07	HII region
502-IG11 a	BCG	338-IG04	BCG, interacting
502-IG11 b	BCG	338-IG08	starburst galaxy
UM 448	BCG or HII galaxy	105-IG11	BCG, high excitation
462	BCG	342-IG13	not BCG, weak [OII]
ESO 505-G12	BCG or HII galaxy	530-G42	no emission lines(?)
CTS 1027	BCG	289-IG08	starburst galaxy
ESO 322-IG32	starburst galaxy	290-G01	starburst galaxy
383-G20	not BCG, strong[NII]	293-G04	irregular galaxy

EMISSION LINE VARIATIONS OF BLRG

M. DIETRICH and W. KOLLATSCHNY
Universitätssternwarte, Geismarlandstr. 11, D-37083 Göttingen, F.R.G.

In late 1989 we started a monitoring campaign of the line profile variations of more than 40 Broad-Line Radio Galaxies (BLRG) at Calar Alto Observatory/Spain. BLRG are the most extreme species of AGN regarding line width and structure of their optical emission line profiles showing FWZI up to 35000 km s^{-1}, eg. 3C332 or Arp102B. Quite often the broad emission line profiles are characterized by a double hump structure. The analysis of the broad emission lines provides information about fundamental parameters of the inner part of the AGN like size, structure and kinematics of the line emitting region. In the following we present line profile variations of a BLRG we are studying.

Spectra of 3C390.3 at different epochs are displayed in Fig. 1 together with the corresponding difference spectrum. Prominent double peaked profiles of the Hα and Hβ emission lines are visible in the difference spectrum. These line profiles were transformed to the velocity space (cf. Fig. 1). In both lines the humps are shifted by $v_{rel} = -2000$ km s^{-1} and $v_{rel} = +7000$ km s^{-1}, respectively.

This work has been supported by DFG grant Ko 857/13-1

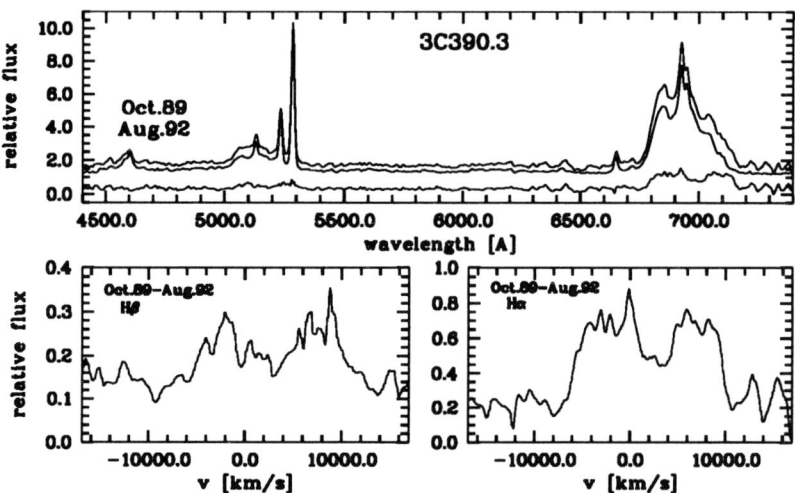

Fig. 1. Spectra of 3C390.3 taken at Aug. 31, 1992 and Oct. 28, 1989. The difference spectrum is also shown. The line profiles of Hβ and Hα of the difference spectrum were transformed to the velocity space and are displayed above

HIGH SPATIAL RESOLUTION 2D SPECTROGRAPHY OF ENLR IN SEYFERT GALAXIES

E. PECONTAL and P. FERRUIT
Observatoire de Lyon 9,av. Charles André 69561 St Genis Laval Cedex FRANCE

Three active nuclei environments have been observed with the integral field spectrograph TIGER operating in the visible domain : Mkn 34, NGC 5929 and M 51. These three objects exhibit linear radio-sources evocating expulsion of plasmons from the nucleus. Long-slit or Pérot-Fabry observations showed that the gas of the galaxies is interacting with the radio-emitter, and models of this interaction have been proposed. These new observations combines the two spatial dimensions of integral field spectrography (with 0.7" FWHM) with the rather large spectral domain of classical spectrography.

We present observational results for two of these objects, NGC 5929 and Messier 51. The first one displays a very simple radio structure with two lobes (Ulvestad and Wilson 1984, ApJ 285,439) which are associated with two line emitting gas components kinematically distincts (Whittle *et al.* 1986, MNRAS 222,189). Our data allow us to disentangle the emission arising from each region and to construct maps of their physical parameters (intensity, velocity fields and line ratios). Figure 1 shows some of these maps. M 51 exhibits a much more complicated structure with an extra-nuclear cloud, the XNC (Ford *et al.* 1985, ApJ 293,132). The intensity distribution of the line emission of the main body of the lines is strongly correlated with the radio map as already found by Cecil (1988, ApJ 329,38).

We plan to use this large set of data to set physical parameters of existing simple bow-shocks models of Taylor *et al.* (1992, MNRAS 255,351).

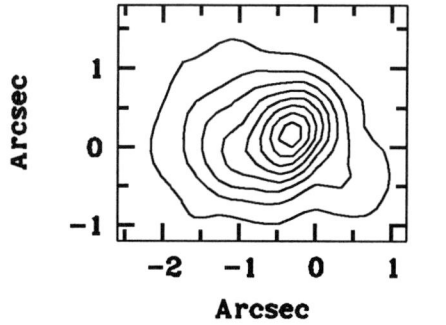 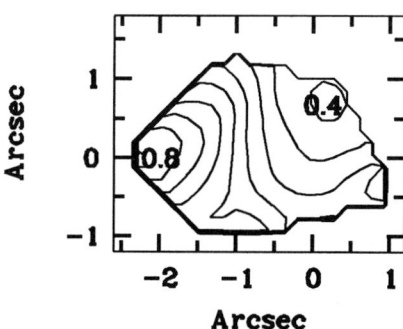

Fig. 1. Left, Hα intensity map of the eastern lobe (25:200:25 ; 10^{-19} W m^{-2} arcsec^{-2}). Right, [NII]6583.4 / Hα ratio (0.4:0.8:0.05 ; unitless) of the same lobe.

KINEMATICS OF THE BLR CLOUDS IN AGNS AND QUASARS

P M GONDHALEKAR
Astrophysics Group
Rutherford Appleton Laboratory
Chilton, OXON, OX11 0QX, England

September 21, 1993

Abstract. The Lyα and the C IV line profiles in the spectra of quasars and Seyfert 1 galaxies can be described by a combination of a narrow and a broad component. Over four orders of magnitude in luminosity the profiles of Lyα and C IV lines in the quasar spectra are very similar but a change is observed in the C IV profile in the spectra of Seyfert 1 galaxies which cover a smaller luminosity range. The correlation of the FWHM of Lyα, C IV and Hβ lines suggests that the Lyα and C IV lines are emitted from clouds in a similar kinematic enviornment but part of C IV line may also be emitted from clouds which emit the Hβ line.

Key words: Quasars, Seyfert 1 galaxies, Line profiles

1. Introduction

The profiles of Lyα and C IV lines in a sample of radio loud (RLQ) and radio quiet (RQQ) quasars and Seyfert 1 galaxies have been analysed. The sample is derived from the archive of *IUE* spectra, and only spectra with a continuum signal-to-noise ratio of ten or higher have been selected.

2. Discussion and Conclusions

The profiles of Lyα and C IV lines in the spectra of both the quasars and the Seyfert 1 galaxies are composed of a narrow (core) and a broad component and these lines are very symmetric in the quasar spectra. Over four orders of magnitude in continuum luminosity the object-to-object variation in the profiles of these lines, in the quasar spectra, is very small, but this is not true of the C IV profile in the spectra of Seyfert 1 galaxies. The FWHM of the Lyα and C IV lines and that of C IV and Hβ lines are correlated. For the Lyα/C IV correlation the Spearman rank correlation coefficient $r = 0.85$ and the slope of the straight line fitted to these data is $s = 0.77 \pm 0.22$. For the C IV/Hβ correlation $r = 0.68$ and $s = 0.68 \pm 0.07$, however, there is *no* correlation between the FWHM of Lyα and Hβ lines.

The high degree of symmetry of both Lyα and C IV lines, and both these lines are optically thick, seems to rule-out unidirectional flow of the BLR clouds. The correlation of the FWHM of Lyα and C IV lines suggests that these lines are emitted from clouds in a similar kinematic enviornment, but the correlation between the FWHM of C IV and Hβ lines seems to indicate that a part of the C IV line may also be emitted from clouds whose kinematic enviornment is similar those which emit the Hβ line and these clouds are not significant Lyα emitters.

INFRARED CORONAL EMISSION LINES AND THE POSSIBILITY OF THEIR LASER EMISSION IN SEYFERT NUCLEI

MATTHEW A. GREENHOUSE
Laboratory for Astrophysics, National Air and Space Museum, Smithsonian Institution, Washington, DC 20560

URI FELDMAN
Solar Terrestrial Relationships Branch, Naval Research Laboratory, Washington, DC 20375

HOWARD A. SMITH
Laboratory for Astrophysics, National Air and Space Museum, Smithsonian Institution, Washington, DC 20560

MARCEL KLAPISCH
Artep Inc., Naval Research Laboratory, Code 4694, Washington, DC 20375

ANAND K. BHATIA
NASA/Goddard Space Flight Center, Greenbelt, MD 20771

and

AVI BAR-SHALOM
Nuclear Research Center of the Negev, P.O. Box 9001, Beer Sheva, Israel

Infrared coronal emission lines are providing a new window for observation and analysis of highly ionized gas in Galactic and extragalactic sources such as Seyfert nuclei and classical novae shells. These lines are expected to be primary coolants in colliding galaxies, galaxy cluster cooling flows, cometary-compact HII regions, and supernova remnants. In this poster, we summarize results discussed in detail by Greenhouse et al. 1993, ApJS, **88**, 23. We discuss approximately 74 infrared ($1 < \lambda$ μm < 280) transitions within the ground configurations $2s^2 2p^k$ and $3s^2 3p^k$ ($k = 1$ to 5) or the first excited configurations $2s2p$ and $3s3p$ of highly ionized ($\chi \geq 100$ eV) O, Ne, Na, Mg, Al, Si, S, Ar, Ca, Fe, and Ni. We present results from detailed balance calculations, critical densities for collisional de-excitation, intrinsic photon rates, branching ratios, and excitation temperatures for the transitions. The temperature and density parameter space for dominant cooling via infrared coronal lines is presented, and the relationship of infrared and optical coronal lines is discussed.

We find that under physical conditions found in Seyfert nuclei, 14 of 70 transitions examined have significant population inversions. Laser gain lengths and corresponding column densities are calculated. We find that several infrared coronal line transitions have laser gain lengths that correspond to column densities which are modeled to exist in Seyfert nuclei. Observations that can reveal inverted level populations and laser gain in infrared coronal lines are also suggested.

NUCLEAR ACTIVITY IN THE SEYFERT GALAXY NGC 1365

M. HJELM, P.O. LINDBLAD and S. JÖRSÄTER
Stockholm Observatory, 133 36 Saltsjöbaden, Sweden

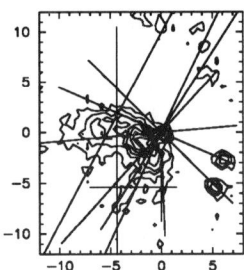 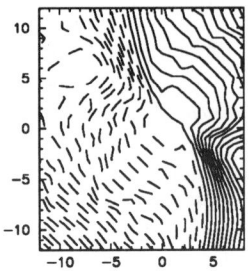 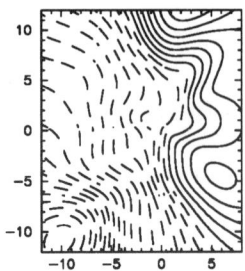

Fig. 1. a: Continuum-subtracted narrow band image in [OIII]λ5007. Superimposed are positions of spectra used in deriving the velocity fields. b: "Anomalous" velocity field. c: Model velocity field, in which the velocity, v, in the cone depends on the distance from the center, r, and on the angular distance, t, from the cone axis as where t_o is half the opening angle of the cone. Density = const $/r^4$. The inclination of the cone axis to the line of sight is $40°$, $t_o = 45°$ and the cones are directed $180°$ away from each other. Units on the axes in all three plots are arcseconds offset from the center. In b and c the dashed lines indicate blueshifted velocities and the velocity interval is $10 km s^{-1}$.

NGC 1365 is a barred Seyfert 1.5 galaxy. Fig 1a is a contour plot of an [OIII]λ5007 image (Jörsäter & Lindblad 1989), showing a plume-like structure pointing towards SE in the direction of the minor axis of the galaxy, which also is the direction of the steepest velocity gradient in [OIII]λ5007. The plume is roughly aligned with a radio feature in the 6 and 20 cm continuum (see Sandqvist et al., this volume). The [OIII]-emission on the NW side is weaker. This plume-like structure is not seen in Hα.

The velocity field of the inner parts of NGC 1365 can be divided into two separate components. The low excitation lines (eg. the Balmer lines) show the galactic rotation. The higher excitation lines (eg. the [OIII]-lines) show "anomalous" velocities that are blueshifted relative to the systemic velocity of the galaxy to the SE of the nucleus and redshifted to the NW (Fig. 1b). We model these "anomalous" velocities as a bi-conical accelerated outflow, in which the density is given by the equation of continuity (steady state) for a chosen velocity law in the cones. Fig 1c shows the velocity field derived from one of our preliminary models. For details and further discussion see Hjelm et al. 1994.

References

Hjelm M., Lindblad P.O. and Jörsäter S., 1994, in prep.
Jörsäter S. and Lindblad P.O.,1989, ESO Workshop on Extra Nuclear Activity, p 39.

COMPOSITE MODELS FOR NARROW LINE REGIONS

STEFANIE KOMOSSA
Astronom. Institut, Ruhr-Universität, D-44780 Bochum

The basic excitation mechanism for the emission lines of Seyfert galaxies is generally believed to be photoionization by radiation emerging from the central power source. In order to deduce the essential physical parameters appropriate to the narrow line region (NLR) on this basis and to overcome some systematic deficiencies of existing photoionization models, we drop the frequent assumption of a constant ionization parameter throughout the NLR and, in particular, allow for a range of densities at fixed radius. We predict *all* observed emission-lines (ranging from CIVλ1549 in the UV- to [SIII]λ9069 in the NIR spectral region) for the sample of Sy 2 galaxies using the code CLOUDY (Ferland 1993). Our models are characterized by (values in parentheses give the considered range for each parameter):
1) the shape and strength of the ionizing continuum (power laws with $\alpha_{uv-x} = -1... - 2.5$, black bodies with $T_{bb} = (1.0...2.5) \times 10^5$K, combinations of both; log $Q = 54, 53$) constrained in the non-EUV part by recent observations.
2) the geometrical and physical properties of the clouds, i.e.
i) distances from the central source ($r = 10^{20...21.5}$ cm for $Q = 10^{54}$ s^{-1}),
ii) cloud column densities ($N_h = 10^{18...24}$ cm^{-2}; for the employed range of densities 10^{18} always corresponds to matter-bounded clouds, 10^{24} to ionization-bounded ones),
iii) total hydrogen density ($n = 10^{2...6}$ cm^{-3}) and density distribution over the whole region and
iv) metal abundances ($0.3...3 \times$ solar).

The major **results** are the following:

By composing the NLR of an ensemble of clouds distributed over a range of radii and with a range of densities at each radius we improve the predictions of high ionization lines like [NeV], CIII] and CIV and, in particular, the hitherto poorly modeled NIR-lines [SIII] and [OII].

To cover the whole observed range for each emission line, various spectral energy distributions incident on the clouds are required, involving both single power laws and black body components in the EUV-part of the continuum.

Solving the 'temperature-problem' on the basis of pure photoionization modelling can only consistently be attained by reducing the metal abundances (as compared to the contribution of a high-density component, matter-bounded clouds or the influence of dust).

No parameter-combination could be found to account on the one hand for the correlations of [NI]λ5200 with other lines (namely HeII) and on the other hand for its observed strength and correlations between further lines.

OPTICAL, HST, AND ROSAT OBSERVATIONS OF BAL QSOS

MICHAEL KOPKO, JR., DAVID A. TURNSHEK, and AND BRIAN R. ESPEY
Department of Physics and Astronomy, University of Pittsburgh

We consider multiwavelength observations of Broad Absorption Line (BAL) QSOs which include optical spectrophotometry supplemented by HST-FOS UV spectrophotometry and/or ROSAT-PSPC x-ray observations. For moderate to high redshift objects, increasing the wavelength coverage into the UV permits observation of BALs due to different ions of the same element for several different elements for the first time. By employing appropriate assumptions, ionic column densities as a function of outflow velocity can be derived from these observations, resulting in constraints on the level of ionization and chemical composition of the BAL region (BALR) gas and the photoionization model itself.

Under the assumption that the BALR gas is photoionized, analysis with the photoionization code CLOUDY developed by Ferland suggests that BALR abundances must be generally enhanced at least 10 to 100 times solar values. The wide range in ionization level observed in some BAL QSOs appears to indicate that for a given outflow velocity a range of ionization parameters applies.

HST FOS UV observations of the low-redshift weak-BAL QSO PG0043+039 show a deficit of flux shortward of $\sim$ 2200 Å which is consistent with intrinsic dust-extinction by SMC-like dust with E(B-V) $\simeq$ 0.1. There is also evidence for significant amounts of FeII emission and a lack of narrow-lined [OIII] emission. The details of this result are reported in a publication by the HST Quasar Absorption Line Key Project Team (Turnshek et al.).

From observations of BAL QSO emission-line profiles, the absence of significant emission components expected from resonance line scattering has generally been used to infer that BALR covering factors are < 0.2. However, given the likelihood that narrow-line [OIII] emission is emitted isotropically, the high incidence of BALs in samples of QSOs with weak-[OIII] emission and strong-FeII emission suggests that the BALR covering factor may be large in some subset of QSOs. The presence of dust in the BAL region could provide a mechanism to destroy photons which were resonance line scattered in the BAL region.

Pointed ROSAT PSPC observations of a sample of 5 moderate-redshift (z $\simeq$ 2) BAL QSOs show that $1.9 < \alpha_{OX} < 2.5$ ($f_\nu \sim \nu^{-\alpha}$) is consistent with the observations. This is well outside the range of 1.3 – 1.8 quoted in the literature for moderate-redshift radio quiet QSOs. This indicates that either significant photoelectric absorption occurs in the BALR or that there is decreased x-ray production along the direction of BALR outflow. The presence of photoelectric absorption would be consistent with enhanced BALR elemental abundances.

WOLF-RAYET STARS AS TRACERS OF THE RECENT HISTORY OF THE STAR FORMATION RATE

G. MEYNET
Geneva Observatory, ch. des Maillettes 51, CH-1290 Sauverny, Switzerland

October 25, 1993

Abstract. Recently Vacca & Conti (1992) have measured the ratio of the luminosity in the broad He II $\lambda 4686$ emission feature to that in the H_β emission line in fourteen starburst galaxies. They related these luminosity ratios to the relative numbers of Wolf-Rayet (type WNL) to O-type stars in these galaxies (higher is the ratio $L(\lambda 4686)/L(H_\beta)$, higher is N_{WNL}/N_O). They found that in general the number ratios are an order of magnitude larger than those expected in region of constant star formation rate. On Fig. 1 the predicted line ratios of our starbursts models (instantaneous burst of star formation at time $t = 0$; initial mass function $dN/dM = CM^{-2}$; stellar models from Meynet et al. 1993; conversion formula between the line ratios and the number ratios of WNL to O-type stars given by Vacca & Conti 1992, with $\eta = 1$) are compared with the observed values given by these authors. This figure shows that a starburst taking place about 2-3 millions years ago can account for the high observed values of $L(\lambda 4686)/L(H_\beta)$. One sees that the effects of the age of the burst (*i.e.* the time elapsed since the burst) and of the metallicity are quite important. It is the hope that in a next future, it will be possible on the base of this kind of luminosity ratios to disentangle the various effects influencing the WR population resulting from a starburst.

References

Meynet, G., Maeder, A., Schaller, G., Schaerer, D., Charbonnel, C.: 1993, *A&AS* in press
Vacca, W.D., Conti, P.S.: 1992, *ApJ*, **401**, 543

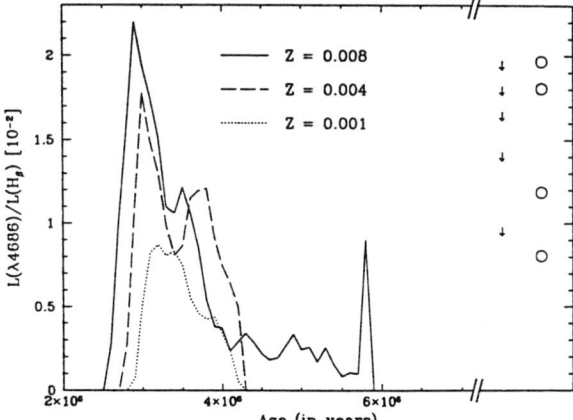

Fig. 1. Evolution with time of the ratio of the luminosity in the broad He II $\lambda 4686$ emission feature to that in the H_β line predicted by our starburst models at different metallicities (see text). Among the 14 galaxies observed by Vacca & Conti, nine have $L(\lambda 4686)/L(H_\beta) < 2.3$. The observed values for these galaxies are given at the right of the figure; the down arrows indicate upper limits.

MODELLING OF DOUBLE-PEAKED EMISSION LINES IN AGN'S

L.S.NAZAROVA
Royal Greenwich Observatory, Cambridge CB3 0EZ, U.K.

Optical spectra of the galaxies NGC3227, Mkn6, Mkn9, Mkn279, Mkn376 with double-peaked emission lines are presented. The data were obtained with the 1024-pixel TV scanner of the 6-m telescope during the 1984-1991 period. The double-peaked structure of the $H\beta$ line in most galaxies shows changes during this period and only in the galaxy Mkn6 is the double-peaked stucture of $H\beta$ constant. In all the galaxies except Mkn6, the blue peak of line coincides with the rest-wavelength of the line (the opposite occurs in Mkn6).

It is suggested that the double-peaked line profiles are due to large scale gas motion in the disk envelope which characterizes the radial and the tangential components of velocity. The double-peaked profiles of $H\beta$ have been modelled with an elliptical geometry for the gas envelope of these galaxies. The shape of the envelope in the models is determined by the tangential component of velocity. The equations of stationary population of levels were solved by Sobolev's probabilistic method. The velocity gradient is determined for the envelope with axially symmetric kinematics. It is also suggested that the rotation axis has an arbitrary inclination to the observer. Comparison of the theoretical and observed profiles of the lines shows that the double-peaked structure of $H\beta$ in the galaxies could be produced in the envelope of the accretion disk which is characterised by Keplerian rotation $V_\theta = const * R^{-0.5}$ and radial component of velocity $V_r = const * R^{-1}$ with decelerated expansion for the galaxies NGC3227, Mkn9, Mkn279, Mkn376 and with accretion for Mkn6.

Key words: data analysis-galaxies: NGC3227, Mkn6, Mkn9, Mkn279, Mkn376-galaxies Seyfert-lines:profile-double-peaked structure

TIME DEPENDENT BLR PHOTOIONIZATION MODELS

P.T. O'BRIEN and M.R. GOAD
Department of Physics & Astronomy, University College London

Abstract. Recent intensive campaigns to monitor the variable broad emission lines of AGN have provided a wealth of observational data. To reliably derive the BLR structure from such data requires a detailed comparison with physically realistic models. To provide such models we have combined photoionization calculations with a BLR modelling code. We find that the line emission is significantly anisotropic for the majority of lines at some radii in our models, particularly where the ionization parameter is large. As both anisotropic line emission and negative responsivity can have a substantial impact on the form of response functions, they must be taken into account when comparing BLR models to monitoring data. Differentiating between the effects of anisotropy and negative responsivity requires comparison of the variability behaviour of lines covering a wide range in ionization state with detailed photoionization models similar to those presented here. Failure to take such physical effects into account will lead to incorrect conclusions regarding the BLR structure.

Based on the form of the anisotropic response functions for our models, we argue that in NGC5548 the high ionization lines (including CIII] $\lambda 1909$) can be explained by a spherical BLR component which has a significant radial ionization parameter gradient. The clouds in this component may be optically thin at the Lyman limit in the inner part of the BLR due to the high ionization parameter. The bulk of the low ionization lines, and in particular almost all of the Balmer line emission, arises from a non-spherical BLR component, possibly an accretion disc.

We have also investigated the accuracy of the linear response approximation by implicitly calculating the emissivity as a function of continuum level. For several lines the response is significantly non-linear over the typical continuum variability range of AGN. The form of the response functions is therefore luminosity dependent. Work is under way to incorporate this effect into the maximum-entropy-method used to recover response functions from monitoring data.

Key words: AGN, Photoionization Models, Emission Lines

The Nuclear Region of the Sbc Spiral Galaxy NGC 5055: A Mildly Active Nucleus.

P. Pişmiş[1], M. Manteiga[2], A. Mampaso[3], E. Recillas-Cruz[4] and G. Cruz-González[1]
[1] *UNAM (México)*
[2] *LAEFF (Spain)*
[3] *IAC (Spain)*
[4] *INAOE (México)*

Abstract.
NGC 5055 is one of a list of nearby large spiral galaxies we have selected for a morphological and kinematic study of their bulges. CCD images are obtained at various spectral bands and in the Hα line using the 2.1 m reflector at San Pedro Martir observatory in México, while long-slit spectra at four different PA's are secured with the IDS spectrograph of the Isaac Newton Telescope of the observatory Roque de los Muchachos at La Palma, Spain. Some of the galaxies on our list for which observations are already performed are, aside from NGC 5055, NGC 3351, 4314, 5383 and 5915. These galaxies are not known to have active nuclei. However, we believe that activity may be a common phenomenon spanning a wide range of energetics, from the most active quasars and radio-galaxies (with jets) down to the mildest cases like our Galaxy or M31. The latter may be designated as MAGN (mildly active galactic nuclei). These ideas are treated in a review by Pişmiş (1987, Rev. Mex. Astron. Astrof. 14). The MAGN are usually nearby, and hence offer the possibility to study them in sufficient detail. It is reasonable to expect that the physical phenomenon underlying activity in galactic nuclei is one and the same, and it is essentially the global parameters such as energy at the nucleus that give rise to the different classes of AGN. Thus by studying the nearby mildly active galaxies one can construct a physically acceptable model supported by observation; such model, based on firmer grounds, can then be applied to all AGN.

NGC 5055 (Sbc) with an adopted distance of 8.2 Mpc has a very bright bulge with a tenuous multiple arm structure around it, extending to 8 arcmin. Our velocity field and morphology of NGC 5055 lead to the following conclusions.

1. The galaxy presents a north-south asymmetry; the southern approaching side has its isophotes closer together than at the farther side. We adopt PA 101° for the line of nodes. Our long-slit spectra cover the position angles: 58°, 101°, 112° and 155°.

2. The rotation curve shows a linear region around the nucleus. At PA 101°, the line of nodes, the amplitude is 270 km s^{-1} between points ± 5 arcsec on either side of the nucleus.

3. There is high concentration of mass at the nuclear region (the bulge of the galaxy). An estimate of the mass for the region where the rotation curve is linear ± 5 arcsec, is 1.5×10^8 M$_\odot$. Burbidge et al (1960) give a total mass of 7.6×10^{10} M$_\odot$. The bulge density is thus found to be 500 times the average density of this galaxy.

Analysis of the general velocity field based on four long-slit spectra shows an asymmetry between the East and West sides, ± 5 arcsec around the nucleus of NGC 5055, suggestive of the existence of non-circular motion. Indeed evidence is found of an outflow on the west side of the Galaxy within a cone emanating from the center towards the NW with a projected radial velocity of around 80 km s^{-1}. Thus this galaxy shows activity at its center and is a candidate for the MAGN group.

Key words: galaxy, spiral, mild activity, MAGN

SUBARCSECOND-SCALE OPTICAL AND RADIO STRUCTURE CORRELATIONS IN SEYFERT GALAXIES

R.W. POGGE
*Department of Astronomy, The Ohio State University,
Columbus, Ohio, USA*

and

M.M. DEROBERTIS
*Department of Physics & Astronomy, York University,
North York, Ontario, Canada*

An unanswered question in the study of Seyferts is the nature of the relationship between the extended radio-continuum and optical emission-line gas. Detailed comparison is difficult as most Seyferts have radio structure on sub-arcsecond scales, while most optical imaging is on 1–2″ scales. Despite this, some basic features have emerged. Extended radio and optical emission regions are generally aligned in projection, but the overall optical emission extends to much greater galactocentric radii. Pedlar et al. (1989) and Whittle et al. (1986) have reported a tendency for optical emission knots to lie *behind* the radio lobes, interpreted in terms of the "cooling length" of gas heated by a radiative bowshock driven into the ISM by a jet. In NGC 1068 (Cecil et al. 1990) and M51 (Cecil 1988), however, ambient gas appears to be piling up *ahead* of a radio lobe at the terminus of the radio jet.

We have obtained subarcsecond seeing CFHT and HST Archive emission-line imaging of Seyfert galaxies, allowing a fair comparison of the radio and optical maps at identical scales. The bright knots of optical emission-line gas are primarily *further* from the nucleus than the radio knots, with a few showing cospatial emission. This is consistent with the radiative shock picture (e.g., Wilson 1989). In one case, Mrk 573, we see a distinctive bow-shaped morphology in the [OIII] emission-line gas. We also see S-shaped distributions of emission-line knots on scales larger than the radio structure. Ionization cones, where present in these galaxies, appear as a diffuse background upon which the bright knots associated with the radio structures are superimposed.

References

Cecil, G., Bland, J., & Tully, R.B. 1990, ApJ, 355, 70
Cecil, G. 1988, ApJ, 329, 38
Pedlar, A. et al. 1989, MNRAS, 238, 863
Whittle, M. et al. 1986, MNRAS, 222, 189
Wilson, A.S. 1989, in Extranuclear Activity in Galaxies (ESO Conference and Workshop Proceedings 32), ed. E.J.A. Meurs & R.A.E. Fosbury, 215.

INFLUENCE OF THE GRAVITATIONAL FIELD ON THE SHAPE OF SPECTRAL LINES IN SPECTRA OF SEYFERT GALAXIES AND QUASARS

L. Č. POPOVIĆ, I. VINCE and A. KUBIČELA
Astronomical Observatory,
Volgina 7, Belgrade
Yugoslavia

On the basis of preliminary examination of the model of Seyfert galaxies and quasars we have concluded that their gravitational field may have an important influence on the shape of spectral lines when the emission cloud is in gravitational field of massive nuclei (mass about $10^7 - 10^8$ $M_\odot$ for Seyfert galaxies, see e.g. Padovani *et al.* 1990), and the emission cloud is large enough (for example about $10^{14} - 10^{15}$ m for Broad Line Region (BLR) of NGC 7469 (Bonatto & Pastoriza 1990) or about 10^{21} m for quasar 3C 257 (Forbes *et al.* 1990)), so that the emitters are in very different gravitational field along the line of sight.

In this preliminary analyses we consider a very simple model of Seyfert galaxy, in wich a massive nuclei is surrounded by circumnucleus cloud with only one sort of emitters and with only one possible transition $m \to n$. The circumnucleus cloud is static and opthicaly thin. We assume that the profile of spectral line, at a given point of the cloud is Lorentzian one.

If we consider that the emitters are in the gravitational field of a mass M at a distance r from the central body, and that the gravitational potential at observer is negligibly small, the transition frequency (ω_0) will be shifted by $\omega' - \omega_0 = \omega_0 GM/(rc^2)$ (see e.g. Weinberg 1972), where G is the gravitational constant and ω' is the shifted frequency. For a cloud of thickness $D = R - R_0$ the emerging line profile is given as a sum of radiation of all emitters along the line of sight. In this case intensity at a given frequency in shape of spectral line profile can be given by

$$S'(\omega, R, R_0, M) = S(\omega)\Phi(\omega, R_0, R, M),$$

where S is the Lorenzian profile and Φ is the correction function that determines the degree of spectral line profile distortion produced by the gravitation field.

As an illustration of the influence of gravitational field on spectral line shape we calculated the shape of H_α line with the following parameters: $M = 10^8$ $M_\odot$, $R_0 = 10^{14}$ m, $R = 10^{16}$ m. As a results we have obtained that the spectral line is shifted by about 0.12 Å toward the red and is broaded about 18%. The blue wing of the deformed spectral line is lower, while the red one is higher than in the unperturbed line profile. This influence is more prominent for spectral lines in infrared and radio than in ultraviolent range of wavelength.

REFERNCES

Bonatto, C. J., and Pastoriza, M. G., 1990, *ApJ.* **353**, 445.
Padovani, P., Burg, R., and Edelson, R. A., 1990, *ApJ.* **353**, 438.
Forbes, D. A., Crawford, C. S., Fabian, A. C. & Johnstone, R. M., 1990, *Mon. Not. R. Astr. Soc.* **244**, 680.
Weinberg, S., 1972, *Gravitation and Cosmology*, John Wiley and Sons, New York – London – Sydney – Toronto.

SIZE OF THE BROAD LINE REGION IN NGC 3227: RESULTS FROM THE LAG CAMPAIGN *

I. SALAMANCA, T. BARIBAUD and D. ALLOIN
Observatoire de Meudon DAEC, 5 Place Jules Janssen, 92195 Meudon Cedex, France

We present the results of a monitoring campaign of the AGN in NGC 3227, conducted at La Palma by the LAG european consortium. The data are long slit spectra, obtained with the 1.5m INT and 4.2m WHT, from January to June 1990. The slit width was 1.5 arcsec. The spectral resolution was about 3 A.

From the observed MgI $\lambda 5167, 5173, 5183$ in the NGC 3227 spectra we deduce that the stellar contribution is at most 40% at $\lambda 5180$ A when the AGN is in his lower state.

To separate the BLR and NLR emissions we have performed a gaussian decomposition of the blends Hα + [NII] + [SII] and Hβ + [OIII]. We have constrained as many parameters as possible, following physical arguments. The same blend modeling was used for all epochs, only the amplitude of the BLR components was allowed to vary.

After having performed the line blend decomposition, subtracted the stellar population contribution and applied corrections for seeing differences, we have obtained the light curves for the BLR Hα and Hβ line emission as well as for the visible continuum (derived under the assumption that the NLR emission does not vary with time). We find out variation amplitudes around 40% over timescales of 6 weeks. **The cross correlation of the line and continuum light curves provides a BLR size of 17 +/- 7 days.**

The Balmer line difference between the highest and lowest levels shows a symmetrical flat-topped profile with FWHM $\approx$ 2900 km/s. Assuming a keplerian broadening of the lines, and that the accretion disc is coplanar with the galactic disc, we conclude that the bulk of the BLR emission occurs at 10^3 Schwarzschild radii and that the central object has a mass of $2 \times 10^8 M_\odot$.

From a small sample of AGN (7) we find that the BLR size is proportional to L^γ, where $\gamma < 0.5$, instead of 0.5 assumed in the models. A correlation analysis indicates a slope of about 0.2.

References

Baribaud T., Salamanca I., Alloin D., Wagner S.,1993, 'Sources of uncertainty in the relative scaling of spectroscopic data', *A&A SS*, in press.
Salamanca I., Alloin D., Baribaud T. et al., 1993, LAG paper, 'Spectroscopic monitoring of active galactic nuclei III. Size of the Broad Line Region in NGC 3227.',*A&A*, in press.

* The Lovers of Active Galaxies (LAG) consortium was set up to apply for international time on the Canaries Observatories and conducted a series of programs related to the physics of AGN.

A UNIFIED VIEW OF NGC 4151

HARTMUT SCHULZ
Radioastronomisches Institut, Universität Bonn, Auf dem Hügel 71, D-53121 Bonn
and Astronomisches Institut, Ruhr-Universität, D-44780 Bochum

Abstract. Using crude constraints from recent HST images, we fit the emission-line spectra of the NLR and VNLR of NGC 4151 with photoionization models. There is no need for an *intrinsic* anisotropy of the distribution of the ionizing flux.

1. Introduction

NGC 4151 is the first Seyfert galaxy in which a 30° wide UV radiation cone was invoked to explain the ionization of the VNLR (very-narrow line region; 5" to 20" SW of the nucleus; Schulz 1988). A NE counterpart (60° wide cone) together with a bipolar NE-SW outflow structure was suggested by Schulz (1990). Recent HST images in [OIII] and Hα light (Boksenberg 1992; Evans et al. 1993) corroborate the geometry predicted from ground-based data by showing an inner bicone ($\pm 2°$) and reveal striking details on the morphology of the emission-line cloud aggregates.

2. Results

The VNLR spectrum can be well represented by a single cloud of fixed density photoionized by a bright and most probably anisotropic continuum that has a 230000 K black-body spectrum in the EUV bump. An alternative fit involving multidensity clouds is consistent with an intrinsically isotropic continuum (Schulz & Komossa 1993). The HST constrained model of the NLR bicone favors the latter continuum model, a mixture of densities ($\log n = 3$ to 4), and a γ-break at least at 9.9 MeV.

3. Acknowledgements

This work was supported by DARA grant 50 OR 9102.

References

Boksenberg A., 1992, ESO Conf. Proc. 44, p.61
Evans I.N. et al., 1993, ApJ., in press
Ferland G., 1993, Univ. Ky. Dep. Phys. & Astron. Int. Report
Schulz H., 1988, A&A, 203, 233
Schulz H., 1990, AJ, 99, 1442
Schulz H., Komossa S., 1993, A&A, in press

IMPLICATIONS OF NONLINEAR LINE RESPONSE IN VARIABLE SEYFERT NUCLEI

JOSEPH C. SHIELDS
Steward Observatory, University of Arizona

GARY J. FERLAND
Department of Physics & Astronomy, University of Kentucky

and

BRADLEY M. PETERSON
Department of Astronomy, Ohio State University

Variable Seyfert nuclei exhibit correlations between emission-line and continuum luminosity consistent with photoionization, although the emission lines (including recombination features such as Lyα) tend to respond nonlinearly to changes in the continuum. A nonlinear response will result if some of the broad-line region (BLR) clouds become fully ionized in hydrogen, such that their emission measure does not then vary in proportion to the incident continuum flux. For thin clouds, emission in transitions from heavy elements may grow or decline with increasing incident flux, reflecting changes in the ionization structure of the cloud. Direct evidence for a fully ionized component within the BLR is provided by correlation analysis of ultraviolet line response in NGC 5548 (Sparke, 1993).

We have completed photoionization calculations with the code Cloudy for a BLR cloud population containing a mix of thick and thin clouds. The results are capable of producing good agreement with the observed slopes of line response relative to the continuum for high ionization lines. In particular, the inverse relation between the C IV $\lambda1549$/Lyα ratio and continuum luminosity observed in NGC 5548 and other Seyferts can be reproduced in this scenario. Understanding this behavior remains problematic if the clouds are all thick (Shields & Ferland, 1993).

A thin cloud population can also act as a "warm absorber" for soft x-rays. One signature of a warm absorber is an inverse relation between apparent absorbing column and luminosity, resulting from changes in the degree of oxygen ionization. Thin clouds with high covering factor can generate significant ultraviolet line emission, and also display warm absorber behavior when the continuuum luminosity is increased only slightly, if the ionizing continuum is relatively hard. Such a scenario may be appropriate for NGC 5548, for which the continuum is approximately described by a power-law $f_\nu \propto \nu^{-1.2}$ from the ultraviolet to soft x-ray region (Clavel et al., 1992).

References

Clavel, J., et al. 1992, *ApJ*, **393**, 113
Shields, J. C., & Ferland, G. J. 1993, *ApJ*, **402**, 425
Sparke, L. S. 1993, *ApJ*, **404**, 570

STAR–FORMATION AND NUCLEAR ACTIVITY IN THREE GALAXIES WITH NUCLEAR RINGS

THAISA STORCHI–BERGMANN*
Instituto de Fisica, UFRGS, C.P. 15051, P. Alegre, RS, Brasil
ANDREW S. WILSON*
Space Telescope Science Institute, Baltimore, MD, USA
and
JACK A. BALDWIN
Cerro Tololo InterAmerican Observatory, La Serena, Chile

We investigate two current problems in active galactic nuclei – the mode of fueling the putative black hole, and the question whether the circumnuclear regions have experienced unusual chemical processing – by studying the kinematics and chemical abundance of the gas in the nuclear region of galaxies with sites of on-going star formation near the active nucleus. We discuss the results for three galaxies with nuclear rings: NGC1097 – for which we recently discovered broad double peaked $H\alpha$ and $H\beta$ emission from its LINER nucleus (Storchi-Bergmann, Baldwin & Wilson 1993, ApJ 410, L11); NGC1672, which also presents a LINER nucleus; and NGC5248, a galaxy with a ring but no nuclear activity, used as a comparison. Narrow–band images obtained with the CTIO 1.5m telescope were used to map the emitting gas. Longslit spectroscopy obtained with the 4m telescope at high spectral resolution (at several positions over the nuclear region) was used to obtain the gas velocity field. In the two galaxies with LINER nucleus, the star-forming rings are located in the turnover of the rotation curves, which show that the gas is rotating faster at the rings than farther out. We conclude that the rings are associated with inner Lindblad ressonances, which may be particularly effective in forcing gas inwards and fuelling the black hole (Wilson et al. 1986, ApJ 310, 121). From the emission line ratios the gas excitation is maped, and it is found that even in NGC5248, the nucleus presents a different excitation, suggesting a very mild LINER activity. Low–dispersion spectroscopy was also obtained in order to calculate the chemical abundances and compare the values obtained for the nuclear gas with those obtained for the HII regions in the ring and beyond the ring (when present). The goal is to check the results of recent studies (Storchi-Bergmann & Pastoriza 1989, ApJ 347, 195; 1990, PASP 102, 1359), based on spectroscopy of the *nucleus*, which indicate an enhanced abundance of nitrogen, up to 5 times solar for the gas of the narrow line region of LINER and Seyfert 2 nuclei.

* Visiting Astronomer at the Cerro Tololo Interamerican Observatory, operated by AURA under contract with the National Science Foundation

HIGH RESOLUTION NIR IMAGING OF THE STARBURST GALAXY NGC 1808

L.E. TACCONI-GARMAN and A. KRABBE
Max-Planck-Institut für extraterrestrische Physik

A. STERNBERG
Tel Aviv University

and

R. GENZEL
Max-Planck-Institut für extraterrestrische Physik

Abstract. We report 0.6″ res. J, H, and K and 1.5″ res. imaging of 2.17 μm HI Brγ and 2.12 μm H_2 1-0 S(1) line emission towards the nucleus of the starburst galaxy NGC 1808. In the K-band data we (partially) resolve the nucleus and see several small knots in the circumnuclear region. Further, our JHK continuum images show that a large fraction of the near infrared light in NGC 1808 is produced in young star forming clusters. The Brγ emission originates from a compact nuclear source and from several distinct emission knots in the circumnuclear region. These knots are spatially well correlated with a family of compact radio sources, but uncorrelated with the optical "hot spots". We propose that the Brγ knots trace the actual sites of starburst activity, while the optical hot spots are just directions of low foreground extinction.

We use our data together with radio and far-infrared continuum emission measurements to constrain the parameters of the *individual* starburst sites in NGC 1808. The data suggest that the starbursts are unsynchronized and prolonged ($5 \times 10^6 - 5 \times 10^7$ yrs). The star formation rates in the active sites range from ~ 0.1 to ~ 0.6 $M_\odot$ yr^{-1}, and the present rapid rate of star-formation in NGC 1808 can be maintained for at most another $\sim 7 \times 10^7$ yrs.

Portions of this work are presently in press (Krabbe, Sternberg, and Genzel 1993), and a second paper is in preparation (Tacconi-Garman et al. 1993).

References

Krabbe, A., Sternberg, A., and Genzel, R.: 1993, 'Near Infrared Spectral Imaging of NGC 1808: Probing the Starburst', *Astrophys. J.*, in press
Tacconi-Garman, L.E., *et al.*: 1993, in preparation.

INFRARED LINE SHAPES IN ACTIVE GALACTIC NUCLEI

Rodger I. Thompson
Steward Observatory, University of Arizona

ABSTRACT. This paper presents the complete spectrum of NGC 4151 from 0.87 to 2.5 µm as well as detail of the Paschen α profile in 3C273 at a resolution of about 5000. Analysis of the several observed Fe II lines yields a most probable electron density of 10^4 and a most likely temperature of 10^4. The Fe II to H ratio is $2 +/- 1 \times 10^{-6}$ which is 5-7% of the available iron if the Fe/H ratio is solar.

1. Observations

The complete 0.87 - 2.5 µm spectrum of NGC 4151 was observed with the Grating Infrared Spectrometer (GRIS) at an average resolution of 5300. Only regions of severe telluric absorption are missing from the spectrum. This spectrum has the advantage of simultaneous and consistent observation of many spectral lines from both the narrow and broad line region plus overlap with the spectral region available to visible CCD spectrometers. The known variability of the continuum and broad line emission from NGC 4151 make the simultaneous observations very important.

2. Results

The major analysis is an examination of the Fe II lines using the new excitation parameters of Pradhan and Zhang (1993). Comparison of the solution of the equilibrium equations with the observed line ratios yields a most probable electron temperature of 10^4 °K and an electron density of 10^4 cm^{-3}. Some observations from Osterbrock et al 1993 were utilized in the analysis. The ratio of Fe II 1.644 µm to Paschen β yields a FeII/H ratio of $2+/-1 \times 10^{-6}$ or about 6% of the iron for a solar Fe/H ratio. This indicates significant grain evaporation or destruction in the Fe II emission region. This appears to be more consistent with fast shocks as the ionization mechanism as opposed to slow shocks. High energy photons could also play a role in destroying the grains. Not all of the line ratios are consistent with the determined density but they may suffer from blends or other observational effects.

The Paschen α emission in 3C273 is consistent with the line shapes in the optical region. There is no evidence in the present spectrum of previously reported H$_2$ emission from the object.

3. References

Osterbrock, D.E., Shaw, R.A., and Veilleux, S. 1992, *Ap.J.*, **352**, 561.

Pradhan, A.K. and Zhang, H.L. 1993, *ApJ (Letters)*, **409**, L77.

FEII, FEI EMISSION LINES FROM ACCRETION DISKS: AN EXPLANATION FOR "FEII PROBLEM" IN AGNS ?

CHUNYAN WEI and FUZHEN CHENG
Center for Astrophysics, University of Science and Technology of China, Hefei 230026, P. R. China

and

JUNHAN YOU
Department of Physics, Shanghai Jiao Tong University, Shanghai 200030, P. R. China

For the solution of the puzzling "FeII problem" in active galactic nuclei(AGNs) (Netzer et al. 1983; Wills et al. 1985), we pay our attention to optical band and suggest: (1)the observed so-called "FeII emission lines" features may be blending of FeII multiples and FeI multiples. Our previous work(Wei et al. 1993) has showed that there are many FeI emission lines whose wavelength lie around the observed "FeII emission lines" features. In fact, FeI emission lines have been observed in the spectrum of PHL 1092(Bergeron et al. 1980; Cheng et al. 1993). (2)the emission lines from accretion disk must be considered besides the emission from broad line region.

Using a Non-LTE α-disk model without external continuum irradiation, we calculate the HI, FeII and FeI emission lines from an accretion disk with one set of the parameters of the disk, $M_{bh} = 10^8 M_\odot$, $\dot{M} = 0.67 M_\odot/yr$, $\alpha = 0.01$. Considering the emission lines from both accretion disk and broad line region, in which the relative intensity of the emission lines from broad line region is take from Wills et al.(1985), $\frac{I(FeII)_{opt}^{BLR}}{I(H\beta)^{BLR}} \simeq 3$, we estimate the ratio of $\frac{I(FeII+FeI)_{opt}}{I(H\beta)}$ is about 6. It nearly reach the average observation value of 7. So the model and results presented here indicate a new way to solve the strong intensity of "FeII problem" in AGNs.

For the observed large changes in the ratio of the intensities of optical to UV FeII emission lines in different objects, our explanation is that the different viewing angle of the disk result in different ratio of observed $I(FeII)_{UV}/I(FeII)_{opt}$ from object to object.

The authors would like to thank Dr. J. Kwan for fruitful discussion.

References

Bergeron, J. and Kunth, D., 1980, *A&A*, **85**, L11.
Cheng, F. Z., et al., 1993, in preparation.
Netzer, H. and Wills, B. 1983, *ApJ*, **275**, 445.
Wei, C. Y., Cheng, F. Z. and You, J. H. et al., 1993, *Kexue Tongbao*, in press.
Wills, B., Netzer, H. and Wills, D., 1985, *ApJ*, **288**, 94.

THE ANISOTROPIC RADIATION FIELD IN NGC 3516

I. YANKULOVA and V. GOLEV
Department of Astronomy, University of Sofia, 5 James Bourchier st., BG – 1126 Sofia, Bulgaria

T. BONEV
Institute of Astronomy, Bulgarian Academy of Sciences, 72 Tsarigradsko chaussee blvd., BG – 1784 Sofia, Bulgaria

and

K. JOCKERS
Max-Planck-Institute for Aeronomy, D – 3411 Katlenburg-Lindau, Germany

We present new narrow-band images of the Extended Emission-Line Region (EELR) in NGC 3516 in light of [O III] λ 4959, Hα + [N II] $\lambda\lambda$ 6548, 84, [O I] λ 6364, He I λ 6678 and [Fe VII] + [Ca V] λ 6087. The observations were carried with the 2-m reflector of the Bulgarian National Astronomical Observatory and the Focal Reducer of the Max-Plank-Institut for Aeronomy. Our [O III] and Hα + [N II] images confirm previously reported EELR features. In contrast, the image in the high-excitation [Fe VII] + [Ca V] line shows a different structure. We identify a biconical morphology over a kiloparsec scale with peak intensities 5.9×10^{-16} ergs cm^{-2} s^{-1} arcsec^{-2} and 3.5×10^{-16} ergs cm^{-2} s^{-1} arcsec^{-2} to north and south of the nucleus, respectively. The total flux of the [Fe VII] + [Ca V] emission in 5″ and 24″ circular apertures centered at the nucleus is $(9.97 \pm 0.38) \times 10^{-14}$ ergs cm^{-2} s^{-1} and $(1.53 \pm 0.15) \times 10^{-13}$ ergs cm^{-2} s^{-1}, respectively, which is in good agreement with measurements of Boksenberg & Netzer (1977) through the 5″ aperture. The cone axis lies at PA $\sim -10°$. The continuum images (Veilleux et al.,1993, Miyaji et al.,1992) indicate a "bar" aligned along PA $\sim -10°$. The velocity extrema regions revealed by Veilleux et al. (1993) are coincident with the peak intensities in our [Fe VII] + [Ca V]. We suppose that our image in [Fe VII] + [Ca V] outlines a Coronal-Line Region (CLR) of NGC 3516, which extends far beyond the classical NLR of the galaxy. Korista & Ferland (1989) have recently shown theoretically that the CLR in Seyferts may be a result of a low-density interstellar medium exposed to and photoionized by a "bare" active nucleus. A typical ISM with $N_e \sim 1-5$ cm^{-3} may produce such an extended CLR as that observed by us.

This research was supported by the Bulgarian National Scientific Foundation grant under contract No. F-109/1991 with the Bulgarian Ministry of Education and Sciences.

References

Boksenberg, A., & Netzer, H.: 1977, *Ap.J.*, **212**, 37
Korista, K. T., & Ferland, G. J.: 1989, *Ap.J.*, **343**, 678
Miyaji, T., Wilson, A. S., & Perez-Fournon, I.: 1992, *Ap.J.*, **385**, 137
Veilleux, S., Tully, R. B., & Bland-Hawthorn, J.: 1993, *A.J.*, **105**, 1318

Poster Contributions:
Polarisation

INTERNAL FARADAY ROTATION IN COMPACT RADIO SOURCES (CRS)

CLAES-INGVAR BJÖRNSSON
Stockholm Observatory
S-133 36 Saltsjöbaden, Sweden

ABSTRACT. Standard inhomogeneous synchrotron models can explain the polarization properties of CRS if proper account is taken for internal Faraday rotation; neither depolarization due to synchrotron self-absorption nor a tangled magnetic field plays an important rôle. The effects of inhomogeneities are important; the polarization properties of flux escaping from a Faraday thick region are quite different from those coming from a homogeneous source. The observed polarization of strong lined CRS indicates (i) a jet with constant opening angle, (ii) a large scale magnetic field whose main component is along the jet axis and (iii) the density of Faraday rotating electrons scales with radial distance in the same way as does the density of synchrotron emitting ones. Furthermore, the preferred values of both the degree of polarization and the polarization angle, as well as their approximate frequency independence, are consistent with the assumption that the Faraday rotating electrons and the synchrotron emitting electrons belong to the same energy distribution; the former being the non/semi-relativistic end of the latter. The source properties of weak lined CRS (BL Lac objects) are harder to infer, mainly because much of the polarized flux remains unresolved even at the VLBI-scale. The weak polarization structure in BL Lac objects might be due to shocks and no Faraday rotation but it can also be accounted for with a jet structure similar to the one inferred for the strong lined CRS. Although an efficient Fermi-type acceleration mechanism can not be excluded, the polarization characteristics of Faraday thick shocks in a jet make them unlikely as acceleration sites.

The polarization properties of the VLBI-core are not those expected from the unresolved inner parts of the jet structure deduced from the resolved emission. Instead it seems likely that the VLBI-core consists of two components; in addition to the inner part of the jet, there is a component which is directly related to the optical blazar emission region. Time variations and the properties of the circular polarization imply that the latter, blazar component, is characterized both by a large scale magnetic field and large Doppler factors. Relativistic streaming in the blazar component is important; for example, the gross features of the initial phases of radio outbursts can be explained in a scenario where the streaming electrons are pitch angle scattered.

SCATTERED LIGHT IN A DISTANT RADIO GALAXY

ANDREA CIMATTI
ESO (Germany) and Dipartimento di Astronomia, Firenze (Italy)

and

SPERELLO DI SEREGO ALIGHIERI
Osservatorio Astrofisico di Arcetri, Firenze (Italy)

Abstract. We present optical polarimetric observations of a radio galaxy at z=2.63 and the results of its spectral modelling made by using stellar and scattered anisotropic nuclear radiation.

1. Polarimetric observations

High redshift radio galaxies (HZRG) [2] provide a unique possibility to study galaxies at early cosmological epochs. An accurate separation of the stellar and non-stellar components is necessary to infer the evolutionary status of these galaxies. In order to derive the ratio between the stellar and the non-stellar radiation we have observed a very distant radio galaxy by CCD imaging polarimetry.
Polarimetric observations of the radio galaxy MRC2025-218 were made at 9 different position angles in R band ($\lambda_{rest} \sim 1900$ Å) at the ESO/MPG 2.2m and ESO 3.6m telescopes. We detect high linear polarization (P=8.3±2.3%, $\theta=93°\pm8°$) with the **E**-vector perpendicular to the radio-optical axis. The polarization properties are consistent with the ones of radio galaxies at 0.5<z<1.2 [1]. The presence of high polarization indicates that the UV radiation is not purely stellar and that a realistic modelling of the spectral energy distribution (SED) must include both stellar and non-stellar radiation.

2. Spectral and polarization modelling

The high perpendicular polarization suggests that the UV radiation is due to scattering of anisotropic nuclear radiation rather than to a jet-induced starburst phenomenon [1]. We have successfully fitted both the degree of polarization and the SED between 1000-6000 Å by using two components : an evolved host galaxy (age $\sim$2 Gyr) and a dust (Galactic grains) scattered quasar spectrum (scattering angle 45°). According to this model, all the UV radiation is due to scattering of the light coming from the obscured quasar nucleus. Our result is consistent with the radio loud quasars – radio galaxies unifying scheme.

References

1. Cimatti A., di Serego Alighieri S., Fosbury R., Salvati M., Taylor D., 1993, *MNRAS*, 264, 421.
2. McCarthy P.J. 1993 *ARAA*, in press

OPTICAL AND VLBI POLARIZATION MEASUREMENTS OF AGN

D. C. GABUZDA
Department of Physics and Astronomy, University of Calgary

and

M. L. SITKO
Department of Astronomy, University of Cincinnati

Abstract. One of the most promising approaches to unravelling the relationship between the optical and radio emission in AGN is to obtain nearly simultaneous optical and VLBI polarization data. We have obtained such data for five AGN (0735+178, OJ 287, 1219+285, 3C 279, and BL Lac). These data suggest that there is a direct link between the optical and radio polarized emission, and that frequently the optical polarization of AGN is associated with the emergence of new VLBI components.

3C 279 is an OVV quasar, while the other four sources listed above are BL Lacertae objects. The continua of these sources and other "blazars" are dominated by nonthermal emission which is variable and highly polarized at optical–radio wavelengths. It is believed that this non-thermal emission is associated with the relativistic jets which are known to exist in these sources, but details of the jet structure and physics are still very uncertain. We use χ_{opt}, χ_c, and χ_j to refer to the optical, VLBI core, and VLBI jet polarization position angles.

In every one of the five AGN for which we have data, there is evidence for a link between χ_{opt} and either χ_c or χ_j. In each of the BL Lacertae objects, the correspondance seems to be between χ_{opt} and χ_c, while χ_{opt} in 3C 279 (which was experiencing a large polarized outburst during our observations) is aligned with χ_j in a newly emerging jet component. These results suggest a much closer connection between the optical and radio emission than has usually been expected; the most likely origin for this connection is cospatiality of the optical and radio emission regions. Our data for OJ 287, for which we have two epochs a year apart, are particularly intriguing. Between the two epochs, χ_j was roughly constant while χ_c rotated by some 60°; this rotation in χ_c was found to be a precursor to the birth of a new VLBI component. Although there is no obvious relation between χ_{opt} and χ_c at either epoch, the difference between these two angles is the same at the two epochs; i.e., χ_{opt} appears to have experienced the same rotation as χ_c. Thus, our results for both OJ 287 and 3C 279 both point towards an association between the optical polarization and the birth and emergence of new VLBI components. We suggest that a significant amount of the optical polarization in AGN is generated in such components, many of which are probably energetic, compact relativistic shocks.

A more thorough presentation and discussion of these data will be given in a paper by Sitko & Gabuzda, to be submitted to the *Astronomical Journal* in October 1993.

POLARIZED OPTICAL AND INFRARED EMISSION FROM HIGH REDSHIFT RADIO GALAXIES

BUELL T. JANNUZI
Institute for Advanced Study, Princeton, NJ, 08540, USA

We have detected highly polarized ($>$ 5%) optical and/or infrared emission (rest frame UV to near infrared) from 5 of the 8 high redshift radio galaxies (HZRG; $z > 0.7$) we have observed. There are now a total of 9 (out of 12 observed) HZRG known to be polarized in spatially integrated measurements (cf. [1]). We have made images of the extended polarized emission from two radio galaxies (3C 265 and 3C 256). Detection of extended polarized emission from a HZRG has previously been reported for 3C 368 [6]. All of the existing polarization observations support the hypothesis [2] that the "alignment effect" (the tendency of the extended UV light to be aligned with the extended radio emission, e.g. [4]) is not solely produced by a burst of star formation, but contains a very significant component produced by the scattering of the light from a hidden active galactic nucleus (AGN). Our modeling of the frequency dependence of the polarized flux from 3C 265 suggests that the most probable scatterer is dust.

Another example of a high redshift object that appears to contain a hidden AGN is the z=2.286 "proto" galaxy/AGN IRAS F10214+4724, one of the bolometrically most luminous objects known [5]. Our spectro-polarimetry confirms the detection by Lawrence *et al.* [3] of highly polarized emission (17%) and reveals that the rest frame UV narrow emission lines are polarized to the same high degree as the continuum. This suggests that the scatterers are cool, i.e. dust. The observed SED of the polarized emission has a slope of -0.9 in $\log(F_\nu)$ vs. $\log(\nu)$. From this we infer an incident radiation spectrum with a slope between -0.9 to -3, considerably different from what would be expected from a young star burst, but similar to some AGNs.

Collaborators in this research are R. Elston, G. D. Schmidt, and P. S. Smith. Complete details of the reported observations will be discussed in papers in preparation (Elston & Jannuzi 1994 and Jannuzi *et al.* 1994). This research was partially supported by a Hubble Fellowship from NASA through grant number HF–1045.01–93A from the Space Telescope Science Institute.

References

1. Cimatti, A., *et al.* 1993, M.N.R.A.S., 264, 421
2. di Serego Alighieri, S., *et al.* 1989, Nature, 341, 307
3. Lawrence, A., *et al.* 1993, M.N.R.A.S, 260, 28
4. McCarthy, P., *et al.* 1987, Ap.J., 321, L29
5. Rowan-Robinson, M., *et al.* 1991, Nature, 351, 719
6. Scarrott, S.M., Rolph, C., & Tadhunter, C.N. *et al.* 1990, M.N.R.A.S., 243, 5p

Multifrequency Polarimetry of CSS Sources

Everton Lüdke, Chidi E. Akujor and Simon T. Garrington

NRAL-Jodrell Bank, Cheshire, SK11 9DL, UK

Compact Steep Spectrum Sources (CSS) have projected linear sizes $\leq$ 20 kpc and overall steep spectrum ($S \sim \nu^{-\alpha}$, $\alpha \geq 0.6$). Interactions with a dense interstellar medium may confine the CSSs and produce the high amounts of Faraday rotation that some distorted sources present [1]. The first results of a detailed study of the multifrequency polarization properties of a sample of CSS sources with MERLIN and the VLA are presented.

Some of the CSSs show symmetric brightness distribution with asymmetry in the polarized brightness. Radiogalaxies tend to be less distorted than quasars, with weaker Faraday effects. Quasars which show no integrated polarized flux in low-resolution observations also show no polarized flux in the components, although they often exhibit features like jets, lobes and hotspots, which are associated with the polarized components of extended quasars (e.g. 3C49). The majority of the sources have magnetic field lines colinear with the jet axis (e.g. 3C138), which is similar to the magnetic field configuration of jets in extended radio sources. Comparison with published data at 3.6 and 2 cm [2] also suggests that the CSSs have larger Faraday dispersions than radio sources with larger linear sizes. This may be the effect of the external hot media denser than in extended sources, thermal gas mixed with source plasma or both, with $\sim$ 300 pc maximum linear scale for thermal irregularities which depolarize the radio radiation.

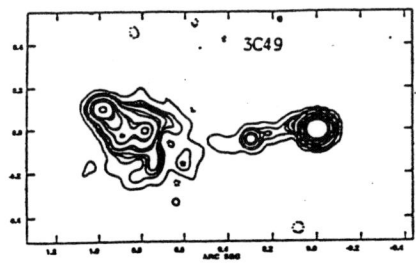 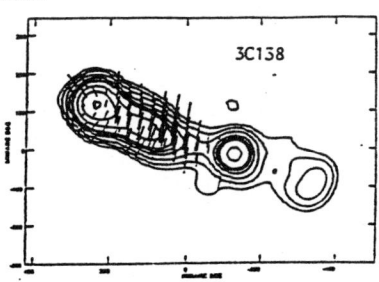

[1] Kato T., Tabara H., Ionue M. and Aizu K., 1987, Nature, 329, 223

[2] van Breugel W.J.M. et al, 1992, A&A, 256, 56

FREQUENCY-DEPENDENT POLARIZATION IN COMPTONIZATION MODELS FOR AGN

JURI POUTANEN

Observatory and Astrophysics Laboratory, P.O.Box 14, SF-00014 University of Helsinki, Finland

Abstract. The angular distribution and the polarization of radiation as a function of the angle and frequency for the two-phase model of accretion disks in AGN are found. The results depend strongly on the temperature of the hot corona.

Key words: accretion disks, polarization, scattering

Two-phase accretion disk models have been used to explain the X-ray emission from galactic black hole candidates (Bisnovatyi-Kogan and Blinnikov, 1977) and AGNs (Czerny and Elvis, 1987; Wandel and Liang, 1991; Haardt and Maraschi, 1993). In such models soft radiation emitted by cool optically thick disk scatters in hot optically thin corona and gives rise to the power-law X-ray spectrum.

Taking into account only these two components I calculate the polarization degree in UV and X-ray region as a function of frequency and inclination angle of the disk. Such calculation in the single scattering approximation were performed by Poutanen and Vilhu (1993). Radiative transfer equation for Compton scattering of polarized light is formulated by Nagirner and Poutanen (1993a,b). For the case of the plane-parallel atmosphere this equation is solved by the iteration method (Sunyaev and Titarchuk, 1985).

The results show that polarization degree decreases when the electron temperature of the corona increases. The exact Compton scattering matrix predicts much smaller degree of linear polarization than the Rayleigh matrix does (Haardt and Matt, 1993). In the energy band where first scattering is important polarization depends on the angular and polarization properties of initial radiation from cool disk. The polarization angle in hard X-ray region is orthogonal to that in UV region. Future polarimetric observations in UV and X-ray regions need the theory of multiple Compton scattering of polarized light taking into account the exact scattering matrix.

References

Bisnovatyi-Kogan, G.S. and Blinnikov, S.I.: 1977, *Astron.Astrophys.* **59**, 111
Czerny, B. and Elvis, M.: 1987, *Astrophys.J.* **321**, 305
Haardt, F. and Maraschi, L.: 1993, *Astrophys.J.* **413**, 507
Haardt, F. and Matt, G.: 1993, *Monthly Not. Roy. Astr. Soc.* **261**, 346
Nagirner, D.I. and Poutanen, J.: 1993a, *Astron.Astrophys.* **275**, 325
Nagirner, D.I. and Poutanen, J.: 1993b, in: R.A. Sunyaev (ed.) *Sov. Sci. Rev. E Astrophys. and Space Phys.* **9**, 1
Poutanen, J. and Vilhu, O.: 1993, *Astron.Astrophys.* **275**, 337
Sunyaev, R.A. and Titarchuk, L.G.: 1985, *Astron.Astrophys.* **143**, 374
Wandel, A. and Liang, E.P.: 1991, *Astrophys.J.* **380**, 84

A TENTATIVE SKETCH FOR BL LACERTAE OBJECTS

HÉLÈNE SOL and LOURDES VICENTE
Observatoire de Paris-Meudon, France

VLBI polarization data show magnetic field configuration parallel to the nuclear jet in quasars and perpendicular to it in BL Lac objects [1]. It appears difficult to account for this contrast within the unified scheme for AGN [2]. To investigate a direct explanation of this peculiarity of BL Lac objects, we study the possibilities of propagation and radiation of beams of particles in transverse ambient magnetic field. High energy streams with kinetic energy density larger than the ambient magnetic one, $E_{kin} > E_{mag}$, can easily propagate with enhancement of the transverse magnetic field at the leading edge of the stream and reconnection of magnetic lines in its wake. Synchrotron radiation in front shocks naturally leads to the observed polarization. Moreover self-polarization, with formation of charge layers and $E \times B$ drift velocity, allows substantial propagation for even lower energy streams with $E_{kin} < E_{mag}$, as long as their density n_o is large enough, typically $\kappa = 4\pi n_o m_i c^2/B^2 > 1$. Such low energy streams are non diamagnetic and do not modify the ambient field. Any high energy tail of the total particle distribution in the jet therefore radiates in the transverse field pattern. This concerns for instance the BL Lac object W Comae if we assume a proton-electron jet with bulk velocity $v_o = 0.1c$ (as the source does not require relativistic beaming so far), an equipartition magnetic field B = 0.02 G and a density of radiating particles $n_r = 0.05 cm^{-3}$ at about 7 pc from the nucleus (knot K3). For $n_r/n_o = 10^{-3}$, one gets the stream density $n_o = 50 cm^{-3}$ which allows good propagation as κ reaches 2×10^3, and still corresponds to a moderate mass outflow of 0.06 M$\odot$ /year for a VLBI jet cross-sectional area of $2pc^2$.

If the ambient magnetic field is roughly aligned with the large scale jet direction, as observed in stellar jets and expected in the case of a large magnetic structure which favors formation of the accretion disc in the plane perpendicular to the field, we reach the idea that BL Lac objects are active galactic nuclei with highly twisted accretion discs. VLBI beams are emitted from the inner edges of the disc along the central black hole rotation axis while the extended radio source comes from the outer misaligned part of the disc driven by the recently accreted matter. This scenario has several consequences such as the possibility of different orientation of the ionizing cone which can strongly modify the observed spectral lines properties and the occurence of large misalignment angles between parsec and kiloparsec radio structures as often observed in BL Lac objects.

References

1. Gabuzda, D.C., Cawthorne, T.V., Roberts, D.H., Wardle, J.F.C., 1992, ApJ, 388, p.40
2. Gopal-Krishna, Wiita, P.J., 1993, Nature, 363, p.142

THE INTRANIGHT VARIABILITY OF OPTICAL POLARIZATION IN PKS 0109+224

L. VALTAOJA
Nordita, Blegdamsvej 17, DK-2100 Copenhagen, Denmark

Abstract. It is found that the statistics of the photopolarimetric intraday data of PKS 0109+224 do not favor the shocked jet scenario as an explanation for intranight variability in this source.

Here we consider the optical variability of PKS 0109+224, which has both flux and polarization varying in different timescales including intranight variations. We have obtained truly simultaneous UBVRI-photopolarimetric observations with Nordic Optical Telescope (NOT). The data includes four long integrations (between 17th and 20th September 1991), which allow us to study the intranight phenomena.

In the optical bands the microvariability seems to be intrinsic. There are basically three kinds of mechanisms to produce it. The models based on phenomena in the accretion disks are difficult to compare with the observations. The so-called Christmas tree models are testable by looking at the correlations between the changes in the Stokes parameters at different colors. (We are working on that.) The shocked jet models are also able to produce short timescale variations. The shock scenario gives us some statistical predictions, which we compare here with the observations.

The polarization (P), the position angle (PA) and the flux vary considerably during a single night, and even the character of the frequency dependence of the polarization (FDP) changes during the night. Statistical analysis of our data gives following results. 1) The observed quantities in different colors (UBVRI) are highly correlated. 2) The degree of polarization and the total flux are only weakly correlated. 3) The changes in the total flux do not correlate with the changes in polarized flux. 4) The strength of FDP does not correlate with P, PA or flux. 5) Largest changes in PA occur at minimum P.

According to the shock scenarios one would expect the total flux and the degree of polarization to depend on each other. In the long timescale data this seems to be the case, but not in the intranight data (2). In addition the appearance of the more polarized shock should cause the changes of total flux to correlate with changes of polarized flux, which was not the case (3), again contrary to the behaviour seen in the long timescale observations. Neither does the non-existing correlation between the stregth of the FDP and the polarization support the shocked jet model (4). Thus, unlike for the long timescale data, the statistical information does not seem to favor the growth and decay of shocks in the jet as a reason for the microvariability. For more details see L. Valtaoja et al., AJ, in press (1993).

Poster Contributions:
Disks Structure and Emission

ON THE ACCRETION DISK MODELS WITH STATIONARY AND NON-STATIONARY SHOCK WAVES

SANDIP K. CHAKRABARTI
TIFR, Bombay (Permanent address) & ICTP, Trieste

An important point which emerged from this meeting is that disks in AGNs are not simply thin, Keplerian type; they show more complex behaviour. Chakrabarti (1990a and references therein) has shown that in an inviscid accretion disk with significant angular momentum, the centrifugal barrier is strong enough to produce axisymmetric standing shock wave. Subsequently, this work was extended to include the non-axisymmetric and viscous disks (Chakrabarti, 1990b). Particularly important are the solutions with viscosity, as they show that as the viscosity is increased, the stable becomes weaker and weaker till it disappears completely. This solution has a unifying character that inviscid *pressure driven* disks have almost constant angular momentum and can have shock discontinuities, but *viscous driven* disks dissipate angular momentum quick enough not to have centrifugal barrier and therefore no shock waves. Chakrabarti & Molteni (1993), using Smoothed Particle Hydrodynamics have shown that shocks are produced in inviscid disks, exactly where they are predicted.

Unlike a Keplerian disk, a disk with a shock has basically two temperature zones. The post shock solution is responsible for the Big Blue Bump and UV excess (Chakrabarti and Wiita, 1992). At the shock location, the disk is 'bulged'; the hard radiation from this region is intercepted by the cooler pre-shock flow. The shock strength and location are sensitive to input specific energy of the flow. This configuration might be responsible for the 'zero-lag' correlated variability of, say, NGC 5548 (Chakrabarti, Haardt, Maraschi & Molendi, AA, submitted) discussed in this meeting. Spiral shocks which may be produced in disks in a binary system can also appear in disks around AGNs; the perturbation may be due to passage of massive objects (Chakrabarti & Wiita, 1993a). They also cause time variations in the double horned pattern from disk line emission (Chakrabarti & Wiita 1993b) as observed in, say ARP 102B. All these observations point that shocks are probably important ingredients in any accretion disk in AGNs

References

Chakrabarti, S. K., 1990a, ApJ, 362, 406.
Chakrabarti, S. K. 1990b, MNRAS, 243, 610.
Chakrabarti, S. K. & Molteni, D. 1993, ApJ, (Nov. 10th).
Chakrabarti, S. K. & Wiita, P.J. 1992, ApJ, 387, L21.
Chakrabarti, S. K. & Wiita, P. J. 1993a, ApJ, 411, 602.
Chakrabarti, S. K. & Wiita, P. J. 1993b, AA, 271, 216.

The Vertical Structures of Accretion Disks in AGN

H. Riffert[1], T. Dörrer[2], R. Staubert[2], H. Ruder[1]
[1] Theoretische Astrophysik, 72076 Tübingen, Germany
[2] Astronomisches Institut, 72076 Tübingen, Germany

Radiation emitted from an accretion disk around a central black hole is the widely accepted model for the observed optical to UV emission from AGN. We have calculated the properties of a standard α -accretion disk (Shakura and Sunyaev, 1973). We present a fully self-consistent model of the structure and the spectrum of such a disk, i.e. the internal vertical density and temperature profiles are calculated simultaneously with the local spectra. Constant density models have been presented by (Ross et al., 1992). The central object is assumed to be a Kerr black hole (BH); relativistic corrections are included. The local energy production is assumed to be entirely due to turbulence. The radiative transfer equation is solved using the Eddington approximation. Inelastic Compton scattering is treated approximately by the Kompaneets equation, and the absorption cross section contains free-free and bound-free processes for hydrogen. The energy transport includes radiation and convection, and the convective flux is calculated in the mixing length theory, taking into account the heating and cooling of the rising elements. Although the convective flux is energetically negligible it has a strong influence on the vertical density structure. We performed several calculations for different parameters such as $\dot{M}$ and α. In regions, where the surface radiation flux is large, we get a strong density inversion because the radiation force per unit mass overcomes the gravitational force. Such a density profile, however, is unstable against convection. Including the convective flux then leads to a monotonic density profile. Figures 1 and 2 show the structure and integrated spectrum for two different disk models.

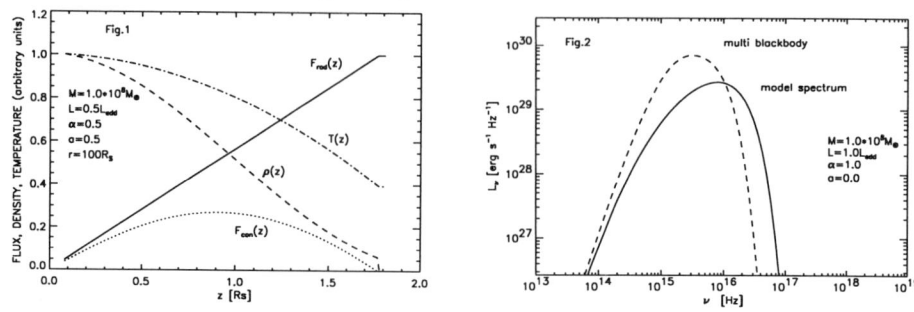

Fig. 1. Vertical disk structure for a fixed radius r, BH mass M, luminosity L, viscosity α, and angular momentum a. ρ, T are normalized to their central values $\rho_0 = 6.3 * 10^{-12}$ g/cm^3, $T_0 = 1.7 * 10^5$ K; the fluxes are in units of $F_0 = 2.1 * 10^{15}$ erg/cm^2/s. The energy production is proportional to the total pressure.

Fig. 2. Integrated spectrum for a disk with an energy production proportional to the gas pressure only.

References

Ross, R.R., Fabian, A.C. and Mineshige, S.: 1992, Mon.Not.R.astr.Soc.**258**,189.
Shakura, N.I. and Sunyaev, R.A.: 1973, Astr.Ap.**24**, 337.

THE NUCLEAR DISK OF NGC4261: HST IMAGES AND WHT SPECTRA

LAURA FERRARESE and HOLLAND C. FORD
Johns Hopkins University and Space Telescope Science Institute
WALTER JAFFE and FRANK VAN DEN BOSCH
Leiden Observatory
and
ROBERT W. O'CONNELL
University of Virginia

Abstract. The properties of the nuclear region of the elliptical galaxy NGC4261 (=3C270), the second brightest radio source in the Virgo cluster, have been studied using high resolution HST images and La Palma WHT spectra.

A 2400s HST Planetary Camera exposure was taken through the F555W filter ($\sim$ Johnson V) and shows a central smooth disk of dust with sharp edges and a point-like nucleus [1]. The disk is elliptical, with axes of $1.''71 \times 0.''74$, corresponding to 121×51 pc at the distance of the Virgo cluster, 14.7 Mpc (Jacoby, Ciardullo, & Ford 1990). The major axis of the disk is roughly parallel to the major axis of the galaxy and perpendicular to the axis of the radio jets (Birkinshaw & Davies 1985). The face-on optical depth of the disk increases from 0.3 at the edge to 0.8 at the centre. The nucleus, located in the centre of the disk, is unresolved ($< 0.''1 = 7$ pc) and has an apparent magnitude m_{5500} of 23.6.

WHT long slit spectra in the wavelength ranges 4730 to 5110 Å and 6380 to 6840 Å, show strong [NII] $\lambda\lambda$ 6548,6584, Hα and [SII] $\lambda\lambda$ 6717,6730 emission lines. The emission line ratios classify the nucleus of NGC 4261 as a LINER. The most striking feature in the spectrum is the presence of broad wings (FWZI $\sim$ 4500 km s^{-1}) on the Hα + [NII] blend and on the [SII] doublet. At least one broad component (FWHM $\sim$ 2500 km s^{-1}) and one narrow component (FWHM $\sim$ 450 km s^{-1}) are required to fit each of the observed line profiles. The symmetry of the broad lines and the identity of their mean velocity with the systemic velocity of the galaxy, suggest that the broad emission regions are either rapidly rotating or confined to a region small compared with the dust distribution.

The mean velocity of the narrow emission lines increases steadily from systemic velocity in the centre, to about 100 km s^{-1} 1 arcsec away. The rotation curve cannot be reproduced by a constant M/L model. Since NGC 4261 is a radio source, we tried to model the rotation curve by adding the contribution of a central point mass, corresponding to a hypothetical black hole, to the stellar potential (for a constant M/L) and to the (insignificant) potential of the disk of dust. The best fit is obtained for a central point mass 10^7 to 10^8 M$_\odot$.

References

1. Jaffe, W. *et al.* 1993, Nature, **364**, 312 and references therein.

PHYSICS OF MAGNETIZED ACCRETION DISKS IN AGN

GEORGE FIELD
Center for Astrophysics and Osservatorio Astrofisico di Arcetri

and

ROBERT ROGERS
Institute of Geophysics and Planetary Physics, Lawrence Livermore National Laboratory

1. Introduction

Field and Rogers (1993) proposed that the accretion disks in moderate - L AGN ($\lesssim 10^{44}$ erg sec^{-1}) are dominated by magnetic stress rather than gas or radiation pressure. A magnetic field parallel to the disk forms loops above and below it where reconnection accelerates electrons to relativistic energies. The nonthermal radiation observed is the synchrotron emission and inverse-Compton scattering by these electrons.

Most authors have assumed that the magnetic fields are weak, so that $B^2/8\pi < p$. If they are accreted along with the gas in the dust torus, and if B there is $\sim 10^{-3}$ G, we find that $B^2/8\pi > p$.

2. Magnetic Field

The accretion disk contains matter whose magnetic field $\mathbf{B}_0$ was originally uniform. The flux threading the original region penetrates the accretion disk. In doing so, it is compressed along with the accreting gas, and is wound up by the differential rotation of the disk, forming spiral sectors of alternating polarity separated by neutral sheets. Reconnection of magnetic loops of opposite polarity at neutral sheets limits the strength of the field as it winds up.

At a radius R from the black hole, the radial extension of an adjoining pair of sectors is ΔR. If the local scale height is H, the strength of the field along the sector is

$$\frac{B}{B_0} = (2\pi)^{2/3} \left(\frac{\rho}{\rho_0}\right)^{2/3} \left(\frac{R}{H}\right)^{1/3} \frac{R}{\Delta R} \quad (1)$$

where we have used the conservation of mass and flux starting from $B = B_0$ and $\rho = \rho_0$.

By balancing the increase in B due to windup with the decrease due to reconnection at speed v_M, we find that $\Delta R = (16\pi v_M/3v_\phi)^{1/2} R$. We have shown that $v_M < v'_M$ where v'_M is the merger speed in the corona, and since according to Petschek (1964) $v'_M \leq v'_A/70$, where v'_A is the Alfvén speed in the corona, equation

(1) implies that

$$\frac{B}{B_0} > 7 \left(\frac{\rho}{\rho_0}\right)^{2/3} \left(\frac{R}{H}\right)^{1/3} \left(\frac{v_\phi}{v'_A}\right)^{1/2} \qquad (2)$$

Plausible values for ρ, ρ_0, H, and v'_A yield $B \geq 10^4 B_0$ at the distance from a 10^8 M$_\odot$ black hole inside of which half the luminosity originates. If $B_0 = 10^{-3}$ G, $B > 10$ G, with $B \sim 10^3$ G more likely.

3. Dynamics

The equation of motion of a steady-state disk is $\nabla \cdot \mathbf{T} = \rho \mathbf{g}$, where the components of the stress tensor for moving matter plus magnetic field are $T_{ij} = \rho v_i v_j + B^2 \delta_{ij}/8\pi - B_i B_j/4\pi$. The R component of the equation of motion implies that $v_\phi = (GM/R)^{1/2}$. The z component gives $\rho = \rho_* \exp(-z^2/H^2)$ where the scale height is $H = v_A R/v_\phi$ and where the Alfvén speed $v_A = B/\sqrt{4\pi\rho}$ is assumed independent of z in the disk. If $\xi \equiv -B_R/B_\phi$ is independent of z, then the ϕ component gives

$$-v_R = \xi \left(\frac{v_A}{v_\phi}\right)^2 \left[1 - \left(\frac{R_i}{R}\right)^{1/2}\right]^{-1} \qquad (3)$$

Finally the conservation of mass implies that the accretion rate $\dot{M} = -4\pi\rho H R v_R$ is independent of R. If v_M, the merger rate, were known, the above equations enable one to determine ρ, v_R, v_ϕ, B_R, and B_ϕ as functions of R. If v_M is a power law, all the other variables are also power laws.

4. Conclusion

We have explored the possibility that AGN accretion disks are magnetically dominated as a result of fields accreted along with interstellar gas. Detailed predictions await the calculation of the merger rate for coronal magnetic loops.

References

Field, G.B. & Rogers, R.D. 1993, ApJ, 403, 94
Petschek, H.E. 1964, in *AAS-NASA Symposium on the Physics of Solar Flares*, (NASA Special Publication SP-50), p. 425

AGN ACCRETION DISKS WITH FINITE TURBULENT PRANDTL NUMBER

O. M. HEINRICH
Institut für Theoretische Astrophysik, INF 561, Heidelberg D-69120, Germany

The present paper reports results of a study of turbulent energy transport in AGN accretion disks. We follow the spirit of the papers by Shakura et al. and Rüdiger et. al in which the concept of an eddy heat conductivity was introduced and developed. For the viscosity we use a standard α- description with the main component of the stress tensor $t_{r\phi}$ proportional to the gas pressure. The dimensionless Prandtl number Pr given as the ratio of turbulent kinematic viscosity and turbulent heat conductivity enters the model as a free parameter. The disk structure depends sensitively on material properties such as opacities and specific heats. In our calculations we have used an equation of state and mean opacities taking into account a list of the most important ionization processes as well as the radiation contributions to thermodynamic quantities. Here are our main conclusions:
- There is a broad range of parameters where turbulent energy transport is of utmost importance.In particular turbulent energy transport is much more efficient than the radiative one in hydrogen ionization zones and in regions with high radiation pressure dominance. The latter fact might seem suprisingly but is caused by the strong radiative contribution to the specific heat c_P. In general we found a flattening of the temperature profile and a steepening of the density profile with almost complete removal of density inversions due to turbulent transport.
- We have calculated surface density- surface temperature relations for various values of α and Pr. In H- ionization zones the value of Pr is of great influence on the shape of this relation. While for $Pr = \infty$ it has the usual S- shape, more complicated structures with the possible occurrence of double S- shapes are found for Pr=O(1).
- Turbulent energy transport changes the shape of the disk. In general an efficient turbulent transport decreases the disk height h. This can lead to a concave disk shape in which irradiation of outer disk parts by inner ones is possible.The height decreasing effect of turbulent energy transport is of great importance for the innermost region of disk models near the Eddington limit. We have calculated a model with $M_{bh} = 10^6 M_\odot$ and $L/L_{edd}=0.9$ in which a decrease of h by a factor of 3 occurs when Pr is changed from ∞ to 1/3. Thus turbulent energy transport does not resolve but considerably weakens the problem of disk thickening.

References

Shakura,N.I.,Sunyaev,R.A. and Zilitinkevich,S.S.: 1978,*Astron. Astrophys.* **Vol. 62**, pp.179-187
Rüdiger,G.,Elstner,D. and Tschäpe,R.: 1988,*Acta Astron.* **Vol. 38**, pp.299-314

STRUCTURE OF OUTER REGIONS OF ACCRETION DISKS IN AGN
Non irradiated, vertically averaged accretion disks

Jean Marc HURÉ, Suzy COLLIN & Guillaume PINEAU DES FORÊTS

DAEC et URA 173 du CNRS
Observatoire de Paris-Meudon - Place Jules Janssen - 92195 Meudon Principal

DESCRIPTION OF THE MODEL

Radial structure of outer regions of α-disks (Shakura & Sunyaev 1973) is investigated in a more sophisticated way than in Collin-Souffrin & Dumont (1990). The vertically averaged equations for the disk structure hold but some of them are reconsidered: the equation of state (atoms, ions and molecules) with a the rigourous treatment of opacities is introduced. The radiative flux is treated as in Hubeny (1990), and finally a rigourous treatment of the self-gravitaty is included.

We have studied the influence of $(M, \dot{M}, \alpha)$ on the global structure, the influence of self-gravity, and the effect of neglecting some opacity sources. Here are the results for the temperature T and the half-thickness H for a "canonical" model (appliable to NGC 5548 for instance): $M = 5 \cdot 10^7 M_\odot$, $L = 1/30 L_{Edd}$ ($\approx 0.04 M_\odot$/yr) and $\alpha = 1$.

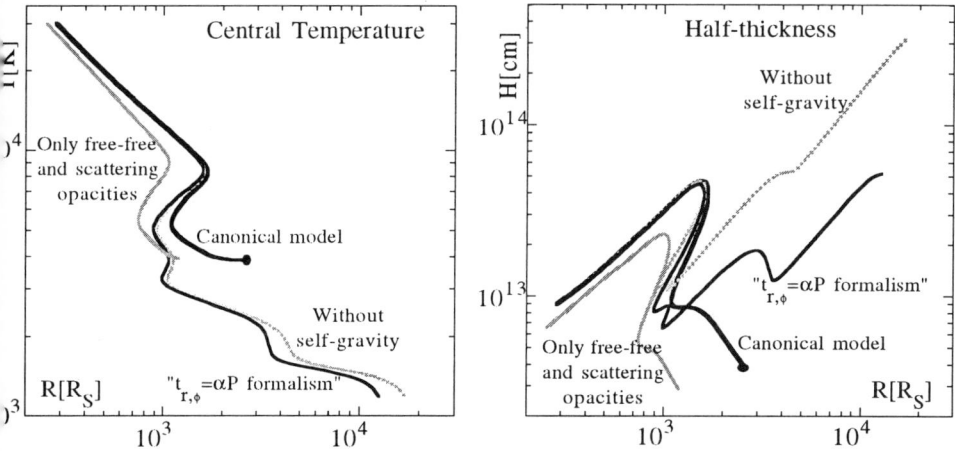

MAIN CONCLUSIONS

Hydrostatic equilibrium cannot be maintained if the gas is too cool to support its own gravity ($\Rightarrow T/\mu|_{min.} \approx m_H/k (2GM/3\alpha)^{2/3}$). The "$t_{r,\phi} = \alpha P$" formalism is not consistent with the equations that include self-gravity. Finally, the opacity is dominated by atomic bound-free and molecular absorption.

Preliminary results on irradiated disks show that the self-graviting dominated region is pushed further away, at $R \approx 10^4 R_S$.

REFERENCES

Collin-Souffrin S. & Dumont A.M., 1990, *Astro. & Astrophys.*, **229**, 292
Hubeny I., 1990, *Ap. J.*, **351**, 632
Shakura N.I. & Sunyaev R.A., 1973, *Astro. & Astrophys.*, **24**, 337

STRUCTURE AND EMISSION LINE SPECTRUM OF AN X-RAY HEATED ACCRETION DISK IN AGN

YUAN-KUEN KO* and TIMOTHY R. KALLMAN
Laboratory for High Energy Astrophysics, NASA/Goddard Space Flight Center, USA.

Abstract. We investigate the structure of an X-ray heated accretion disk in active galactic nuclei. It is found that X-ray heating can prevent the disk to be disrupted by its self-gravity under sufficient X-ray heating. The disk size can be two orders of magnitute larger than that limited by self-gravity of the disk without X-ray heating. An accretion disk corona will be formed by X-ray heating and can be a site for line emission. We present such emission line spectra which range from optical to hard X-ray energies and compare with the observational data.

1. Results

1.1. THE SELF-GRAVITY RADIUS OF AN X-RAY HEATING DOMINATED DISK

See Fig. 1. This is based on a disk where radiation pressure dominates gas pressure and electron scattering opacity dominates free-free opacity. (with $M = 10^8 M_\odot$ and $f\eta$ is the ratio of illuminating X-ray luminosity and $\dot{M}c^2$.)

1.2. EMISSION SPECTRUM

We made non-LTE and self-consistent calculations of the temperature, density, ionization, ion level populations and emitted spectrum in the accretion disk corona above the optically thick disk photosphere. Fig.2 shows the UV line ratios (relatiave to CIVλ1550) compared with the observational data (Wu, Boggess & Gull 1983, Laor et al. 1993).

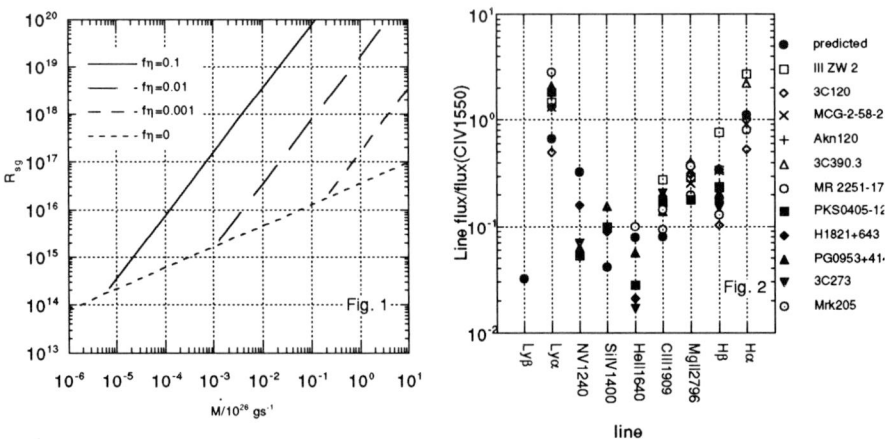

* Also Center for EUV Astrophysics, University of California, Berkeley, USA.

T. J.-L. Courvoisier and A. Blecha: *Multi-Wavelength Continuum Emission of AGN*, 484.
© 1994 IAU. Printed in the Netherlands.

HOT ACCRETION DISKS WITH MAGNETIC FIELD AND THERMAL CYCLO-SYNCHROTRON RADIATION

MASAAKI KUSUNOSE and ANDRZEJ A. ZDZIARSKI
Nicolaus Copernicus Astronomical Center, Bartycka 18, 00-716 Warsaw, Poland

We study the structure of hot, two-temperature accretion disks around black holes, including the effects of thermal cyclo-synchrotron radiation and magnetic viscosity. This work is an extension of previous work by Björnsson & Svensson (1991a, b, 1992) and Kusunose & Mineshige (1992), which did not include those effects. Magnetic field, B, is assumed to be randomly oriented and determined by prescribing the ratio $\alpha = P_{\text{mag}}/P_{\text{gas}}$ or $\alpha = P_{\text{mag}}/(P_{\text{gas}} + P_{\text{rad}})$, where P_{mag}, P_{gas}, and P_{rad} are the pressures of magnetic field, gas, and radiation, respectively. We find those effects do not change the qualitative properties of the disks, i.e., there are two critical accretion rates related to production of $e^{\pm}$ pairs, $\dot{M}_{\text{cr}}^U$ and $\dot{M}_{\text{cr}}^L$ ($\dot{M}_{\text{cr}}^U > \dot{M}_{\text{cr}}^L$), that affect the number of local and global disk solutions, as recently found for the case with $B = 0$ (Björnsson & Svensson 1991a, b, 1992). However, a critical value of the α-viscosity parameter above which those critical accretion rates disappear becomes smaller than $\alpha_{\text{cr}} = 1$ found in the case of $B = 0$, for $P_{\text{mag}} = \alpha(P_{\text{gas}} + P_{\text{rad}})$. If $P_{\text{mag}} = \alpha P_{\text{gas}}$, on the other hand, α_{cr} is still about unity. Moreover, when Comptonized cyclo-synchrotron radiation dominates Comptonized bremsstrahlung, radiation from the disk obeys a power law with the energy spectral index of ~ 0.5, in a qualitative agreement with X-ray observations of AGNs and Galactic black hole candidates. The spectral index is weakly dependent on the mass accretion rate.

We also find that the hot disk solutions for $P_{\text{mag}} = \alpha(P_{\text{gas}} + P_{\text{rad}})$ are extended to the effectively optically thick region, where they merge with the standard cold disk solutions, as was found for the case with $B = 0$ by Kusunose & Mineshige (1992). Finally, we find a region in the disk parameter space with no solutions due to the inability of Coulomb heating to supply enough energy to electrons.

References

Björnsson G., & Svensson R. 1991a, in Structure and Emission Properties of Accretion Disks (IAU Colloq. 129), ed. C. Bertout, S. Collin, J.-P. Lasota, & J. Tran Thanh Van (Gif-sur-Yvette: Éditions Frontières), 379
Björnsson G., & Svensson R. 1991b, in Relativistic Hadrons in Cosmic Compact Objects, ed. A. A. Zdziarski & M. Sikora (Berlin: Springer), 53
Björnsson G., & Svensson R. 1992, Astrophysical Journal, 394, 500
Kusunose, M., & Mineshige, S. 1992, Astrophysical Journal, 392, 653

GALACTIC DYNAMO MODELS

N.P. MOORE, J.S. PANESAR and A.H. NELSON
Dept. of Physics and Astronomy
University of Wales College of Cardiff
PO Box 913
Cardiff CF2 3YB

Abstract. We present the results of 3-d numerical simulations of galactic dynamos. Using reasonably justifiable parameters, kinematic dynamo models are shown to reproduce the gross features of galaxy magnetic fields. The central field morphology is suggestive of an embryonic jet generator.

In recent years there have been considerable advances in the mapping of magnetic fields in spiral galaxies. The observations show that the large-scale or systematic field lines generally lie parallel to the spiral arms, with magnetic intensity decreasing away from the plane of the disc. Towards the centre, the field grows in strength and becomes dominated by the vertical component. The currently favoured theory which explains the existence and form of the large-scale galactic field involves a dynamo process (Moffatt, 1978). Here we present simulations of 3-d galactic dynamos. The model and results are described in more detail in Panesar & Nelson (1992).

The simulations evolve an initially weak seed field according to the dynamo equation

$$\frac{\partial \mathbf{B}}{\partial t} = \nabla \times (\mathbf{v} \times \mathbf{B}) - \nabla \times (\eta \nabla \times \mathbf{B}) + \nabla \times (\alpha \mathbf{B}) \quad (1)$$

v is the velocity field of the galaxy including differential rotation and a shock induced by a stellar density wave. The second term in (1) is the Ohmic diffusion term. The α coefficient in the last term is proportional to the helicity of the flow. This is the dynamo term where the field is amplified by cyclonic turbulence. As the field strength increases the α-effect diminishes, representing the tendency for the magnetic field to suppress turbulence.

The evolved field patterns are in agreement with the principal observed morphological characteristics as detailed above. In particular the results show a propensity for strong vertical fields at the centre of the disc. The field here is helical in nature with the pitch of the helix decreasing with height, i.e. we have an embryonic jet generator based on the torsional Alfvén wave model for jets.

References

Moffatt, H.K., 1978, *Magnetic Field Generation in Electrically Conducting Fluids*, Cambridge University Press.
Panesar, J.S., and Nelson, A.H., 1992, *Astron. Astrophys.* **264**, 77-85.

VARIABILITY OF BLACK HOLE ACCRETION DISKS

MICHAEL A. NOWAK
CITA, 60 St. George St., Toronto, Ont. M5S 1A7

and

ROBERT V. WAGONER
Stanford, Dept. of Physics and CSSA, Stanford, CA 94305-4060

Abstract. We consider two forms of variability in black hole accretion disks: harmonically oscillating modes trapped near the disk inner edge, and noisy fluctuations throughout the entire disk. We apply the former to AGN disks, and the latter to the X-ray power spectra of GX339-4.

Key words: accretion disk, harmonic oscillations, X-ray power spectra, GX339-4

The mechanism governing harmonic oscillations in a disk is similar to that governing the g-modes of helioseismology, accept that the bouyancy frequency is replaced by the epicyclic frequency (*i.e.* angular momentum conservation). The modes, which are trapped near the epicyclic frequency maximum near the disk inner edge, oscillate predominantly in the vertical direction, have radial extents on the order of $2\ GM/c^2$, and have frequencies on the order of $0.03\ c^3/GM$ (a period of 10^6 seconds for $10^9\ M_\odot$). Detailed discussion of these modes can be found in [1, 2]. We recently have considered the stochastic excitation of these modes by turbulence, and we have made simple estimates of the resulting disk luminosity fluctuations [3]. For AGN parameters, R-band modulation may be on the order of several tenths of a precent, and UV modulation may be several percent. Observed mode periods are tens of days.

To explain the X-ray power spectra of galactic black hole candidates, we have constructed a simple kinematic model of accretion disk variability which is based upon viscous and thermal instabilities [4]. We assume local, exponentially growing fluctuations that grow on either a viscous time scale, $\tau_V^{-1} \sim \alpha \mathcal{L}^2\ \Omega$, or a thermal time scale, $\tau_H^{-1} \sim \alpha\Omega$ ($\alpha \sim 0.1$ is the usual Shakura-Sunyaev α parameter, Ω is the local Keplerian rotation frequency of the disk, $\mathcal{L} \sim 0.1$ is the ratio of the disk luminosity to the Eddington luminosity). The viscous instabilities can be supressed by a mass loss from the disk, whereas the thermal instabilities can be supressed by an energy loss. The X-ray power spectra of GX339-4 in its very high state is modelled as a transition from viscous to thermal instabilities. This constrains the temperature and total energy of a stabilizing "wind", the energetics of which are found to be consistent with Compton cloud models of GX339-4. Furthermore, the computed X-ray power spectra is consistent with the observations.

References

1. Nowak, M. A., and R. V. Wagoner 1991, ApJ, 378, 656.
2. Nowak, M. A., and R. V. Wagoner 1992, ApJ, 393, 697.
3. Nowak, M. A., and R. V. Wagoner 1993, ApJ, 418, to be Published.
4. Nowak, M. A. 1994, ApJ, to be Published.

THE STRUCTURE AND STABILITY OF TWO-TEMPERATURE ACCRETION DISK

MYEONG-GU PARK *
Department of Astronomy and Meteorology
Kyungpook National University
Taegu 702-701, KOREA

Abstract. The structure and stability of the two-temperature accretion disk around compact object with central soft-photon source is studied.

Key words: Accretion Disk, Black Holes, AGN, QSO

1. Summary

The structure and stability of the gas pressure dominated, thin accretion disk, cooled by Comptonization of the central soft photons is studied. Steady-state solutions have two branches: High-temperature (HT) solutions have very different ion and electron temperatures and correspond to the classic solutions of Shapiro, Lightman, and Eardley. Low-temperature (LT) solutions have same ion and electron temperature, which is very close to the Compton temperature of soft photons.

The linear analysis, allowing for the surface density perturbation and dynamics in the vertical direction, shows that LT disk is stable while HT disk is not. LT disk is stable because ions and electrons are locked to the Compton temperatre of the soft photons. HT disk generally has 4 local modes: (1) Heating mode grows in thermal time scale, $(5/3)(\alpha\omega)^{-1}$, where ω is Keplerian frequency. (2) Cooling mode decays in Compton time scale, $(2/5)(T_e/T_i)(\alpha\omega)^{-1} \ll (\alpha\omega)^{-1}$. (3) Lightman-Eardley mode decays in viscous time scale, $(8/11)(\Lambda/H_0)^2(\alpha\omega)^{-1}$, where Λ is the wavelength of the perturbation and H_0 is the disk height. (4) Vertical oscillatory modes oscillate in Keplerian time scale, $(3/8)^{1/2}\omega^{-1}$ with growth rate $\propto (H_0/\Lambda)^2$. Including dynamics in the vertical direction does not change the stability behavior in general, adding only the oscillatory modes which gradually grow as H_0/Λ increases.

Non-linear behaviour of the disk is followed by numerical integration. Cooling function covering both effectively optically thin and thick cases is used. Only the ion temperature perturbation is important and the disk either expands or collapses vertically, depending on the sign of the perturbation. When expanding, the ion temperature becomes very high while the electron temperature very low, resulting in runaway behavior due to the decreased Coulomb coupling, especially so if the ion velocity effect is considered. When collapsing, the disk approaches LT solutions and stabilizes at Compton temperature.

* This work was supported by Korea Science and Engineering Foundation grant 923-0200-009-1.

REPROCESSING OF UV AND LINE EMISSION IN AGN ACCRETION DISCS

EVLABIA ROKAKI
QMW, Univ. of London, Dept of Physics, Mile End Road, London E1 4NS, UK

and

CATHERINE BOISSON
DAEC, Observatoire de Paris-Meudon, Meudon, France

It is commonly admitted that AGN contain a massive black hole fuelled most likely by an accretion disc. Several spectral features of the AGN, as the continuum excess in the UV and the broad line spectrum, involving different physical processes of emission (thermal for the UV continuum, photoionisation for the line spectrum) have been proposed as signatures of the disc. Physical parameters of the nucleus (as the mass of the black hole, M, the disc inclination, i, and accretion rate, $\dot{M}$) are better determined when these two spectral features are modelled simultaneously. Here, we present results from the disc modelling (see [1]) of the UV and broad Hβ emission of the 22 Seyfert 1 galaxies in a complete AGN sample selected in a hard X-ray survey [2].

The comparison of the Eddington, $L_{Edd}(= 1.26\,10^{38}\,M/M_\odot$ erg s^{-1}) and bolometric, L_{Bol}, luminosities shows that all objects radiate well below the Eddington limit ($L_{Bol} = (0.2-2.5)\,10^{-1}L_{Edd}$). The derived values of i span the range of $15°-45°$ and are consistent with the AGN unified scheme for the Seyferts 1 nuclei. An interesting parameter, derived from the line fitting is also the size of the region which emits the bulk of the line emission $R_{H\beta}$. This parameter (multiplied by c) represents the time lag of the line light curve with respect to that of the continuum. A relation $R_{H\beta} \propto L_{Bol}^{0.53}$ is derived.

The contribution of the UV emission of the disc (gravitationally heated), L_{UV}, to the overall emission of the nucleus is significantly scattered and this is most likely to be due to the heavy absorption of the UV emission from some objects in the sample. When these objects are excluded the disc contribution seems independent of the bolometric luminosity with $L_{UV} \sim 0.15\,L_{Bol}$.

For the non-absorbed objects we also compute the UV spectra assuming both radiative, (X-ray) and gravitational heating of the disc as is described in [3]. The X-ray source is assumed spherical with a radius equal to the inner radius of the disc (= $3\,R_{gr}$), and with a luminosity equal to that observed between 2 and 10 KeV. From the new fits of the UV spectra we find that 5-50 % of the UV emission can be produced by the X-ray radiation. Having selected the UV-brightest objects of the sample and assuming the most conservative value for the X-ray luminosity, this fraction represents only the lower limit of the reprocessed UV emission. More realistic values of the reprocessed UV component demand the study of the UV and X-ray variability.

References

1. Rokaki, E., Boisson, C., and Collin-Souffrin, S., 1992, A&A 253, 57
2. Piccinotti G., Mushotzky R.F., Boldt E.A., et al., 1982 ApJ 253, 485
3. Rokaki, E., Collin-Souffrin, S. and Magnan C., 1993, A&A 272, 8

SIMULTANEOUS IMPLOSIVE ACCRETION AND JET FORMATION IN QUASARS: CORRELATION OF OPTICAL OUTBURSTS WITH VLBI JETS

M.M. ROMANOVA
Space Research Institute, Russian Academy of Sciences, Moscow, Russia, 117810

and

R.V.E. LOVELACE
Department of Astronomy and Applied Physics, Cornell University, Ithaca, NY 14853, U.S.A.

A model and simulation code have been developed for time-dependent axisymmetric disk accretion onto a compact object including for the first time the influence of an ordered magnetic field and magnetically driven outflow of energy and angular momentum in ($\pm z$) directions (see also Lovelace et al., 1993). It was shown that the system behaviour crucially depends on the amplitude of the poloidal magnetic field fluctuation $B_{\rm p}$, compared to the critical value $B_{\rm cr} \sim (\alpha^2 T^{1/2} \sigma / R^{3/2})^{1/2}$, where $T(r,t)$ is the temperature, $\sigma(r,t)$ the surface density of the disk, R the radial distance, α the alpha coefficient of Shakura-Sunyaev disk model. If the fluctuation is small, $B_{\rm p} < B_{\rm cr}$, then it diffuses outwards with decreasing the amplitude and eventually disappears. In the opposite case $B_{\rm p} > B_{\rm cr}$, a soliton-like structure forms in the disk density, temperature, and magnetic field and propagates implosively inward. In this case the radial accretion speed $u(r,t)$ is shown to be the sum of the usual viscous contribution and magneitic contribution $\sim R^{3/2} B_{\rm p}^2 / \sigma$. The essential part of angular momentum and energy is going to the jet from the region of fluctuation. Compression of matter in the propagating wave leads to enhancement of magnetic field and more effective angular momentum outflow. This leads in turn to accelerated accretion and subsequent enhancement of magnetic field. It gives the implosive nature of the process, which can be observed as: simultaneous burst in the radiation and outflow. The model is pertinent to the formation of discrete components observed in VLBI jets which appear to originate at times of optical outbursts at some quasars.

References

Lovelace R.V.E., Romanova M.M. and Newman W.I. 1993. Preprint : of the Space Research Institute of the Russian Academy of Sci.,: N 1875, pp.1-17.

HYBRID ACCRETION-DISKS IN AGN

and the AGN contribution to the XRB

AMRI WANDEL
The Hebrew University, Jeruslem, Israel.

Abstract. The hybrid accretion-disk (HAD) model links the two characteristic components of AGN spectra – the UV bump and the X-ray power-law – in the framework of one physical model. The radially stratified hybrid disk is a self consistent combination of a thin, cool accretion disk at large radii, with an inner hot two-temperature disk. Its spectrum consists of three components, corresponding to the three radial disk regions: a blackbody thermal spectrum from the outer cool disk, a Comptonized soft photon power-law spectrum from the intermediate region, and a thermal Comptonized bremsstrahlung spectrum from the inner region. The dependence of the hybrid disk spectrum on the accretion rate and on other parameters is discussed and applied to AGN spectral evolution, and in particular to explaining the cosmic X-ray background by AGN.

Key words: accretion-disk, active nuclei, quasars, x-rays

0.1. RADIAL DISK STRUCTURE

The thin accretion disk (Sakura and Sunyaev 1973) is unstable in the inner, radiation- pressure dominated region. The HAD model (Wandel and Liang 1991) removes this instability by assuming that in the inner region the disk heats up reaching a stable configuration - the two temperature hot disk (Lightman and Eardley 1974). For high accretion rates the hot disk has two parts - an outer part, where the cooling is dominated by inverse Compton scattering of the soft photons from the cool disk, and an inner region, which is shaded from most of the soft photon flux, and is dominated by Comptonized bremsstrahlung . Unlike the vertically stratified cold-hot disk models, the radial configuration involves no additional free parameters and is thermally self consistent.

0.2. SPECTRUM

The HAD spectrum consists of three components, corresponding to the three hybrid disk regions: modified black body from the outer cool disk, Comptonized soft photon power-law from the intermediate region, and thermal Comptonized bremsstrahlung from the inner region.

0.3. THE HAD-AGN MODEL OF THE XRB

Wien pair equilibrium leads to electron temperatures in a narrow range of 50-100keV in the inner region, which makes the HAD in AGN an attractive model for the X-ray background (Wandel 1992, in Proc. MPE conf. on AGN and the XRB). Convolving the HAD AGN spectrum with the X-ray luminosity function of AGN (Boyle et.al. 1993) gives a good fit to the hard XRB (Wandel 1993).

SPECTRUM OF A MAGNETIZED ACCRETION DISK AGN MODEL

YI WANG
*Department of Astronomy and Astrophysics, and Department of Physics,
Harvard University, Cambridge, MA 02138, U.S.A*

Abstract. Using the magnetized accretion disk AGN model of Field and Rogers (1993, ApJ, 403, 94), the spectral emission from far infrared to γ-ray is calculated. The resulting spectrum closely resembles that of Seyfert 1 galaxies. The effect of optically thin dust above the accretion disk is investigated.

Key words: AGN, spectrum

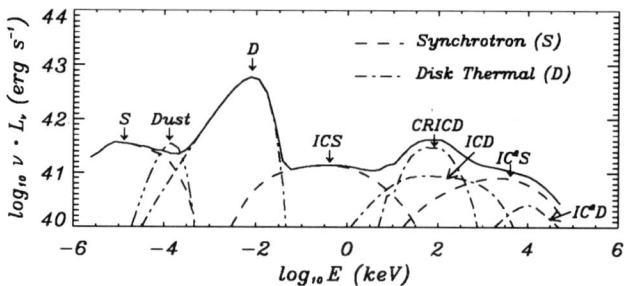

Fig. 1. The νL_ν plot of the radially integrated spectrum. A far-infrared cutoff exists just below 10^{-6} keV due to synchrotron self-absorption. A UV-bump and a large γ-ray tail are present. Here, IC represents Inverse-Compton emission, IC^2 represents second order IC emission, and CR represents a Compton Reflected component.

In the model of Field and Rogers (1993), gravitational energy released by gas spiraling inward in the accretion process is converted to magnetic energy and radiated away. Loops in the corona form due to magnetic buoyancy. Reconnection occurs and it accelerates particles to relativistic energies. The radially integrated spectrum is shown in figure 1.

The dust emission from an optically thin ($\tau = 0.1$) layer of dust suspended above the accretion disk exceeds emission from thermal and synchrotron spectra near 10 μm (figure 1). Optically thick dust in the torus contributes significantly to the infrared emission. Thus the predicted 10% variability in the infrared can be greatly reduced by dust. The sharp cutoff in dust spectrum at high temperature almost always produces an infrared minimum at a few μm.

Acknowledgement:
I am grateful to Prof. George Field at CfA and Dr. Robert Rogers at LLNL for their guidance and assistance. Support from the Rowland Fund from Department of Physics at Harvard University is acknowledged.

ON ILLUMINATION EFFECT IN AGN DISKS

INSU YI
Harvard-Smithsonian Center for Astrophysics
60 Garden St., Cambridge, MA 02138, USA

We consider illumination effects in a weakly self-gravitating ($4\pi G\Sigma \sim \Omega^2 z$) outer disk region ($R \sim 10^{17}$ cm) around a supermassive blackhole of $M_8 = M/10^8$ $M_{sun} = 1$ with relatively high mass accretion rates ($\dot{M}_{25} = \dot{M}/10^{25} g/s = 1$). (e.g., Shlosman and Begelman 1989, APJ, **341**, 685; Shore and White 1982, APJ, **256**, 390).
We adopt an effective gravitational viscosity which is characterized by $\nu_{eff} \sim (G/M^3)^{1/2}\Sigma^2 R^{9/2}$ (Lin and Pringle 1987, MNRAS, **225**, 607). The illumination is dominant over the local viscous dissipation for $R > 3GM/2c^2\epsilon\eta = 2.2 \times 10^{15} M_8(\epsilon\eta)^{-1}_{0.01}$ cm (ϵ is the fraction of energy emitted by the central hard X-ray source, η is the efficiency of accretion). We consider the region where $P = P_{gas} \gg P_{radiation}$ and $\tau_{ff} \gg \tau_{es}$. We calculate the vertical structure and self-consistently determine the disk parameters assuming that the illuminating photon energies are of a single value, $E = 10$ keV. We find $H \propto R^{0.6}$ and $T_c \propto R^{-0.8}$ (cf. for a standard α disk $H \propto R^{8/9}$ and $T_c \propto R^{-3/4}$). For a fixed set of model parameters ($\dot{M}$, M, R, and E), we vary the fraction ($\epsilon\eta$) of accretion energy, which is converted into illumination flux, from 0.001 to 0.1. We calculate expected model spectra from a radial zone $R = 10^{17} - 3 \times 10^{17}$ cm. When the illuminating photon energy is fixed, as the fraction increases from 0.001 to 0.1, the peak photon energy increases from $10^3 K$ to $3.2 \times 10^3 K$. Although the hard X-ray photons do not penetrate the AGN disks into the central region, the surface heating and resulting redistribution of energy affects the disk structure and emission. For a fixed set of accretion parameters, with $\epsilon\eta = 0.01$, we vary the characteristic energy E of the illuminating hard-X-ray flux. Due to energy dependence of X-ray absorption depth, the vertical structure is significantly affected by varying incident photon energies even when the illumination flux is kept constant.
The results described above are interestingly related to the following analytically derived radial structure of accretion disks for the same region, $T_c \approx [3.7 \times 10^5 K] M_8^{3/17} \dot{M}_{25}^{4/17} R_{15}^{-13/17} (\epsilon\eta)_{0.01}^{2/17}$, $\Sigma \approx [2.3 \times 10^7$ $g/cm^2] M_8^{1/2} \dot{M}_{25}^{1/3} R_{15}^{-3/2}$, and $H \approx [5.3 \times 10^{12}$ $cm] M_8^{-11/34} \dot{M}_{25}^{-5/51} R_{15}^{25/34} (\epsilon\eta)_{0.01}^{2/17}$. The results of numerical calculations are roughly in qualitative agreement with the above analytic results.

Poster Contributions: Statistical Studies and Evolution

CM–WAVELENGTH FLUX VARIABILITY: CONSTRAINTS ON AGN MODELING

M.F. ALLER, H.D. ALLER and P.A. HUGHES
Astronomy Department, University of Michigan

1. Introduction

To study whether the radio properties of BL Lacertae type objects and QSOs differ, we initiated a program in 1979 to monitor the total flux density and linear polarization at 14.5, 8.0 and 4.8 GHz of the strongest then known BL Lac objects which met the Hewitt-Burbidge criteria (Ledden, private communications) plus 3 subsequently identified high declination BL Lacs (Biermann et al. 1981). Results based on the behavior of the 45 brightest sample members are compared here with the properties of the QSOs in the flux-limited Pearson-Readhead sample (Pearson and Readhead 1988).

2. Results

We find that: the average degree of variability is higher and the highest amplitude variations are exhibited by the BL Lac sources; the average spectral indices for the BL Lacs are near 0.0, but dramatic steepening to values near 1.0 ($S_\nu \propto \nu^{+\alpha}$) sometimes occurs during large outbursts; large events in both classes are often separated by intervals of several years, but well-resolved events are more common in BL Lacs than in QSOs (where the flux outbursts frequently represent simultaneous contributions from several components that are sometimes resolved in the polarization) – these BL Lac events often show rapid declines; and few BL Lacs exhibit a well-defined long-term stable position angle characteristic of many of the QSOs. While the data cannot be used to unambiguously distinguish the two types of objects, these characteristic properties suggest intrinsic differences. We attribute the high amplitude flux changes with rapid drop-offs in the BL Lacs to more pronounced energy loss along their jets. The lower degrees of ordering of the magnetic field in BL Lacs compared to QSOs suggests that here the polarized flux is commonly dominated by the contribution from the evolving compact components rather than by the well-ordered underlying quiescent flows in the jets.

This work was supported in part by grant AST 9120224 from the NSF.

References

Biermannn, P. et al. 1981, ApJ, **247**, L53.
Pearson, T. J. and Readhead, A. C. S. 1988, ApJ, **328**, 114.

K-CORRECTION BIASES AND THE QUASAR LUMINOSITY FUNCTION

A. C. BAKER and P. C. HEWETT
Institute of Astronomy, Madingley Road, Cambridge, CB3 0HA England

Abstract.
By characterising the range of quasar UV-optical spectral indices and any correlation with it e.g. luminosity or line parameters, we hope to remove one more bias from the quasar luminosity function (QLF).

Although the rest-frame quasar UV-optical spectrum is well-fit by a power law ($f(\nu) \propto \nu^\alpha$) with a mean spectral index $\alpha_{\rm UVO} \sim 0.3$ (Francis et al. 1993 ApJ 407 519), this is not a sufficient description for the purposes of calculating the QLF. The QLF is 'blurred' by the *range* in spectral index (Warren et al. 1994 ApJ in press) and a flux-limited sample reaches deeper into the blue QLF, mimicing faster 'evolution'.

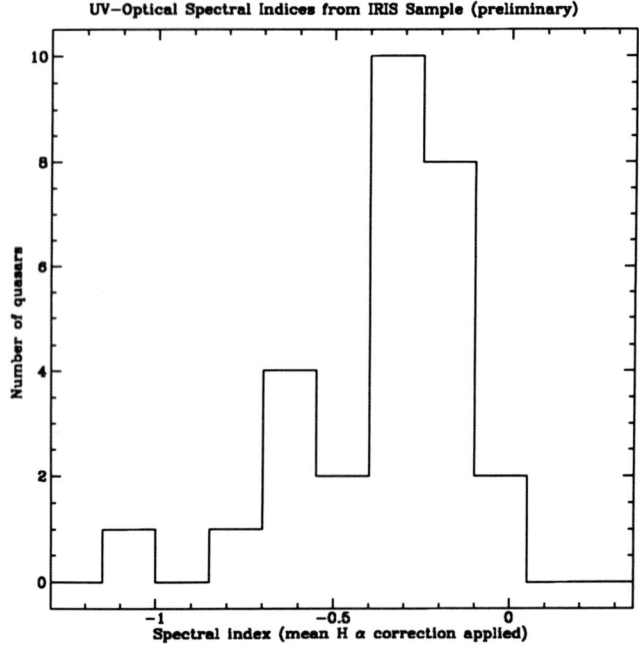

To combine with our existing B_J magnitudes, we have obtained JHK' photometry using IRIS on the AAT for over 100 quasars, selected from the LBQS to have a wide range in other properties. Preliminary $\alpha_{\rm UVO}$ for the first 29 objects in the sample are shown in Figure 1; note the range $-1 < \alpha_{UVO} < 0$.

NUCLEAR ACTIVITY IN INTERACTING GALAXIES

TAPAN K. CHATTERJEE
Facultad de Ciencias, Fisico-Matematicas, Universidad A. de Puebla, (A.P. 1316), Puebla, Mexico

Dynamical studies of galactic collisions, conducted previously (Chatterjee, 1992, 1993a,b), indicated that most of the mergers take place in two to three shrinking orbital periods. We extend this line of research work to study induced nuclear activity. We study the binary evolution of a spiral galaxy perturbed by a compact elliptical galaxy of comparable mass and find that each time the perturber penetrates the disk of the spiral, the disk is subjected to an appropriate perturbation, causing inflow of gas towards its nucleus due to loss of angular momentum; there it could activate an inert black hole (consistent with previous studies; e.g. Naguchi, 1988). However a new feature that we find is that repeated episodes of disk penetration by the perturber occur in gradually shorter timescales, causing an overlap of the activity timescale and the dynamical timescale. In fact if the elliptical is very compact and the spiral has a massive bulge, subsequent dynamical timescales reduce by more than an order of magnitude. This periodic increase in the activity of the nucleus is of a secular nature (in contrast to a reactivation process), and could lead to the evolution of the spiral along the following lines, Starburst → Seyfert 2 → Seyfert 1, as interpenetrations follow.

References

Chatterjee, T.K., 1992: Astrophys. Space Sci. 196, 283
Chatterjee, T.K., 1993a: Astrophys. Space Sci. 199, 35
Chatterjee, T.K., 1993b: Astrophys. Space Sci. 199, 189
Naguchi, M., 1988: Astron. Astrophys. 203, 259

SPATIAL DISTRIBUTION OF QUASARS IN THE LBQS SAMPLE

YAOQUAN CHU and JIANHUI TAO
Center for Astrophysics
University of Science and Technology of China
Hefei, Anhui 230026, China

The information of spatial distribution of quasars would be very important in building the model of formation of AGN. Following the development of high efficiency detectors and related technique, quasar samples with less selection effect were and will gradually available. As we know that the power spectrum analysis gives a more robust determination of large scale distribution of objets. We develop a new methods to determine the power spectrum in a three dimension ball, which is more suited to the case of quasars survey. The details of our approach can be found in Tao and Chu (1993).

The basic simplified formula for a part of the ball is:

$$P = \langle \frac{|b_{lk}^m - nI_{lk}^m|^2}{J_m^l} \rangle / n - 1$$

where

$$b_{lk}^m = \sum_{i=1}^{N} \frac{1}{r_i \sqrt{r_0}} e^{ik(r_i - R)} Y_l^m(\theta_i, \phi_i), \quad I_{lk}^m = \int W(\Omega) \frac{1}{r \sqrt{r_0}} e^{ik(r-R)} Y_l^m(\theta, \phi) dV$$

$$J_l^m = \int_{\Delta\Omega} |Y_l^m|^2 d\Omega$$

We use these methods to analysis the space distribution of quasars in the sample of Large Bright Quasar Survey (LBQS), The LBQS survey cover about 800 degree on the sky which contain about 1000 quasars with redshift $0.2 < z < 3.3$.

We have calculate the power spectrum for all 18 fields in LBQS sample and find no strong evidence for clustering. For all fields, $P \approx 0.2$ and no any tendency to decrease at the scale of 15 Mpc have been found. We also check the influence of the survey border, the value of q_0, and the error in determining of redshift, and find no one of these seriously effect our statistical results.

References

1. Tao J. and Chu, Y., 1993, Scientia Sinica, in press
2. Foltz C. B. et. al. AJ. **94** (1978), p. 1423; **98** (1989), p. 1959; **101** (1991), p. 1121.(LBQS)

ON THE UNIFICATION OF RADIO GALAXIES AND QUASARS

KRZYSZTOF T. CHYZY
Astronomocal Observatory of the Jagiellonian University
ul. Orla 171, 30-244 Krakow, Poland

A comparison of the projected linear size evolution of extended quasars and radio galaxies are often used as a test of the radio galaxy - quasar unification schemes. If both the mentioned categories of radio sources differ to an observer only due to the various viewing directions, then their radio linear sizes are expected to evolve with redshift in the same way (Gopal-Krishna & Kulkarni, 1992). However, apart from the simplest linear size parameter L we can still determine two independent parameters assessing the asymmetry of the radio structure: Q - the arm lengths ratio; M - misalignment, which measures the apparent bending, and defined as the ratio of the displacement of the core from the source axis to the linear size. The asymmetry parameters Q and M can also be a powerful tool in the consistency test for the orientation based unification scheme as their evolutionary patterns should be the same for radio galaxies and quasars. Contrary to the linear size, they are not sensitive to the simple homological rescaling of the whole structure and hence to the age or expansion velocity of the structure.

In order to derive the evolutionary behavior of the asymmetry of radio structures the dependence of the median values of Q and M parameter on redshift and radio luminosity has been estimated in the form $\propto (1+z)^n$ and $\propto P^\beta$ respectively for 152 powerful radio galaxies and 173 steep-spectrum quasars (Chyzy & Zieba, 1993). The striking result of the comparison of the fitted parameters is the stronger evolution of the asymmetry Q and M for radio galaxies than for quasars in concordance with the faster decrease of radio galaxy linear sizes. Moreover, linear size and bending are more closely correlated with radio luminosity for radio galaxies than for quasars.

The possible causes underlying for the evolution of geometrical parameters seem to be similar for radio galaxies and quasars. For both the types of objects with increasing redshift there is a fast decrease in their sizes, increase in bending and slower increase in arm asymmetry. However, this epoch dependency is far stronger for radio galaxies than for quasars, which resembles an analogous tendency in correlation with luminosity. As radio morfology is closely connected with the state of galactic surroundings we may conjecture that the evolutionary state of galactic environment associated with quasars and radio galaxies slightly differs. A part of this effect may be caused by somewhat distinct redshift range for these objects.

References

Chyzy, K.T., Zieba, S., 1993, *Astron. Astrophys.* **267**, L27.
Gopal-Krishna, Kulkarni, V.K., 1992, *Astron. Astrophys.* **257**, 11.

PRINCIPAL COMPONENT ANALYSIS OF MULTIWAVELENGTH PROPERTIES OF SEYFERT GALAXIES

DEBORAH DULTZIN–HACYAN and CARLOS RUANO
Instituto de Astronomía UNAM, Apdo 70-264 Mexico DF 04510, Mexico

Abstract. A multidimensional statistical analysis of observed properties of Seyfert galaxies has been carried out using Principal Component Analysis (PCA) applied to X-ray, optical, near and far IR and radio data for all the Seyfert galaxies types 1 and 2 for the catalog by Lipovtsky et al. (1987).

Key words: Seyfert galaxies, multiwavelength emission, multidimensional statistics

1. Analysis and Results

The catalog by Lipovetsky et al. (1987) provides the largest list of Seyferts (nearly 1000) with multiwavelength data. The effects of incompleteness of the sample as well as other details of the statistics and interpretation will be discussed elsewhere (Dultzin-Hacyan & Ruano, in preparation). PCA is a statistical method which permits to determine the minimum number of independent or uncorrelated variables underlying a larger number of observed variables, see e.g. Kendall (1957). Dimensional analysis is a powerful tool in studies that require classification and is very important for understanding correlations among different variables.

Our main result is that the emission of Seyfert 1 galaxies is well represented by only one eigenvector; i.e., the observed quantities generate a one–dimensional space of variables. On the other hand, in the case of Seyfert 2 galaxies, the space of variables generated by the PCA is three–dimensional. Several details of the analysis lead us to the following interpretation of this result: In the case of Seyferts 1, the main process at the origin of radiation is the release of energy of gravitational origin by accretion unto a supermassive black hole. All the considered luminosities (even if they are the result of re–processing of this energy) are thus coupled to the intensity of this process. In the case of Seyferts 2, appart from accretion energy, there are two other important processes which we may identify with stellar and interstellar radiation from the underlying galaxy. A detailed analysis reveals that the variance in luminosity related to radiation of stellar origin in no case exceeds $\sim 13\%$ for Seyferts 1. In contrast, for Seyferts 2 the radiation of stellar origin can account for $\sim 46\%$ of the variance in certain luminosities.

References

Lipovetsky, V.A., Neisvestny, S.I. & Neizvestnaya, O.M. 1987, Communications of the Special Astrophysical Observatory No. 55, Acad. Sci. URSS.
Kendall, M.G. 1957, "A Course of Multivariate Analysis" Griffin & Co., London

OPTICAL DIFFERENCES BETWEEN RADIO-LOUD AND RADIO-QUIET QSOS

PAUL J. FRANCIS
Steward Observatory, University of Arizona
Tucson AZ 85721, U.S.A. *

Abstract. We analyse the rest-frame UV spectra of a complete sample of optically selected radio-loud and radio-quiet QSOs. Our results are:
  1: Broad absorption-line QSOs (BALQSOs) are all radio quiet, *but* they are strongly clustered toward the top end of the radio-quiet population in radio power.
  2: Radio-loud QSOs have higher equivalent-width, narrower high ionisation emission-lines than radio-quiet QSOs.

1. Observations

We observed 250 of the brightest QSOs in the optically selected Large Bright QSO Survey with the MMT and VLA. A clearly bimodal distribution of radio-to-optical flux ratios was found.

2. Results

2.1. ABSORPTION LINES

None of the radio-loud QSOs showed broad absorption lines (BALs). However, the BAL QSOs were not uniformly distributed through the radio-quiet population; they were strongly clustered towards the highest radio-to-optical flux ratios of the radio-quiet population (99.99% confidence). This result can be explained if, for radio-quiet QSOs, the BAL covering factor is proportional to the radio flux.

2.2. EMISSION-LINES

We studied the Lyman-α (1216), C IV (1549) and C III] (1909) emission-lines. In radio-loud QSOs they had $\sim$ 30% higher mean equivalent-widths (97 % confidence) and were 30% narrower (96 % confidence) than in radio-quiet QSOs. This is evidence for more extended emission-line regions in radio-loud QSOs.

3. Further Reading

See Francis, Hooper & Impey, 1993, Astronomical Journal 106, 417 and references therein for more details and discussion.

---

* Current Address: School of Physics, University of Melbourne, Parkville, Victoria, Australia 3052

ON THE WAY AGN'S TURN OFF

D. FRIEDLI

Geneva Observatory, CH-1290 Sauverny, Switzerland

and

Steward Observatory, University of Arizona, Tucson AZ 85721, USA

Observations of nearby galaxies indicate non-negligible (dark) mass in their nuclei, interpreted either as very dense clusters or Massive Black Holes (MBH's). The latter hypothesis is supported by the widespread idea that MBH's can be the engine powering Active Galactic Nuclei (AGN's), and that interaction- or bar-induced central mass accretion can feed MBH's with large scale, plentiful fuel. However, there are fewer AGN's at the present time than at high redshifts, although many if not all bright galaxies must harbour relics of central active MBH's. How can we explain the fact that some AGN's are now turned off? Is it only due to the exhaustion or evaporation of the available fuel, and/or to the lower rate of interactions at the present epoch?

An alternative possibility stemming from the MBH - bar - fueling connections is suggested. 1) Interactions do not exclusively induce starbursts and/or AGN's. In strongly barred galaxies, significant large scale, bar-induced mass accretion occurs as well [1], and mass transport to the very centre can be produced by a system of embedded bars with different pattern speeds [2]. 2) Since mass concentrations in the nucleus lead to the generation of chaos, MBH's are not innocuous for the global galactic dynamics. The self-consistent response of the potential acts to minimise this chaos and the central region tends to become nearly spherical. When the mass of the MBH reaches about 2% of the stellar disc mass, the bar is destroyed [3]. 3) The bar dissolution results in a drastic reduction of both the angular momentum transfer and the dissipation rate, and gas fueling switches off.

Thus, the existence of these turned off AGN's could be explained in part by the complete dissolution of the bars by supermassive black holes. Except in the unlikely event that the fuel is completely exhausted or evaporated, sporadic accretion of small amounts of local matter remains possible at any time. Finally, since the bar dissolution will usually occur in evolved thick bars, MBH's candidates for the upper end of the black hole mass function should preferably be searched in S0 or Sa galaxies, some of which may result from dissolved thick bars [1]. Such galaxies would be excellent candidates for turned off quasars or Seyferts.

References

[1] Friedli, D., Benz, W., 1993, A&A 268, 65
[2] Friedli, D., Martinet, L., 1993, A&A, 277, 27
[3] Friedli, D., 1993, in: Mass-Transfer Induced Activity in Galaxies, Lexington Conference, ed. I. Shlosman. Cambridge University Press, Cambridge, in press

MULTIFREQUENCY SPECTRA OF EXOSAT BLAZARS

K. K. GHOSH and S. SOUNDARARAJAPERUMAL
Indian Institute of Astrophysics, Vainu Bappu Observatory, India

BL Lacs, Highly Polarized Quasars (HPQs) and Optically Violent Variables (OVVs) are classed as blazars. Based on the multifrequency spectral properties of these objects we have classified the blazars into two groups which are as follows:

- Group I - BL Lacs

  BL Lac objects are soft X-ray sources (Hardness ratio is less than 5) with steep and flat spectral indices in the X-ray and ultraviolet bands, respectively. They are relatively less polarized (in the optical band) and low luminous objects ($M_{abs} < -23$). Radio to X-ray spectra of BL Lacs can be fitted with a single parabolic curve and these objects radiate most of their energies in the X-ray band (radio through X-ray frequencies).

- Group II - HPQs/OVVs

  HPQs and OVVs are hard X-ray sources (Hardness ratio is greater than 5) with flat and steep spectral indices in the X-ray and ultraviolet bands, respectively. These are highly polarized and highly luminous ($M_{abs} > -23$) objects. Radio to X-ray continuum emission spectra of HPQs/OVVs can be fitted with two parabolic curves with the spectral break between UV and X-ray bands and these objects radiate most of their energies in the infrared band.

We have found a new bimodal nature of the distribution of blazars in the radio and the X-ray luminosity plane. Also the spectral indices between UV and X-ray bands and the UV spectral indices are correlated and anticorrelated for BL Lacs and HPQs/OVVs, respectively. The above results suggest that either BL Lacs and HPQs/OVVs are two different classes of objects, or they are same type of objects with different alignments of the jets to the line-of-sight. Inhomogenious jet model of Ghisellini et al. (1985, A&A, 146, 204) can explain the steep and flat X-ray spectra of BL Lacs and HPQs/OVVs, respectively. However, the above model which suggests that the UV radiation is due to synchrotron mechanism from both the types of objects, can not explain the relatively flat and very steep UV spectral indices of BL Lacs and HPQs/OVVs, respectively. Also the model of Ghisellini et al. (1985) is unable to explain certain other detail results of blazars. There are other competetive models of blazars but, at present, any single model is unable to explain all the observed properties of blazars.

RADIO TO X-RAY ENERGY DISTRIBUTION OF BL LACERTAE OBJECTS

P. GIOMMI, S. G. ANSARI and A. MICOL
ESIS, Information Systems Division of ESA, ESRIN, Frascati, Italy

Abstract.
We have constructed radio to X-ray energy distributions for a large number of BL Lacertae objects using archival data. We find that Radio to optical spectra of RBLs and XBLs are very similar. Large differences are seen at higher frequencies where RBLs very frequently show a cutoff near the optical band while XBLs usually do not show any turnover before UV/X-ray energies. Our data is consistent with a picture where RBL and XBL are from the same parent population, the XBL simply being those (rare) objects where the break in the energy distribution is located at high energy.

1. Summary

We present multifrequency spectra of about 110 Radio and X-ray selected BL Lacertae objects constructed using non-simultaneous archival data. The data have been collected using the European Space Information System (ESIS, Giommi & Ansari 1993) and are from several radio catalogues, the IRAS faint source catalogue, the Veron & Veron (91) catalogue, the *Einstein* and EXOSAT databases.
The radio to optical spectrum of Radio Selected (RBL) and X-ray Selected (XBL) BL Lacs are very similar. The difference between these two classes of objects is limited to the optical/X-ray part of the spectrum. The classical radio discovered BL Lacs are characterized by an energy distribution with a sharp cutoff in the IR/optical band while in most of the X-ray discovered objects the turnover is located near the UV/X-ray band or at higher frequencies. For a given radio flux this can give rise to a difference in the X-ray flux of factor of 100 or more. The population of BL Lac objects probably includes a wide range of energy cutoffs. From the radio selected sample we see that the probability that the cutoff falls in the X-Ray band (or beyond) is small and is of the order of 10 %. The LogN-LogS of RBL (Stickel et al. 1991) gives $N(S > 1Jy) = 10^{-3} deg^{-2}$. Since the typical radio flux of a XBL with X-ray flux of $5 \times 10^{-13} - 10^{-12} erg\ cm^{-2}\ s^{-1}$ is of the order of 10 mJy, the expected density of these objects is $N(S > 10\ mJy) = 10^{-3} \times (1000\ mJy/10\ mJy)^{1.5} = 1.0\ deg^{-2}$ (if the Euclidean slope extends to 10 mJy). Assuming that about 10% of these sources have a cutoff at sufficiently high energy to be detected in the EMSS, the expected density of BL Lacs in the EMSS is $\approx 0.1 deg^{-2}$ which is about the value measured (Maccacaro et al. 1989, Giommi et al. 1989). RBL and XBL are known to populate different regions of the α_{ox} vs α_{ro} diagram. We find that this is probably the consequence of the different position of the energy cutoff.

References

Giommi, P., et al. 1989, in "BL Lac Objects", L. Maraschi, T. Maccacaro, and M.-H. Ulrich eds., Springer-Verlag p. 231.
Giommi, P., & Ansari 1993, Proc of XXVII ESLAB Symposium, Noordwijk, The Netherlands.
Maccacaro, T., Gioia, I. M., Schild, R. E., Wolter, A., Morris, S. L. and Stocke, J. T. 1989, in "BL Lac Objects", L. Maraschi, T. Maccacaro, and M.-H. Ulrich eds., Springer-Verlag p. 222.
Stickel, M., Fried, J.W., Kuehr, H., Padovani, P., and Urry, C.M., 1991, APJ, 374, 431.

OPTICAL PROPERTIES OF SOFT X-RAY SELECTED BRIGHT NEW ROSAT AGN

D. GRUPE, K. BEUERMANN and K. REINSCH
Universitäts-Sternwarte Göttingen, FRG

and

H.-C. THOMAS and H.H. FINK
MPI f. Astrophysik, Garching, FRG, and MPI f. Extraterrestrische Physik, Garching, FRG

Abstract. Optical identification of bright sources from the ROSAT All-Sky Survey has led to the discovery of 40 new Seyfert galaxies with very soft X-ray spectra. Nearly half of these are narrow-line Seyfert 1 (Hβ-FWHM < 2000 km/s) with strong optical FeII emission, suggesting the presence of an unseen hard X-ray component. The remainder are Sy1.5 and a few broad-line Sy1.

Key words: Soft X-ray, narrow-line Seyfert 1

Our Sample results an from optical identification program of bright soft X-ray sources (PSPC countrate > 0.5 cts/s; galactic latitude |b| > 20°, HR1 < 0.0) from the ROSAT All-Sky Survey. Of a total of 314 sources, 122 were not previously known. There are 60 known and 48 new AGN in this soft X-ray selected sample. For these we started a program to obtain better X-ray spectra via pointed ROSAT PSPC observations and to acquire optical spectra of higher resolution ($\sim$5Å/mm). From a quick classification of the identification spectra we found that most of our soft X-ray selected AGN show rather "narrow" broad permitted lines and qualify as "Narrow-Line Seyfert 1" (nlS1) type AGN ($\sim$ 50% nlS1, $\sim$ 30% S1.5). For comparison the known AGN with hard X-ray spectra (HR1 > 0.0) are mostly Seyfert 1 type galaxies with decidedly broad permitted lines.

Our X-ray observations suggest the absence of intrinsic absorption within the individual AGN. From our optical spectra we find, that the AGN with relatively narrow permitted lines (nlS1) always show very strong FeII emission and weak emission from the NLR. On the contrary, the Seyfert 1.5 galaxies show weak or absent FeII emission but strong lines from the NLR. The FeII emission of the Seyfert 1 with broad lines is visible but not as strong as the FeII emission of the nlS1s. This leads to the following conclusions:

1) The continuum "seen" by the NLR gas is weaker in nlS1 than in Sy1.5 and Sy1.
2) The continuum "seen" by the nl-BLR is harder than the very soft X-ray spectra detected by the PSPC (FeII requires deeply penetrating hard ionizing photons).
3) The BLR of nlS1 is probably at a larger distance to the central object than the BLR of a typical Seyfert 1 galaxy. Alternatively broader components are intrinsically present but obscured. Nevertheless, soft X-rays are able to escape. If the latter model applies, we should expect polarized light from the hidden broad wings. So far, no such study has been performed for nlS1.

THE REDSHIFT DISTRIBUTION OF Ly α FOREST LINES IN SPECTRA OF QSOs[1]

Huang Keliang[1] and Zhou Hongnan[2]

[1]Department of Physics, Nanjing Normal University, Nanjing 210024, China

[2]Department of Astronomy, Nanjing University, Nanjing 210008, China

Numerous narrow absorption lines in the region of wavelength shorter than $1216(1+z_{em})$ (z_{em} is the emission redshift), i.e. so-called Ly α forest lines, detected in QSO spectra are usually thought to be produced in intervening primeval clouds. The study of Ly α clouds may reveal how matter distributes in space and how it evolves with time at the early universe and provide valuable information about the large scale structure of the universe and its evolution. Based on intermediate resolution (1 ~ 2 A) spectra, many authors (e.g. Lu et al. 1991) deduced that the evolutionary index $\gamma \simeq 2$, (dN/dz ~ $(1+z)^\gamma$, dN/dz is the number of clouds per unit redshift interval at redshift z). It means that Ly α clouds have strong cosmological evolution. In recent years, there appear high-resolution ($< 30 km/sec$) spectra of QSOs. High resolution spectra may provide more information than medium resolution spectra. Hence, it is necessary to study the evolution of Ly α clouds, using the spectra with higher resolution. Carswell et al. (1987) found $\gamma=1$ in the redshift interval 1.9-3.8. But Rauch et al. (1992) found $\gamma=2.1$ for the line sample with $logN(HI) \geq 13.75$. It is more interesting that Giallongo (1991) found a differential evolution: γ is depended on the equivalent width W of line and no evolution for the strong line sample with 0.5>W>0.3. However, these studies involved very few QSOs (three or four). In this paper, we use a larger sample of QSOs to study the evolution of Ly α clouds.

So far, there are spectra of Ly α forest region of five QSOs with high-resolution published: QSOs 2000-330 ($z_{em} = 3.78$; Carswell et al. 1987), 0420-388 ($z_{em} = 3.12$; Atwood et al. 1985), 2206-199N ($z_{em} = 2.56$; Pettini et al. 1990), 1101-264 ($z_{em} = 2.14$; Carswell et al. 1991) and 0014+813 ($z_{em} = 3.38$; Rauch et al. 1992). Recently We obtained spectral data of other two QSOs: 1331+170 ($z_{em} = 2.08$; York et al. 1993, Kulkarni et al. 1993) and 1225+317 ($z_{em} = 2.22$; Huang et al. 1993). By use of the data of these seven QSOs, we study the evolution of Ly α clouds. Following Giallongo (1991), we take W= 0.14 as the threshold of equivalent width to select lines. Besides, we do not consider those Ly α lines for which their observing wavelengths are shorter than $1025 (1+z_{em})$. These lines may confuse with Ly β of other clouds. In our sample, there are 378 Ly α lines with 0.5> W > 0.14. The redshift interval the sample covered is 1.60-3.78.

The value of γ can be easily derived by use of the maximum likelihood method (Murdoch et al. 1986, Lu et al. 1991). Many authors studied the evolution of strong lines with W>0.36 when they analysed the data of medium-resolution spectra. Giallongo (1991) discussed the cases of differential evolution with various intervals of equivalent width. Following these authors, we also discuss the cases with different intervals of W, as a comparison.

The following table is the result of analysing our sample. In the table, nt is the total number of lines in the intervals of W, Q the Q-test and K-S the Kolmogorov-Smirnov test. The values in parenthesis are the results after removing the lines within 8 Mpc of the QSOs to eliminate the influence of the inverse effect.

	W>0.36	0.5>W>0.3	0.3>W>0.2	0.2>W>0.14
γ	1.93±0.48 (2.49±0.48)	2.00±0.53 (2.29±0.58)	1.66±0.55 (1.97±0.61)	2.41±0.60 (2.56±0.67)
nt	171 (162)	138 (125)	126 (111)	114 (97)
Q	0.24 (0.19)	0.25 (0.22)	0.31 (0.28)	0.22 (0.22)
K-S	0.10 (0.25)	0.49 (0.72)	0.54 (0.38)	0.86 (0.95)

---

[1]supported partly by the National Natural Science Foundation of China

1. We can see $\gamma \simeq 2$ for all the intervals of equivalent width. We did not find the differential evolution Giallongo (1991) found. The data of high-resolution spectra show that Ly α clouds indeed have strong cosmological evolution, as that concluded by use of medium-resolution data.

2. Although the values of γ are derived, the statistical test is disappointing. Considering the inverse effect, we obtained only slightly better statistic. This implies that the density distribution of Ly α lines of individual QSOs strongly departs from the grobal distribution. In fact, we obtain quite different values of γ if the data of individual QSOs is seperately used. Therefore, in order to understand the evolution of Ly α clouds better, we need more data of high-resolution spectra.

3. Lu et al.(1991) found a broken power-law form of redshift distribution of Ly α clouds: $\gamma=4.60$ for $z<2.32$ and $\gamma=1.71$ for $z>2.32$. It means that the evolution at lower redshift is faster than that at higher redshift. The number of Ly α lines at lower redshift predicted by this law is even more less than that observed by Hubble Space Telescope for 3C 273 (Bahcall et al. 1991; Morris et al.1991). We examined this broken power-law form by use of the sample in this paper. The sample could not be fitted well by the law.

Reference

Atwood,B., Baldwin,J.A. and Carswell, R.F., 1985 Ap.J. **292**,58
Bahcall,J.N., Jannuzi,B.T., Schneider,D.P., Harting,G.F., Bohlin,R. and Junkharinen,V., 1991 Ap.J. **377**,L5
Carswell,R.F., Webb,J.K., Baldwin,J.A. and Atwood,B.,1987 Ap,J. **319**,709
Carswell,R.F., Lanzetta,K.M., Parnell,H.C. and Webb,J.K.,1991 Ap.J. **371**,36
Giallongo,E. 1991 M.N.R.A.S. **251**,541
Huang,K. et al. (in preparation)
Kulkarni,V.P., Welty,D.E., York,D.G., Huang,K. L., Green, R. F. and Bechtold,J., 1993 Ap.J. (in press)
Lu,L., Wolfe,A.M. and Turnshek,D.A. 1991 Ap.J. **367**,19
Morris,S.L., Weymann,R.J., Savage,B.D. and Gilliland,R.L.,1991 Ap.J. **377**,L21
Murdoch,H.S., Hunstead,R.W., Pettini,M. and Blades,J.C.,1986 Ap.J. **309**,19
Pettini,M., Hunstead,R.W., Smith,L.J. and Mar,D.P.,1990 M.N.R.A.S. **246**,545
Rauch,M., Carswell,R.F., Chaffee,F.H., Foltz, C.B., Webb,J.K., Weymann,R.J., Bechtold, J., and Green, R.F., 1992 Ap. J. **390**,387
York,D.G., Green,R.F., Huang,K.L., Bechtold,J., Welty,D.E. and Carlson,M., 1993 Ap.J. (in press)

DO LUMINOUS QSOS HAVE SOFTER UV SPECTRA ?

THORSTEN KÖHLER
Hamburger Sternwarte, Gojenbergsweg 112, D-21029 Hamburg, Germany

We present a comparison of well defined continuum slope distributions for QSOs found in the LBQS (Chaffee, F.H. et al. 1991, Astron. J., 102, 461 and references therein) and in the Hamburg−ESO−survey (hereafter HES) (Reimers, D. 1990, The messenger 60, 13).

The method for the determination of power−law indices of the LBQS QSO sample is described by Francis (Francis, P.J. 1993, ApJ 407, 519): transformation of all spectra in the quasar rest−frame; defining three pairs of continuum windows in the quasar rest−frame avoiding emission lines: 1430−1480 Å and 2150−2230 Å , 2150−2230 Å and 3020−3100 Å , 3020−3100 Å and 4150−4250 Å; fitting power−laws ($f_\nu \propto \nu^\alpha$) to all QSO−spectra containing one of these pairs.

For a direct comparison we applied the same method to the HES long slit QSO spectra. The measured mean continuum slopes are :

$$z = 0.2 \text{ to } 0.5 : < \alpha_{LBQS} > = +0.3 , < \alpha_{HES} > = -0.18$$
$$z = 0.7 \text{ to } 1.1 : < \alpha_{LBQS} > = -0.8 , < \alpha_{HES} > = -0.95$$
$$z = 1.6 \text{ to } 1.9 : < \alpha_{LBQS} > = -0.5 , < \alpha_{HES} > = -1.03$$

Since blue objects are easier to find we conclude that the dominance of softer spectra in the HES sample is real and not effected by selection effects.

Francis (1993, see above) has pointed out in principal that it is not possible to obtain the true mean continuum slope of the QSO population in a given redshift range without taking into account the survey brightness limits, a dispersion in the continuum slopes and individual K−correction. For this reason we simulated a survey on a virtual QSO population: Each QSO gets an absolute magnitude by the LF (Hewett, P.C., Foltz, C.B., Chaffee, F.H. 1993, ApJ 406, L43), a random z in an initial z−range and an individual α scattered randomly around α_0 (with a gaussian distribution of half width $\Delta\alpha = 0.5$). At last an apparent brightness for each QSO is calculated using its individual α for the K−correction. Adopting the survey limits of HES ($B = 14$ to 17.5) and LBQS ($B = 16.5$ to 18.85) together with the redshift ranges given above we can obtain virtual QSO subsamples with mean continuum slopes now comparable to the measured values (see above).

We can show that a linear correlation between α_0 and M_{abs} ($\alpha_0 = b + a \cdot M_{abs}$, $a = 0.4 \pm 0.1$, $b = 9.4 \pm 0.05$) is valid over a z range from 0.2 to 1.9 covering a dynamic range of nearly 5 magnitudes in each bin (see above). Our result is in strong contrast to Zheng & Malkan (Zheng, W.,Malkan ,M. 1993,ApJ 415, 517). Since their QSO sample is not separated in redshift bins evolution effects are very likely. Francis (1993, see above) has shown that high−redshift QSOs have intrinsically harder slopes than low redshift QSOs.

A POSSIBILITY OF REVEAL OF PARENT POPULATIONS FOR THE OBJECTS WITH ACTIVE NUCLEI ON THE BASIS OF COMPARING THEIR SPATIAL CORRELATION FUNCTIONS

B.V. KOMBERG
Astrospace Center, Lebedev's Physical Institute,
Profsoyuznaya St, 84/32, Moscow, 117810, Russia

and

A.V. KRAVTSOV
Sternberg Astronomical Institute, Moscow State University,
Universitetsky Pr, 13, Moscow, 117234, Russia

For a model of the World with $q_0 = 0.5$ and $H_0 = 100h^{-1}$ there was received a 2-point spatial correlation function (CF) (The sample of 300 objects with $z < 0.045$ and $\delta > 0$ was used [1]) for Seyfert Galaxies with the following parameters : $r_0 = (9 \pm 2)h^{-1} Mpc, \gamma = -(1.7 \pm 0.2)$.

CF for QSOs was studied by using of the Catalogue [2]. In order to decrease a heterogeneity influence of QSO sample the CF was estimated by *method of normalisation to the large scales* [3]. A CF obtained for a sample of 875 QSOs with $0.5 < z < 1$ has $r_0 = 4.5h^{-1} Mpc, \gamma = -1.8$. One can see (taking into account the evolution of structure) that the CF of these QSOs is located between the CF of galaxies and groups of galaxies.

The results of CF calculations for QSO subsamples with different redshifts gives the law of evolution of the CF as: $\xi(r,z) = \xi(r,0)(1+z)^{-3}$.

A division of the whole sample from the Catalogue [2] in subsamples with different absolute magnitudes is, to some extent , analogous to division of QSOs in radio-loud QSS which are brighter and fainter radio-quiet QSG. We obtained that a CF amplitude for QSS subsample with $M = -26$ is 3-4 times higher than one for QSG subsample with $M = -24$.

An evolutional match of the CF of far QSS with the CF of near radio-galaxies and evolutional match of QSG's CF with the CF of near SyG is entirely corresponds to the hypothesis that radio-galaxies and SyG are the final stages of QSS and QSG evolution respectively.

References

1. Lipovetsky,V.A., et al. (1989).*A Catalogue of galaxies Sy-types*, SAO, USSR
2. Veron-Cetty,M.-P., and Veron,P. (1991). ESO Sci. Rep. No.10.
3. Shaver,P.A. (1984). Astron. Astrophys. **136**, L9.

ACTIVITY AS THE RESULT OF MERGING

V. M. KONTOROVICH, A. V. KATS and D. S. KRIVITSKY
Institute of Radio Astronomy of the Ukrainian Academy of Sciences
4 Chervonopraporna str., 310002 Kharkiv, Ukraine

The observed correlation between activity and mergers of galaxies may be explained by the compensation of the angular momentum by merging. This leads to accretion on the galaxy centre [2].

The statistical description of the merger process is based on the generalized Smoluchowsky kinetic equation for the galaxy mass and angular momentum distribution $f(M, \mathbf{S}, t)$ [1]. The model allows to find the luminosity function of active objects $\varphi(L, t)$ connected with $f(M, \mathbf{S}, t)$ by the quadratic in f integral relation. It supposes the luminosity-mass excess Δm relation: $L = B \cdot \Delta m$ (mass Δm is able to fall to the centre). The simplest calculation scheme results in the function $\varphi \propto 1/L$, close to the observed one, if the asymptotic expression of the mass function MF $f(M, t) \equiv \int f(M, \mathbf{S}, t) d\mathbf{S} \propto M^\alpha$ with $\alpha = -(u+2)/2$, where u is defined by the dependence of coalescence coefficient $U \propto M^u$ on mass. MF with $\alpha = -(u+2)/2$ corresponds to the approximate conservation of the number of massive galaxies if their interaction with small ones (masses $\sim M_*$) prevails.

At $u > 1$ (this value is typical for galaxies in the wide mass interval) the "explosive" evolution occurs — analog of the phase transition of gel formation [4], when the power-type tail of MF is formed during a finite time interval. Accordingly an "explosive" formation of active objects, i.e. the epoch of quasar formation, takes place [3]. On large enough masses $M_f \geq 10^{14} M_\odot$ the maximum value of the impact parameter is already limited by the mean free pass length or mean distance between the galaxies. So the mass dependence of the merger cross-section σ and U disappears ($u \to 0$) and the "explosion" stops. This model (assuming an early emergence of small-mass galaxies $M_* \sim 10^6 M_\odot$) enables to explain the abrupt disappearance of quasars at $z = z_{cr} \geq 3$.

The authors are grateful to ISF and the organizers of the IAU Symposium 159 for being given the opportunity to take part in the conference thanks to the Soros Travel Grant for one of the authors (V.M.K.).

References

1. Kats, A. V., and Kontorovich, V.M., 1990, JETP, 70,1; Pis'ma v. Astron. Zh., 17, 229; 1992, Astron. Astrophys. Trans., 2, 183.
2. Kontorovich, V. M., Kats, A. V. and Krivitsky, D. S., 1992, JETP Lett., 55, 1.
3. Stockmayer, W. H., 1943, J. Chem. Phys., 11, 45.
4. Toomre, A., and Toomre, J., 1972, Astrophys. J., 178, 623.

THE CFA SEYFERT SAMPLE AT 8.4 GHZ

MAREK J. KUKULA & ALAN PEDLAR
NRAL, University of Manchester, Jodrell Bank, Nr Macclesfield, Cheshire SK11 9DL, UK.

S. BAUM & C. O'DEA
STScI, 3700 San Martin Drive, Baltimore MD21218, USA.

and

S. UNGER
RGO, Madingley Road, Cambridge CB3 0EZ, UK.

Increasingly, the evidence from optical and infra–red wavebands suggests that the difference between Seyfert 1 and Seyfert 2 nuclei is due largely to orientation effects rather than intrinsic differences between the two classes, but the evidence from radio observations has been less clear–cut. We have observed the CfA Seyfert Sample at 8.4 GHz using the VLA in A– and C–configurations. At this frequency our A–array maps have a resolution of 0.25″ – much higher than those achieved in previous surveys – whilst the 3″ C–array beam is ideal for measuring the total radio flux of the active nucleus. The 1–sigma noise in both sets of observations was 70 μJy.

Space densities for the CfA Seyferts have been derived by Huchra & Burg (1992) and Edelson (1987). The results of our work are consistent with these previous studies: the total space density of Seyferts with 8.4 GHz luminosities between $10^{20.6}$ and $10^{22.6}$ WHz^{-1} is 3.5×10^{-5} Mpc^{-3}. For Seyfert 1s the density is 1.0×10^{-5} Mpc^{-3} and the density of Seyfert 2s is approximately 2.5 times larger.

We find that the mean 8.4 GHz luminosities are $10^{21.3 \pm 0.2}$ WHz^{-1} for Type 1 and $10^{21.4 \pm 0.2}$ WHz^{-1} for Type 2, and a variety of statistical tests all show that there is no significant difference between the luminosity functions of the two types.

Although the mean radio luminosities of Seyfert 1s and 2s are the same, their radio structures are very different. In the high–resolution A–array maps 45% of the Type 2 objects are extended, with double, triple and linear structures. However, at the same resolution only 8% of the Seyfert 1s are resolved, and only one nearby object, NGC4151, contains multiple radio components. This difference could be explained by orientation effects: if type 1s are seen 'head on' (so that their radio axes lie close to the line of sight) and type 2s 'side on' then any extended radio structure in Seyfert 1s will be foreshortened. We find that the individual compact components in Seyfert 2 nuclei have, on average, only 75% of the luminosity of those in Seyfert 1s. This suggests that foreshortening in Seyfert 1s has resulted in multiple radio components being superimposed and appearing as a single compact source.

Our observations therefore support the idea that Seyfert 1s and 2s have the same intrinsic luminosities and radio structures, and that the apparent differences between them are due to their orientation relative to the line of sight.

THE INFLUENCE OF SELECTION EFFECTS ON THE PROPERTIES OF BL LACS

M.J.M. MARCHÃ [*] and I.W.A. Browne
NRAL, University of Manchester, Jodrell Bank, Nr Macclesfield, Cheshire SK11 9DL, UK.

The fact that the recognition of BL Lacs always requires optical confirmation, regardless of whether the objects were first selected in the radio, or X-ray frequencies means that deep surveys will miss some objects simply because the optical emission from the host galaxy outshines that of the BL Lac. In particular, the deeper the survey, the more difficult it will become to recognize low luminosity BL Lacs in the nuclei of luminous galaxies. This recognition effect will modify the intrinsic distribution of objects, and influence their statistical properties in general.

We have developed a method to quantify this recognition problem which enables us to predict source counts, redshift distributions, $< V/V_{max} >$, and percentage of missed objects for flux limited samples. Many of the predicted features appear in the observed distributions of X-ray and radio selected BL Lacs. In particular, comparisons between the predicted and observed properties of BL Lacs selected from the EMSS X-ray survey suggest that this BL Lac sample is only 80% complete, unless there is a lower limit cutoff in BL Lac luminosities. Although the redshift dependance of the recognition problem produces spurious cosmological evolution, the predicted value of $< V/V_{max} >$ for the EMSS sample of BL Lacs is not sufficient to account for the strong negative cosmological evolution found for these objects. However, the fact that the recognition problem does mimic some negative evolution means that such strong intrinsic cosmological evolution is not required for the BL Lacs of this sample.

As deeper surveys become available we predict that the above selection effect will become more significant. It should therefore be taken into account before comparisons between samples are made, and conclusions about the statistical properties of BL Lacs are reached.

---

[*] M.J.M.M. acknowledges the grant from "Programa Ciência" of the Junta Nacional de Investigação Científica e Tecnológica (Portugal).

THE QED (QUASAR ENERGY DISTRIBUTIONS) ATLAS

J. MCDOWELL
Harvard-Smithsonian Center for Astrophysics

Abstract. While the energy distributions of optically and radio selected quasars have the same, reproducible, mean shape in the infrared to ultraviolet region, the strength of the infrared and ultraviolet components can vary by over a decade from object to object.

Key words: Quasars

We have constructed an Atlas of Quasar Energy Distributions (QED Atlas, Elvis et al 1993 in preparation), consisting of starlight-subtracted spectral energy distributions (SEDs) from the radio to the x-ray of 47 quasars observed with the Einstein Observatory. From 100 to 0.1 microns the SEDs consist of two components, the ultraviolet 'big bump' and a broader infrared component, separated by an inflection at a median wavelength of 1.5 microns. We find:

- IRAS upper limits are included using the Kaplan-Meier estimator; this lowers the mean by a factor of three at 100 microns relative to the results of Sanders et al (1989, ApJ 347, 29).
- Flat spectrum radio-loud quasars are brighter at 60-100 microns than other types, but the mean is not otherwise dependent on type.
- The range in UV and IR component strength is a factor of 10 to 20.
- The wavelength of the IR peak is different in different objects, so the mean energy distribution is not a good representation of an individual object.

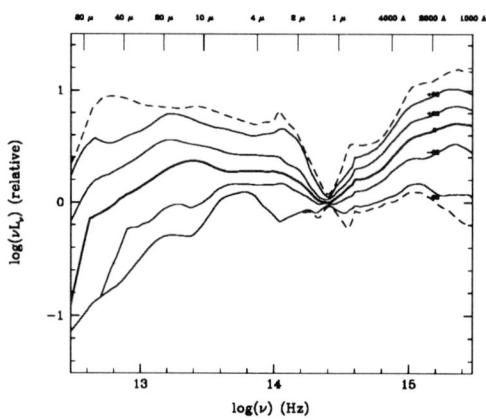

Fig. 1. The dispersion in SED shapes, normalized at 1.5 microns. A dashed line shows the sample envelope at each wavelength; the solid lines show the median and the 68 and 90 percentile contours on each side.

A ROSAT SURVEY OF A SPATIALLY AND MAGNITUDE COMPLETE QUASAR SAMPLE

K. MOLTHAGEN, H.J. WENDKER
Hamburger Sternwarte, Gojenbergsweg 112, D 21029 Hamburg, Germany

and

U.G. BRIEL
Max Planck Institut f. extraterrestrische Physik, Gießenbachstr., D 85740 Garching, Germany

September 10, 1993

The field studied is part of the Hamburg Quasar Survey (Engels et al. 1988, PASPC 2, 143). It contains a spatially and magnitude (down to $\sim 18^m\!.5$) complete AGN sample. Several follow–up observations were made: a deep radio continuum survey at 21 cm and 74 cm, a HI survey and a ROSAT survey consisting of 48 overlapping PSPC pointings with $\overline{t_{exp}} = 2500\,s$.

28 AGN with $16^m\!.7 \leq m_B \leq 19^m\!.0$ and $0.2 < z < 2.5$ (Engels and Hagen, priv. comm.) lie within our region of interest. Remarkable (but not very significant) is a gap between $z = 1.2$ and $z = 1.5$, which can hardly be due to missing bright emission lines since the candidates were selected by a blue continuum. The AGN distribution on the sky appears to be somewhat peculiar, too. An apparent clustering is visible for which no obvious reason, either technical or astronomical, can be found.

Considering their X–ray properties, our AGN sample behaves normally. No unexpected features are found, the only peculiar object is OK492, a known OVV. The tentative identification of some more X–ray sources weakened the contrast between the AGN–rich part and the AGN–poor part somewhat, but it is still clearly visible, the contrast being now 3:1.

The gap in z has not been filled, either. This was not to be expected, because most of the additional AGN are found identifying RASS sources (s. Bade, 1993, PhD thesis, Univ. Hamburg). More fainter AGN have to be identified to test the reality of these two effects.

The fractions of stars and AGN among our X–ray bright sample follow roughly the EMSS (Stocke et al., 1991, ApJ Suppl. **76**, 813). One third of this sample is fainter than $m_B \approx 19^m$.

Most sources (about 80%) show indications of variability. This is confirmed by a comparison of our X–ray bright sample with the RASS source list of the same region.

The merging of the indivdual pointings to increase the net observing time to about 10 ksec. is underway. For a full reference, see Molthagen, Wendker and Briel, 1993 (in preparation).

SOME PECULIARITIES OF VARIABLE OPTICAL SPECTRA OF 11 LOW-REDSHIFT QUASARS

I.I. PRONIK
Crimean Astrophysical Observatory, Ukraine

The flux of narrow 5007 Å [OIII] forbidden line during last 20 years is accepted as a bench-mark for calibration of the continuum and broad-lines fluxes in AGN. But one can not get away from the problem of the forbidden lines variability in these objects. The first report were published by Bardin et al. (1967). Some of the results are reviewed by V.Pronik, I.Pronik (1988,1992). There were 4 years monitoring of[OIII] fluxes variability in the Seyfert galaxy NGC 1275 nucleus (Pronik et al.,1990). Now we reexamine the published data of W.Zheng et al. (1986,1987,1988), discussed earlier in supposition of forbidden lines constant in spectra of 11 low-redshifted QSOs. Maximal calibration coefficients adopted by W.Zheng et al in supposition of [OIII] lines constant were in the interval 2.0-3.6. On the other hand the fluxes errors are about 15%. Disagreement is very high. This is one of the arguments in the forbidden lines variability rightness.

The considered relation $log\, L_{[OIII]} - log\, L_\beta$ has a jumping character. Theire smooth parts of the relation have coefficients of correlations equal to 0.72 – 0.98. Single QSO variation shows the similar behaviour as that of the objects of different luminosities. The degree of one QSO [OIII] lines flux variation are 25 – 100% during 4 – 9 years. QSOs groups formed one smooth relation divided mainly by ratio $I_{[OIII]}/I_\beta$. When $I_{[OIII]}/I_\beta > 1.6$, $I_{[OIII]} \sim I_\beta$; when $I_{[OIII]}/I_\beta < 1$, $I_{[OIII]} \sim I_\beta^{1.4}$

Untill now there is no general theory explaining the forbidden lines variations in AGN in a time scale of years and months. Same attempts were published by Bochkarev (1987) and Fabrika (1987). V.Pronik and I.Pronik (1988,1992) proposed for the regions of neutral hydrogen the mechanism of ionization of oxygen and other heavy elements by soft X-rays. For regions of ionized hydrogen they proposed usual well known collisional exitation mechanism in HII regions of high electron temperatures ($T_e > 28000$ K) and high electron concentrations ($n_e > 10^7 cm^{-3}$).

References

Bardin B., Chopinet M., Duflot-Augarde R.: 1967, *Comp. Rendus Acad. Sci.* Paris, **Vol. 265**, Serie B, P.1149.
Bochkarev N.: 1987, *Procced. IAU Sym. 121*, eds. E.Khachikian et al., Reidel Publ. Comp., P.223.
Fabrika S.: 1987, In"Active nuclei and stars cosmology", ed. D.Martinov, Moscow Univ. Press, P.107.
Pronik I., Merkulova N., Metik L.: 1990, *Astrophys.a.Space Sci*, **Vol. 171**, P.91.
Pronik I., Pronik V.: 1992, *Astronomical a.Astrophys.Trancac*, **Vol.3**, P.57.
Pronik V. Pronik I.: 1988, *Russian Astron.J*, **Vol.65**, P.478.
Zheng W.: 1988, *Astrophys.J.*, **Vol.324**, P.801.
Zheng W., Burbidge E.M.: 1986, *Astrophys.J.*, **Vol.306**, P. L 67.
Zheng W., Burbidge E.M., Smith H.E., Cohen R.D., Bradley S.E.: 1987, *Astrophys.J.*, **Vol.322**, P.164.

A LOWER LIMIT TO THE EXCESS OF COMPANIONS AMONG SEYFERT GALAXIES

P. RAFANELLI and M. VIOLATO
Department of Astronomy, University of Padova, Italy

It is not yet clear which is the role played by interaction on Seyfert activity. Presently there are on this topic contradictory statistical results, based on the study of more or less rich samples of Seyfert galaxies. Petrosian (1982), Dahari (1985) and Mac Kenty (1989, 1990) found an excess of Seyferts among galaxies which have nearby companions and an excess of galaxies with nearby neighbors among Seyfert galaxies. The contrary result was found by Bushouse (1987) and by Fuentes Williams & Stocke (1988). In order to find more reliable conclusions, taking advantage of the increased number of Seyfert galaxies identified in the last years, we have applied the statistical procedures used by Dahari (1985) to a sample of 287 Seyfert-1 (S1) and 195 Seyfert-2 (S2) (namely to all known S1 and S2 galaxies with $z \leq 0.11$ and $\delta \geq -23°$). A subsample has been extracted carefully excluding objects, identified as Seyfert on the basis of their morphology (e.g. on the basis of the presence of a companion) and brighter than $m_B = 15.5$, the magnitude at which the cumulative number of S1 and S2 galaxies becomes flat. The possible companions of our Seyfert galaxies have been identified on the blue POSS prints by visual inspection looking for objects not more distant than 3 diameters from the Seyfert galaxy and not more than 3 magnitudes fainter. It results that $\sim 35\%$ of both the S1 and S2 galaxies have a neighbor. This percentage is reduced to $\sim 14\%$ for both classes of galaxies excluding the possible optical companions, the upper limit to the number of which has been derived starting from the counts of galaxies published by Shane and Wirtanen (1967), assuming that the probability of finding an optical companion is a Poisson probability. The same procedure applied to a control sample of 281 normal galaxies indicates that only $\sim 3\%$ of the galaxies have a close companion. In conclusion Seyfert galaxies show a larger percentage (14%) of objects with companion than normal galaxies do (3%). This excess, which is the same for S1 and S2 galaxies, has been derived overestimating the number of optical companions and its value is then a lower limit.

References

Bushouse, H.A.: 1987, *ApJ*, **320**, 49
Dahari, O.: 1985, *ApJS*, **57**, 643
Fuentes Williams, T., Stocke, J.T.: 1988, *AJ*, **96**, 1235
MacKenty, J.W.: 1989, *ApJ*, **343**, 125
MacKenty, J.W.: 1990, *ApJS*, **72**, 231
Petrosian, A.R.: 1982, *Astrofizika*, **18**, 548 (English trans. *Astrophysics*, **18**, 312)
Shane, C.D., Wirtanen, C.A.: 1967, *Publ. Lick Obs.*, **Vol. 22**, Part I, University of Chicago, Chicago Illinois, 647

ARE REDSHIFTS OF SEYFERT GALAXIES COSMOLOGICAL?

A.K. SAPRE and P.S. PARIHAR
School of Studies In Physics, Pt. Ravishanker Shukla University, Raipur 492010, India

Abstract. We have studied the Hubble diagram for an almost complete sample of 48 Seyfert galaxies. We find that if Malmquist bias is taken into account, the correlation of $[\log(cz), V]$ pairs for Seyfert galaxies is statistically significant and that the slopes of the two regression lines are consistent with those expected if the redshifts of Seyfert galaxies are cosmological.

1. The Hubble Diagram of Seyfert Galaxies

The Hubble diagram, i.e., $[\log(cz), V]$ plot for Seyfert galaxies is generally considered to be not very enlightening (Weedman 1976) to show that the redshifts of Seyfert galaxies are proportional to their distances. This is mainly because such a diagram turns out to be a scatter diagram. In this paper we have studied the Hubble diagram for an almost complete sample of 48 Seyfert galaxies (Mackenty 1990) taking Malmquist bias into account. For this sample we find that the regression coefficient for the linear regression of $\log(cz)$ on V is $b_1 = 0.133 \pm 0.024$ and that for the linear regression of V on $\log(cz)$ is $b_2 = 2.889 \pm 0.525$. Though the correlation coefficient $r = 0.62$ is statistically significant the two regression coefficients are not statistically consistent with the expected values of 0.2 and 5.0 respectively if the redshifts of Seyfert galaxies are cosmological. To minimise the effect of Malmquist bias we have considered only those Seyfert galaxies which lie to the brighter side of the best-fit line $V = 5\log(cz) - 5.73$. For this sub-sample of 25 Seyfert galaxies we find that the regression coefficient for the linear regression of $\log(cz)$ on V is $b_1 = 0.157 \pm 0.018$ and that for the linear regression of V on $\log(cz)$ is $b_2 = 3.704 \pm 0.655$. The correlation coefficient $r = 0.76$ is statistically significant at 99% confidence level. We find that the departure of the two regression coefficients from the theoretically expected slopes of 0.2 and 5.0 respectively is not statistically significant, at 95% confidence level, implying that the redshifts of Seyfert galaxies are cosmological in nature.

References

Mackenty, J. W. 1990, Ap. J. Suppl. Series 72, 231.
Weedman, D. W. 1976, Q. J. R. Astr. Soc. 17, 227.

AGN AS A RESULT OF EVOLUTION OF BINARY GRAVIMAGNETIC ROTATORS

OLGA K.SIL'CHENKO
Sternberg Astronomical Institute, Universitetskij pr., 13, 119899 Moscow, Russia

and

VLADIMIR M.LIPUNOV
Sternberg Astronomical Institute, Universitetskij pr., 13, 119899 Moscow, Russia
Faculty of Physics, Moscow University, Leninskie Gory, 117234 Moscow, Russia

Supermassive black holes in AGN can form directly as the results of collapse of supermassive magnetic and rotating stars (gravimagnetic rotators, [5], [4]). In this case naturally we can consider the binary magnetic rotator systems as progenitors of active galactic nuclei. One of the most important consequence of this scenario is an existence of moving galactic nuclei [3], so called "nomadic" nuclei [6].

The most striking picture has been found in the nearest big spiral galaxy **M 31**. Its nucleus is not a Seyfert one; it is known to be a very quiet. But the stellar kinematics reveals a presence of a "dead" black hole with a mass of some $10^7 M_\odot$ in the dynamical center of the galaxy coinciding with the center of the very smooth isophotes; and the compact continuum source is shifted by $0.2'' \div 0.5''$ ($1 \div 2\,pc$) [1], [2]! **M 31** is a spiral galaxy with a very regular structure; it has no signs of merging. The only proposed hypothesis is a lot of dust distributed in such a way that a true center is fully closed and a small region in 0.5'' from it is opened. But it is a quite improbable picture: the central isophotes are regular and the nucleus is point-like. We think that our suggestion is more available: in **M 31** we see the supermassive binary system in the center; one of the components (more massive, surely the black hole) is invisible and situated in the dynamical center, and the other is seen as a continuum point-like source. This system may be a prototype of an active galactic nucleus; the only reason why the nucleus of **M 31** is not a Seyfert one is evidently a lack of gas in the very center of the galaxy.[1]

References

1. Dressler, A., Richstone, D.O.: 1988, *Astrophys. J.* **324**, 701.
2. Ford, H.C., Caganoff, S., Kriss, G.A., Tsvetanov, Z., Evans, I.N.: 1992, *BAAS* **24**, 818.
3. Lipunov, V.M.: 1979, *Astron. Tzirk.* **1065**, 1 (in Russian).
4. Lipunov, V.M.: 1987, *Astrophys. Space Sci.* **132**, 1.
5. Ozernoy, L.M.: 1966, *Soviet Astronomy* **10**, 241.
6. Sil'chenko, O.K. and V.M.Lipunov: 1985, *Astrophys. Space Sci.* **117**, 293.

---

[1] After the preparation of this poster was complete, we have received an issue of Hubble Telescope News (Press Release No.STSCI-Pr93-18) where a discovery of a double nucleus in M 31 is reported. A second, week point-like luminous source is found in the geometrical center of the galaxy. So, our prediction is fully confirmed.

BURIED QUASARS IN RADIOGALAXIES?

C.J. SIMPSON, M.J. WARD, D.L. CLEMENTS and S. RAWLINGS
Astrophysics, University of Oxford, Keble Road, Oxford OX1 3RH, United Kingdom

and

A.S. WILSON
Space Telescope Science Institute, 3700 San Martin Drive, Baltimore MD 21218

Abstract. We report on the results of a near-infrared imaging survey of low-redshift radiogalaxies. We find that one of our 13 objects harbours an unresolved source at K, which we interpret as a quasar-like central engine seen through ~ 40 magnitudes of visual extinction.

We have undertaken J and K-band imaging of a sample of 13 low-redshift ($z < 0.11$) narrow- and absorption-line radiogalaxies. Radial surface brightness profile fitting was then performed with a two-component model of unresolved central source and de Vaucouleurs $r^{1/4}$ law to determine whether we see a heavily-reddened quasar continuum, as predicted by current 'unified schemes' (see e.g. Antonucci 1993 for a review). We have also performed an identical analysis on two low-redshift broad-line radiogalaxies, and find that our estimates of the reddening agree well with those obtained from the Balmer decrements of the broad lines.

We detect such a source in one of our objects, the giant radiogalaxy PKS 0634-205 and we determine the magnitude of this source to be $m_K = 15.67 \pm 0.20$. We can estimate the unobscured nuclear continuum at $2.2\,\mu$m as the [O III] $\lambda 5007$ line is a good indicator of quasar power (its equivalent width is nearly constant in quasars; Miller *et al.* 1992), and we thus derive an obscuration of $A_V = 39 \pm 12$ mag. We estimate an upper limit for the remainder of our sample to be $m_K \approx 17.5$, which typically implies $A_V > 25$ mag. Although our sample is relatively small, we suggest that the reason why we do not see any sources with inferred extinctions in or below this range is because the dusty torus favoured in the present paradigm is 'hard-edged', i.e. its optical depth varies very rapidly with viewing angle at some critical value.

Having estimated the reddening, we can correct the K-band and soft X-ray points and determine the number of ionizing photons, assuming a typical quasar SED. This figure ($\sim$ a few $\times 10^{55}$ photons s^{-1}) agrees well with the value determined from longslit spectroscopy of extended emission-line gas (see Hansen *et al.* 1987 for narrow-band images) and this agreement adds extra credence to our conclusion that PKS 0634-205 is a quasar firing into the plane of the sky.

References

Antonucci, R., 1993, *Ann. Rev. Astron. Astrophys.*, in press
Hansen, L., Nørgaard-Nielsen, H.U. & Jørgensen, H.E., 1987, *Astron. Astrophys. Suppl.*, **71**, 465
Miller, P., Rawlings, S., Saunders, R. & Eales, S., 1992, *Mon. Not. R. astr. Soc.*, **254**, 93

A COMPARISON OF THE CENTIMETRE-TO-SUBMILLIMETRE CONTINUUM SPECTRA OF BL LACERTAE OBJECTS AND FLAT SPECTRUM RADIO QUASARS

J.A. STEVENS and S.J. LITCHFIELD
Centre for Astrophysics, University of Central Lancashire, Preston PR1 2HE, UK

E.I. ROBSON
Joint Astronomy Centre, 660 N. A'ohōkū Place, University Park, Hilo, Hawaii 96720, USA

W. K. GEAR
Royal Observatory, Blackford Hill, Edinburgh EH9 3HJ, UK

and

D.H. HUGHES
Department of Physics, Nuclear Physics Laboratory, Keble Road, Oxford OX1 3RH, UK

Abstract. A comparison of the centimetre to submillimetre continuum spectra of 22 BL Lacertae objects and 24 flat-spectrum radio quasars (FSRQ) has been conducted in order to search for systematic differences between the two classes. The same overall spectral shape is found for all sources and it is concluded that the same basic physical model applies to the continuum emission over this frequency range in both cases. There is clear evidence, however, for the BL Lacs to have flatter high frequency spectra and this difference is reconciled with an intrinsic difference in the underlying jets of the two classes.

The quasi-simultaneous multifrequency continuum spectra (see Gear et al. 1993) of each object was analysed by fitting two-point spectral indices to the 5–37 GHz points and the 37–270 GHz points. In addition, a weighted fit was made to the 375, 270, 230 and 150 GHz data which describes the optically thin spectral index. A 1-D Kolmogorov-Smirnov (KS) test showed no difference in the 5–37 GHz two-point spectral indices of BL Lacs and FSRQs. However, significant differences were found between the 37–270 GHz two-point spectral index populations (98.8%) and between the optically thin spectral index populations (99.5%), with the BL Lacs having the flatter slopes.

Multi-epoch data, which were available for many sources, showed that the derived spectral indices were highly variable. The optically thin spectral indices were thus separated into two categories; those describing the sources in a high state and those describing the sources in a low state. Re-applying the K-S test, a separation in the populations was still found when the sources are in the low state, but this disappears when the sources are flaring. It is concluded that the observed difference in spectral indices between BL Lacs and FSRQs is due to an intrinsic difference in the underlying jets.

References

Gear, W. K. et al. 1993, Monthly Notices of the RAS, in press

SPECTROPHOTOMETRICAL MODELS FOR AGN

DORU MARIAN SURAN and NEDELIA ANTONIA POPESCU
Astronomical Institute of the Romanian Academy,
75212 Bucharest 28 ROMANIA

Abstract. We compare the observational colour diagrams with the position of the lensed quasars, mean cluster memberships for $z < 1$ and galactic evolutive tracks in order to determine the influence of these effects in AGN evolution.

1. OBSERVATIONAL DATA

We have used the following catalogs: Catalog of Quasars and Active Nuclei (Veron Cetty, Veron 1991); Catalog of lensed quasars (Suran, Popescu 1992); Mean evolutive colours for galaxies in high redshift clusters (Suran, Popescu 1987); Atlas of synthetic spectra of galaxies (Rocca-Volmerange, Guideroni 1988).

2. RESULTS AND CONCLUSIONS

The comparison between the different data in the two colour diagrams (U-B, B-V) and in the cosmological diagrams (U-B, z), (B-V, z) reveals three distinct regions: AGN with $z < 0.8 - 1.2$; quasars with $0.8 - 1.2 < z < 2.5 - 3$; quasars with $z > 2.5 - 3$. Also, different effects as lensing, cluster and cosmological effects have been discussed.

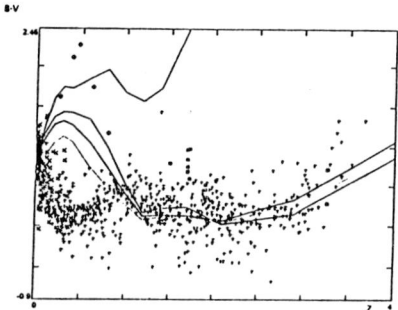

Fig. 1. The cosmological diagram (B-V,z). (+): quasars, (x): Seyfert gal., (squares): lensed quasars, (diamonds): clusters, lines: differents tracks of galactic evolution from burst to Im (Rocca–Volmerange and Guideroni 1988).

References

Rocca-Volmerange,B., Guideroni,B. 1988 Astr.Astrophys.Suppl.Ser.,75,93;
Suran,M.D., Popescu,N.A. 1987, Internal CSEN Report No. 9047;
Suran,M.D., Popescu,C.C. 1992,Rom.Astron.J.,2,1;
Veron-Cetty,M.P., Veron,P. 1991, ESO Report,No.10.

Subject Index

This index is based solely on the title of the papers.

AGN
  Continuum 5
  Evolution 424, 504, 524
  Formation 521
  Gamma Rays 347
  Spatial Distributions 512

BL Lac Objects
  CM/MM Emission 523
  IR Emission 17, 400
  Models 473
  Multiwavelength 506
  Optical Emission 17
  Selection Effects 515
  UV Emission 17
  Variability 393, 394
  X-Ray Emission 377, 384

Blazars
  Energy Distribution 221, 320
  High-Energy Emission 29
  IR Emission 321, 392
  MM Emission 328
  Multiwavelength 155, 505
  Optical Emission 321, 392
  X-Ray Emission 383

Central Regions 233, 351, 375, 405, 431, 441, 454, 479

Cosmology 293, 361, 509, 520

Disks
  Emission 261, 484, 492
  MHD Models 480, 485, 492
  Models 349, 477, 482, 491, 493
  Reprocessing 489
  Structure . . . 478, 479, 483, 484, 488
  Variability 487

Emission
  Auger Process 356
  Dust 355
  Electron Energy Distribution . . 358
  Free-Free 344
  Neutrons 345
  Synchrotron 320, 358, 485

  Thompson Cascade 343

Galaxies
  Emission Lines 442
  Interactions . . . 372, 499, 513, 519
  Origin of the Magnetic Field . . . 486

IR
  High Resolution 461
  Origin 336
  Spiral Galaxies 337
  Ultraluminous Galaxies 332

Jets
  and Clouds 346
  and Stellar Winds 421
  Emission 257
  Gamma Rays 381
  IR Emission 329
  MHD Models 249, 341, 360
  Models . 233, 397, 428, 432, 433, 490
  Multiwavelength Emission . 333, 397
  Particle Acceleration 29
  Radio Emission 428
  Structure 257, 341
  VLBI 187

Lines
  Ly_α Forest 509
  BAL 450
  BLR 399, 444, 446, 453
  BLR and Stars 437
  BLR Structure 163, 457
  from Disks 463
  IR 335, 447, 462
  Kinematics 441, 446
  Models 452, 453
  Narrow Emission 372
  Ne Emission 356
  NLR 445, 449
  Profile 395, 456, 462
  Variability . . . 403, 439, 444, 459
Luminosity Function 498

Objects
  0235+16 397
  3C 273 . . . 193, 323, 327, 333, 382
  3C 279 . . . . 323, 413
  3C 345 . . . . . . 390, 391, 413, 432
  CEN A 323
  CYG A 375

E 1615+061 317
M 87 329
MRK 335 408
MRK 421 404
MRK 474 366
NGC 1068 422
NGC 1365 . . . . 431, 448
NGC 1808 461
NGC 3227 . . . . 429, 457
NGC 3516 464
NGC 3783 . . . . 403, 407
NGC 4051 335
NGC 4151 . 289, 323, 379, 401, 402,
    405, 441, 458
NGC 4261 479
NGC 5055 454
NGC 5548 . . . . 177, 325
NGC 5682 366
NGC 5929 440
NGC 5930 440
NGC 6814 127
OJ 287 . . . . 404, 409
PG 1116+215 197
PKS 0109+224 474
PKS 2155-304 . . . . . . 319, 322
The Milky Way 351
Observations
    ASCA 383
    C-GRO 39, 49, 379
    EUVE 105, 322, 325
    EXOSAT 505
    GINGA 73, 330, 383
    GRANAT 63, 379
    HST   83, 177, 289, 318, 440, 450, 479
    IUE 159, 177, 318, 330
    MERLIN 430
    Multiwavelength . 159, 323, 327, 332,
        404, 409
    ROSAT   53, 197, 318, 330, 338, 366,
        368–370, 372, 374, 377, 384, 450,
        508, 517
    SEST 410, 431
    VLA 431
    VLBI . 181, 187, 423, 425, 469, 490
Optical
    High Resolution 455

Particle Acceleration 379
Plasmas 346, 357
Polarisation
    Faraday Rotation 467

IR 470
Models 472
Multiwavelength 471
Optical 469, 470, 474
VLBI 469

Quasars
    and Fundamental Constants . . . 361
    Emission 203
    Energy Distribution 516
    Extreme 25, 271
    Formation 279
    High z 318
    IR Emission 350
    Lines 399
    MM Emission 324
    Multiwavelength 318, 450
    Optical Emission 518
    Optically-Quiet 419
    Radio-Loud 285, 503, 523
    Radio-Quiet . . . 285, 324, 414, 503
    Spatial Distribution 500
    UV Emission 511
    Variability 518
    X-Ray Emission . 123, 368, 369, 373

Radio 326
    GHz-Peaked Spectrum . . . 424, 425
    High Resolution . 189, 389, 390, 419,
        422, 423, 425–427, 429–431, 455
    Outbursts 397, 420
    Spiral Galaxies 337
Radio Galaxies
    Connection to Quasars 21, 522
    Polarisation 468, 470
    X-Ray Emission 385

Seyfert Galaxies
    Emission 203
    Gamma Ray Emission 379
    IR Emission 355
    Lines 445
    Multiwavelength 330, 502
    Optical Emission 275
    Radio Emission 430, 514
Starbursts . . . 351, 365, 438, 460, 461
    Wolf-Rayet 451
Surveys
    Optical 275, 368
    X-Rays 217, 374, 517

Unified Models . 289, 301, 427, 458, 501

Variability
　Accretion Disks 253
　CM 497
　Emission Lines 163, 173, 395
　High Energy 113
　IR 392, 400
　MM 145, 193, 410
　Models 239
　Multiwavelength . 131, 401, 407, 413
　Optical . 391–394, 408, 411, 412, 414, 415
　Periodicity 127, 390
　Radio . 145, 396, 398, 410, 411, 420
　Supernovae 371
　UV 331, 403, 408
　VLBI 181, 389
　X-Rays 193, 371, 382, 402

X-Rays
　Cosmic Background 378
　Einstein Sources 370
　Emission 213
　Luminous Galaxies 311
　Optical Properties 508
　Origin 33, 380
　Reflection 213, 348
　Soft Spectrum 197, 217, 317, 338, 373, 382
　X-ray Loud AGN 365

Author Index

Akujor Chidi E. **419**, 471
Alef W. 389
Alexander Tal **437**
Alighieri Sperello di Serego . . **21**, 468
Alkemade F. J. M. 325
Aller H. D. 233, 319, 497
Aller M. F. 233, 319, **497**
Alloin D. 457
Altschuler D. R. **389**
Ansari S. G. 506
Antonucci Robert **301**
Apalkov Yu. 63
Appl Stefan **341**
Appleton P. N. 311
Araújo F. X. De 356
Ardeberg A. 419
Aretxaga Itziar **438**
Axon Dave J. 429

Bååth L. B. **181**
Babadzhanyants M. K. **390**
Bade N. **365**, 372
Baker A. C. **498**
Baldwin J. 415
Baldwin Jack A. 460
Ballet J. 63
Balonek T. 327
Bałucińska-Church M. **317**
Band D. L. **318**
Bar-Shalom Avi 447
Baribaud T. 457
Baring Matthew G. **343**
Bartel N. **420**
Barvainis Richard **344**
Baum S. 514
Begelman Mitchell C. 127
Belokon' E. T. **390**
Berezinsky V. S. **349**
Berlin A. B. 326
Bersanelli M. 321
Bertsch D. L. 49
Beuermann K. 508
Bhatia Anand K. 447
Bi Hongguang **366**
Bica Eduardo 355
Biermann Peter L. 350
Birkinshaw M. 385
Björnsson Claes-Ingvar **467**

Blanco P. R. 318
Blecha A. **319**
Bloom Steven D. 155
Bochkarev N. G. **439**
Bock H. 404
Bodo G. **360**
Boisson Catherine 489
Boksenberg A. 441
Boldt Elihu 378
Bonatto Charles 355
Bondi M. **420**
Bonev T. 464
Borgeest U. 391
Bouchet L. 63
Bouchet P. 319
Bower G. A. **440**
Bowman Mark **421**
Bowyer C. S. 322
Bratschi P. 319
Bregman Joel N. 5
Briel U. G. 517
Brinkmann W. **53**
Brinks Elias **422**
Britzen S. **423**
Browne I. W. A. 515
Brunner H. . . **368**, 369, 374, 377, 382
Bühler P. **369**
Burbidge E. M. 318
Burbidge G. **293**
Byun Y. I. 442

Calamai G. 335
Camenzind Max **257**, 341
Campbell J. 423
Cappi M. **370**
Carilli C. L. 375
Carini M. T. 319
Carone T. 325
Carone T. E. 322
Carrasco Luis 320, 358
Carson J. E. 389
Carvalho Joel C. **424**
Catchpole R. M. **441**
Ceca R. Della 370
Celotti Annalisa 221
Chagelishvili G. 360
Chakrabarti Sandip K. **477**
Chatterjee Tapan K. **499**
Cheng Fuzhen 463
Chernyakova M. 63
Chiang J. 49

Chu Yaoquan	**500**	Ferreira J.	249
Chun M. S.	**442**	Ferruit P.	445
Churazov E.	**63**	Fichtel C. E.	49
Chuvaev K. K.	401	Field George	**480**
Chyzy Krzysztof T.	**501**	Fierro J. M.	49
Cimatti Andrea	21, **468**	Filippenko A. V.	319
Claret A.	63	Filippenko Alexei V.	275
Clavel J.	**131**	Fink H.	317, 372, 382
Clements D. L.	522	Fink H. H.	330, 338, 373, 384, 508
Cohen R. D.	318	Finoguenov A.	63
Collin Suzy	483	Fiore F.	317
Conway R. G.	333	Fiore Fabrizio	127
Courvoisier T. J.-L.	319, 327, 330, 331, 346, 369, 382	Ford H. C.	289
		Ford Holland C.	479
Courvoisier Thierry J.-L.	**203**	Francis Paul J.	**503**
Cruz–González G.	454	Franco J.	371
Cruz-González Irene	**320**, 358	Friedli D.	**504**
Czerny B.	**261**	Friedrich P.	368, **374**
		Friedrich S.	382
Dallacasa D.	**425**	Fruscione A.	**322**
de Bruyn A. G.	426	Fujimoto R.	383
Dennison B.	389		
De Robertis M. M.	455	Gabuzda D. C.	**469**
Dermer C. D.	29	Garrington Simon T.	471
Dietrich M.	395, **444**	Gear W. K.	193, 324, 523
Dingus B. L.	49	Gear Walter K.	155
Donahue M.	319	Gehrels Neil	323
Done Chris	127	Genzel R.	461
Dörrer T.	368, 374, 478	Ghisellini Gabriele	**33**, 221
Dultzin-Hacyan Deborah	**502**	Ghosh K. K.	**505**
Dunlop J. S.	311, 324	Ghosh T.	426
		Giannuzzo E.	335
Edelson Rick	**113**	Gilfanov M.	63
Efimov Yu. S.	404	Giommi P.	**506**
Ekejiuba Ifeanyi E.	**345**	Glass I. S.	319
Elvis Martin	**25**	Goad M. R.	453
Engels D.	**372**	Golev V.	464
Espey Brian R.	450	Gondhalekar P. M.	**446**
Esposito J. A.	49	Gontier A.-M.	423
Evans I. N.	289	Gopal–Krishna	414
		Graham D.	389, 426
Fabian A. C.	217, 380	Graham D. A.	187
Falomo R.	17, **321**	Grebenev S.	63
Fanti C.	425	Greenhouse Matthew A.	447
Fanti R.	420, 425	Gregorini L.	420
Fedorenko V.	**346**	Grupe D.	**508**
Feigelson E. D.	319	Gurvits L. I.	389
Feldman Uri	447		
Ferland Gary J.	459	Haardt Francesco	33
Fernandes R. Cid	371	Haehnelt M. G.	**279**
Ferrarese Laura	**479**	Hagen-Thorn V. A.	**392**

Harris D. E.	375	Kollgaard R. I.	319
Hartman R. C.	49	Komberg B. V.	**512**
Heckman T. M.	440	Komossa Stefanie	**449**
Heidt J.	319, 391, 404	Kondo Yoji	408
Heidt Jochen	**393**	Kontorovich V. M.	**513**
Heinrich O. M.	**482**	Kopko Michael, Jr.	**450**
Heise J.	325	Korista K. T.	177
Henri G.	347	Kovalev Y. A.	326
Hewett P. C.	498	Kovalev Y. Y.	**397, 428**
Hintzen Paul M. N.	408	Krabbe A.	461
Hjelm M.	**448**	Kravtsov A. V.	512
Ho Luis C.	**275**	Krichbaum T. P.	**187**, 432
Hongnan Zhou	509	Kriss G. A.	289
Howard Sethanne	323	Krivitsky D. S.	513
Hoyle F.	293	Krolik J. H.	357, 440
Hughes D. H.	311, **324**, 523	Krolik Julian H.	**163**
Hughes P. A.	**233**, 319, 497	Kron R. G.	412
Hunt L. K.	335	Kubičela A.	456
Hunter S. D.	49	Kukula M. J.	426, 430
Huré Jean-Marc	**483**	Kukula Marek J.	**514**
Hutchings J. B.	427	Kurfess J. D.	**39**
		Kusunose Masaaki	**485**
Inoue H.	**73**		
		Lüdke Everton	**471**
Jörsäter S.	431, 448	Lagunov I.	63
Jaffe Walter	479	Lainela Markku	**396**
Jannuzi Buell T.	399, **470**	Lamer G.	368, 369, **377**
Jockers K.	464	Lapidus I.	349
Johnson N.	327	Larionov G. M.	397
Jourdain E.	63, **348**	Laurent P.	63
Junkkarinen V. T.	318	Lawrence A.	332
Junor W.	193	Lawson A. J.	**123**
		Leach C. M.	193
Köhler Thorsten	**511**	Lebrun F.	63
Königl A.	322	Leiter Darryl	**378**
Kaastra J. S.	**325**	Lichti G. C.	**327**
Kahn S. M.	322	Lightman Alan P.	379
Kallman Timothy R.	484	Lin Sung-Nan	408
Kanbach G.	49	Lin Ting-Chang	408
Kaspi Shai	**399**	Lin Y. C.	49
Kats A. V.	513	Lindblad P. O.	431, 448
Kellang Huang	**509**	Linde J. v.	391
Kembhavi Ajit	**217**	Lipunov Vladimir M.	521
Kidger M.	335	Litchfield S. J.	**328**, 523
Kii T.	383	Lorenzetti D.	400
Kikuchi S.	**394**	Lovelace R. V. E.	490
Klapisch Marcel	447	Lyutyi V. M.	401
Kniffen D. A.	49		
Ko Yuan-Kuen	**484**	Maccacaro T.	370
Kohmura Y.	383	Macchetto F.	**83**
Kollatschny W.	**395**, 444	Machalski J.	398

Maciołek-Niedźwiecki Andrzej . . **379**
Madejski Greg M. **127**
Magdziarz P. **398**
Makino F. 330, **383**
Makishima K. 383
Malkan Matthew A. 408
Mampaso A. 454
Mandrou P. 63
Mannheim Karl **285**
Mannucci F. 335
Manteiga M. 454
Mantovani F. 420
Maoz Dan **399**
Maraschi Laura **221**
Marchã M. J. M. **515**
Marcowith A. **347**
Marscher Alan P. **155**
Marshall H. L. 322
Marshall Herman L. **105**
Massaro E. **400**
Matheson T. 319
Mathur Smita **271**
Matsuoka M. 317
Matt G. **380**
Mattox J. R. 49
Mayer–Hasselwander H. A. 49
McConnell M. 327
McDowell J. **516**
McHardy I. M. **193**, 402
Meisenheimer K. 329, 333
Mewe R. 325
Meyer Donald **381**
Meynet G. **451**
Michelson P. F. 49
Micol A. 506
Miley G. K. 426, 440
Miller H. R. 319
Molendi S. 197
Molthagen K. **517**
Moon H. K. 442
Moore N. P. **486**
Mulchaey J. S. 440
Mundell Carole G. **429**
Mushotzky Richard F. 127

Nair A. D. 411
Narlikar J. V. 293
Nazarova L. S. 439, **452**
Neff S. G. 427
Nelson A. H. 486
Nesci R. 400

Netzer Hagai **213**, 399, 437
Neumann M. **329**, 333
Nicolson G. D. 420
Niemeyer Martina **350**
Nizelski N. A. **326**
Noble J. C. 319
Nolan P. L. 49
Nowak Michael A. **487**

O'Brien P. T. **453**
O'Connell Robert W. 479
O'Dea C. 514
Odenwald Sten **323**
Ohashi T. 383
Oknyanskij V. L. **401**
Orr A. **330**
Otani C. 330, 383
Otterbein K. 382
Ozernoy Leonid M. **351**

Paciesas W. 327
Padrielli L. 420
Paltani S. **331**, 346
Palumbo G. G. C. 370
Panesar J. S. 486
Papadakis I. 193
Papadakis I. E. **402**
Parihar P. S. 520
Park Myeong-Gu **488**
Pastoriza M. G. **355**, 415
Patnaik A. R. 419
Pecontal E. **445**
Pedlar A. 189, 426, **430**
Pedlar Alan 429, 514
Pelletier G. **249**, 347
Perley R. A. 375
Perola G. C. 317, 400
Peterson B. M. **177**
Peterson Bradley M. . . . 408, 415, 459
Petrini D. **356**
Pişmiş P. 454
Pian E. 17
Pietrini P. **357**
Pineau des Forêts Guillaume . . . 483
Piro L. 317
Pogge R. W. **455**
Popescu Nedelia Antonia 524
Popović L. Č. **456**
Porcas R. W. 419
Potekhin A. Y. 361
Poutanen Juri **472**

Pronik I. I.	**518**	Shields Joseph C.	**459**
		Shrader Chris R.	408
Qian S. J.	432	Sikora Marek	127
		Sil'chenko Olga K.	**521**
Radecke H.-D.	49	Sillanpää A.	**404**
Rafanelli P.	**519**	Simpson C. J.	**522**
Rawlings S.	522	Sitko M. L.	469
Recillas–Cruz E.	454	Skillman Evan D.	422
Rees Martin J.	**239**	Smith A. G.	327, 411
Reichert G. A.	318, **403**	Smith Howard A.	447
Reinsch K.	508	Smith P. S.	319
Riffert H.	**478**	Smith Paul S.	399
Rigopoulou D.	**332**	Soffitta P.	317
Robson E. I.	193, 324, 327, 328, **523**	Sol Hélène	**473**
Rogers Robert	480	Solomos Nicolaos H.	**405**
Rokaki Evlabia	**489**	Soundararajaperumal S.	505
Romanova M. M.	**490**	Spinoglio L.	400
Romney J. D.	420	Sreekumar P.	49
Roques J. P.	63, 348	Standke K. J.	187
Röser H.-J.	329, **333**	Stanga R. M.	335
Ross R. R.	380	Staubert R.	193, 327, 368, 369, 374, 377, **382**, 478
Rosso F.	249		
Rothschild R. E.	318	Steffen W.	**432**
Rozyczka M.	371	Stenholm L. G.	**336**
Ruano Carlos	502	Steppe H.	193, 327
Ruder H.	478	Sternberg A.	461
		Stevens J. A.	328, **523**
Sadun A.	327	Stickel M.	329
Sagar Ram	414	Stirpe Giovanna M.	**407**
Saikia D. J.	**189**, 426	Stone Remington P. S.	408
Salamanca I.	**457**	Storchi–Bergmann T.	355, 415
Salas Luis	320, **358**	Storchi–Bergmann Thaisa	**460**
Salvati M.	**335**	Studt J.	372
Sandqvist Aa.	**431**	Sun Wei–Hsin	**408**
Sapre A. K.	**520**	Sung E. C.	442
Sargent Wallace L. W.	275	Sunyaev R.	63
Scarpa R.	17	Suran Doru Marian	**524**
Schaeidt S.	365	Swanenburg B. N.	327
Schalinski C.	327		
Schalinski C. J.	187, 423	Tacconi–Garman L. E.	**461**
Schartel N.	**373**	Takalo L. O.	404, **409**
Schilizzi R. T.	426	Tao Jianhui	500
Schlickeiser R.	**29**	Tashiro M.	**383**
Schneid E. J.	49	Team The Faint Object Camera	441
Schramm K. J.	**391**	Team *ASCA*	383
Schrijver C. J.	325	Team *Ginga*	383
Schulz Hartmut	**458**	Tenorio–Tagle G.	371
Serlemitsos Peter	127	Teräsranta H.	404, **410**
Shakhovskoy N.	404	Terlevich Elena	422
Shapovalova A. I.	439	Terlevich R.	371
Shevchenko Ivan I.	**173**	Terlevich Roberto	438

Terlevich Roberto J. 422
Thomas H.-C. **384**, 508
Thompson D. J. **49**
Thompson Rodger I. **462**
Tornikoski M. 410, **411**
Tovmassian H. M. **337**
Trèvese D. **412**
Treves A. 17
Trotter A. S. 389
Trussoni E. 360
Tsvetanov Z. **289**
Turner M. J. L. 123, 327
Turner T. Jane 127
Turner Tracey J. 408
Turnshek David A. 450

Ulrich M.-H. **197**, 327, 382
Unger S. 514

Valtaoja E. 404, 410, 411
Valtaoja Esko **145**, 396
Valtaoja L. **474**
van den Bosch Frank 479
Varshalovich D. A. **361**
Vicente Lourdes **433**, 473
Vince I. 456
Vio R. 159
Violato M. 519
von Montigny C. 49, 327

Wagner S. 319
Wagner S. J. 391
Wagoner Robert V. 487
Wallinder F. H. **253**
Walter R. 330, **338**, 373
Wamsteker W. **159**, 330, 335
Wandel Amri **491**
Wang Yi **492**
Ward M. J. **311**, 522
Webb James R. **413**
Wei Chunyan **463**
Weiler K. W. 420
Wendker H. J. 517
Wiita Paul J. **414**
Williams O. R. 327
Wilson A. S. 440, 522
Wilson Andrew S. 460
Winge Cláudia **415**
Witzel A. 187, 423, 432
Worrall D. M. **385**

Yankulova I. **464**
Yi Insu **493**
You Junhan 463

Zanna G. Del 335
Zdziarski Andrzej A. 379, 485
Zensus J. A. 187, 432
Zentsova A. 346
Zhang Yun Fei 155